Physics at the Terascale

*Edited by Ian C. Brock and
Thomas Schörner-Sadenius*

Related Titles

Huang, K.

Quantum Field Theory
From Operators to Path Integrals

2010
ISBN 978-3-527-40846-7

Stock, R. (ed.)

Encyclopedia of Applied High Energy and Particle Physics

2009
ISBN: 978-3-527-40691-3

Martin, B., Shaw, G. P.

Particle Physics

2008
ISBN: 978-0-470-03294-7

Griffiths, D.

Introduction to Elementary Particles

2008
ISBN: 978-3-527-40601-2

Frampton, P. H.

Gauge Field Theories

2008
ISBN: 978-3-527-40835-1

Belusevic, R.

Relativity, Astrophysics and Cosmology

2008
ISBN: 978-3-527-40764-4

Iliadis, C.

Nuclear Physics of Stars

2007
ISBN: 978-3-527-40602-9

Physics at the Terascale

Edited by
Ian C. Brock and Thomas Schörner-Sadenius

WILEY-VCH Verlag GmbH & Co. KGaA

The Editors

Prof. Dr. Ian C. Brock
Physics Institute
University of Bonn
Germany
brock@physik.uni-bonn.de

Dr. Thomas Schörner-Sadenius
DESY Hamburg
Germany
thomas.schoerner@desy.de

All books published by **Wiley-VCH** are carefully produced. Nevertheless, authors, editors, and publisher do not warrant the information contained in these books, including this book, to be free of errors. Readers are advised to keep in mind that statements, data, illustrations, procedural details or other items may inadvertently be inaccurate.

Library of Congress Card No.:
applied for

British Library Cataloguing-in-Publication Data:
A catalogue record for this book is available from the British Library.

Bibliographic information published by the Deutsche Nationalbibliothek
The Deutsche Nationalbibliothek lists this publication in the Deutsche Nationalbibliografie; detailed bibliographic data are available on the Internet at http://dnb.d-nb.de.

© 2011 WILEY-VCH Verlag GmbH & Co. KGaA, Boschstr. 12, 69469 Weinheim, Germany

All rights reserved (including those of translation into other languages). No part of this book may be reproduced in any form – by photoprinting, microfilm, or any other means – nor transmitted or translated into a machine language without written permission from the publishers. Registered names, trademarks, etc. used in this book, even when not specifically marked as such, are not to be considered unprotected by law.

Typesetting le-tex publishing services GmbH, Leipzig
Printing and Binding Fabulous Printers Pte Ltd, Singapore
Cover Design Adam Design, Weinheim

Printed in Singapore
Printed on acid-free paper

ISBN Print 978-3-527-41001-9

ISBN ePDF 978-3-527-63498-9
ISBN ePub 978-3-527-63497-2
ISBN oBook 978-3-527-63496-5
ISBN Mobi 978-3-527-63499-6

For Rosi and Outi

Foreword

The past decades have seen tremendous progress in the amount and quality of data delivered by experiments at the highest energies at different colliders: Sp$\bar{p}$S, LEP, HERA, Tevatron. The corresponding measurements confirmed in many respects the expected: by finding long sought-for particles like the W and Z bosons, the top quark and the tau neutrino; by establishing the final word on important questions of the Standard Model like the number of interacting neutrino generations; by providing almost ultimate experimental precision as in the case of the Z mass; or by opening up new realms as in the case of the investigation of QCD in studies of the proton structure. In addition, immense work has been invested in new accelerator and detector technologies.

The 1980s and 1990s brought the consolidation of the Standard Model. However, neither was the Higgs boson found, nor did we observe any signs of supersymmetry or any other exciting extension of the Standard Model that might explain some of the open questions we are facing. Now, with the start-up of the Large Hadron Collider – the biggest scientific enterprise ever undertaken by humans – a new sense of something between mild excitement and wild enthusiasm can be felt. If our expectations and our physical intuition are not completely wrong, we can be sure that in the coming years discoveries will be made that have the potential to fundamentally change our view of the microcosm. They may well even shed some light on questions relating to the very large: to the content, structure and evolution of the universe.

Compared to previous experiments, collaborations at the LHC, and in the future also at an e^+e^- collider, enter a new domain of size and complexity, both on the technological and on the social side. Experiments with 10^8 electronic channels, 40 m in length, and with far more than 2000 collaborators require also new approaches in design and construction, in running and maintenance, and in management, communication and coordination.

This exciting moment also seems the appropriate time to document the achievements of the past and to discuss the opportunities that are in front of us. Doing this in a comprehensive manner and in a style appealing to both students and more senior physicists should provide a work which will find its place not only on many bookshelves, but also in the hands of many interested readers. One of the attractions of the book is that it discusses, in addition to the physics and technology, also the social, political and financial environment of today's high energy physics.

The German Helmholtz Alliance "Physics at the Terascale" is one of the instruments designed to prepare better the German particle physics community for future challenges. It aims at strengthening cooperation between the different experimental groups and between experimentalists and theorists as well as increasing the impact of the German high energy physics community on the physics at the Terascale. It is only natural that the impulse for the present book should come from within the Terascale Alliance, demonstrating the broad and deep coverage of the field in Germany.

We hope that at some point a new volume of the same kind as the present will become necessary, then treating the achievements of the LHC in much the same fashion as this volume discusses the era of LEP, HERA and the Tevatron.

Rolf-Dieter Heuer, Director-General of CERN
Joachim Mnich, DESY High Energy Physics and Astroparticle Physics Director

Contents

Foreword *VII*

Preface *XIX*

List of Authors *XXIII*

The Authors *XXVII*

List of Abbreviations *XXXV*

Part One The Physics *1*

1 Setting the Scene *3*
Ian C. Brock and Thomas Schörner-Sadenius
1.1 From the 1970s into the Twenty-first Century *3*
1.1.1 Weak Neutral Currents *4*
1.1.2 November Revolution *4*
1.1.3 Third Generation *5*
1.1.4 τ Lepton *5*
1.1.5 B Mixing and CP Violation in b-Quark Systems *6*
1.1.6 Gluon Discovery *6*
1.1.7 W and Z Discoveries *7*
1.1.8 LEP and the Standard Model *8*
1.1.9 HERA and Proton Structure *8*
1.1.10 Top-Quark Discovery *10*
1.1.11 Searches for Higgs Particles and Supersymmetry *11*
1.1.12 Tau-Neutrino Discovery *11*
1.2 Problems of the Standard Model *12*
1.3 Other Topics Connected to High Energy Physics *13*
1.3.1 Neutrino Physics *13*
1.3.2 Astroparticle Physics *16*
1.3.3 Heavy Ion Physics *19*
1.3.4 Spin Physics *20*
Further Reading *21*

2 The Standard Model: Our Picture of the Microcosm 23
Markus Diehl and Wolfgang Hollik

- 2.1 Introduction 23
- 2.2 Local Gauge Invariance 24
- 2.3 Formulation of QCD 27
- 2.4 Formulation of the Electroweak Standard Model 28
- 2.4.1 Gauge Fields 30
- 2.4.2 Fermion Fields and Their Gauge Interactions 31
- 2.4.3 The Higgs Field and Spontaneous Symmetry Breaking 31
- 2.4.4 Yukawa Interactions: Fermion Masses and Mixing 34
- 2.5 Renormalisation 37
- 2.6 Electroweak Parameters and Observables 39
- 2.7 Some Remarks on Quantum Chromodynamics 41
- 2.8 Symmetries 42
- 2.9 Mass Scales and Effective Theories 44
- References 46

3 Electroweak and Standard Model Tests: the Quest for Precision 47
Klaus Mönig and Georg Steinbrück

- 3.1 The Standard Model at Born Level 47
- 3.2 The Gain from Additional Precision 49
- 3.2.1 Sensitivity to New Physics in Loops 49
- 3.2.2 Sensitivity to Born-Level Effects at High Mass Scales 51
- 3.3 Measurements 51
- 3.3.1 Parameters Measured on the Z Resonance 52
- 3.3.2 W Boson Mass and Couplings 55
- 3.3.3 Two-Fermion Processes Off Resonance 58
- 3.3.4 Low Energy Data 60
- 3.4 Constraints from Precision Data 60
- 3.4.1 Status of Theoretical Predictions 61
- 3.4.2 Standard Model Constraints 62
- 3.4.3 Constraints on New Physics 65
- 3.4.4 Expectations from LHC and ILC 68
- References 69

4 Hard QCD: Still Going Strong 73
Sven-Olaf Moch and Klaus Rabbertz

- 4.1 Introduction 73
- 4.2 The Strong Coupling 75
- 4.3 Perturbative QCD at Colliders 76
- 4.4 Hard Parton Scattering 78
- 4.5 Parton Luminosity 80
- 4.6 Fragmentation Functions and Event Shapes 83
- 4.7 Jet Production 86
- 4.8 Gauge-Boson Production 89
- 4.9 Jet Shapes 91

4.10	Tests of the QCD Gauge Structure	93
4.11	Outlook	94
	References	94

5 Monte Carlo Generators and Fixed-order Calculations: Predicting the (Un)Expected 97

Stefan Gieseke and Zoltán Nagy

5.1	Fixed-Order Born-Level Calculations	97
5.2	Next-to-Leading Order Calculations	98
5.3	Next-to-Next-to-Leading Order Calculations	99
5.4	Leading-Order Parton Showers	99
5.4.1	Shower Evolution	100
5.4.2	The Splitting Operator	102
5.5	Implementations and Shower Schemes	104
5.5.1	Angular-Ordered Shower	104
5.5.2	Partitioned Dipole Shower with Leading-Colour Approximation	106
5.5.3	Antenna Dipole Shower with Leading-Colour Approximation	107
5.6	Matching Parton Showers to Fixed-Order Calculations	108
5.6.1	Born-Level Matching	108
5.6.2	Next-to-Leading Order Matching	110
5.7	Hadronisation	111
5.7.1	Independent Fragmentation	112
5.7.2	Lund String Fragmentation	113
5.7.3	Cluster Hadronisation	114
5.8	The Underlying Event	115
5.8.1	Relevance of the Underlying Event at Hadron Colliders	116
5.8.2	Measuring the Underlying Event	116
5.8.3	Modelling the Underlying Event in Monte Carlo Event Generators	117
	References	118

6 The Higgs Boson: Still Elusive After 40 Years 123

Markus Schumacher and Michael Spira

6.1	The Higgs Boson Mass	123
6.2	Higgs Boson Decays	126
6.3	Higgs Boson Production at the LEP Collider	127
6.4	Higgs Boson Production at Hadron Colliders	129
6.5	Past and Present Searches at LEP and Tevatron	131
6.5.1	Searches at LEP	132
6.5.2	Searches at the Tevatron	134
6.6	Prospects for Higgs Boson Searches at the LHC	136
6.7	Implications of Observation or Exclusion	139
	References	140

7 Supersymmetry 143

Herbert Dreiner and Peter Wienemann

7.1	Introduction	143
7.2	Supersymmetry Transformations and Fields	143

7.3	Superfields and Superpotential 145
7.4	Discrete Symmetries 146
7.5	R-Parity Conservation (P_6 Model) vs. R-Parity Violation (B_3 Model) 147
7.5.1	Model Building 148
7.5.2	Phenomenology 149
7.6	Measuring Supersymmetry 151
7.6.1	Searches at the LHC 151
7.6.2	Measuring SUSY Properties at the LHC 152
7.6.3	Measuring SUSY Properties at the ILC 155
7.6.4	SUSY Parameter Determination 157
7.7	Summary and Conclusions 159
	References 160

8 Quark Flavour Physics 163
Gudrun Hiller and Ulrich Uwer

8.1	Flavour Within the Standard Model 164
8.1.1	The CKM Quark Mixing Matrix 164
8.1.2	The Unitarity Triangle 166
8.2	Flavour and New Physics 169
8.2.1	Flavour Changing Neutral Currents 169
8.2.2	Null Tests 172
8.3	B-Meson Key Measurements 172
8.3.1	Production of B Mesons 173
8.3.2	Mixing of Neutral B Mesons 174
8.3.3	Measurement of the CP Violating Mixing Phases 178
8.3.4	Search for Very Rare Decays 184
8.4	Flavour at the Terascale – Outlook 185
	References 186

9 Top Quarks: the Peak of the Mass Hierarchy? 187
Peter Uwer and Wolfgang Wagner

9.1	Introduction 187
9.2	Top-Quark Pair Production in Hadronic Collisions 190
9.2.1	Cross-Section Calculations 190
9.2.2	Top–Antitop Cross-Section Measurements 192
9.2.3	Spin Correlations in Top-Quark Pair Production 193
9.2.4	Forward–Backward Charge Asymmetry 195
9.3	Single Top-Quark Production 195
9.3.1	Cross-Section Calculations 196
9.3.2	First Observation of Single Top Quarks 197
9.4	Top-Quark Decay 199
9.4.1	W-Boson Helicity in Top-Quark Decays 199
9.4.2	Measurement of $\mathcal{R}_b$ 201
9.5	Top-Quark Mass 201
9.5.1	Mass Measurement Techniques 203
9.6	The Top Quark as a Window to New Physics 204

9.6.1	Top–Antitop Resonances	204
9.6.2	Non-Standard Model Sources of Single Top Quarks	204
9.6.3	Search for Non-Standard Models Top-Quark Decays	205
	References	206

10 Beyond SUSY and the Standard Model: Exotica 209
Christophe Grojean, Thomas Hebbeker and Arnd Meyer

10.1	Alternative Higgs	209
10.2	Technicolour, Composite Higgs and Partial Compositeness	211
10.3	Extra Dimensions, Strings and Branes	213
10.4	Grand Unified Theories	216
10.5	Extra Gauge Bosons	218
10.6	Leptoquarks	218
10.7	Unexpected Physics: Hidden Valley, Quirks, Unparticles…	219
10.8	Model-Independent Search for New Physics	220
	References	221

11 Forward and Diffractive Physics: Bridging the Soft and the Hard 225
Jochen Bartels and Kerstin Borras

11.1	Introduction	225
11.2	Cross Sections in pp and ep Scattering	225
11.3	Parton Densities, Small-x and BFKL Dynamics	228
11.4	Saturation	231
11.5	Diffractive Final States	233
11.6	Multiple Scattering, Underlying Event and AGK	235
11.7	Necessary Instrumentation at the LHC	237
	References	239

Part Two The Technology 241

12 Accelerators: the Particle Smashers 243
Helmut Burkhardt, Jean-Pierre Delahaye and Günther Geschonke

12.1	Introduction	243
12.1.1	Why Are Existing Accelerators So Big?	244
12.1.2	Energy and Luminosity	244
12.2	LEP	246
12.3	Tevatron	250
12.3.1	Performance	251
12.3.2	Where Do the Protons and Antiprotons Come from?	252
12.4	HERA	253
12.5	LHC	255
12.6	Linear Collider	258
12.6.1	Acceleration	259
12.6.2	Superconducting Accelerating System	260
12.6.3	Normal-Conducting Acceleration System	260
12.6.4	Achieving the Luminosity	260
	References	263

13	**Detector Concepts: from Technologies to Physics Results** *265*	
	Ian C. Brock, Karsten Büßer and Thomas Schörner-Sadenius	
13.1	Introduction *265*	
13.2	Technical Concepts *265*	
13.3	Infrastructure *269*	
13.4	Organisation *270*	
13.5	ALEPH, DELPHI, L3 and OPAL at LEP *271*	
13.5.1	ALEPH *272*	
13.5.2	DELPHI *273*	
13.5.3	L3 *274*	
13.5.4	OPAL *274*	
13.6	H1 and ZEUS at HERA *275*	
13.6.1	H1 *276*	
13.6.2	ZEUS *277*	
13.7	CDF and DØ at the Tevatron *278*	
13.7.1	CDF *278*	
13.7.2	DØ *279*	
13.8	ATLAS and CMS at the LHC *280*	
13.8.1	ATLAS *281*	
13.8.2	CMS *282*	
13.9	ILD – a Detector Concept for the International Linear Collider *283*	
13.9.1	Requirements and Rationales *284*	
13.9.2	The ILD Detector Concept *284*	
13.9.3	The Challenge of Push–Pull Operations *286*	
	References *288*	
14	**Tracking Detectors: Following the Charges** *291*	
	Jörn Große-Knetter, Rainer Mankel and Christoph Rembser	
14.1	Introduction *291*	
14.2	Gaseous Detectors *292*	
14.2.1	Working Principle *292*	
14.2.2	Operation Modes of Gaseous Detectors *293*	
14.2.3	Gas Mixtures in Gaseous Detectors *295*	
14.2.4	Gaseous Detectors at Modern Experiments *296*	
14.2.5	Micro-Pattern Gas Detectors *298*	
14.3	Semiconductor Detectors *299*	
14.3.1	Silicon as Detector Material *300*	
14.3.2	Strip Detectors *301*	
14.3.3	Hybrid Pixel Detectors *302*	
14.3.4	Active Pixel Detectors *303*	
14.3.5	Radiation Tolerance *304*	
14.4	Track Reconstruction *306*	
14.4.1	Track Pattern Recognition *306*	
14.4.2	Track Fitting *307*	
14.5	Alignment *309*	

14.6	Tagging of Heavy Flavours *310*	
	References *311*	

15 Calorimetry: Precise Energy Measurements *313*
Felix Sefkow and Christian Zeitnitz

15.1	Introduction *313*	
15.2	Basic Principles of Particle Detection *313*	
15.2.1	Energy Loss of $e^{\pm}$ and Photons *314*	
15.2.2	Interaction of Hadrons with Matter *315*	
15.3	Particle Showers *316*	
15.3.1	Electromagnetic Cascades *316*	
15.3.2	Hadronic Cascades *318*	
15.4	Calorimeters: Response and Resolution *319*	
15.4.1	Response and Resolution of a Sampling Calorimeter for Electromagnetic Particles *320*	
15.4.2	Homogeneous Calorimeters *322*	
15.4.3	Response and Resolution of Hadron Calorimeters *322*	
15.4.4	Spatial Resolution *325*	
15.5	New Concepts *325*	
15.5.1	Dual-Readout Calorimetry *326*	
15.5.2	Particle-Flow Calorimetry *327*	
15.6	Summary *330*	
	References *330*	

16 Muon Detectors: Catching Penetrating Particles *333*
Kerstin Hoepfner and Oliver Kortner

16.1	Sources of Muons *333*
16.2	Energy Loss of Muons and Muon Identification *334*
16.3	Measurement of Muon Momenta *336*
16.4	Muon Identification in ATLAS and CMS *337*
16.5	ATLAS and CMS Muon Chambers *339*
16.5.1	Drift-Tube Detectors *340*
16.5.2	Resistive-Plate Chambers *341*
16.5.3	Cathode-Strip Chambers *342*
16.6	Muon Track Reconstruction and Identification *343*
	References *345*

17 Luminosity Determination: Normalising the Rates *347*
Ian C. Brock and Hasko Stenzel

17.1	Outline *347*
17.2	Luminosity Determination in e^+e^- Machines *348*
17.3	Luminosity Determination at HERA *351*
17.4	Luminosity Determination at Hadron Colliders *353*
17.4.1	Luminosity from Counting Inelastic Events *353*
17.4.2	Luminosity from Elastic Scattering *358*
17.4.3	Luminosity from W/Z Production *359*
	References *360*

18	**Trigger Systems in High Energy Physics Experiments** *363*
	Eckhard Elsen and Johannes Haller
18.1	Introduction *363*
18.2	Elements of a Trigger System *365*
18.2.1	Clocked Readout and Triggering *366*
18.2.2	Central Trigger Logic *367*
18.2.3	Buffering *368*
18.2.4	Dead-Time *368*
18.2.5	Multi-level Trigger Systems and Event Building *370*
18.2.6	Trigger Rates and Downscale *371*
18.3	Trigger Systems in Modern HEP Experiments *372*
18.3.1	Trigger Strategies *372*
18.3.2	Example HERA: the Trigger System of the H1 Experiment *374*
18.3.3	Example LHC: the Trigger System of the ATLAS Experiment *375*
18.3.4	Example LC: Trigger-Less Data Acquisition at the ILC *378*
18.4	Trigger Systems and HEP Data Analysis *379*
18.5	Summary and Outlook *381*
	References *381*

19	**Grid Computing in High Energy Physics** *383*
	Wolfgang Ehrenfeld and Thomas Kreß
19.1	Introduction *383*
19.2	Access to the Grid *385*
19.3	Tier-0 Grid Layer *386*
19.4	Tier-1 Grid Layer *386*
19.5	Tier-2 Grid Layer *386*
19.6	Tier Centres' Hardware Components *387*
19.7	Tier-3 Grid layer *388*
19.8	User Analysis on the Grid *390*
19.9	National Analysis Facility *392*
19.10	Last Steps of a Typical HEP Analysis *393*
19.11	Cloud Computing – the Future? *394*
19.12	Data Preservation *395*
	References *396*

Part Three   The Organisation *399*

20	**The Sociology and Management of Terascale Experiments: Organisation and Community** *401*
	R. Michael Barnett and Markus Nordberg
20.1	Introduction *401*
20.2	Performance and Instruments of Funding *401*
20.3	Technology, Project Structures and Organisation *405*
20.4	From Data Analysis to Physics Publications – the Case of ATLAS *408*
20.5	Budget and Time Considerations *410*
20.6	Conclusions *412*
	Further Reading *413*

21	**Funding of High Energy Physics** *415*	
	Klaus Ehret	
21.1	Outline *415*	
21.2	High Energy Physics – an International Effort between Accelerator Laboratories and Universities *415*	
21.3	Funding and Interplay with Politics *416*	
21.3.1	Accelerator Laboratories for High Energy Physics *417*	
21.3.2	Constitution of International Collaborations *419*	
21.3.3	Project Funds for University Groups *420*	
21.4	Federal Structure of Science Policy and Funding in Germany *421*	
21.4.1	Federal Ministry of Education and Research – BMBF *421*	
21.4.2	Institutions for Basic Scientific Research in Germany *422*	
21.4.3	Project Funding for Basic Research *423*	
21.5	European Research Area and EC Funding *426*	
21.5.1	European Research Council – ERC *426*	
21.5.2	Seventh Research Framework Programme – FP7 *427*	
21.6	Strategic Decision-Making *428*	
21.7	Funding of the LHC and Its Experiments *429*	
21.8	Summary and Outlook *430*	
22	**The Role of the Big Labs** *431*	
	Albrecht Wagner	
22.1	Why Does Particle Physics Need Large Laboratories? *431*	
22.2	Examples of Large Laboratories *432*	
22.3	Complementarities of Universities and Large Laboratories *433*	
22.4	Key Functions and Assets of Large Laboratories *434*	
22.4.1	Research and Development of Accelerators and Experiments *434*	
22.4.2	Construction *435*	
22.4.3	Operation *436*	
22.4.4	Computing *437*	
22.4.5	User Support *438*	
22.4.6	Management *438*	
22.4.7	Scientific Staff at Large Laboratories *438*	
22.5	Collaborations and Their Individual Members *439*	
22.6	Organisational Models for Particle Physics Facilities *440*	
22.7	Access to Large Laboratories and Their Facilities *441*	
22.8	Strategic Planning for Different Laboratories and Regions *441*	
22.9	Decision Process and the Role of Politics *442*	
22.10	Possible Future Developments *444*	
22.11	Summary and Outlook *446*	
23	**Communication, Outreach and the Terascale** *449*	
	James Gillies and Barbara Warmbein	
23.1	Why Communicate? *450*	
23.2	The Place of Communication Within an Organisation *453*	
23.3	Audiences and Tools – the Basics *454*	

23.4 How to Engage with Your Audience 455
23.5 Communication at the Terascale 459
Appendix: CERN Strategic Communication Plan 2009–2013, Summary 462

Index 465

Preface

The fundamental physics that can only be studied at the highest achievable energies and luminosities is a driving motivation for particle physics. With the start of the LHC this physics now enters the *Terascale* regime. Most of the colleagues who contributed to this volume have been working on the LHC, its experiments or theoretical foundations, for many years – be it on the hardware, on the preparation of the data analyses and the necessary tools, or on particle theory and phenomenology. Some even contributed to the initial discussions about a proton–proton collider project as far back as 1984.

In 2007 the Helmholtz Alliance "Physics at the Terascale" was approved in Germany. It is a research network supported by the Helmholtz Association and comprises the research centres DESY and KIT (GF), 18 German universities, and the Max Planck Institute for Physics. Its stated aim is "Within the framework of the world-wide investigation of the fundamental properties of matter using accelerators at the highest energies, to sustainably concentrate and advance the expertise and strengths of the participating institutes." As both of us are heavily involved in the Alliance (ICB: Scientific Manager from 2007 to 2010, TSS: Head of the Analysis Centre at DESY), when looking for authors for the various chapters in the book it seemed natural mostly to approach our colleagues from the many Alliance partners. We also tried to find authors who are currently actively involved in the physics and detectors and thus also fulfil a further aim of the Alliance, which is to give more responsibility and visibility to the younger members of the community.

Although first ideas for a book about the status of our field of high energy particle physics were circulated on the eve of the start of LHC operation, it was not difficult to assemble a team of competent and highly motivated colleagues, who were eager to share their knowledge of the Physics at the Terascale and particularly that of the LHC. Not even the overlap in time between the writing of the manuscript and the start-up of LHC operations – which naturally means a lot of work both in terms of the understanding of the detectors and the beginning data analyses – could stop them, although it did slow them down a bit!

The aim of this book is to provide a comprehensive overview of "Physics at the Terascale" which naturally emphasises physics at the LHC, both in terms of the theoretical foundations and the experimental status. Consequently, the first part of the volume contains chapters dealing with the Standard Model and its different building blocks like the electroweak theory and QCD, with several of its extensions and with important particular aspects like Higgs boson or top-quark physics.

Physics at the LHC requires massive experimental devices – accelerators and detectors. Therefore, a complete coverage of high energy physics also demands a discussion of the instruments we are dealing with. This aspect is covered in the second part of the book, in which accelerators, detectors and relevant detector technologies used at the LHC are introduced, together with the important topics of triggering and computing.

The third part of the book is somewhat unusual for a physics textbook. The LHC is often called the largest experiment ever undertaken by mankind. Projects of the size of the LHC take a long time, involve many hundreds or even thousands of people and are extremely expensive. We therefore felt that the physics discussions in the first two parts of the book should be supplemented by a discussion of the funding of high energy physics and its organisation and management. The chapters in the third part of the book deal with these issues. They are written from different viewpoints: active members of a collaboration, a (former) lab director and someone intimately involved with the financing of particle physics. We hope that these different viewpoints provide an insight into the how and why of particle physics organisation. They may even help you to understand better why decisions get made in the way they do. Last, but by no means least, the final chapter in the book considers communication in high energy physics; what it means, the specific audiences that one has to consider and some guidelines as to how one should approach it.

It is clear that such a book cannot discuss all the elaborate concepts of particle physics in great depth; in order to cure this shortcoming, all chapters include a list of references or links to further information, which allow the interested reader to gain deeper insight and find more facts. All chapters are designed as independent units which should be self-contained and understandable for the average graduate student of the field. However, it is probably advisable to read at least the chapters in one part of the book in the order in which they are presented.

A large number of persons have contributed, in one way or the other, to this book project and it is a pleasure to thank them all. These are, first and foremost, the authors of the individual chapters. All of them were extremely eager to contribute and very cooperative. Consequently the biggest and almost only problem for us editors during the preparation of the volume was to achieve a reasonable balance between the page number limit given by the publisher and the seemingly never ending wealth of detail the authors wanted to write down.

Besides the authors, a number of other people are needed to successfully finish a large book project: Andrea Fürstenberg took on the incredibly tedious and time-consuming task of checking all the references and adding missing information; Michaela Grimm took care of copyright issues for some of the figures. Our graphics editor, Katarina Brock, spent many hours editing all the figures and providing a unified layout. The quality of the figures in the book is in no short measure due to her skills. Caren Hagner, Dieter Horns, Thomas Peitzmann and Caroline Riedl gave invaluable input to Chapter 1 as did Wolfgang Walkowiak for Chapter 8. Wolfram Zeuner's contribution to Chapter 13 cannot be overestimated.

The Helmholtz Alliance "Physics at the Terascale" kindly granted substantial financial support which is gratefully acknowledged. We are also grateful to Wiley-VCH for publishing this work, and especially to Anja Tschörtner who carried us through this project, enduring our many questions and several requests for yet another delay of the deadline.

Finally, our heartfelt thanks go to our families. They not only endured, but also massively supported our work on this volume. Reading, editing and correcting the chapters occupied many of our days, evenings, weekends and even vacations for the best part of the first half of 2010. Without their understanding and patience Wiley would have had to wait even longer for the book to be completed.

While we have made every effort to avoid mistakes in the volume, we certainly cannot rule them out. For this purpose we have set up a web page on which we will list corrections and any other relevant information connected with the book:

http://www.terascale.de/terascale_book.

Please send any errors you find to us by email: *brock@physik.uni-bonn.de, thomas.schoerner@desy.de*

Bonn and Hamburg
August 2010

Ian C. Brock and
Thomas Schörner-Sadenius

List of Authors

R. Michael Barnett
Mailstop 50R-6008
Lawrence Berkeley National Laboratory
1 Cyclotron Road
Berkeley
CA 94720
USA
barnett@lbl.gov

Joachim Bartels
DESY
Notketstr. 85
22607 Hamburg
Germany
joachim.bartels@desy.de

Kerstin Borras
DESY-CMS
Notkestr. 85
22607 Hamburg
Germany
kerstin.borras@desy.de

Ian C. Brock
Universität Bonn
Physikalisches Institut
Nußallee 12
53115 Bonn
Germany
brock@physik.uni-bonn.de

Karsten Büßer
DESY-FLC
Notkestr. 85
22607 Hamburg
Germany
karsten.buesser@desy.de

Helmut Burkhardt
CERN
1211 Genève 23
Switzerland
helmut.burkhardt@cern.ch

Jean-Pierre Delahaye
CERN
1211 Genève 23
Switzerland
Jean-Pierre.Delahaye@cern.ch

Markus Diehl
DESY
Notkestr. 85
22607 Hamburg
Germany
markus.diehl@desy.de

Herbert Dreiner
Universität Bonn
Physikalisches Institut
Nußallee 12
53115 Bonn
Germany
dreiner@th.physik.uni-bonn.de

Wolfgang Ehrenfeld
DESY
Notkestr. 85
22607 Hamburg
Germany
wolfgang.ehrenfeld@desy.de

Klaus Ehret
DESY-PT
Notkestr. 85
22607 Hamburg
Germany
klaus.ehret@desy.de

Eckhard Elsen
DESY-FLC
Notkestr. 85
22607 Hamburg
Germany
eckhard.elsen@desy.de

Günther Geschonke
CERN
1211 Genève 23
Switzerland
gunther.geschonke@cern.ch

Stefan Gieseke
Karlsruhe Institute of Technology (KIT)
Fakultät für Physik
Institut für Theoretische Physik (IThP)
Postfach 6980
76128 Karlsruhe
Germany
gieseke@particle.uni-karlsruhe.de

James Gillies
CERN
1211 Genève 23
Switzerland
James.Gillies@cern.ch

Christophe Grojean
CERN-TH
1211 Genève 23
Switzerland
christophe.grojean@cern.ch

Jörn Große-Knetter
II. Physikalisches Institut
Universität Göttingen
Friedrich-Hund-Platz 1
37077 Göttingen
Germany
jgrosse1@uni-goettingen.de

Johannes Haller
II. Physikalisches Institut
Universität Göttingen
Friedrich-Hund-Platz 1
37077 Göttingen
Germany
johannes.haller@desy.de

Thomas Hebbeker
III. Physikalisches Institut A
Physikzentrum
RWTH Aachen
52056 Aachen
Germany
hebbeker@physik.rwth-aachen.de

Gudrun Hiller
TU Dortmund
Theoretische Physik T3
44221 Dortmund
Germany
ghiller@physik.uni-dortmund.de

Wolfgang Hollik
Max-Planck-Institut für Physik
(Werner-Heisenberg-Institut)
Föhringer Ring 6
80805 München
Germany
hollik@mppmu.mpg.de

Kerstin Hoepfner
III. Physikalisches Institut A
Physikzentrum
RWTH Aachen
52056 Aachen
Germany
kerstin.hoepfner@physik.rwth-aachen.de

Oliver Kortner
Max-Planck-Institut für Physik
(Werner-Heisenberg-Institut)
Föhringer Ring 6
80805 München
Germany
kortner@mppmu.mpg.de

Thomas Kreß
III. Physikalisches Institut B
Physikzentrum
RWTH Aachen
52056 Aachen
Germany
thomas.kress@physik.rwth-aachen.de

Rainer Mankel
DESY-CMS
Notkestr. 85
22607 Hamburg
Germany
rainer.mankel@desy.de

Arnd Meyer
III. Physikalisches Institut A
Physikzentrum
RWTH Aachen
52056 Aachen
Germany
arnd.meyer@physik.rwth-aachen.de

Klaus Mönig
DESY-Zeuthen
Platanenallee 6
15738 Zeuthen
Germany
klaus.moenig@desy.de

Sven-Olaf Moch
DESY-Zeuthen
Platanenallee 6
15738 Zeuthen
Germany
sven-olaf.moch@desy.de

Zoltán Nagy
DESY-FH/CMS
Notkestr. 85
22607 Hamburg
Germany
zoltan.nagy@desy.de

Markus Nordberg
CERN
1211 Genève 23
Switzerland
markus.nordberg@cern.ch

Klaus Rabbertz
KIT-Karlsruher Institut für Technologie
Institut für Experimentelle Kernphysik
Campus Süd
Postfach 69 80
76128 Karlsruhe
Germany
klaus.rabbertz@kit.edu

Christoph Rembser
CERN
1211 Genève 23
Switzerland
christoph.rembser@cern.ch

Thomas Schörner-Sadenius
DESY
Notketstr. 85
22607 Hamburg
Germany
thomas.schoerner@desy.de

Markus Schumacher
Physikalisches Institut
Hermann-Herder-Str. 3
79104 Freiburg
Germany
markus.schumacher@physik.uni-freiburg.de

Felix Sefkow
DESY-FLC
Notkestr. 85
22607 Hamburg
Germany
felix.sefkow@desy.de

Michael Spira
Theory Group LTP
Paul-Scherrer-Institut
5232 Villigen PSI
Switzerland
Michael.Spira@psi.ch

Georg Steinbrück
Institut für Experimentalphysik
Universität Hamburg
Luruper Chaussee 149
22761 Hamburg
Germany
georg.steinbrueck@desy.de

Hasko Stenzel
II. Physikalisches Institut
Universität Giessen
Heinrich-Buff-Ring 16
35392 Giessen
Germany
hasko.stenzel@cern.ch

Peter Uwer
Humboldt-Universität
Department Physik
Newtonstr. 15
12489 Berlin
Germany
Peter.Uwer@physik.hu-berlin.de

Ulrich Uwer
Physikalisches Institut
Universität Heidelberg
Philosophenweg 12
69120 Heidelberg
Germany
ulrich.uwer@physi.uni-heidelberg.de

Albrecht Wagner
DESY
Notkestr. 85
22607 Hamburg
Germany
albrecht.wagner@desy.de

Wolfgang Wagner
Fachgruppe Physik
Bergische Universität Wuppertal
Gaußstr. 20
42097 Wuppertal
Germany
wagner@physik.uni-wuppertal.de

Barbara Warmbein
DESY-PR/FLC
Notkestr. 85
22607 Hamburg
Germany
barbara.warmbein@desy.de

Peter Wienemann
Universität Bonn
Physikalisches Institut
Nußallee 12
53115 Bonn
Germany
wienemann@physik.uni-bonn.de

Christian Zeitnitz
Fachgruppe Physik
Bergische Universität Wuppertal
Gaußstr. 20
42097 Wuppertal
Germany
zeitnitz@physik.uni-wuppertal.de

The Authors

R. Michael Barnett is a senior physicist at Lawrence Berkeley National Laboratory. He is the Education and Outreach Coordinator of the ATLAS Experiment at the Large Hadron Collider. He is head of the international Particle Data Group. Barnett received his Ph.D. from the University of Chicago, and worked at Harvard and SLAC before coming to Berkeley. He is co-founder of two US national educational projects, QuarkNet and the Contemporary Physics Education Project. As a theoretical physicist, he has focused on supersymmetry, QCD, Higgs bosons, and c and b quark physics. Barnett was the first Chair of the US LHC Users Organisation.

Jochen Bartels studied physics at the Universities of Tübingen and Hamburg, and he held postdoc positions at Fermilab and at CERN. He then became professor at Hamburg University. From 1999 until 2008 he was Editor in Chief/Theory of the European Physical Journal C (EPJC); between 2008 and 2009 he was head of the physics department at Hamburg University. His current research interests include high energy QCD and the AdS/CFT correspondence.

Kerstin Borras obtained her Ph.D. at the University of Dortmund working on the construction and calibration of the liquid-argon calorimeter for the H1 experiment. After a fellowship at the Rockefeller University in New York she became a staff member at DESY, presently leading the DESY–CMS group. She has participated in different experiments like H1 and ZEUS, CDF and CMS. She has coordinated the operation of calorimeters and was convener for physics analysis working groups for forward physics and diffraction.

Ian C. Brock studied physics at the University of Oxford where he obtained his D.Phil. in 1983. During his career he has worked on a whole series of experiments mostly at e^+e^- machines: TASSO, Crystal Ball, CLEO, L3, ZEUS, CLEOc and ATLAS. He was employed by Carnegie Mellon University from 1982 to 1996. Since then he has occupied a full professorship at the University of Bonn. He specialises in the physics of heavy quarks and has worked with many detector technologies, from wire chambers and silicon to crystals and luminosity monitors. From 2007 until early in 2010 he was the Scientific Manager of the Helmholtz Alliance "Physics at the Terascale".

Helmut Burkhardt studied physics at the University of Hamburg where he obtained his Ph.D. in experimental physics at DESY in 1982. He then moved to CERN where he worked on LEP and SPS operation as engineer in charge and machine coordinator from 1990 to 1998. He is now a senior staff member in the CERN accelerator physics group, mostly concentrating on the commissioning and optimisation of the experimental conditions in the LHC.

Karsten Büßer studied physics at the University of Hamburg. He started his scientific career in hadron physics (EDDA, COSY) and switched later to high energy particle physics (OPAL, TESLA, ILC). He is currently working at DESY in Hamburg on detector concepts and the machine–detector interface of future linear colliders. In addition he is the Administrative Coordinator of the Helmholtz Alliance "Physics at the Terascale".

Jean-Pierre Delahaye got his Ph.D. at the University of Grenoble in 1971. A CERN staff member from 1974, he was nominated PS Division Leader in charge of CERN accelerators and experimental areas up to 25 GeV in 2000. Since 1994, he has been responsible for the CLIC study of a Linear Collider in the multi-TeV energy range.

Markus Diehl studied physics in Göttingen, Paris, Heidelberg and Cambridge. He held postdoctoral positions in Palaiseau, Saclay, Hamburg, Stanford and Aachen and is currently a staff member in the theory group at DESY. His main interest is in Quantum Chromodynamics.

Herbi Dreiner studied at the Universities of Bonn and Wisconsin, Madison. He has since held positions at DESY, Oxford University, ETH Zurich and the Rutherford Laboratory. He is currently a professor at the University of Bonn. His interests are in searches for physics beyond the Standard Model.

Wolfgang Ehrenfeld studied physics at the University of Oldenburg, at King's College London and at the University of Hamburg. He received this diploma and Ph.D. from the University of Hamburg for work on the OPAL experiment and the TESLA project. Since then he has worked on the BABAR and ATLAS experiments. Currently he is working at DESY for the ATLAS experiment. Besides searching for supersymmetry his main focus is on Grid computing and large-scale user analysis.

Klaus Ehret studied physics at the Universities of Ulm and Heidelberg. He has worked on the ARGUS and HERA-B experiments. For some years he has been active in the project management organisation of the German Federal Ministry of Education and Research at DESY, responsible for the funding of high energy physics activities at German universities. In addition to that he searches for WISPs (Weakly Interacting Sub-eV Particles) with the ALPS experiment at DESY.

Eckhard Elsen received his Ph.D. in particle physics at the University of Hamburg and habilitated at the University of Heidelberg. He has been working on QCD and electroweak physics at the experiments JADE and H1, DELCO and BABAR and OPAL. He led the trigger activities for the H1 experiment before becoming the spokesperson. Most recently he has engaged in accelerator physics with emphasis on e^+e^- linear colliders, ILC. He is a professor at the University of Hamburg.

Günther Geschonke studied physics at the Technical University Munich. At CERN he has worked on the RF system of LEP, in the last years of the running of LEP 2 as group leader of the LEP RF group. Since 2001 he has been a member of the CLIC study team and project leader of the CLIC test facility CTF3.

Stefan Gieseke got his Ph.D. at the University of Hamburg. Moving to Cambridge in 2001, he became one of the main authors of the newly developed Monte Carlo event generator HERWIG++. This is still his main interest after moving to Karlsruhe in 2004 where he has become leader of a young investigator group on Monte Carlo event generator development in 2008 at the KIT.

James Gillies is head of communication at CERN. He holds a Doctorate in physics from the University of Oxford and began his research career working at CERN in the mid-1980s. In 1993, he left research to become Head of Science with the British Council in Paris. After managing the Council's bilateral programme of scientific visits, exchanges, bursaries and cultural events for two years, he returned to CERN in 1995 as a science writer. He has been head of the organisation's communication group since 2003 and is co-author of "How the Web was Born", a history of the Internet published in 2000 and described by the London Times as being among the year's ten best books for inquisitive minds.

Christophe Grojean got his Ph.D. at the University Paris XI, Orsay, and has worked at CEA-Saclay where he holds a permanent research staff position. He worked for two years at the University of California at Berkeley as a postdoc and spent one year at the University of Michigan at Ann Arbor as a visiting professor. He is currently working in the theory unit of the physics department at CERN. His topic of research concerns various aspects of physics beyond the Standard Model.

Jörn Große-Knetter got his Ph.D. at the University of Hamburg in 1997 working on the ZEUS experiment. Subsequently he joined the ZEUS and ATLAS groups at the University of Oxford, followed by a fellowship at CERN and several years at the University of Bonn, both on the ATLAS experiment. He habilitated in Bonn in 2008. He is now head of the detector laboratory of the particle physics group at the University of Göttingen. In the field of detector development he has worked on strip and pixel semiconductor detectors for the ATLAS experiment.

Johannes Haller got his diploma and Ph.D. in physics from the University of Heidelberg working on the OPAL and H1 experiments. He then moved to CERN as a research fellow in the ATLAS trigger group. As a junior professor at the University of Hamburg his current focus is on the ATLAS trigger system, LHC data analysis and fits of the SM and beyond.

Thomas Hebbeker obtained his Ph.D. at the University of Hamburg working on the CHARM neutrino experiment. Later he worked at CERN and became professor at Humboldt University, Berlin, in 1994. Since 2001 he has been professor at the RWTH Aachen. He was a member of the L3 collaboration at LEP for many years and in 1994 joined the CMS experiment, where he worked on the muon chamber construction and now focuses on the data analysis, in particular the search for new physics. He is, in addition, involved in the DØ experiment at the Tevatron and the Pierre Auger Observatory where he studies cosmic rays at high energies.

Gudrun Hiller   received her physics diploma and Ph.D. from the University of Hamburg/DESY. After positions at SLAC/Stanford, Munich University and CERN she is now a professor at TU Dortmund. She is interested in the theory and phenomenology of fundamental interactions, mostly beyond the Standard Model and with emphasis on flavour.

Kerstin Hoepfner   studied in Berlin and obtained her Ph.D. in particle physics while at CERN working on the CHORUS neutrino oscillation experiment. After a Leopoldina Fellowship at the Technion Haifa, Israel, she accepted a postdoctoral position at DESY, Hamburg, working on the HERA-B experiment to coordinate a vertex detector upgrade. In 2001 she joined the CMS experiment and moved to the RWTH Aachen, where she now holds the position of senior scientist. For six years she coordinated the construction of 1/4 of the CMS muon barrel system, which is now providing data at the LHC. After detector commissioning she transitioned to analysis with an emphasis on the search for new particles and is currently leading the search for new heavy vector bosons.

Wolfgang Hollik   studied physics at Würzburg University and received his Ph.D. in 1989. He was a postdoctoral researcher in Würzburg until 1983, then a scientific assistant at Hamburg University until 1989 (habilitation 1989). Further steps in his career were Scientific Associate at CERN until 1990, staff member at the Max Planck Institute for Physics in Munich until 1993, professor of theoretical physics at Karlsruhe University until 2002. Since 2002, he has been a director at the Max Planck Institute of Physics in Munich, and honorary professor at Technical University Munich.

Oliver Kortner   is a senior scientist at the Max Planck Institute for Physics in Munich. From 1993 to 1998 he studied physics at the Ludwig-Maximilians University Munich. He then worked on the Crystal Barrel experiment before he joined ATLAS. Here he investigated the shower production of highly energetic muons in matter and developed a test stand for the precision muon drift tube chambers. In 2002 Oliver Kortner joined the ATLAS group of the Max Planck Institute for Physics in Munich where he continued to work on the muon system. From 2007 to 2009 he served as ATLAS muon combined performance coordinator, and from 2009 to 2010 as muon calibration coordinator. He is the head of the ATLAS muon calibration centre in Munich, one of three calibration centres world-wide. In parallel to his convenorships he contributed to the preparation of the Higgs boson searches and the inclusive muon cross-section measurement with the ATLAS detector. In June 2010 he was reelected as ATLAS muon combined performance coordinator.

Thomas Kreß   received his diploma and Ph.D. from the University of Heidelberg working on non-perturbative QCD aspects within the OPAL collaboration. After postdoctoral work at CERN for the University of California, Riverside, on

Bose–Einstein correlation effects in W^+W^- physics and computing support for OPAL he joined RWTH Aachen in 2002 as head of the physics department's IT division and member of the CMS collaboration.

Rainer Mankel studied physics at the University of Dortmund, where he obtained his Ph.D. in 1987. During his career he worked on the experiments Split-Field Magnet (CERN-ISR), ARGUS, HERA-B, ZEUS and CMS, and held positions at DESY and the Humboldt University of Berlin. His main research interest focuses on tracking, alignment, offline computing, QCD and heavy-flavour physics. In 1999 he became a staff member of the experimental particle physics group at DESY; from 2008 to 2010 he was convener for alignment and calibration of the CMS experiment.

Arnd Meyer obtained his doctoral degree at the University of Hamburg working on charmonium physics within the H1 experiment. After postdoctoral work at Fermilab and building the CDF data-acquisition system, he joined RWTH Aachen and coordinated data-taking at the second large Tevatron experiment, DØ. He is now devoting most of his research time to CMS. His main physics interests are searches for new phenomena and supersymmetry.

Sven-Olaf Moch studied physics and mathematics in Heidelberg and Hamburg. He has obtained a Ph.D. in theoretical physics at DESY in Hamburg and held positions at Nikhef in Amsterdam and Karlsruhe. Since 2002 he has been a staff member at DESY in Zeuthen. His research interests are centred around Standard Model phenomenology, precision predictions in Quantum Chromodynamics, large-scale computer algebra and mathematical aspects of quantum corrections at higher orders.

Klaus Mönig studied physics at the University of Wuppertal and received his Ph.D. in 1990 working on the DELPHI experiment at LEP. Working at CERN he continued with DELPHI and was an active member of the LEP Electroweak Working Group. In 1998 he became a leading scientist at DESY and started to work on the preparation of the International Linear Collider. In 2006 he joined the ATLAS experiment at the LHC as leader of the DESY-ATLAS group. His main physics interest has always been precision measurements of the electroweak Standard Model.

Zoltán Nagy received his Ph.D. at the University of Debrecen. He held postdoctoral positions at Durham University, University of Oregon, University of Zürich and CERN. Currently he is at DESY and a member of the Analysis Centre of the Helmholtz Alliance "Physics at the Terascale". His main interest is in higher-order calculations and Monte Carlo development in perturbative QCD. He is the author of the NLOJet++ program.

Markus Nordberg is the Resources Coordinator of the ATLAS project at CERN, where his responsibilities include budget planning, resource allocation and reporting for the ATLAS project. He has a degree both in physics and in business administration and has served as Visiting Senior Research Fellow at the Centrum voor Bedrijfseconomie, Faculty ESP-Solvay Business School, University of Brussels. Dr. Nordberg is a member of the Academy of Management,

Strategic Management Society and is a member of the Association of Finnish Parliament Members and Scientists, TUTKAS.

Klaus Rabbertz obtained his Ph.D. in 1998 at the RWTH Aachen for research performed within the H1 experiment at the electron–proton collider HERA. As a CERN research fellow he worked within the OPAL experiment at the e^+e^- collider LEP. Since 2002, he has been a member of the CMS collaboration at the LHC as a senior scientist for the University of Karlsruhe. From 2007 to 2008, he was convener of the CMS working group on QCD. Currently, his main research topics are QCD and jets with the first LHC data.

Christoph Rembser got enthusiastic about high energy physics as a CERN summer student in 1989. Since then he has worked on the ZEUS and OPAL experiments, gaining expertise in various types of detectors and searching for new physics phenomena at the Gigascale. After a short intermezzo at the University of Erlangen, teaching about detectors and learning about astroparticle physics, he returned to CERN in 2005. Christoph is convinced that he will see new particles in the ATLAS Transition Radiation Tracker which he helped design and build and which he is happy to see in operation.

Thomas Schörner-Sadenius studied physics at the Universities of Hamburg and Munich. He held postdoc positions in Munich, at CERN and in Hamburg, working on a number of different experiments (OPAL, H1, ATLAS, ZEUS, CMS). In 2008 he joined DESY where he is currently the leader of the Analysis Centre of the Helmholtz Alliance "Physics at the Terascale".

Markus Schumacher studied at the Rheinische Friedrich-Wilhelms-Universität Bonn, where he obtained his diploma degree (1996) and Ph.D. degree (1999) working at the OPAL experiment at CERN. After a fellowship at DESY working for ILC and a postdoc position at Bonn he held a professorship at Siegen University for two years. Since 2008 he has occupied a professorship at the Albert-Ludwigs-Universität in Freiburg. In 2001 he joined the ATLAS collaboration. His main research field is the investigation of electroweak symmetry breaking and the search for Higgs bosons in the Standard Model and beyond.

Felix Sefkow studied physics in Hamburg and Paris. He obtained his doctoral degree in the ARGUS collaboration and worked as a CERN and DESY fellow on the ALEPH and H1 experiments before becoming assistant professor at Zurich. At present, he is a staff scientist at DESY and spokesperson of the CALICE collaboration developing calorimeters for a future linear collider.

Michael Spira studied physics at the RWTH Aachen and graduated in theoretical particle physics. Since then he has worked at different institutes, that is, DESY Hamburg, University of Hamburg and CERN. He is now a staff member of the theory group at the Paul Scherrer Institute in Villigen (Switzerland).

Georg Steinbrück studied physics in Heidelberg and at the University of Oklahoma where he received his Ph.D. in 1999. He continued his involvement in the DØ experiment at Fermilab as a postdoctoral researcher at Columbia University. At DØ he worked on electroweak physics and on the impact parameter trigger. Since 2003, he has been a member of the scientific staff at the University of Hamburg. As part of the CMS collaboration, he is working on silicon detector research and development and on top-quark physics.

Hasko Stenzel   studied physics in Heidelberg where he received his Ph.D. in 1996 working on the ALEPH experiment at LEP. As a postdoc at MPI Munich he joined ATLAS and worked on the hadron calorimeter and did QCD data analyses for ALEPH. In 2001 he became a staff member of the University of Giessen and worked on the HERMES experiment at HERA before his return in 2005 to ATLAS, where he is now involved in the luminosity detectors.

Peter Uwer   studied physics at the RWTH Aachen where he also received his Ph.D. in theoretical physics in 1998. He worked as a scientist in Saclay, Karlsruhe and at CERN before he became professor for theoretical particle physics at the Humboldt University, Berlin, in 2008. His main research interests are QCD and top-quark physics.

Ulrich Uwer   studied physics in Aachen (RWTH). He obtained his doctoral degree in 1994 with the precision determination of the properties of the Z boson. After a research fellowship at CERN and postdoc positions at DESY and the Humboldt University Berlin he became professor of physics at the University of Heidelberg in 2001. His research focusses on tests of the quark mixing in the electroweak Standard Model and precise measurements of the CP violation in heavy meson decays as a possibility to search for new phenomena.

Albrecht Wagner   studied physics at the Universities of Munich, Göttingen and Heidelberg. He was professor for physics at the Universities of Heidelberg (1984–1991) and Hamburg (1991–2006). He was Director of Research of DESY (1991–1999) and Chair of the Board of Directors (1999–2009). Since 2008 he has chaired the Council of the University of Hamburg.

Wolfgang Wagner   studied physics at the University of Bonn and the Ohio State University in Columbus where he graduated as an M.Sc. in 1996. At the Max Planck Institute for Physics in Munich he worked for the HERA-B experiment at DESY and earned his Ph.D. from the Ludwig-Maximilians-Universität of Munich in 2000. After graduation he started a postdoc position at the University of Karlsruhe and joined the CDF experiment at the Fermilab Tevatron where he spent two years as a visiting researcher. During his time at CDF he worked on several analyses in the field of top-quark physics. He has been at the University of Wuppertal since 2008, where he is currently associate professor. At Wuppertal he contributes to the operation of the ATLAS pixel detector and prepares for analyses of ATLAS data on top-quark properties and the search for the Higgs boson.

Barbara Warmbein   did not study physics. But she likes to talk about it. A journalist with a degree in literature, she caught the particle physics bug during a science journalism internship at the CERN Press Office. After a couple of years of being scientific editor at the European Space Agency she became part of the PR team for the International Linear Collider. Barbara is based at DESY and at CERN.

Peter Wienemann   studied physics at RWTH Aachen from which he also obtained his Ph.D. Later he held research positions at DESY and Freiburg University. At present he is working at the University of Bonn. His main research interests are physics beyond the Standard Model and tracking detectors.

Christian Zeitnitz studied physics at the University of Hamburg where he obtained his Ph.D. in 1992. He was involved in detector projects – mainly calorimeter related – at the H1, DØ and ATLAS experiments. In addition he worked on b-quark and Higgs boson physics studies at ALEPH and DØ. His current focus is on the search for the Higgs boson at the LHC, the upgrade of the ATLAS detector for high luminosity and R&D projects in the framework of the CALICE collaboration. He is a professor at the University of Wuppertal.

List of Abbreviations

ACFA	Asian Committee for Future Accelerators
AFS	Andrew File System – global file system
AGK	Abramovsky–Gribov–Kancheli
ALEPH	LEP experiment
ALICE	LHC experiment
ASIC	application-specific integrated circuit
ATLAS	LHC experiment
BABAR	PEP II experiment
Belle	KEK-B experiment
BFKL	Balitsky–Fadin–Kuraev–Lipatov
BMBF	Bundesministerium für Bildung und Forschung – German Federal Ministry for Education and Research
BRAHMS	RHIC experiment
BSM	physics "Beyond the Standard Model"
CA	Certificate Authority
CAF	CERN Analysis Facility
CCD	charged-coupled device
CDF	Tevatron experiment
CELLO	PETRA experiment
CERN	Conseil Européenne pour la Recherche Nucléaire, Geneva, Switzerland – European Organisation for Nuclear Research
CESR	Cornell Electron Storage Ring, Ithaca, USA – symmetric e^+e^- collider which ran mostly at $\Upsilon(4S)$ centre-of-mass energies
CGC	colour glass condensate
CKM	Cabibbo–Kobayashi–Maskawa
CLEO	CESR experiment
CLIC	Compact Linear Collider – planned e^+e^- linear accelerator, $\sqrt{s}$ up to 3 TeV
CMS	LHC experiment
CSC	cathode-strip chamber
DØ	Tevatron experiment
DELPHI	LEP experiment

DEPFET	depleted field effect transistor
DESY	Deutsches Elektronen-Synchrotron, Hamburg, Germany – German particle physics laboratory
DFG	Deutsche Forschungsgemeinschaft – German Research Association
DGLAP	Dokshitzer–Gribov–Lipatov–Altarelli–Parisi
DORIS	Doppelringsynchrotron, DESY – symmetric e^+e^- collider which ran at Υ centre-of-mass energies
DT	drift tube
ECFA	European Committee for Future Accelerators
ENC	equivalent noise charge
ERA	European Research Area
ERC	European Research Council
EWSB	electroweak symmetry breaking
FALC	Funding Agencies for the Large Collider
Fermilab	Fermi National Accelerator Laboratory, Batavia, USA
FF	fragmentation function
FP7	Seventh Framework Programme for Research and Technology of the European Union
FSP	Forschungsschwerpunkt
GAN	Global Accelerator Network
GDE	Global Design Effort
GEM	gas electron multiplier
GIM	Glashow–Iliopoulos–Maiani
GPD	generalised parton distribution
GUT	grand unified theory
H1	HERA experiment
HEPAP	High Energy Physics Advisory Panel
HERA	Hadron-Elektron-Ringanlage, DESY – ep collider, $\sqrt{s} = 300\text{–}318\,\text{GeV}$
HERA-B	HERA experiment
HERMES	HERA experiment
HGF	Helmholtz-Gemeinschaft – Helmholtz Association
ICFA	International Committee for Future Accelerators
ILC	International Linear Collider – planned e^+e^- accelerator, $\sqrt{s} = 0.5\text{–}1.0\,\text{TeV}$
ILD	ILC experiment
IP	interaction point
ITRP	International Technology Recommendation Panel
JADE	PETRA experiment
KEK	Japanese national high energy physics laboratory
KEK-B	Asymmetric e^+e^- collider running mostly at $\Upsilon(4S)$ centre-of-mass energies
L3	LEP experiment
LEP	Large Electron–Positron Collider, CERN – e^+e^- accelerator, $\sqrt{s} = 90\text{–}209\,\text{GeV}$
LHC	Large Hadron Collider, CERN – pp accelerator, $\sqrt{s} = 7\text{–}14\,\text{TeV}$

LHCb	LHC experiment
LoI	Letter of Intent
LSP	lightest supersymmetric particle
Lustre	a high performance file system
MAPS	monolithic active pixel sensor
Mark-J	PETRA experiment
MC	Monte Carlo
MDT	monitored drift tube
mip	minimum ionising particle
MoU	Memorandum of Understanding
MPG	Max-Planck-Gesellschaft – Max Planck Society
MPGD	micro-pattern gas detector
NAF	National Analysis Facility, DESY
NLO	next-to-leading order
NNLO	next-to-next-to-leading order
OPAL	LEP experiment
PDF	parton distribution function
PEP	Positron–Electron Project, SLAC – e^+e^- collider which ran at $\sqrt{s} = 29$ GeV
PEP II	Positron–Electron Project, SLAC – asymmetric e^+e^- collider which ran mostly at $\Upsilon(4S)$ centre-of-mass energies
PETRA	Positron-Elektron-Tandem-Ringanlage, DESY – e^+e^- collider which ran at centre-of-mass energies from 13 GeV to 46 GeV
PHENIX	RHIC experiment
PHOBOS	RHIC experiment
PLUTO	PETRA experiment
PMT	photomultiplier
PS	Proton Synchrotron, CERN – proton accelerator used mainly as an injector
QCD	Quantum Chromodynamics
QED	Quantum Electrodynamics
QGP	quark–gluon plasma
R&D	research and development
RHIC	Relativistic Heavy Ion Collider, Brookhaven
RPC	resistive-plate chamber
SE	storage element
SiD	ILC experiment
SLAC	Stanford Linear Accelerator Laboratory, Stanford, USA
SLC	Stanford Linear Collider – e^+e^- collider used for physics at mass of the Z boson with polarised beams
SM	Standard Model
Sp$\bar{\text{p}}$S	Proton-antiproton collider version of SPS
SPS	Super Proton Synchrotron, CERN – mainly used as a proton accelerator for fixed target experiments, part of LHC and LEP injection chains
STAR	RHIC experiment

SUSY	supersymmetry
TASSO	PETRA experiment
TCO	total cost of ownership
Terascale	physics that is relevant at centre-of-mass energy scales of 1 TeV and above
Tevatron	$p\bar{p}$ collider, Fermilab – $\sqrt{s} = 1.96\,\text{TeV}$
TGC	thin-gap chamber
TPC	time projection chamber
VOMS	Virtual Organisation Management Service
WIMP	weakly interacting massive particle
WLCG	Worldwide LHC Computing Grid
WMAP	Wilkinson Microwave Anisotropy Probe
WN	worker node
ZEUS	HERA experiment

Part One The Physics

1
Setting the Scene
Ian C. Brock and Thomas Schörner-Sadenius

In this chapter we introduce the basic features of the field of particle physics at the Terascale and give a short historical perspective. Given that we are experimentalists, we concentrate more on the landmark experimental measurements and leave the discussion of theoretical developments to the later chapters on the Standard Model (SM), supersymmetry (SUSY), physics beyond the Standard Model and so on. We also briefly cover a few topics that are otherwise not addressed in the rest of the book, such as the connection between particle physics and astrophysics, neutrinos and spin physics.

Throughout the book we use the usual particle physics units, that is, $\hbar = c = 1$, and use energy units for momenta and masses, for example $m_\mu = 0.105$ GeV and $p_T > 10$ GeV.

1.1
From the 1970s into the Twenty-first Century

It is difficult to know where to start when writing an introduction to both a book and the field of particle physics. We decided that the 1970s would be the appropriate time, as this was when the Standard Model of particle physics started to establish itself as the theory of fundamental particles and their interactions; it was also the decade when one of us (ICB) entered the field.

The 1970s saw a whole slew of fundamental discoveries and theoretical developments, to name just a few:

- the discovery of weak neutral currents;
- the discovery of the J/ψ meson and further excited charmonium states;
- the discovery of the τ lepton;
- the discovery of the b quark;
- the discovery of the gluon at the end of the decade.

Within theory notable developments include:

Physics at the Terascale, First Edition. Edited by Ian C. Brock and Thomas Schörner-Sadenius.
© 2011 WILEY-VCH Verlag GmbH & Co. KGaA, Weinheim.
Published 2011 by WILEY-VCH Verlag GmbH & Co. KGaA.

- the proof that local gauge theories are renormalisable;
- the development of Quantum Chromodynamics (QCD), the theory of the strong interaction;
- the recognition that CP violation could be explained within the framework of the Standard Model, if there are at least three generations of quarks and leptons. In other words the Cabibbo–Kobayashi–Maskawa (CKM) matrix contains a non-trivial phase for three or more generations.

Although the pace of discoveries slowed in the last three decades, both the experimental measurements and the theoretical developments have been essential in establishing the Standard Model as the theory of fundamental particles and interactions as well as exposing weaknesses in the models and indicating directions for future accelerators, detectors and theory.

1.1.1
Weak Neutral Currents

The combination of a theory of the weak interactions which relied on the local gauge principle and the Higgs mechanism led to the formulation of the Standard Model by Glashow, Weinberg and Salam in the mid-1960s. Together with the proof of the renormalisability of such theories by 't Hooft and Veltman in 1971, the prediction of the existence of a neutral partner for the $W^{\pm}$ bosons (responsible for charged-current interactions) became a hot topic for the experimentalists.

Groups in the USA and at CERN looked for bubble chamber events in which a group of hadrons appear from nowhere! These were supposed to be due to reactions such as $\nu p \rightarrow \nu X$. A major difficulty in extracting a signal from such events was that neutron-induced interactions look very similar. Very detailed studies of neutron production in the detector surroundings were necessary before it was possible to convince both the collaborations and the community at large that such neutral-current events actually exist. First evidence was announced in 1973 by the Gargamelle collaboration and by June 1974 three different collaborations all showed clear evidence for weak neutral currents.

This discovery marked the beginning of a huge experimental and theoretical activity in the field of electroweak unification at CERN and around the world. By comparing the charged-current and neutral-current cross sections it was possible to determine the weak mixing angle, θ_W. This yielded a prediction for the mass of the W boson, which in turn led to the idea of building a proton–antiproton collider in order to be able to discover the W and Z bosons well before the start of the e^+e^- collider, LEP.

1.1.2
November Revolution

The discovery of the J/ψ meson in 1974 and the ψ' shortly thereafter have rightly been named the "November Revolution". Quite remarkably the J/ψ meson was observed in two very different experiments at the same time: e^+e^- collisions at a centre-of-mass energy of 3.1 GeV using the SPEAR storage ring at SLAC; a fixed-target experiment looking at the e^+e^- final state in p–Be collisions at Brookhaven. The identification of the resonances as bound states of a new quark, the charm quark, meant that there were now four quarks and four leptons, which could be classified into two generations of quarks and leptons. This provided a natural explanation for the non-existence of so-called *flavour changing neutral currents* (FCNC) via the GIM (Glashow, Iliopoulos, Maiani) mechanism. It is fair to say that this established the Standard Model as a serious theory for the interactions of fundamental particles and was also very instrumental in making particle physicists really believe that quarks were "real" particles rather than just abstract mathematical concepts.

This nice simple, symmetric picture was relatively short-lived, as the tau lepton was discovered only one year later through the decay chain where one τ lepton decays to an electron and accompanying neutrinos, while the other decays to a muon+neutrinos. Such an event signature – an $e\mu$ final state and missing energy – provided a clear signature for the existence of a third generation of leptons.

1.1.3
Third Generation

With the discovery of charm and then the τ lepton, it was natural to see if even more quarks existed. The highest energies could be reached with proton accelerators. This time the $\mu^+\mu^-$ final state was used to look for signs of new resonances. The location was Fermilab, and protons with energies up to 400 GeV were used. A clear signal for at least one resonance, with hints of a further two, was seen at an invariant mass of around 9.5 GeV, ushering in the existence of the fifth quark. Early in 1978, groups at DESY using the e^+e^- collider DORIS were able to separate the $\Upsilon(1S)$ and the $\Upsilon(2S)$. CESR, a new storage ring at Cornell University, extended the list of resonances to $\Upsilon(3S)$ and $\Upsilon(4S)$ a couple of years later.

The spectroscopy of both the $c\bar{c}$ and $b\bar{b}$ resonances has since been investigated in quite some detail. Masses and branching fractions can be compared to potential models, in order to study the strong interaction at intermediate energy scales. For many applications a non-relativistic quark model is sufficient, which simplifies the models considerably.

1.1.4
τ Lepton

Studies of the τ lepton produced a wealth of physics information that is not really covered in this book. Just to give a few examples:

- the measurement of the leptonic branching fractions of τ decays clearly shows the need for colour and yields a precise determination of the strong coupling, α_s;

- the hadronic mass spectra of the τ decay products yield important information on resonances in the 1 GeV range. This information is also useful for evaluating the running of the electromagnetic coupling, α, from a low energy scale to the mass of the Z boson;
- measurements of the τ lifetime and the branching fractions to electrons and muons are important tests of lepton universality.

1.1.5
B Mixing and CP Violation in b-Quark Systems

In the same way that the neutral kaon flavour eigenstates mix to form the mass eigenstates K^0 and $\overline{K}^0$, the neutral D and B hadrons should also mix. However, as the phase space for the decay is so much larger, the lifetimes of the two states are almost identical. One has to look for other signatures of such mixing, for example the observation of a $B^0 B^0$ event coming from the decay of $\Upsilon(4S)$. Such mixing was first observed by the ARGUS experiment at DESY in 1986. The observation was confirmed later by the CLEO experiment at Cornell University. Mixing in the D-hadron system is expected to be much smaller and was not observed until the B-factories KEK-B and PEP-II had been taking data for a number of years.

The machines that run at the centre-of-mass energy of the $\Upsilon(4S)$ observe mixing in the B_d^0 system; the Tevatron experiments have recently also observed mixing in the B_s^0 system, with the expected much higher oscillation frequency.

It is also possible for CP violation to occur in the B-hadron system. Theoretical studies showed that the most promising channel was the decay to the CP eigenstate $J/\psi K_S^0$. It was, however, necessary to measure the number of B^0 and $\overline{B}^0$ as a function of the time difference between their decay and the decay of the CP eigenstate in order to observe an effect. This made it necessary to build asymmetric machines, with different energies for the electron and positron beams. The BABAR and Belle experiments started taking data in 1999 and 3 years later produced clear evidence for CP violation. While the level of CP violation can be explained within the framework of the Standard Model, it is by far not enough to explain the matter–antimatter asymmetry in the universe, the origin of which is one of the big questions for both particle and astroparticle physics as well as cosmology.

These topics are discussed in much more detail in Chapter 8.

1.1.6
Gluon Discovery

One of the main goals of the PETRA (DESY) and PEP (SLAC) accelerators was to discover the top quark. Although they ultimately failed in this goal, they did discover the gluon! The discovery put the theory of the strong interaction, QCD, on a much stronger footing and was the result of a very productive interplay between experimentalists and theorists. There is some controversy over which of the PETRA collaborations discovered the gluon first. The European Physical Society credited four members of the TASSO collaboration with the discovery, for which they received the EPS prize in 1995. However, they also awarded a complementary prize to the four PETRA collaborations: JADE, MARK J, PLUTO and TASSO in recognition of their combined efforts.

A 1976 paper from J. Ellis, M. Gaillard and G. Ross had suggested that 3-jet events with hard gluon bremsstrahlung should be observable in e^+e^- collisions. First data were taken by the PETRA experiments at centre-of-mass energies of 13 and 17 GeV in 1978. Later that year the energy was increased up to 30 GeV. One of the keys to the discovery was to investigate how the jet shapes changed as a function of energy. Do the jets get wider and is the topology of the jet-widening consistent with a general increase of the intrinsic transverse momentum of the particles in the jet? Or, is the broadening confined to a plane, as would be expected if hard gluons are emitted? By summer 1979 there was clear evidence that the latter was the case, and all experiments had textbook pictures of events in which three clear jets were seen. More data allowed the spin of the gluon to be determined in the following year.

1.1.7
W and Z Discoveries

As discussed above, the relative rate of neutral-current and charged-current interactions of neutrinos could be used to measure the weak mixing angle and make a prediction for the masses of the W and Z bosons. At CERN, the SPS came into operation in 1976 with beam energies of 350–400 GeV, insufficient to produce W or Z bosons in fixed-target experiments. In the same year, D. Cline, C. Rubbia and P. McIntyre proposed changing the SPS into the Sp$\bar{p}$S, that is, a proton–antiproton collider with enough energy and intensity to produce the weak bosons and detect their production in collider detectors.

While the proposal was certainly controversial, one may even say audacious, it led to the W and Z being discovered at least 6 years before the Z boson could have been produced at LEP. It is also clear from all accounts of this period that without the drive, enthusiasm and skill of Carlo Rubbia the Sp$\bar{p}$S would not have been approved or built. And without the invention of stochastic cooling by Simon van der Meer it would not have been possible to produce enough "cooled" antiprotons to achieve the necessary luminosity.

Both the experiments and the collider were built in an impressively short time. Machine and detectors were both ready by summer 1981, just three years after the project had been approved. The luminosity increased rapidly and the collider run at the end of 1982 yielded enough luminosity (18 nb^{-1}) to see clear W-boson events in both the UA1 and UA2 detectors (time was shared between fixed-target and collider running). A further increase in luminosity in 1983 (118 nb^{-1}) led to Z-boson decays being observed, and the electroweak Standard Model was established as the correct description of electromagnetic and weak processes.

1.1.8
LEP and the Standard Model

One of the major achievements of LEP was undoubtedly the precise determination of the number of light neutrino families, which also put severe constraints on possible extensions of the Standard Model which contain other weakly interacting light particles.

LEP also saw clear evidence for the self-coupling of both gluons and the vector bosons, W and Z, one of the key predictions of non-Abelian theories like the electroweak Standard Model and QCD.

The measurements at LEP, as well as input from SLC and Tevatron, all feed into a global fit of all Standard Model parameters that impressively demonstrates the validity of the Standard Model and also gives a prediction for the mass of the Higgs boson.

As these measurements are discussed in detail in Chapter 3, we will not go into them further here.

1.1.9
HERA and Proton Structure

The idea for an electron–proton collider had been around since the early 1970s. It was decided to build the accelerator at DESY, and the project was officially launched in 1984. After the start of the machine in 1990, first physics results came from the running in 1992. Protons with energies of 820 or 920 GeV were brought into collision with electrons or positrons of 27.5 GeV, which implies a centre-of-mass energy of 300–320 GeV.

The early running already brought two surprises: at the small distance scales probed by HERA, many more gluons and quark–antiquark pairs from the sea inside the proton were observed than expected; a proton could participate in a hard interaction while remaining intact in a much larger fraction of events than it was reasonable to expect at such high momentum transfers. For the physics at the LHC, such a large number of gluons enhances many cross sections substantially and so is very relevant for "Physics at the Terascale". The origin of the "diffractive" events is still not fully understood and has spawned a whole series of studies and theoretical models (see Chapter 11 for a discussion of diffractive physics).

The lasting legacy of HERA though is certainly the very precise measurements of the structure functions of the proton; the clear demonstration of electroweak unification through the measurement of both neutral-current and charged-current cross sections over a very wide range of squared four-momentum transfer, Q^2, and the precise measurements of the strong coupling, also showing a clear running of the coupling as a function of the energy scale within a single experiment.

In recent years the two collider experiments, H1 and ZEUS, have started to produce combined results in the same spirit as the LEP and Tevatron experiments. This has led to an impressive improvement in the precision of the structure func-

H1 and ZEUS

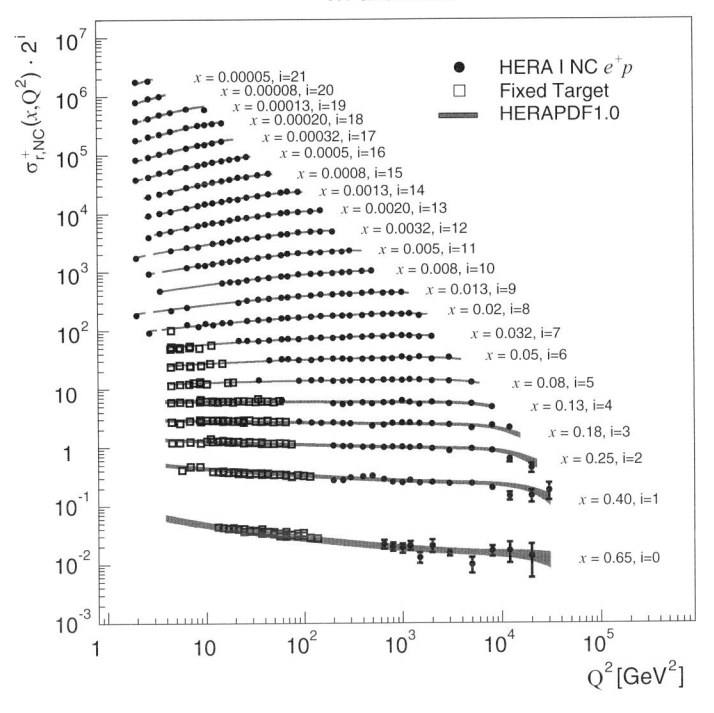

[Figure: Plot of $\sigma_{r,NC}^+(x,Q^2) \cdot 2^i$ vs Q^2 [GeV^2], showing HERA I NC e^+p data, Fixed Target data, and HERAPDF1.0 curves, for x values from $x=0.00005$ ($i=21$) down to $x=0.65$ ($i=0$).]

Figure 1.1 Combined measurements of the ZEUS and H1 collaborations of the proton structure function (adapted from H1 Collaboration and ZEUS Collaboration (F.D. Aaron et al.), Combined Measurement and QCD Analysis of the Inclusive $e^{\pm}p$ Scattering Cross Sections at HERA. JHEP 1001:109 (2010)).

tion measurements, which form the basis for cross-section predictions for the LHC (and any other hadron collisions).

Given that both of us spent the last 10–15 years of our careers working in either the H1 or the ZEUS collaborations we cannot resist showing the comparison of the cross-section measurements in neutral-current processes with the parton distribution function (PDF) extracted from the HERA data, Figure 1.1. The plot shows in fact the reduced cross section, σ_r, as a function of Q^2, over a wide range of x, where the so-called Bjorken x can be interpreted as the fraction of the proton's longitudinal momentum that participates in the hard interaction. The reduced cross section is closely related to the structure function F_2. As discussed at the end of Chapter 17 it even appears feasible to use W-boson production to measure the LHC integrated luminosity, thanks to next-to-next-to-leading-order QCD calculations and the accurate determinations of the structure functions. It is always interesting to see the Nobel prizes of the past being used for "bread and butter" physics in the next generation of colliders!

The neutral-current and charged-current cross sections for electrons and positrons are shown in Figure 1.2. At low Q^2, the neutral-current cross section is dom-

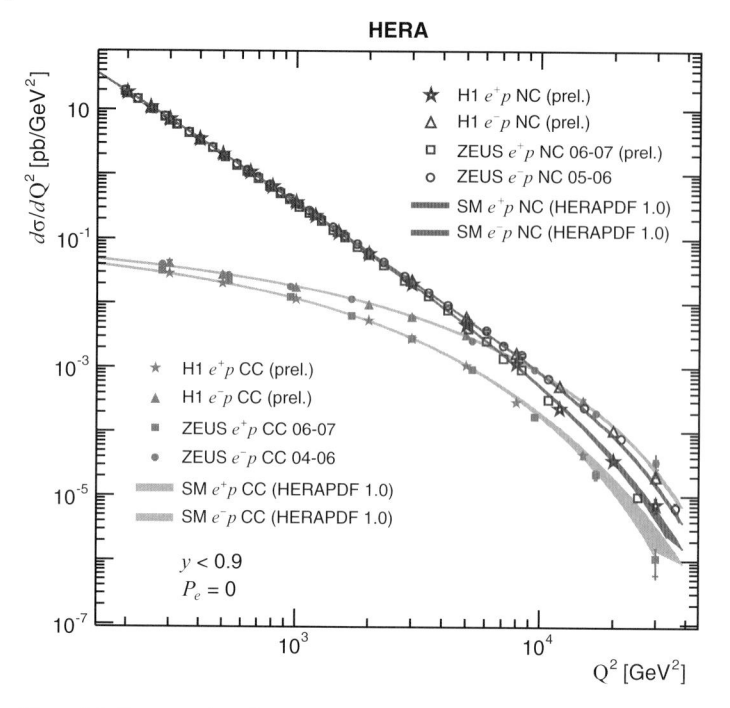

Figure 1.2 Electroweak unification at HERA. Neutral-current and charged-current cross sections for both electrons and positrons as a function of Q^2 are shown (adapted from https://www.desy.de/h1zeus/combined_results/index.php?do=nc_cc).

inated by photon exchange. At high $Q^2 \sim M_W^2$ both weak and electromagnetic process contribute with similar strength, as can be clearly seen in the figure – a real textbook demonstration of electroweak unification.

When it comes to measurements of the strong coupling, α_s, at HERA, theory lags behind experiment. The experimental precision is substantially better than the theoretical uncertainties in the extraction of α_s. Progress here is slow, as higher-order QCD calculations are notoriously difficult to perform and higher-order Monte Carlo simulations for ep collisions are not at the top of the priority list for most theorists.

1.1.10
Top-Quark Discovery

After the discovery of the b quark in 1977 and the τ lepton in 1975, it was clear that a sixth quark, the top quark, should exist. Studies of the properties of B-hadron decays using the CESR accelerator at Cornell provided further evidence that the b quark was a member of a weak isospin doublet.

One of the main physics goals of the subsequent e^+e^- colliders, PETRA, PEP, TRISTAN and LEP was therefore to discover the top quark. Despite heroic efforts, fine energy scans to look for $t\bar{t}$ resonances and pushing the accelerator energies to the maximum possible, no direct evidence for the top quark was found.

After the failures to find the top quark at PEP, PETRA and TRISTAN, one of the first indications that the top quark was very heavy came from the discovery of $B^0\bar{B}^0$ mixing by the ARGUS experiment at DESY in 1986. Such a large mixing could be explained by a large top-quark mass, as terms such as $m_t^2 - m_c^2$ appear in the calculation (here, m_t and m_c are the masses of the top and the charm quark, respectively).

With the high-precision data collected by the four LEP experiments in the early 1990s it was possible to make a prediction for the value of the top-quark mass with an error of around 20 GeV. This prediction was stunningly confirmed when the top quark was finally discovered by the CDF and DØ collaborations in 1995 with a mass in excellent agreement with the prediction from LEP. This topic is discussed in more detail in Chapter 9.

1.1.11
Searches for Higgs Particles and Supersymmetry

The Higgs mechanism discussed in Chapters 2 and 6 was invented almost 50 years ago to explain how mass could be generated within the framework of spontaneous symmetry breaking in local gauge theories. It is the only mechanism thought up so far that has withstood the test of time. However, one of its key predictions is that at least one massive scalar boson should exist. Again here, measurements of the free parameters in the Standard Model lead to a prediction for the Higgs mass, which will certainly be within the reach of future colliders. However, no clear evidence for a Higgs particle has been found so far. Indeed no fundamental scalar particles have ever been discovered, so a Higgs boson would be a major new aspect of the Standard Model of particle physics.

Supersymmetry is the favoured theoretical model for physics beyond the Standard Model. It is discussed in detail in Chapter 7. Unbroken supersymmetry would imply a partner particle for every Standard Model particle with the same mass. Such particles clearly do not exist, hence supersymmetry must be broken and the supersymmetric particles must in general be significantly heavier than their Standard Model counterparts. Unsuccessful searches for such particles have been made at all colliders over the past decades. As the LHC can cover the scale of electroweak symmetry breaking, evidence for supersymmetry, if it exists, should finally be discovered there!

1.1.12
Tau-Neutrino Discovery

After the discovery of the top quark, and ignoring for now the Higgs boson, the only other missing particle in the Standard Model was the tau neutrino. Decays of the τ lepton indicated that a partner neutrino almost certainly existed. However, direct interactions of the tau neutrino are very difficult to observe. First one has to produce such neutrinos. This can happen through the decay of tau leptons or D_S mesons; producing an intense beam of tau leptons in a fixed-target experiment requires both high intensity and high energy. A handful of events which demonstrate directly the existence of the tau neutrino via a charged-current interaction which produced a tau lepton were observed in the DONUT experiment at Fermilab in 2000.

1.2
Problems of the Standard Model

So far, the history of the Standard Model (see also Chapter 2) as told above is a story of successes. However, the Standard Model also has numerous shortcomings which are one of the main motivations for the construction of future collider facilities like the LHC. In this section, the most striking of the deficiencies are mentioned briefly; more details can be found in Chapters 2, 7 and 10.

First of all, the Standard Model has a number of conceptional shortcomings: one fundamental problem is the so-called hierarchy problem – the question why the scale of electroweak symmetry breaking (or alternatively the expected Higgs mass) is $\mathcal{O}$ (100 GeV), when in principle this scale should receive corrections of the order of the largest scale relevant in the Standard Model (like the Planck scale or the grand unification scale). This discrepancy can either be explained by a fine tuning of tree-level and loop contributions to the Higgs mass (hence also the term "fine-tuning problem"), or by the introduction of a new symmetry (like supersymmetry, see Chapter 7). The onset of "new physics" should then be within the reach of the LHC. Putting it another way, despite the invention of the Higgs mechanism, the puzzle of electroweak symmetry breaking (i.e. the question how gauge bosons acquire their observed masses) is far from being understood.

A second conceptual (and also aesthetic) problem is the large number of parameters of the Standard Model: in a fundamental theory one would expect explanations of the values that certain parameters take. However, in the Standard Model, the values of about 30 parameters have to be put in by hand (masses, mixing angles, couplings). In addition, from a phenomenological point of view, the values of some of the parameters are rather puzzling. There is, for example, a huge spread in fermion masses (from meV for neutrinos to more than 170 GeV for the top quark) and no obvious mechanism for the generation of these masses. Similarly, the different mixing behaviour of quarks (almost diagonal) and neutral leptons (very large mixing) is a challenge.

Talking about neutrino masses, also the question of the particle character of the neutrino is open: due to its zero electric charge, the neutrino is special among the fermions and might, eventually, be its own antiparticle (a "Majorana" neutrino instead of a "Dirac" particle like the other fermions). This question is still unanswered (see Section 1.3.1).

There are further far-reaching questions to the Standard Model, of which we only mention two. In the Standard Model the quantisation of electric charge can only be explained if magnetic monopoles exist – which so far have not been observed.

Furthermore, the Standard Model does not provide charge and mass unification (the three fundamental interactions do not unify at some large unification scale), and it is unclear why atoms are neutral when quarks and leptons belong to different multiplets. However, a truly fundamental theory should include such a unification of forces.

Also from the cosmological side, the Standard Model is challenged: first, it is obvious that the Standard Model does not contain a theory for the description of the interactions that govern the large-scale structure of the universe – there is no renormalisable quantum theory of gravity. Second, and of more direct consequence for particle physics, the Standard Model does not provide a candidate particle to explain the large dark-matter content of the universe of close to 25%. The dark matter is necessary to account for, among other features, the rotation curves of galaxies and is also favoured by measurements of the cosmic microwave background (CMB). Finally, also the baryon asymmetry as observed in the universe still awaits an explanation.

Current efforts in high energy physics are focused on solving at least some of these questions and a few definite experimental answers are expected from the Large Hadron Collider. An attractive theoretical alternative to or extension of the Standard Model which provides satisfactory answers to many of the questions is supersymmetry which is discussed in more detail in Chapter 7.

1.3
Other Topics Connected to High Energy Physics

This book focuses on recent and near-future collider-based high energy physics experiments, their organisation, construction, operation, and results. However, high energy particle physics is of course a much wider field, and there are numerous topics which, because of space limitations, cannot be covered in detail. In this section, a number of these topics are briefly touched upon. At the end of the chapter we include a few suggestions for further reading.

1.3.1
Neutrino Physics

Neutrinos as elementary particles were first suggested by W. Pauli in the 1930s in order to explain the spectrum of nuclear β decay; the first experimental observation (of the electron neutrino) was in 1956 by Cowan and Reines; the muon and tau neutrinos were discovered in 1962 and 2000, respectively. However, the neutrino is still a mystery.

Firstly, for some time now the neutrino mass has been known to be extremely small, but distinctly different from zero (the evidence for neutrino oscillations which gives rise to this knowledge is discussed below). This fact is theoretically interesting since it (somewhat contrary to intuition) points to a new large fundamental mass scale and thus to new physics beyond the Standard Model of particle physics. Consequently, a number of experiments have been devised to precisely measure the masses of the neutrinos and to determine their mass hierarchy. Three different methods are used: measurement of the β decay energy spectrum, the detection of neutrinoless double-β decay and cosmological measurements of the cosmic microwave background.

The last method relies on determining the influence of neutrino masses on structure formation in the universe and on primordial nucleosynthesis. Although the model uncertainties are rather large, an upper limit for the sum of the three neutrino masses of less then 0.28 eV has been derived.

In the case of β decay measurements, a precise measurement of the endpoint in the decay energy spectrum is aimed for. The relevant experiments, like the Mainz–Troitsk experiment, have achieved precisions of the order of 2 eV, and new experiments like KATRIN aim at a measurement with a precision of about 0.2 eV for the electron neutrino[1].

Neutrinoless double-β decay experiments search for extremely rare decays of certain isotopes which can only take place if the neutrino has mass and is at the same time its own antiparticle. Past or running experiments to be mentioned here are the Heidelberg–Moscow collaboration, CUORICINO and EXO-200[2]. The question whether the neutrino (the only neutral fermion in the Standard Model!) is its own antiparticle or not (Majorana neutrino instead of Dirac neutrino) is in itself of fundamental interest. Not only would neutrinoless double-β decay imply lepton number violation by 2 units; Majorana-type neutrinos also have implications for the question of CP violation in the lepton sector of the Standard Model and, as a consequence, for leptogenesis.

Strong evidence for non-zero neutrino masses and some information about the mass hierarchy of neutrinos has been obtained from neutrino oscillation observations. In 1998, following results from for example IMB and Kamiokande, the Super-Kamiokande experiment reported on the observation of discrepancies between data and predictions for atmospheric neutrino fluxes. The collaboration investigated the zenith-angle distribution of the ratio of muons to electrons from atmospheric neutrino interactions in the low-energy (few GeV) regime and found a value significantly smaller than two that would be expected from pion and muon decay. This pointed to a lower-than-expected ratio of atmospheric muon to electron neutrinos, $(\nu_\mu + \bar{\nu}_\mu)/(\nu_e + \bar{\nu}_e)$, for neutrinos which travelled a long distance after their production (large zenith angles). The findings were confirmed by other experiments like Soudan 2 and MACRO. Clear evidence for muon neutrino disappearance has now also been seen in accelerator-based experiments like K2K and MINOS.

---

1) Further developments in this direction will be made by the MARE, MIBETA and MANU experiments.
2) In the future also GERDA, CUORE, NEMO-3, SNO+ and COBRA will play a role.

In parallel, the Super-Kamiokande collaboration discovered a deficit (with respect to the Standard Solar Model prediction)[3] of charged-current reactions induced by solar electron neutrinos, which was confirmed by SNO. On the other hand, using neutral-current interactions SNO could show that the *total* neutrino flux is according to expectations.

Quickly oscillations between different neutrino flavour eigenstates composed of mixtures of mass eigenstates (which propagate with different velocities) were suggested as a possible explanation, requiring the neutrinos to have non-zero masses. In the case of disappearing atmospheric muon neutrinos the transition from muon to tau neutrinos, $\nu_\mu \to \nu_\tau$, was held to be mainly responsible[4]. For the solar ν_e disappearance, oscillations $\nu_e \to \nu_{\mu,\tau}$ were assumed. At reactor-based experiments like KAMLAND also the behaviour of anti-electron neutrinos is investigated (KAMLAND has reported a significant disappearance signal).

Considering transitions between all three lepton generations (electron, muon, tau), the oscillation scenario has room for five mixing parameters – namely the three mixing angles θ_{ij} ($i,j = 1,2,3$) and two independent squared mass differences between the three mass eigenstates, Δm_{ij}^2. Different experiments have different sensitivities to these quantities, and numerous measurements have been performed. Here only a few main results are summarised:

- Together, atmospheric and accelerator neutrino experiments (Super-Kamiokande, Soudan, MACRO, K2K, Minos) suggest almost pure $\nu_\mu \to \nu_\tau$ oscillations with large mixing angle θ_{23} of about 45°.
- Solar neutrino experiments provide, via electron–neutrino disappearance, the highest sensitivity to the mixing angle θ_{12}, the best values[5] currently being of the order of 33°; in addition, they provide access to the sign of the squared mass difference Δm_{12}^2.
- In addition to the neutrino mass results mentioned above, the Super-Kamiokande results are suggestive of a minimum neutrino mass of the order of 0.05 eV.
- The highest precision for $|\Delta m_{13}^2|$ ($|\Delta m_{12}^2|$) is achieved by the K2K and MINOS accelerator experiments (KAMLAND reactor-based experiment). The currently quoted best-fit values (assuming $|\Delta m_{13}^2| \cong |\Delta m_{23}^2|$) are $|\Delta m_{13}^2| = 2.40 \times 10^{-3}$ eV^2 and $|\Delta m_{12}^2| = 7.65 \times 10^{-5}$ eV^2.

---

3) Historically, the Homestake experiment was the first to report on a neutrino deficit; it was later followed by a number of other radio-chemical experiments like GALLEX, GNO and SAGE which made use mainly of gallium and chlorine.

4) Recently, the OPERA collaboration reported on the first observation of tau neutrinos from oscillations $\nu_\mu \to \nu_\tau$ – before that, only ν_μ disappearance had been observed by atmospheric and accelerator neutrino experiments.

5) The preferred solution to the solar/reactor neutrino disappearance is the so-called LMA-MSW interpretation, with a large mixing angle $\theta_\odot$ and taking into account matter effects in the sun (the "Mikheyev–Smirnov–Wolfenstein" effect). It is considered a striking feature of neutrino physics that both the atmospheric and the solar mixing angles are large, in contrast to the quark mixing in the CKM matrix (see Chapter 8).

- The quantities describing the 1–3 mixing are under intense investigation: as of now, the mixing angle θ_{13} seems to prefer low values – the question is how low a value is realised, and whether or not it is compatible with zero. First limits were derived by the CHOOZ experiment, and future reactor and accelerator experiments will further investigate this via anti-electron neutrino disappearance[6] and the appearance of electron neutrinos in a muon-neutrino beam[7].

Future emphasis will be on determining in more detail the masses of the eigenstates, the mass hierarchy (or equivalently the sign of the squared mass difference Δm_{23}^2), and on determining the as of now only weakly constrained mixing angle θ_{13}. Furthermore, the question is still open whether CP violation in the neutrino sector exists or not.

All in all, neutrino physics is an extremely active field with many ongoing and planned facilities, and very promising prospects for the coming years.

1.3.2
Astroparticle Physics

Astroparticle physics aims at measuring elementary particles of astronomical origin, and at understanding their origin, production, and acceleration mechanism. In doing so, astronomical and cosmological questions can be addressed. Various different elementary particles are used for astroparticle studies, and the relevant experiments are well adapted to the instrumental challenges (here only a small fraction of all experiments and results can be mentioned).

Neutrinos of all energies (from a few keV to the highest measured energies) are used for astroparticle physics studies. The existence of neutrinos of the highest energies (several TeV and above) is a sign of hadronic acceleration processes in the universe. Because of their low reaction cross sections, neutrinos allow cosmic sources to be observed which otherwise would remain hidden by dense matter distributions or at large distances. The small cross section, on the other hand, makes them hard to detect. Sources that are held responsible for neutrino production in the universe are both galactic (supernova remnants, pulsars, nebulae, binary systems) and extra-galactic (active galactic nuclei, other point sources on the diffuse neutrino background). First-generation high energy neutrino telescopes showed that large-scale facilities may be used to measure neutrinos via the Cerenkov light emitted by reaction products in clean water (Lake Baikal, Antares) or ice (AMANDA) using a lattice of photomultiplier tubes. Currently, kilometre-scale experiments (for example IceCube at the geographic south pole, km3net, …) are being constructed or designed that will significantly increase the sensitivity.

Medium-energy neutrinos (typical energies of 1–1000 GeV) are mainly interesting for atmospheric neutrino oscillation studies and as backgrounds to low-background experiments (like, for example, the search for proton decay). Low-energy

---

6) DoubleCHOOZ, Daya Bay, RENO.
7) MINOS, T2K, planned: NOVA.

neutrinos mostly have solar or supernova origins. Besides their use for neutrino studies (mass, oscillation, etc. – see Section 1.3.1), solar neutrinos allow the solar core region, the processes taking place there and the related neutrino fluxes to be studied (test of the Standard Solar Model).

Like neutrinos (and unlike charged particles), photons traverse the universe undeflected and point back to their production origin, thus allowing the identification of point sources in the sky. Very high or ultra high energy (VHE/UHE) photons with energies of several TeV are assumed to stem from even more energetic primary particles that are probably accelerated via diffuse shock acceleration in the expanding blast waves of supernova remnants. They therefore open a window on the "accelerator sky". There are numerous gamma-ray observatories, typically based on imaging air Cerenkov telescopes, which have in the past decade identified a rich and diverse collection of VHE sources. Examples are H.E.S.S., MAGIC, and VERITAS. Their main observation is a photon energy spectrum that falls with a power of the energy.

Figure 1.3 shows the energy spectrum of cosmic rays (mostly protons) as measured by numerous experiments. The figure shows only the high energy part of the cosmic ray spectrum above 10^{13} eV, scaled by $E^{2.7}$. The plot shows that cosmic rays with energies of more than 10^{20} eV have been observed. Three distinct regimes are visible in Figure 1.3: up to about 3×10^{15} eV (where the so-called "knee" is located), particles are assumed to be accelerated via the above-mentioned shock acceleration mechanism in our galaxy. Between the "knee" and the "ankle" at about 10^{18}–

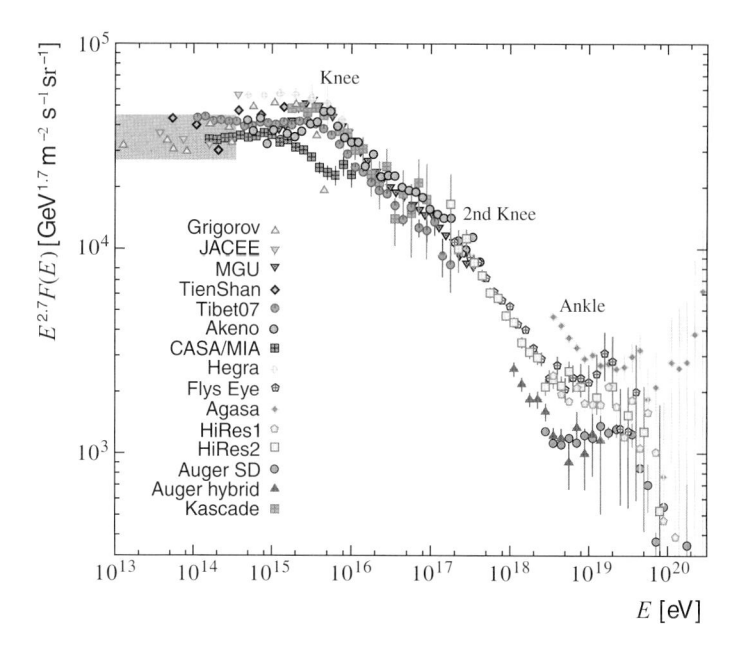

Figure 1.3 Overview of cosmic-ray energy measurements at high energies, including the "knee" and the "ankle" (Source: PDG 2008).

10^{19} eV, the particles are still assumed to be galactic in origin, although the acceleration mechanism is not understood. Beyond the "ankle", particles are assumed to be extra-galactic in origin, and extend in energy up to the so-called Greisen–Zatsepin–Kuzmin (GZK) cutoff at which they start to interact with the cosmic microwave background and thus lose energy. This prominent feature of the spectrum was quite recently observed by the HiRes and Pierre Auger experiments.

Cosmic ray experiments for the highest energy particles are typically based on the reconstruction of cascades of secondary particles produced when the primary impinges on the upper atmosphere. These cascades, or air showers, can be studied using either nitrogen fluorescence in the atmosphere or by sampling shower particles on the ground (or by combining the two methods like in the Pierre Auger Observatory which uses water Cerenkov detectors overlooked by fluorescence telescopes). The experiments are also able, by means of shower-shape analysis, to distinguish between various primary particles, for example proton-induced versus neutrino-induced events.

Astroparticle physics is a large and diverse field which employs numerous different methods and measurements. Only a "multi-messenger" approach, combining information from neutrinos, VHE gamma rays and cosmic rays of highest energies will, in the end, be able to give a full and detailed explanation of the sources of all cosmic particles and their acceleration mechanism. Through this, we will gain deeper insights into astrophysical and cosmological questions, like that of the dark matter in the universe.

A substantial fraction of the matter content of our universe is constituted from unknown particles ("dark matter", see Section 1.2). Weakly interacting massive particles (WIMPs) are a popular dark-matter candidate. These WIMPs are supposed to be located mainly in the galactic halo. One possibility is that they are then swept up by the sun as the solar system moves about, and will occasionally scatter elastically with nuclei in the sun, eventually becoming gravitationally bound. After sufficient time, an equilibrium between WIMP capture and annihilation (for example $\chi\chi \rightarrow W^+W^-$ or $\chi\chi \rightarrow b\bar{b}$ if the WIMP is a supersymmetric neutralino – see Chapter 7) in the sun might build up, allowing the annihilation products and their decay products (and here mainly neutrinos because of their small interaction cross section) to leave the sun and be detected on earth using high energy neutrino telescopes like IceCube. The corresponding signal can be easily predicted, given a certain WIMP mass, and measurements can thus be used to confirm or rule out models or regions in the supersymmetry parameter space; for certain regions the strongest limits presently come from data from the neutrino telescopes Super-Kamiokande, AMANDA, IceCube, and soon also from ANTARES.

Another approach is followed by experiments which try to detect the elastic collisions of the WIMPs in the galactic halo with a detector on earth, as the earth moves through the halo. The aim is to measure the nuclear recoil of the produced neutrinos (here a clear signature for WIMPs would be annual variations of the observed signal)[8]. It should be noted that these indirect searches for dark-matter candidates are complementary to direct, accelerator-based searches.

1.3.3
Heavy Ion Physics

The investigation of heavy ion collisions is a substantial part of the LHC programme, both in the dedicated ALICE experiment and in the omni-purpose experiments ATLAS and CMS. While the LHC prospects are not discussed in this book, some discussion is given here on the current status of the field.

First indications for the creation of a so-called *quark–gluon plasma* (QGP) were obtained in the 1980s and 1990s at the CERN SPS, leading to the announcement of indirect evidence for a new state of matter by CERN in 2000. In recent years, heavy ion collisions using mainly gold and copper atoms have been investigated at the Brookhaven RHIC collider: here, four experiments have been or are still taking data: BRAHMS, PHOBOS, PHENIX, and STAR[9].

The physics questions investigated centre around the specific features of the high temperature/high energy environment. One topic is the behaviour of the gluon density in the nucleon in the limit of extremely small momentum fractions, x, where presumably perturbative QCD is not applicable and saturation effects might set in. The term *colour glass condensate* (CGC) has been created to describe the behaviour of gluons in this kinematic regime. The CGC is already relevant for proton–proton collisions but is still more important in collisions of nuclei, as here the projected area density of gluons should be higher and thus effects of gluon saturation should be stronger. In addition, the knowledge of the low-x gluon distribution determines the initial state of the created matter in heavy ion collisions. Therefore, knowledge about gluon saturation will be crucial for the interpretation of heavy ion data. However, the current data do not allow the details of the gluon density behaviour in this region to be pinned down (although RHIC data can be described using QCD predictions with some assumptions on the gluon at small x).

Arguably the most striking observation at RHIC is the observation of a strong global anisotropy of the azimuthal particle distributions, called elliptic flow. The creation of elliptic flow requires early equilibration of the produced matter, the absolute value points to an extremely low viscosity. Thus, the system produced at RHIC appears not at all as a weakly interacting plasma, but rather as a strongly coupled, close-to-perfect liquid. Detailed features of elliptic flow favour an evolution of the system undergoing a phase transition from the quark–gluon plasma, and the relevance of parton degrees of freedom is clearly visible in scaling properties of the data.

---

8) The DAMA collaboration, which performs dark-matter searches using scintillation techniques, has reported on a controversial 8.2σ signal for rather low mass dark-matter candidates with surprisingly high cross section.
9) The first two experiments finished data-taking in 2006; PHENIX and STAR are still taking data (or are being upgraded).

Another interesting question is that of the relative abundances of hadrons produced from the hot, dense system. It is found that the abundances can be described by a statistical model. In small systems (e. g. proton–proton collisions) such models are also applicable if one accounts for the fact that strangeness conservation provides a very strong constraint, leading to the so-called canonical suppression of strangeness. For large numbers, these constraints can be relaxed such that central heavy ion collisions should exhibit a strangeness enhancement compared to proton–proton collisions. This is indeed observed experimentally by CERN and RHIC experiments. Although the good agreement of the abundances with the statistical model does not prove thermal behaviour, it is striking that the temperature parameters extracted coincide with the predicted values of the transition temperature.

A further issue is the effect of the dense quark–gluon medium on final-state properties. The extended quark–gluon cloud will influence strongly interacting particles, and measurements of these influences will in turn allow conclusions on the properties of the medium. One effect observed by the RHIC experiments is that of a significant reduction of parton energies: in contrast to photons, which traverse the strongly interacting dense medium unaffected, quarks and gluon (and consequently also hadrons like protons or pions or even hadronic jets) lose a significant fraction of their energy in strong interactions before leaving the QGP (*jet quenching*). These effects are also observed in particle correlations: in hard collisions, a pair of back-to-back particles balanced in transverse momentum (quarks, gluons, or also photons) is typically created. Depending on the position of the hard interaction inside the QGP, one of the particles might have a longer path inside the plasma, allowing it to lose more of its energy and thus breaking the momentum balance. The use of different particles and particle correlations (single hadrons, hadron–hadron, photon–hadron, jet–jet, jet–photon) allows different aspects of the energy-loss process and the details of the QGP to be studied.

Concerning the future of heavy ion physics (beyond the LHC), it is very much open, and decisions about which direction to take (higher energy versus better detectors etc.) will only be possible in the light of the first LHC heavy ion data.

1.3.4
Spin Physics

It is common knowledge that the spin of the proton is 1/2, and for a long time this spin was assumed to be due to the spins of the (three) valence quarks in the nucleon which are also spin-1/2 particles. However, in 1988 the European Muon Collaboration (EMC) at CERN initiated the so-called *spin crisis* or *spin puzzle* when it discovered that the quark-spin contribution to the nucleon spin was far below 50%. Later experiments at CERN and SLAC confirmed these findings.

For more than a decade, various experiments were performed to solve this puzzle and to measure precisely the various contributions to the nucleon spin, namely the contributions of the spins of the (valence and sea) quarks and the gluons, $\Delta\Sigma$ and ΔG, and the orbital angular momentum of these two contributions, L_q and L_g. Among these experiments are HERMES and COMPASS (polarised deep inelastic scattering), various experiments like CLAS at Jefferson Lab (JLab), and STAR, PHENIX and BRAHMS at the heavy ion collider RHIC (in the proton–proton mode). A distinguishing feature of all these spin experiments is that they require at least one polarised ingredient – beam or target. The most precisely measured value is $\Delta\Sigma$ for which HERMES has determined a value of 0.33 ± 0.011 (theo.) ± 0.025 (exp.) ± 0.028 (evol.). Numerous spin physics results have been obtained in recent years. Important examples are the determination of polarisation-dependent structure functions, the decomposition of the different quark flavour contributions using identified pions and kaons in polarised deep inelastic scattering, the confirmation of the opposite orientation of the up versus the down valence-quark spins, and determinations of the gluon-spin contribution ΔG (which turns out to be very small).

Figure 1.1 shows the proton parton distribution function measured by the HERA experiments H1 and ZEUS as a function of the longitudinal momentum fraction, x. In contrast to this *longitudinal* PDF, spin physics experiments have also determined the transverse structure of the nucleon. Transverse-momentum-dependent parton distribution functions are recognised as a tool to study spin–orbit correlations, hence providing experimental observables for studying orbital angular momentum via the measurement of certain azimuthal asymmetries.

Generalised parton distributions (GPDs) can be accessed via measurements of azimuthal spin or charge asymmetries in exclusive reactions for which the complete spectrum of produced particles is known. The GPDs give a three-dimensional representation of the nucleon which is often referred to as *nuclear tomography* (two spatial transverse dimensions and one longitudinal momentum dimension). With GPDs, it becomes possible to visualise the transverse position of quarks while scanning different longitudinal momentum slices. In addition, certain moments of the GPDs can in principle give constraints on the total angular momentum carried by quarks in the nucleon.

The future of spin physics offers a rich picture: in the near future, new polarised data will be taken with COMPASS and at JLab (CLAS and Hall A). After 2012, JLab will be upgraded. In the far future, an electron–ion collider (EIC) is foreseen in the US. There are two competing concepts – eRHIC at Brookhaven and eLIC at JLab. Probably at least one of these will be realised. There are also plans for an electron–nucleon collider (ENC) at GSI with $\sqrt{s} = 40$ GeV and also involving polarised beams. And finally, there are plans to provide polarised electrons also for the LHeC, the electron–proton machine foreseen for the LHC tunnel.

Further Reading

There are very many introductory particle physics books as well as ones on more specialised topics. The "best" book on a given topic is largely a matter of taste – we do not even agree on which book we like best. We list here a few introductory books as well as ones on those topics briefly discussed in this chapter, but otherwise not covered in this book.

Griffiths, D. (2008) *Introduction to Elementary Particles*, 2nd revised edn, Wiley-VCH Verlag GmbH.

Berger, C. (2007) *Elementarteilchenphysik – von den Grundlagen zu den modernen Experimenten*, Springer, Berlin.

Cahn, R. and Goldhaber, G. (2009) *The Experimental Foundations of Particle Physics*, 2nd edn, Cambridge University Press.

Particle Data Group, Amsler, C. et al. (2008) Review of Particle Physics, *Phys. Lett. B*, **667** 1, http://pdg.lbl.gov/ – contains a whole series of up-to-date articles on all aspects of particle physics as well as astroparticle physics.

Giunti, C. and Kim, C.W. (2007) *Fundamentals of Neutrino Physics and Astrophysics*, Oxford University Press.

Zuber, K. (2004) *Neutrino Physics*, IOP Publishing, Bristol.

Winter, K. (2008) *Neutrino Physics*, second edition, Cambridge University Press.

Longair, M. (1994) *High Energy Astrophysics*, vols. 1 and 2, Cambridge University Press.

Grupen, C. (2005) *Astroparticle Physics*, Springer, Berlin.

Perkins, D.H. (2003) *Particle Astrophysics*, 2nd edn, Oxford University Press.

Bartke, J. (2009) *Introduction to Relativistic Heavy Ion Physics*, World Scientific.

Vogt, R.L. (2007) *Ultrarelativistic Heavy-Ion Collisions*, Elsevier, Amsterdam.

Burkardt, M., Miller, C.A., and Nowak, W.-D. (2010) Spin-polarized high-energy scattering of charged leptons on nucleons, *Rep. Prog. Phys.*, **73**, 016201, http://arxiv.org/abs/0812.2208.

2
The Standard Model: Our Picture of the Microcosm

Markus Diehl and Wolfgang Hollik

2.1
Introduction

The Standard Model reflects our current knowledge of fundamental particles and their interactions at scales of about 1 fm and below. Historically, it has emerged as a theory unifying the electromagnetic, weak and strong forces [1, 2].

- The electromagnetic interaction is described by Quantum Electrodynamics (QED), which generalises Maxwell's classical theory of electromagnetism to a relativistic quantum field theory. The principle of gauge invariance – already present at the classical level – proves to be a key ingredient in the quantum theory.
- The weak interaction is responsible for instance for the β decay of the neutron or the decay $\pi^+ \to \mu^+ \nu_\mu$. In the Standard Model Lagrangian, weak and electromagnetic interactions are unified and characterised by an $SU(2) \times U(1)$ gauge symmetry. This electroweak symmetry is spontaneously broken by the vacuum, and electromagnetic and weak interactions become very different at low momentum scales (see Section 2.4.3).
- The strong interaction is described by Quantum Chromodynamics (QCD). This is the theory of quarks and gluons interacting because of their colour charge, which is associated with an $SU(3)$ gauge symmetry. Strongly interacting particles like the proton, neutron or pion emerge as bound states. QCD provides a basis both for the quark model of hadrons and for the parton model (see Section 2.7).

With these ingredients the Standard Model applies not only to the physics of elementary particles, but also underlies phenomena at larger distance scales, both in atomic physics (with QED providing a precise description of the hydrogen spectrum) and in nuclear physics. In nuclear physics the strong force gives rise to the binding of protons and nucleons in nuclei, while the strong, weak and electromagnetic interactions are responsible for nuclear α, β and γ decay, respectively.

Physics at the Terascale, First Edition. Edited by Ian C. Brock and Thomas Schörner-Sadenius.
© 2011 WILEY-VCH Verlag GmbH & Co. KGaA, Weinheim.
Published 2011 by WILEY-VCH Verlag GmbH & Co. KGaA.

2 The Standard Model: Our Picture of the Microcosm

To date, the Standard Model gives a highly successful description of the physics governing the microcosm. Nevertheless there are strong arguments calling for an extension of this description, as already mentioned in Section 1.2. An obvious shortcoming of the Standard Model is that it treats neutrinos as massless, whereas neutrino oscillation experiments provide clear evidence for small but non-zero neutrino masses. It also fails to account for a number of phenomena of the macrocosm, such as gravitation, dark matter and the baryon asymmetry of the universe. Finally, there are conceptual issues suggesting that the Standard Model should be embedded in a more fundamental theory.

The following three sections sketch the construction of the Standard Model, emphasising its key principles of gauge invariance and spontaneous symmetry breaking in a renormalisable quantum field theory. Some important properties of the Standard Model are discussed in subsequent sections of this chapter.

2.2 Local Gauge Invariance

This section reviews the general principle of non-Abelian gauge invariance, which dictates the structure of the interactions between fermions and vector bosons as well as the vector boson self-interactions. It is the generalisation to non-Abelian symmetry groups of the Abelian gauge symmetry found in QED.

The Lagrangian for a free fermion field ψ with mass m has a global U(1) symmetry, that is,

$$\mathcal{L}_0 = \overline{\psi} \left(i\gamma^\mu \partial_\mu - m \right) \psi \tag{2.1}$$

is invariant under the phase transformation

$$\psi(x) \to \psi'(x) = e^{i\alpha} \psi(x) \tag{2.2}$$

for arbitrary real numbers α. One can derive QED by extending the global symmetry to *local* transformations, where $\alpha \to \alpha(x)$ is an arbitrary real function. This necessitates the presence of a vector field A_μ and the "minimal substitution" of the derivative in $\mathcal{L}_0$ by the covariant derivative

$$\partial_\mu \to D_\mu = \partial_\mu + ieA_\mu , \tag{2.3}$$

where e is later identified as the elementary charge. The resulting Lagrangian is invariant under the local gauge transformations

$$\psi(x) \to \psi'(x) = e^{i\alpha(x)} \psi(x) ,$$
$$A_\mu(x) \to A'_\mu(x) = A_\mu(x) - \frac{1}{e} \partial_\mu \alpha(x) , \tag{2.4}$$

which form the electromagnetic gauge group U(1). As an immediate consequence, the invariant Lagrangian describes an interaction of the vector field A_μ with the electromagnetic current $J^\mu_{em} = e\bar\psi\gamma^\mu\psi$,

$$\mathcal{L} = \bar\psi\left(i\gamma^\mu D_\mu - m\right)\psi = \mathcal{L}_0 - J^\mu_{em} A_\mu \, . \tag{2.5}$$

The vector field itself is not yet dynamic since a kinetic term for it is missing. Such a term can easily be added by invoking the well-known expression from classical electrodynamics,

$$\mathcal{L}_A = -\frac{1}{4} F_{\mu\nu} F^{\mu\nu} \, , \tag{2.6}$$

where the field strength tensor $F_{\mu\nu} = \partial_\mu A_\nu - \partial_\nu A_\mu$ is invariant under the local gauge transformation (2.4). A_μ thus becomes the photon field obeying Maxwell's equations.

The preceding construction applies to a fermion carrying positive electric charge e. For a fermion with charge Qe the phase factor in (2.2) is replaced by $e^{i\alpha Q}$ and the covariant derivative in (2.3) by $D_\mu = \partial_\mu + iQeA_\mu$. The electromagnetic current is then $J^\mu_{em} = Qe\bar\psi\gamma^\mu\psi$.

In summary, the three basic steps that yield QED as the gauge theory of the electromagnetic interaction are:

1. identifying the global symmetry of the free Lagrangian;
2. making the symmetry local by substituting ∂_μ with the covariant derivative D_μ, which contains a vector field;
3. adding a kinetic term for the vector field.

These steps can be extended to the case of non-Abelian symmetries as follows:

1. The non-interacting system is described by a multiplet $\Psi = (\psi_1, \psi_2, \ldots, \psi_n)^T$ of fermion fields with mass m. An example with $n = 3$ is the triplet $(q_1, q_2, q_3)^T$ of fields describing the three possible colour states of a quark. The non-interacting Lagrangian

$$\mathcal{L}_0 = \bar\Psi\left(i\gamma^\mu\partial_\mu - m\right)\Psi \tag{2.7}$$

with $\bar\Psi = (\bar\psi_1, \ldots, \bar\psi_n)$ is invariant under global transformations

$$\Psi(x) \to U(\alpha^1, \ldots, \alpha^N)\Psi(x) \, , \tag{2.8}$$

with a unitary $n \times n$ matrix U parametrised by N real parameters $\alpha^1, \ldots, \alpha^N$. The set of matrices $U(\alpha^1, \ldots, \alpha^N)$ forms a group, G, under multiplication, which is in general non-Abelian. One can write

$$U(\alpha^1, \ldots, \alpha^N) = \exp(i\alpha^a T^a) \, , \tag{2.9}$$

where here and in the following, repeated group indices $a = 1, \ldots, N$ are to be summed over. Since U is unitary, the $n \times n$ matrices $T^1, \ldots, T^N$ are Hermitian. They are called the generators of the group and satisfy commutation relations

$$[T^a, T^b] = if^{abc}T^c \, , \tag{2.10}$$

where the structure constants f^{abc} are real numbers characteristic of the group. Conventionally, the generators are normalised as

$$\text{Tr}(T^a T^b) = \frac{1}{2}\delta^{ab}. \tag{2.11}$$

Physically important examples for G are the groups $SU(n)$ of $n \times n$ unitary matrices with determinant equal to 1. In particular, $SU(2)$ has $N = 3$ generators given by

$$T^a = \sigma^a/2, \tag{2.12}$$

where σ^a are the Pauli matrices; while $SU(3)$ has $N = 8$ generators

$$T^a = \lambda^a/2 \tag{2.13}$$

conventionally expressed in terms of the so-called Gell-Mann matrices λ^a.

2. The global symmetry can be extended to a local symmetry by converting the constants α^a in (2.9) into real functions $\alpha^a(x)$ and simultaneously introducing a covariant derivative in the free Lagrangian (2.7) via the replacement

$$\partial_\mu \to D_\mu = \partial_\mu - ig W_\mu. \tag{2.14}$$

One has thus introduced a vector field W_μ together with a coupling constant g, which are analogues of A_μ and e in QED[10]. Since D_μ acts on the n-dimensional column Ψ, the vector field is an $n \times n$ matrix. It can be expanded in terms of the generators as

$$W_\mu(x) = T^a W_\mu^a(x). \tag{2.15}$$

In this way, a set of N fields $W_\mu^a(x)$, the gauge fields, enters the Lagrangian (2.7) and induces an interaction term

$$\mathcal{L} = \mathcal{L}_0 + \mathcal{L}_{\text{int}}, \quad \mathcal{L}_{\text{int}} = g \overline{\Psi} \gamma^\mu W_\mu \Psi = -J^{a,\mu} W_\mu^a, \tag{2.16}$$

which describes the interaction of N currents $J^{a,\mu} = -g\overline{\Psi}\gamma^\mu T^a \Psi$ with the gauge fields W_μ^a. The local gauge transformation under which $\mathcal{L}$ is invariant is constructed from the matrix $U(\alpha^1(x), \ldots, \alpha^N(x))$ as

$$\Psi \to \Psi' = U\Psi,$$

$$W_\mu \to W'_\mu = U W_\mu U^{-1} - \frac{i}{g}(\partial_\mu U) U^{-1}. \tag{2.17}$$

The gauge transformation for the vector field looks more familiar when written for the components and expanded for infinitesimal $\alpha^a(x)$:

$$W_\mu^a \to W_\mu'^a = W_\mu^a + \frac{1}{g}\partial_\mu \alpha^a + f^{abc} W_\mu^b \alpha^c. \tag{2.18}$$

---

10) Note the different signs in the Abelian and non-Abelian covariant derivatives in (2.3) and (2.14). The convention where $D_\mu = \partial_\mu + ig W_\mu$ is also used in the literature. This sign change has no physical significance as it corresponds to changing the overall phase of the non-Abelian vector field W_μ.

The term with α^a corresponds to (2.4) in the Abelian case, whereas the term with f^{abc} is of pure non-Abelian nature.

The Lagrangian for a multiplet $\Phi = (\phi_1, \ldots \phi_n)^T$ of scalar fields is obtained by an analogous construction, with the minimal substitution

$$(\partial_\mu \Phi)^\dagger (\partial^\mu \Phi) - m^2 \Phi^\dagger \Phi \rightarrow (D_\mu \Phi)^\dagger (D^\mu \Phi) - m^2 \Phi^\dagger \Phi \tag{2.19}$$

and the same covariant derivative D_μ as in (2.14). This finds its application in the Higgs sector of the Standard Model (see Section 2.4.3).

3. The kinetic term for the W fields can be obtained from a generalisation of the electromagnetic field strength tensor $F_{\mu\nu}$ in (2.6),

$$F_{\mu\nu} = T^a F^a_{\mu\nu} = \partial_\mu W_\nu - \partial_\nu W_\mu - ig[W_\mu, W_\nu], \tag{2.20}$$

with the N components

$$F^a_{\mu\nu} = \partial_\mu W^a_\nu - \partial_\nu W^a_\mu + g f^{abc} W^b_\mu W^c_\nu. \tag{2.21}$$

Under the gauge transformation in (2.17) the field strength transforms as

$$F_{\mu\nu} \rightarrow F'_{\mu\nu} = U F_{\mu\nu} U^{-1}. \tag{2.22}$$

As a consequence, the trace $\text{Tr}(F_{\mu\nu} F^{\mu\nu})$ is gauge invariant and provides the so-called Yang–Mills Lagrangian

$$\mathcal{L}_W = -\frac{1}{2}\text{Tr}(F_{\mu\nu} F^{\mu\nu}) = -\frac{1}{4} F^a_{\mu\nu} F^{a,\mu\nu} \tag{2.23}$$

as the non-Abelian analogue of the kinetic term (2.6) of the photon field. The part of $\mathcal{L}_W$ that is quadratic in the W fields describes the free propagation of the gauge bosons, whereas the cubic and quartic terms describe self-interactions between the vector bosons that are completely determined by the gauge symmetry:

$$\begin{aligned}\mathcal{L}_W = &-\frac{1}{4}\left(\partial_\mu W^a_\nu - \partial_\nu W^a_\mu\right)\left(\partial^\mu W^{a,\nu} - \partial^\nu W^{a,\mu}\right) \\ &- \frac{g}{2} f^{abc} \left(\partial_\mu W^a_\nu - \partial_\nu W^a_\mu\right) W^{b,\mu} W^{c,\nu} \\ &- \frac{g^2}{4} f^{abc} f^{ade} W^b_\mu W^c_\nu W^{d,\mu} W^{e,\nu}.\end{aligned} \tag{2.24}$$

Note that the Lagrangians in (2.6) and (2.24) describe *massless* vector bosons. Mass terms $\frac{1}{2} m^2 A_\mu A^\mu$ and $\frac{1}{2} m^2 W^a_\mu W^{a,\mu}$ for the gauge fields are not invariant under a gauge transformation and thus break the gauge symmetry.

2.3 Formulation of QCD

Quantum Chromodynamics is the gauge theory of the strong interaction. It is formulated using the principles of the previous section for the specific case of the symmetry group $SU(3)_c$, where the subscript c indicates that the quantum number associated with the symmetry is called "colour". The elementary fermions of the theory are quarks in three different colour states, described by triplets of fermion fields $q = (q_1, q_2, q_3)^{\mathrm{T}}$ for each quark flavour $q = u, d, s, c, b, t$. The covariant derivative acting on the quark triplets reads

$$D_\mu = \partial_\mu - i g_s \frac{\lambda^a}{2} G_\mu^a . \qquad (2.25)$$

It involves the strong coupling constant g_s, which in analogy to the fine-structure constant in QED is often expressed in terms of $\alpha_s \equiv g_s^2/(4\pi)$. The gauge fields G_μ^a (with $a = 1, \ldots 8$) describe the gluon in its eight different colour states. Introducing the gluon field strengths

$$G_{\mu\nu}^a = \partial_\mu G_\nu^a - \partial_\nu G_\mu^a + g_s f^{abc} G_\mu^b G_\nu^c , \qquad (2.26)$$

one obtains the Lagrangian of QCD according to the rules of the preceding section as

$$\begin{aligned}\mathcal{L}_{\mathrm{QCD}} &= \sum_q \bar{q} \left(i \gamma^\mu D_\mu - m_q \right) q + \mathcal{L}_G \\ &= \sum_q \bar{q} \left(i \gamma^\mu \partial_\mu - m_q \right) q - \sum_q J_q^{a,\mu} G_\mu^a - \frac{1}{4} G_{\mu\nu}^a G^{a,\mu\nu} ,\end{aligned} \qquad (2.27)$$

where the summation index, q, runs over the six quark flavours. The Lagrangian contains the interaction of the quark currents

$$J_q^{a,\mu} = -g_s \bar{q} \gamma^\mu \frac{\lambda^a}{2} q \qquad (2.28)$$

with the gluon fields, as well as triple and quartic gluon self-interactions as specified in (2.24). The quark mass, m_q, appears in QCD as a free parameter and is different for each quark flavour. Its origin is of electroweak nature and will be discussed in Section 2.4.4.

Despite the apparent simplicity of the Lagrangian (2.27), the dynamics of QCD is highly non-trivial and gives rise to a wide range of phenomena. A few of these will be discussed in Section 2.7, and a detailed account of the applications of QCD to collider processes is given in Chapter 4.

2.4
Formulation of the Electroweak Standard Model

The elementary fermions in the Standard Model appear in three families (or generations) of leptons and quarks. The electroweak symmetry group is $SU(2) \times U(1)$. The symmetry transformations act differently on left-handed and right-handed fermion fields, which are projected out from a usual Dirac field ψ by

$$\psi_L = \frac{1 - \gamma_5}{2} \psi , \quad \psi_R = \frac{1 + \gamma_5}{2} \psi . \qquad (2.29)$$

2.4 Formulation of the Electroweak Standard Model

The $SU(2)$ part of $SU(2) \times U(1)$ is called the weak isospin group, with associated quantum numbers I and I_3. Left-handed fields have $I = 1/2$ and form doublets

$$\begin{pmatrix} \nu_e \\ e \end{pmatrix}_L, \quad \begin{pmatrix} u \\ d \end{pmatrix}_L, \quad \begin{pmatrix} \nu_\mu \\ \mu \end{pmatrix}_L, \quad \begin{pmatrix} c \\ s \end{pmatrix}_L, \quad \begin{pmatrix} \nu_\tau \\ \tau \end{pmatrix}_L, \quad \begin{pmatrix} t \\ b \end{pmatrix}_L, \tag{2.30}$$

which transform with a unitary matrix $U = \exp(i\alpha^a \sigma^a/2)$, where $a = 1, 2, 3$. The right-handed fields

$$e_R, u_R, d_R, \quad \mu_R, c_R, s_R, \quad \tau_R, t_R, b_R \tag{2.31}$$

have $I = 0$, that is, they are singlets that are invariant under the weak isospin transformation. Note that there are no right-handed neutrinos in the Standard Model, where neutrinos are treated as strictly massless. Given that the actual neutrino masses are so small, this simplification is valid for most situations.

For each of the preceding doublets or singlets, the $U(1)$ part of $SU(2) \times U(1)$ corresponds to multiplication by a phase factor $e^{i\alpha Y/2}$, where Y is called weak hypercharge. The electric charge of a particle is given by the Gell-Mann–Nishijima relation

$$Q = I_3 + \frac{Y}{2}. \tag{2.32}$$

For definiteness, from now on the group underlying the electroweak interaction will be called $SU(2)_L \times U(1)_Y$.

The electromagnetic gauge group $U(1)_{em}$ appears as a subgroup of $SU(2)_L \times U(1)_Y$, obtained by combining a hypercharge transformation with a particular isospin transformation. For the isospin doublets (2.30) the corresponding transformation matrix is

$$e^{i\alpha\sigma_3/2} e^{i\alpha Y/2} = \begin{pmatrix} e^{i\alpha/2} & 0 \\ 0 & e^{-i\alpha/2} \end{pmatrix} e^{i\alpha Y/2}, \tag{2.33}$$

which together with the relation (2.32) gives a phase factor $e^{i\alpha Q}$ for each component of the doublet field. For the isospin singlets in (2.31) one simply has $e^{i\alpha Q} = e^{i\alpha Y/2}$. The assignment of quantum numbers to the lepton and quark fields is given in Table 2.1.

The structure just described can be embedded in a gauge invariant field theory of the unified electromagnetic and weak interactions by interpreting $SU(2)_L \times U(1)_Y$ as the group of gauge transformations under which the Lagrangian is invariant. The full symmetry must be broken down to the electromagnetic gauge symmetry, since the photon is massless while the gauge bosons $W^\pm$ and Z have non-zero mass. In the Standard Model, the required symmetry breaking is achieved by the Higgs mechanism and with minimal particle content, namely a single Higgs field which is a doublet under $SU(2)_L$.

The full electroweak Lagrangian consists of gauge, fermion, Higgs and Yukawa terms,

$$\mathcal{L}_{ew} = \mathcal{L}_{WB} + \mathcal{L}_F + \mathcal{L}_H + \mathcal{L}_Y, \tag{2.34}$$

Table 2.1 The weak isospin I_3 and hypercharge Y for the left-handed and right-handed leptons and quarks, as well as their electric charge Q. The first three rows correspond to the first, second and third fermion generation.

	ν_{eL}	e_L	e_R	u_L	d_L	u_R	d_R
	$\nu_{\mu L}$	μ_L	μ_R	c_L	s_L	c_R	s_R
	$\nu_{\tau L}$	τ_L	τ_R	t_L	b_L	t_R	b_R
I_3	$+\frac{1}{2}$	$-\frac{1}{2}$	0	$+\frac{1}{2}$	$-\frac{1}{2}$	0	0
Y	-1	-1	-2	$+\frac{1}{3}$	$+\frac{1}{3}$	$+\frac{4}{3}$	$-\frac{2}{3}$
Q	0	-1	-1	$+\frac{2}{3}$	$-\frac{1}{3}$	$+\frac{2}{3}$	$-\frac{1}{3}$

which are discussed in turn in the following subsections. A complete list of the corresponding Feynman rules can for example be found in [3, 4].

The gauge group of the Standard Model is $SU(3)_c \times SU(2)_L \times U(1)_Y$, and the complete Lagrangian is obtained as the sum of the QCD and electroweak Lagrangians (2.27) and (2.34), minus the free quark term $\sum_q \bar{q}(i\gamma^\mu \partial_\mu - m_q)q$, which appears both in $\mathcal{L}_{QCD}$ and $\mathcal{L}_{ew}$ and must not be double counted. It should be understood that each of the quark fields in (2.30) and (2.31) is a colour triplet $q = (q_1, q_2, q_3)^T$ as discussed in Section 2.3. The colour indices of the quark fields are just summed over in the electroweak Lagrangian; for ease of notation they are omitted in the present section.

2.4.1
Gauge Fields

The gauge field Lagrangian is constructed by applying the method of Section 2.2 separately to each factor of the group $SU(2)_L \times U(1)_Y$. One thus has three vector fields W_μ^a associated with the three generators of $SU(2)_L$ and one vector field B_μ associated with the hypercharge group $U(1)_Y$. The corresponding field strength tensors are

$$W_{\mu\nu}^a = \partial_\mu W_\nu^a - \partial_\nu W_\mu^a + g_2 \epsilon^{abc} W_\mu^b W_\nu^c,$$
$$B_{\mu\nu} = \partial_\mu B_\nu - \partial_\nu B_\mu, \qquad (2.35)$$

where the totally antisymmetric symbol ϵ^{abc} appears as the structure constant of $SU(2)$. Likewise, there are two independent gauge coupling constants, denoted by g_2 for the non-Abelian factor $SU(2)_L$ and by g_1 for the Abelian factor $U(1)_Y$. From the field strength tensors one constructs the pure gauge field Lagrangian

$$\mathcal{L}_{WB} = -\frac{1}{4} W_{\mu\nu}^a W^{\mu\nu,a} - \frac{1}{4} B_{\mu\nu} B^{\mu\nu}, \qquad (2.36)$$

which is invariant under the gauge transformations given in (2.4) and (2.17).

2.4.2
Fermion Fields and Their Gauge Interactions

The kinetic part of the fermion Lagrangian, together with the interaction terms between fermions and gauge bosons, can readily be obtained from the appropriate covariant derivatives for isospin doublets and singlets. One finds

$$\mathcal{L}_F = \sum_{j=1}^{6} \overline{\psi}_L^j i\gamma^\mu \left(\partial_\mu - ig_2 \frac{\sigma^a}{2} W_\mu^a + ig_1 \frac{Y}{2} B_\mu \right) \psi_L^j$$

$$+ \sum_{j=1}^{9} \overline{\psi}_R^j i\gamma^\mu \left(\partial_\mu + ig_1 \frac{Y}{2} B_\mu \right) \psi_R^j \,. \quad (2.37)$$

In the first line ψ_L^j stands for the left-handed doublets of (2.30), and in the second line ψ_R^j represents the right-handed singlets of (2.31). The terms with ∂_μ and B_μ in the first line are understood to contain the two-dimensional unit matrix in isospin space.

Note that there are no fermion mass terms at this stage: since $\overline{\psi}\psi = \overline{\psi}_L\psi_R + \overline{\psi}_R\psi_L$ they would mix left-handed and right-handed fields and thus explicitly break gauge invariance. Instead, fermion masses will be generated dynamically by gauge-invariant Yukawa interactions with the Higgs field.

2.4.3
The Higgs Field and Spontaneous Symmetry Breaking

The Higgs sector of the Standard Model consists of an isospin doublet of complex-valued scalar fields,

$$\Phi(x) = \begin{pmatrix} \phi^+(x) \\ \phi^0(x) \end{pmatrix} \,. \quad (2.38)$$

This doublet has hypercharge $Y = 1$, so that the appropriate covariant derivative is

$$D_\mu = \partial_\mu - ig_2 \frac{\sigma^a}{2} W_\mu^a + ig_1 \frac{1}{2} B_\mu \,. \quad (2.39)$$

Following the Gell-Mann–Nishijima relation (2.32), ϕ^+ carries electric charge +1, whereas ϕ^0 is electrically neutral. In accordance with the prescription in (2.19), the Higgs Lagrangian reads

$$\mathcal{L}_H = (D_\mu \Phi)^\dagger (D^\mu \Phi) - V(\Phi) \,, \quad (2.40)$$

with a gauge invariant self-interaction given by the potential

$$V(\Phi) = -\mu^2 \Phi^\dagger \Phi + \frac{\lambda}{4}(\Phi^\dagger \Phi)^2 = \frac{\lambda}{4}\left(\Phi^\dagger \Phi - \frac{2\mu^2}{\lambda} \right)^2 - \frac{\mu^4}{\lambda} \,. \quad (2.41)$$

The last term on the r.h.s. is a constant, which will be dropped in the following.

In the vacuum, that is, in the ground state of the theory, the potential energy of the field must be minimal. For positive μ^2 and λ the potential $V(\Phi)$ has the form of a "Mexican hat" (when plotted on the z axis as a function of $|\phi^+|$ and $|\phi^0|$ on the x and y axes) and takes its minimum for any field configuration with $\Phi^\dagger \Phi = 2\mu^2/\lambda$. Among these configurations, the vacuum selects the one with

$$\langle \Phi \rangle = \frac{1}{\sqrt{2}} \begin{pmatrix} 0 \\ v \end{pmatrix} \quad \text{with} \quad v = \frac{2\mu}{\sqrt{\lambda}}, \tag{2.42}$$

where the brackets $\langle\,\rangle$ denote the vacuum expectation value. Obviously (2.42) is not invariant under a transformation with $U = \exp(i\alpha^a \sigma^a/2)$ or a phase factor $e^{i\alpha Y/2}$, so that the $SU(2)_L \times U(1)_Y$ gauge symmetry is broken by the vacuum. One speaks of "spontaneous" symmetry breaking (in analogy with a ferromagnetic system, whose Hamiltonian is rotationally symmetric, but whose magnetisation in the ground state singles out a direction in space). However, since the charged field component in (2.42) is $\langle \phi^+ \rangle = 0$ and the electrically neutral component $\langle \phi^0 \rangle = v/\sqrt{2}$ is invariant under an electromagnetic gauge transformation, the $U(1)_{em}$ symmetry is preserved by $\langle \Phi \rangle$, as advertised earlier.

The vacuum is a state containing no particles, and particle excitations correspond to deviations of a field from its vacuum expectation value. To make this explicit one can rewrite the Higgs doublet as

$$\Phi(x) = \frac{1}{\sqrt{2}} \begin{pmatrix} \phi_1(x) + i\phi_2(x) \\ v + H(x) + i\chi(x) \end{pmatrix}, \tag{2.43}$$

where the fields H and χ, ϕ_1, ϕ_2 are real-valued and have zero vacuum expectation values. The Higgs potential then reads

$$V = \mu^2 H^2 + \frac{\mu^2}{v} H \left(H^2 + \chi^2 + \phi_1^2 + \phi_2^2 \right) + \frac{\mu^2}{4v^2} \left(H^2 + \chi^2 + \phi_1^2 + \phi_2^2 \right)^2, \tag{2.44}$$

and one finds that the field H describes an electrically neutral scalar particle, the Higgs boson, with a mass

$$M_H = \sqrt{2}\mu. \tag{2.45}$$

By contrast, no mass terms appear for χ, ϕ_1 and ϕ_2. Small values of these fields correspond to the directions in field space along which $V(\Phi)$ retains its minimal value. This is in agreement with the Goldstone theorem, which states that a continuous symmetry of the Lagrangian that is broken by the vacuum goes along with massless fields, called Goldstone bosons (see e.g. Section 11.1 of [5]). For a local symmetry, these fields can however be eliminated from the Lagrangian by a suitable choice of gauge, which implies that they do not describe physical degrees of freedom (they are sometimes called would-be Goldstone bosons). Indeed, by a local

2.4 Formulation of the Electroweak Standard Model

$SU(2)_L$ transformation the field in (2.43) can be brought into the simple form

$$\Phi(x) = \frac{1}{\sqrt{2}} \begin{pmatrix} 0 \\ v + H(x) \end{pmatrix}, \qquad (2.46)$$

which involves only the physical Higgs field, H. In the corresponding gauge, called the unitary gauge, the Higgs potential (2.44) reads

$$V = \frac{M_H^2}{2} H^2 + \frac{M_H^2}{2v} H^3 + \frac{M_H^2}{8v^2} H^4. \qquad (2.47)$$

Note that both the cubic and the quartic self-interactions of the Higgs field are proportional to M_H^2 for a given v.

Inserting (2.43) into the kinetic term of the Lagrangian (2.40), the couplings of the Higgs to the gauge fields can be obtained, as well as mass terms for the vector bosons in the non-diagonal form

$$\frac{1}{2}\left(\frac{g_2}{2}v\right)^2 \left(W_\mu^1 W^{1,\mu} + W_\mu^2 W^{2,\mu}\right)$$
$$+ \frac{1}{2}\left(\frac{1}{2}v\right)^2 \left(W_\mu^3, B_\mu\right) \begin{pmatrix} g_2^2 & g_1 g_2 \\ g_1 g_2 & g_1^2 \end{pmatrix} \begin{pmatrix} W^{3,\mu} \\ B^\mu \end{pmatrix}. \qquad (2.48)$$

The physical content becomes apparent by a transformation from the fields W_μ^a and B_μ (in terms of which the original gauge symmetry is manifest) to the "physical" fields

$$W_\mu^\pm = \frac{1}{\sqrt{2}} \left(W_\mu^1 \mp i W_\mu^2 \right) \qquad (2.49)$$

and

$$\begin{pmatrix} Z_\mu \\ A_\mu \end{pmatrix} = \begin{pmatrix} \cos\theta_W & \sin\theta_W \\ -\sin\theta_W & \cos\theta_W \end{pmatrix} \begin{pmatrix} W_\mu^3 \\ B_\mu \end{pmatrix}. \qquad (2.50)$$

In terms of these fields the mass term (2.48) is diagonal and has the form

$$M_W^2 W_\mu^+ W^{-\mu} + \frac{1}{2} (Z_\mu, A_\mu) \begin{pmatrix} M_Z^2 & 0 \\ 0 & 0 \end{pmatrix} \begin{pmatrix} Z^\mu \\ A^\mu \end{pmatrix} \qquad (2.51)$$

with

$$M_W = \frac{g_2}{2} v, \quad M_Z = \frac{\sqrt{g_1^2 + g_2^2}}{2} v. \qquad (2.52)$$

The angle in the rotation (2.50) is called the weak mixing angle and is given by

$$\cos\theta_W = \frac{g_2}{\sqrt{g_1^2 + g_2^2}} = \frac{M_W}{M_Z}. \qquad (2.53)$$

The field A_μ remains massless and can thus be identified with the gauge boson of the electromagnetic $U(1)_{\text{em}}$ symmetry, that is, with the photon.

The coupling of the Higgs boson to the physical fields $W_\mu^\pm$, Z_μ and A_μ can be obtained from (2.51) by replacing $M_W \to M_W(v+H)/v$ and $M_Z \to M_Z(v+H)/v$, which gives rise to vertices HZZ, HWW and $HHZZ$, $HHWW$. Expressing the pure gauge Lagrangian (2.36) in terms of the physical fields one obtains their kinetic terms, as well as couplings involving three or four gauge bosons, which give rise to vertices $WW\gamma$, WWZ and $WWWW$, $WW\gamma\gamma$, $WWZZ$, $WWZ\gamma$.

The Higgs mechanism gives masses to W and Z by breaking the $SU(2)_L \times U(1)_Y$ symmetry with a non-zero vacuum expectation value. The original Lagrangian is, however, gauge invariant. This leaves an imprint on the structure of radiative corrections and turns out to be essential for showing that the theory can be renormalised and is thus self-consistent at the quantum level.

2.4.4
Yukawa Interactions: Fermion Masses and Mixing

In the Standard Model, not only the masses of W and Z but also the fermion masses are generated by the spontaneous breaking of the $SU(2)_L \times U(1)_Y$ gauge symmetry. This is achieved by gauge-invariant Yukawa interactions between the Higgs and the fermion fields. For a single generation of leptons and quarks, the Yukawa term in the Lagrangian has the compact form

$$\mathcal{L}_{Y,\text{gen}} = -Y_\ell \overline{L}_L \Phi \ell_R - Y_d \overline{Q}_L \Phi d_R - Y_u \overline{Q}_L \Phi^c u_R + \text{h.c.}, \tag{2.54}$$

where $L_L = (\nu_L, \ell_L)^T$ and $Q_L = (u_L, d_L)^T$ stand for the left-handed lepton and quark doublets, and ℓ is the charged lepton. The charge conjugate of the Higgs doublet Φ is

$$\Phi^c = i\sigma_2 \Phi^* = \begin{pmatrix} \phi^{0*} \\ -\phi^- \end{pmatrix}, \tag{2.55}$$

where ϕ^- denotes the complex conjugate of ϕ^+. Written out in components (2.54) reads

$$\begin{aligned}\mathcal{L}_{Y,\text{gen}} = &- Y_\ell \left(\overline{\nu}_L \phi^+ \ell_R + \overline{\ell}_R \phi^- \nu_L + \overline{\ell}_L \phi^0 \ell_R + \overline{\ell}_R \phi^{0*} \ell_L \right) \\ &- Y_d \left(\overline{u}_L \phi^+ d_R + \overline{d}_R \phi^- u_L + \overline{d}_L \phi^0 d_R + \overline{d}_R \phi^{0*} d_L \right) \\ &+ Y_u \left(\overline{u}_R \phi^+ d_L + \overline{d}_L \phi^- u_R - \overline{u}_R \phi^0 u_L - \overline{u}_L \phi^{0*} u_R \right). \end{aligned} \tag{2.56}$$

The vacuum expectation value of ϕ^0 gives rise to fermion mass terms, where the masses are related to the Yukawa couplings by

$$m_f = Y_f \frac{v}{\sqrt{2}} \tag{2.57}$$

2.4 Formulation of the Electroweak Standard Model

for $f = \ell, d, u$. In the unitary gauge (2.46), the Yukawa Lagrangian becomes particularly simple:

$$\mathcal{L}_{Y,\text{gen}} = -\sum_{f=\ell,d,u} \left(m_f \overline{\psi}_f \psi_f + \frac{m_f}{v} H \overline{\psi}_f \psi_f \right). \tag{2.58}$$

One thus obtains Yukawa interactions between the massive fermions and the physical Higgs field H, with coupling constants proportional to the fermion masses.

In the case of three generations, a more intricate structure arises because the Yukawa terms mix fermions of different generations. The most general gauge invariant Yukawa interaction is given by

$$\mathcal{L}_Y = -\sum_{ij} \left(Y_\ell^{ij} \overline{L}_L^i \Phi \ell_R^j + Y_d^{ij} \overline{Q}_L^i \Phi d_R^j + Y_u^{ij} \overline{Q}_L^i \Phi^c u_R^j \right) + \text{h.c.}, \tag{2.59}$$

where the indices i, j denote the three generations. In this notation u^i denotes the u-type quarks $u^1 = u, u^2 = c, u^3 = t$ and d^i the d-type quarks $d^1 = d$, $d^2 = s, d^3 = b$. The Yukawa couplings Y_f^{ij} are now matrices in generation space with complex-valued entries. The fermion mass terms are obtained from (2.59) by replacing Φ with its vacuum expectation value (2.42), which gives

$$-\frac{v}{\sqrt{2}} \sum_{ij} \left(Y_\ell^{ij} \overline{\ell}_L^i \ell_R^j + Y_d^{ij} \overline{d}_L^i d_R^j + Y_u^{ij} \overline{u}_L^i u_R^j \right) + \text{h.c.} \tag{2.60}$$

The terms involving quark fields can be diagonalised with the help of four unitary matrices $V_{L,u}, V_{R,u}, V_{L,d}, V_{R,d}$, which transform from flavour eigenstates to mass eigenstates,

$$\tilde{q}_L^i = \sum_j V_{L,q}^{ij} q_L^j, \quad \tilde{q}_R^i = \sum_j V_{R,q}^{ij} q_R^j \quad (q = u, d) \tag{2.61}$$

and yield mass matrices $M_u = \text{diag}(m_u, m_c, m_t)$ and $M_d = \text{diag}(m_d, m_s, m_b)$,

$$M_q = \frac{v}{\sqrt{2}} V_{L,q} Y_q (V_{R,q})^\dagger \quad (q = u, d). \tag{2.62}$$

When expressed in terms of the quark masses and the mass eigenstates, the Yukawa interactions of the physical Higgs field with the quarks retain the form given in (2.58), because Φ depends only on the sum $v + H(x)$. Furthermore, introducing the mass eigenstates into the fermion Lagrangian $\mathcal{L}_F$ in (2.37) preserves the form of both the kinetic terms and the interaction terms with the neutral gauge bosons, as both of these have the structure $\sum_i \overline{q}_L^i \dots q_L^i$ or $\sum_i \overline{q}_R^i \dots q_R^i$ and are hence invariant under the unitary transformations (2.61). The same holds for the QCD Lagrangian (2.27). The only modification occurs in the flavour changing quark interaction with the $W^\pm$ bosons in $\mathcal{L}_F$, where the insertion of the mass eigenstates into $\sum_i \overline{u}_L^i \dots d_L^i$ introduces the unitary quark mixing matrix (also called the Cabibbo–Kobayashi–Maskawa or CKM matrix)

$$V_{\text{CKM}} = V_{L,u} (V_{L,d})^\dagger. \tag{2.63}$$

A unitary 3 × 3 matrix has nine real-valued parameters, of which five can be eliminated by redefining the phases of the quark fields (see e. g. Section 20.3 of [5]). Four independent physical parameters remain: three rotation angles and one complex phase. A more detailed discussion of V_{CKM} is given in Chapter 8.

In analogy to (2.61) and (2.62), the matrix Y_ℓ^{ij} in (2.60) can be diagonalised by unitary transformations $V_{\text{L},\ell}$ and $V_{\text{R},\ell}$ that yield mass eigenstates for the charged leptons. Since no neutrino fields appear in (2.60), one is free to additionally transform the left-handed neutrinos with $V_{\text{L},\nu} = V_{\text{L},\ell}$. The analogue of the CKM matrix (2.63) for leptons is then the unit matrix. Generation mixing in the lepton sector only appears if neutrino masses are introduced into the theory.

In terms of fermion mass eigenstates and the physical vector boson fields, the interaction term in the fermion Lagrangian (2.37) can be written as

$$\mathcal{L}_{\text{F,int}} = -J_{\text{em}}^\mu A_\mu - J_{\text{NC}}^\mu Z_\mu - J_{\text{CC}}^\mu W_\mu^+ - \left(J_{\text{CC}}^\mu\right)^\dagger W_\mu^-, \tag{2.64}$$

where

$$J_{\text{em}}^\mu = g_1 \cos\theta_W \sum_f Q_f \overline{\psi}_f \gamma^\mu \psi_f,$$

$$J_{\text{NC}}^\mu = -\frac{g_2}{2\cos\theta_W} \sum_f \overline{\psi}_f \left(g_V^f \gamma^\mu - g_A^f \gamma^\mu \gamma_5\right) \psi_f,$$

$$J_{\text{CC}}^\mu = -\frac{g_2}{\sqrt{2}} \left(\sum_i \overline{\nu}^i \gamma^\mu \frac{1-\gamma_5}{2} \ell^i + \sum_{ij} \overline{u}^i \gamma^\mu \frac{1-\gamma_5}{2} V_{\text{CKM}}^{ij} d^j \right) \tag{2.65}$$

are the electromagnetic current, the neutral weak current and the charged weak current, respectively. The sum over f in J_{em} and J_{NC} runs over all fermions, and the sums over i and j in J_{CC} run over the three generations. For ease of notation, the tilde indicating transformed fields $\tilde{\nu}^i$, $\tilde{\ell}^i$, $\tilde{u}^i$, $\tilde{d}^i$ has been omitted in (2.65). The coupling constants of the neutral current are determined by the electric charge, Q_f, and the isospin, I_3^f, of the left-handed fermion, f_L,

$$g_V^f = I_3^f - 2Q_f \sin^2\theta_W, \quad g_A^f = I_3^f. \tag{2.66}$$

Note that only the *charged* currents change the quark or lepton flavour. In the Standard Model, flavour changing *neutral* currents only appear at the level of loop corrections and are generally small (see Chapter 8).

Comparing J_{em} in (2.65) with its standard form in QED (see Section 2.2) one can identify $g_1 \cos\theta_W$ with the elementary charge e. Together with (2.53) one thus has

$$e = g_1 \cos\theta_W = g_2 \sin\theta_W = \frac{g_1 g_2}{\sqrt{g_1^2 + g_2^2}}. \tag{2.67}$$

The relations given in this and the previous subsection allow the original set of parameters $g_2, g_1, \lambda, \mu^2, Y_f^{ij}$ in the Lagrangian to be replaced with the equivalent

Table 2.2 Masses of the particles in the Standard Model (except for the Higgs boson mass). Values are taken from [6, 7] and rounded for the sake of this overview. The masses of u, d, s are in the $\overline{MS}$ scheme at the scale $\mu = 2\,\text{GeV}$, and those of c and b in the $\overline{MS}$ scheme with the scale equal to the quark mass. The top-quark mass has been extracted from the kinematic reconstruction of t decays [7] (see Section 9.5 for a discussion).

e	μ	τ	W	Z	
511 keV	106 MeV	1.78 GeV	80.4 GeV	91.2 GeV	

u	d	s	c	b	t
1.5–3.3 MeV	3.5–6.0 MeV	104 MeV	1.27 GeV	4.20 GeV	173 GeV

set of more physical parameters e, M_W, M_Z, M_H, m_f, V_{CKM}^{ij}. They are all experimentally known, with the notable exception of the Higgs boson mass, M_H. For reference, the measured fermion and gauge-boson masses are given in Table 2.2.

2.5 Renormalisation

From the Lagrangian discussed in the previous sections, one can derive the Feynman rules of the Standard Model, with quadratic terms in the fields giving propagators, and cubic and quartic terms specifying the vertices in Feynman graphs. The lowest order in the perturbative expansion of a process amplitude is given by tree graphs, where the momenta of all lines are fixed by the momenta of the external particles. Higher-order corrections involve loop graphs, where the momenta of internal lines in the graphs are integrated over, as in the examples of Figure 2.1. The powers of coupling constants, that is, the order in the perturbative expansion, increase with the number of loops.

The momentum integrals of loop corrections to propagators or vertices diverge for large loop momenta. To obtain a meaningful theory one has to first *regularise* these divergences, for example by calculating in a number $D = 4 - \epsilon$ of dimensions where the loop integrals converge. For instance, the graph in Figure 2.1a then

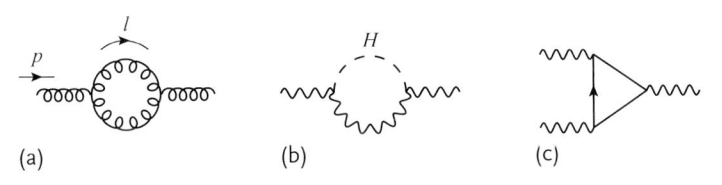

Figure 2.1 Examples of loop graphs for (a) the gluon propagator, (b) the W or Z propagator and (c) the vertex between three gauge bosons.

involves the integral

$$\int d^{4-\epsilon}l \frac{1}{l^2(p-l)^2} \,. \tag{2.68}$$

The would-be divergences are then absorbed in the quantities that appear in the Lagrangian, such as coupling constants and masses. After this procedure, called *renormalisation*, the regulator is removed again ($\epsilon \to 0$). In physical terms, renormalisation means that the effects of quantum fluctuations with momenta much larger than the scale of a given problem can be absorbed into the parameters of the theory.

One has to choose a renormalisation scheme, which prescribes exactly which part of the loop corrections are absorbed into the renormalised quantities. A common choice in the electroweak theory is the on-shell scheme (see e. g. [3, 4]), where propagators have their poles for the squared particle momentum equal to the squared renormalised mass, $p^2 = m_{\text{pole}}^2$. Another frequently used scheme is the $\overline{\text{MS}}$ scheme, which is physically less intuitive but greatly simplifies higher-order calculations.

In the $\overline{\text{MS}}$ scheme (and most other schemes) the renormalised couplings and masses depend on a renormalisation scale μ, which, roughly speaking, specifies the lower limit of loop momenta that have been absorbed in the renormalised parameter. The "running" of the parameter with μ is described by renormalisation group equations, which in the simple case of one coupling and one mass read

$$\mu \frac{d}{d\mu} g(\mu) = \beta(g(\mu)) \,, \quad \mu \frac{d}{d\mu} m(\mu) = m(\mu) \gamma_m(g(\mu)) \,. \tag{2.69}$$

The renormalisation group functions β and γ_m have an expansion in the coupling and can be determined from the relevant loop graphs. If the Standard Model is renormalised in the $\overline{\text{MS}}$ scheme, all couplings and masses, as well as the CKM matrix elements (which are closely connected to the Yukawa couplings), become dependent on μ. With increasing μ the $U(1)_Y$ coupling g_1 increases, whereas the $SU(2)_L$ and $SU(3)_c$ couplings g_2 and g_s decrease.

An expansion in the coupling constants is of course only useful if these are small enough to allow for a truncation of the perturbative series. The electroweak couplings g_1 and g_2 are small at all relevant scales (g_1 becomes large only for $\mu \gg 10^{15}$ GeV, where the Standard Model cannot be expected to be valid anymore). This is not true for the strong coupling, g_s, which has profound consequences for QCD as discussed in Section 2.7.

The triangle graph in Figure 2.1c is characterised by the three fermion currents that couple to the external gauge bosons. Analysing this graph for the specific case of two vector currents $\bar{\psi} \gamma_\mu \psi$, one axial-vector current $\bar{\psi} \gamma_\mu \gamma_5 \psi$ and a massless fermion around the loop, one finds that it is not possible to find a renormalisation prescription for which *both* the vector and the axial-vector current are conserved. From Noether's theorem one would expect these currents to be conserved for a massless fermion, since they are associated with the $U(1)$ symmetries under phase transformations $\psi \to e^{i\alpha}\psi$ in the vector case and $(\psi_L \to e^{i\alpha}\psi_L, \psi_R \to e^{-i\alpha}\psi_R)$ in the axial-vector case. The breaking of an apparent symmetry in the Lagrangian by loop effects is called an "anomaly", and the case just discussed is referred to as the axial or triangle anomaly (see e. g. Chapter 19 of [5]).

Since the electroweak currents (2.65) involve both vector and axial-vector components, the triangle anomaly could potentially break the gauge symmetry of the Standard Model, which would prevent a consistent renormalisation of the theory. However, it turns out that the symmetry-breaking terms cancel in the sum over all fermions running around the loop. The cancellation happens separately for each generation of quarks and leptons, where the quarks come with a factor three in the loop because they have three different colour states. The consistency of the Standard Model thus hinges on the detailed quantum number assignment for the fermions under the full gauge group $SU(3)_c \times SU(2)_L \times U(1)_Y$ – this is suggestive of a deeper underlying structure.

2.6 Electroweak Parameters and Observables

The input parameters of the electroweak theory have to be determined from experiment. As discussed in Section 2.4.4, a convenient choice is the set of physical parameters given by the particle masses, the elements of the CKM matrix and the electromagnetic coupling e. The latter is commonly expressed in terms of the fine-structure constant $\alpha = e^2/(4\pi)$, a low-energy parameter known with extremely high precision. Apart from the Higgs boson mass, M_H, and the flavour sector (the fermion masses and mixing angles), only three independent quantities are required for fixing the input for the gauge sector and the fermion–gauge interactions. These are α and the vector boson masses M_W, M_Z, which correspond to g_1, g_2 and v.

A further very precisely measured low-energy parameter is the Fermi constant G_F, which is the effective four-fermion coupling constant in Fermi's theory of weak decays (see Section 2.9). From the muon lifetime one obtains $G_F = 1.16637(1) \times 10^{-5}$ GeV^{-2} [6]. Consistency between Fermi's theory and the Standard Model at energies much smaller than M_W requires the identification (see Figure 2.2)

$$\frac{G_F}{\sqrt{2}} = \frac{g_2^2}{8M_W^2} = \frac{e^2}{8\sin^2\theta_W M_W^2} = \frac{e^2}{8\sin^2\theta_W \cos^2\theta_W M_Z^2}, \qquad (2.70)$$

which connects the vector boson masses to the parameters α, G_F, $\sin^2\theta_W$ and gives the W–Z mass correlation

$$M_W^2\left(1 - \frac{M_W^2}{M_Z^2}\right) = \frac{\pi\alpha}{\sqrt{2}G_F} \stackrel{\text{def}}{=} A^2, \quad A = 37.2805 \text{ GeV}. \qquad (2.71)$$

The relation (2.70) can be further exploited to express the normalisation of the neutral-current couplings (2.65) in terms of the Fermi constant,

$$\frac{g_2}{2\cos\theta_W} = \left(\sqrt{2}G_F M_Z^2\right)^{\frac{1}{2}} . \tag{2.72}$$

In this way the vector and axial-vector couplings of each fermion species to the Z are determined and can be used to calculate a variety of observables at the Z resonance, like the Z width, partial widths, and a series of asymmetries (see Section 3.3.1). Finally, one obtains the vacuum expectation value of the Higgs field as

$$v = (\sqrt{2}G_F)^{-\frac{1}{2}} = 246.22 \text{ GeV} . \tag{2.73}$$

The preceding relations were obtained using the Standard Model Lagrangian without taking into account that parameters like masses and couplings are renormalised by radiative corrections. In this sense these relations have tree-level accuracy. This turns out to be insufficient when confronted with experimental data, which have achieved extraordinary precision in the era of LEP and the Tevatron (see Chapter 3). Terms beyond the lowest order in perturbation theory must therefore be included. In other words, high experimental precision makes observables sensitive to the quantum structure of the theory.

The loop contributions to the electroweak observables contain all particles of the Standard Model spectrum, in particular also the Higgs boson. The higher-order terms thus induce a dependence of observables on the Higgs boson mass. In this way, M_H is indirectly accessible through precision measurements, although the Higgs particle itself has not yet been discovered.

The implementation of higher-order terms can be written in a compact way for the W–Z mass correlation:

$$M_W^2 \left(1 - \frac{M_W^2}{M_Z^2}\right) = \frac{A^2}{1 - \Delta r} . \tag{2.74}$$

G_F and hence A are defined through the muon lifetime. The loop corrections to muon decay are summarised in the quantity $\Delta r = \Delta r(m_t, M_H)$, which at one-loop order depends logarithmically on the Higgs boson mass and quadratically on the top-quark mass.

The radiative corrections to the vertices between fermions and the Z boson can also be written in a compact way, replacing the lowest-order couplings (2.66) by effective couplings

$$g_V^f = \sqrt{\rho_f}\left(I_3^f - 2Q_f \sin^2\theta_{\text{eff}}^f\right) , \quad g_A^f = \sqrt{\rho_f}\, I_3^f . \tag{2.75}$$

These couplings are specific for each fermion f and express the higher-order contributions in terms of a form factor $\rho_f(m_t, M_H)$ and an effective weak mixing angle, appearing as $\sin^2\theta_{\text{eff}}^f(m_t, M_H)$. Again, their dependence on m_t is quadratic, whereas the one on M_H is only logarithmic. Nevertheless, the effective leptonic $\sin^2\theta_{\text{eff}}^\ell$ is one of the most constraining observables for M_H (see Section 3.4).

2.7
Some Remarks on Quantum Chromodynamics

Among all Standard Model interactions, the dynamics of QCD are most decisively determined by the behaviour of its coupling, α_s. The renormalisation group running drives the strong coupling to large values at low momentum scales, for example from $\alpha_s(m_Z) \approx 0.118$ to $\alpha_s(m_\tau) \approx 0.34$ [6], and somewhere below 1 GeV the perturbative expansion in the coupling breaks down.

Generally, one speaks of perturbative or non-perturbative effects, quantities, and so on depending on whether perturbation theory can be applied or not, and associates a scale of order 1 GeV with the onset of non-perturbative dynamics in QCD. Certain non-perturbative quantities (for instance hadron masses, but in general not scattering amplitudes) can be calculated in lattice QCD, where the Feynman path integral is numerically evaluated on a discrete lattice that approximates the space-time continuum. In many situations non-perturbative dynamics can, however, only be modelled (see e.g. Chapter 5).

Experiments have shown that quarks and gluons do not appear as physical states but are confined within hadrons, which carry zero colour charge and have a typical size of the order of 1 fm. The increase of the strong coupling at large distances makes confinement plausible, although a rigorous derivation of this phenomenon has not been given yet. In the opposite limit of small distances, quarks and gluons interact with a small coupling: one says that they become "asymptotically free".

A key to calculating QCD processes involving a high momentum scale is the concept of factorisation (see Section 4.3). For suitable observables in suitable kinematics one can split the dynamics into a perturbatively calculable short-distance subprocess, which involves quarks and gluons, and non-perturbative quantities like parton distributions, which quantify the structure of hadrons at the quark-gluon level. To lowest order in α_s, the result typically reproduces what is obtained in the parton model.

Important features of the low-energy dynamics in QCD can be understood from approximate symmetries. The masses of u and d quarks (see Table 2.2) are much smaller than the scales governing non-perturbative QCD interactions, so that it is useful to consider as a starting point the theory where m_u and m_d are zero. To less accuracy, the same can be done for the strange quark.

If the light quark masses are set to zero, the QCD Lagrangian has a global symmetry under $SU(2)$ or $SU(3)$ transformations of the doublet $(u, d)^T$ or triplet $(u, d, s)^T$ of quark fields[11]. The quantum number associated with the $SU(2)$ group is called strong isospin. Hadrons can be grouped into corresponding multiplets, with the proton and neutron, for instance, forming an $SU(2)$ doublet and being part of an $SU(3)$ octet (see Section 14 of [6]). This provides a successful scheme for understanding the pattern of hadron masses and of strong decays. The scheme can be refined by treating the effect of the strange quark mass as a perturbation to

---

11) This $SU(3)$ symmetry, which relates quarks of different flavours, is of course distinct from the $SU(3)_c$ gauge symmetry of QCD, which acts in colour space.

the $SU(3)$ symmetry. The quark model, in which baryons are represented as bound states of three quarks and mesons as quark–antiquark systems, implements the $SU(3)$ flavour symmetry in a simple way and historically has been a key ingredient in the development of QCD. The $SU(2)$ and $SU(3)$ symmetries also apply to quantities like parton densities: in the isospin limit one finds for instance that the distribution of u quarks in the proton is the same as that of d quarks in the neutron.

In fact, the QCD Lagrangian for massless quarks has an even larger symmetry, described by the group $SU(3)_L \times SU(3)_R$ for three flavours, because the triplets $(u_L, d_L, s_L)^T$ and $(u_R, d_R, s_R)^T$ of left-handed and right-handed fields can be transformed independently. This "chiral symmetry" is spontaneously broken by a non-zero vacuum expectation value

$$\sum_{q=u,d,s} \langle \bar{q}q \rangle = \sum_{q=u,d,s} \langle \bar{q}_L q_R + \bar{q}_R q_L \rangle \tag{2.76}$$

of quark field bilinears, called "chiral condensate". The remaining symmetry is a single $SU(3)$ acting on $q = q_L + q_R$ as discussed in the previous paragraph. Since the spontaneously broken symmetry is global (and not local as in the Higgs case), the Goldstone mechanism gives a set of massless bosons. In the real world, chiral symmetry is also *explicitly* broken by the quark mass terms $m_q(\bar{q}_L q_R + \bar{q}_R q_L)$. Treating them as a perturbation to the $SU(3)$ symmetric Lagrangian, one finds that the Goldstone bosons acquire masses. They are then called pseudo-Goldstone bosons and can be identified as the $SU(3)$ octet of pions, kaons and the η, whose masses are much smaller than those of the other hadrons. Chiral perturbation theory implements this idea in a systematic fashion and has become a very successful tool to describe strong interaction dynamics at scales small compared to 1 GeV [8]. Its results are even more precise when reduced to the $SU(2)_L \times SU(2)_R$ symmetry of u and d quarks, the pseudo-Goldstone bosons of which are the pions.

Charm, bottom and top are often referred to as "heavy quarks". Their masses are not small compared to 1 GeV, and setting them to zero does therefore not give a good approximation for non-perturbative quantities. On the contrary, m_b is much larger than this scale and therefore the limit of infinite quark mass is a valid starting point to describe bound states containing b quarks (see Chapter 8). The c quark is a borderline case, but with limited accuracy the heavy-quark limit also gives useful results for bound states in the charm sector.

2.8
Symmetries

This section gives a brief overview of various symmetries in the Standard Model, which play a prominent role both in its structure and in the phenomena it gives rise to. Explanations and more detail on the following can be found for example in [9].

- The local symmetry under $SU(3)_c \times SU(2)_L \times U(1)_Y$ determines the gauge interactions. The vacuum expectation value of the Higgs field breaks this symmetry down to the gauge group $SU(3)_c \times U(1)_{em}$ of QCD and QED, giving the W and Z bosons their masses as shown in Section 2.4.3.
- Poincaré symmetry, that is, the invariance under space-time translations, rotations and boosts, leads to conservation of energy, momentum and angular momentum.
- The discrete symmetries parity (P), time reversal (T) and charge conjugation (C) have a more involved pattern. QCD and QED conserve each symmetry individually. P transforms a left-handed fermion into a right-handed fermion; C transforms a left-handed fermion into a right-handed antifermion[12]. The electroweak gauge group $SU(2)_L \times U(1)_Y$ thus breaks P and C since it acts differently on left-handed and right-handed fields, but it conserves T and the combined symmetry CP.

  Finally, CP and T are broken by the non-zero complex phase in the CKM matrix, that is, by the Yukawa couplings between fermions and the Higgs boson. Note that interactions involving only gauge bosons and the Higgs are CP conserving (a feature that is not generally found in extensions of the Standard Model).

  By virtue of a general theorem, the combined symmetry CPT is conserved in any local relativistic field theory.
- The QCD Lagrangian (2.27) is incomplete in the sense that it does not contain all possible gauge invariant terms that can be constructed from quark and gluon fields. The missing term has the form $\theta g_s^2 \epsilon^{\mu\nu\rho\sigma} G^a_{\mu\nu} G^a_{\rho\sigma}/(8\pi)^2$ and changes sign under P and CP transformations, thus violating the associated symmetries. Experimental limits on the electric dipole moment of the neutron give a stringent bound $|\theta| < 10^{-10}$ [6]. The difficulty of explaining the smallness of this parameter is called the "strong CP problem" (as opposed to the CP violation of the weak interactions encoded in the CKM matrix). A class of proposed solutions implies the existence of very light spin-zero particles, called axions.
- The QCD Lagrangian for the light quarks u, d, s has an approximate global symmetry $SU(3)_L \times SU(3)_R$. It is spontaneously broken by the vacuum to a single $SU(3)$, and explicitly broken by the quark masses, leading to a flavour $SU(3)$ octet of pseudo-Goldstone bosons. The amount of explicit symmetry breaking is smaller in the subsector of u and d quarks, where $SU(2)_L \times SU(2)_R$ is spontaneously broken to the $SU(2)$ group of the strong isospin, with the isotriplet of pions as pseudo-Goldstone bosons.

  The QCD Lagrangian for massless quarks also has an axial $U(1)$ symmetry, under which all fields q_L are multiplied with a common phase factor and all q_R by the complex conjugate factor. This symmetry is *not* realised in the mass spectrum as it implies mass degenerate pairs of hadrons with opposite parity, a situation sometimes called the "$U(1)$ problem". In fact, the axial symmetry is broken at the quantum level by the triangle anomaly (see Section 2.5). This is intimate-

---

12) Charge conjugation does not change the helicity of a particle, and a negative-helicity fermion is left-handed, whereas a negative-helicity antifermion is right-handed.

ly related to the structure of the vacuum, with a crucial role being played by the term $g_s^2 \epsilon^{\mu\nu\rho\sigma} G^a_{\mu\nu} G^a_{\rho\sigma}/(8\pi)^2$ already mentioned above.
- The Standard Model Lagrangian is invariant under a global common phase transformation of all quark fields, and under a corresponding transformation of all lepton fields. The associated charges are baryon number ($B = 1/3$ for quarks and $B = -1/3$ for antiquarks) and lepton number ($L = 1$ for leptons e^-, ν_e, and so on and $L = -1$ for antileptons $e^+, \bar{\nu}_e$, etc). The Feynman rules of the Standard Model conserve these quantum numbers.

The lepton and baryon number currents are affected by triangle anomalies, and both B and L can be broken by non-perturbative effects. These effects are in general tiny, except for extreme conditions like in the early universe, where they are relevant for the generation of the baryon asymmetry [10]. The anomalies of B and L cancel in $(B - L)$, which is an exact symmetry in the Standard Model.

2.9
Mass Scales and Effective Theories

The Standard Model contains a wide range of scales, including the quark and lepton masses, the boson masses M_W, M_Z, M_H and the scale v of electroweak symmetry breaking, and the scale of non-perturbative QCD interactions. An important tool to manage multi-scale problems are effective field theories (see e. g. [11] for an introduction). In an effective field theory, particles with masses much larger than the momentum scale, p, of a physical problem do not appear as explicit degrees of freedom (one often speaks of "integrating out" their fields in the path integral formalism). Rather, their effects are incorporated into the coupling constants, masses, and possibly into new operators of the Lagrangian for the effective theory. Typically, one "matches" the effective theory to the full theory by requiring that a suitable quantity be the same to a given accuracy when calculated in the two theories.

A simple example for this is Fermi's theory of charged-current weak interactions. Here the effects of W exchange at scales $p \ll M_W$ are described by a four-fermion interaction, which can be obtained from the corresponding graphs in the Standard Model by approximating $p^2 - M_W^2 \approx -M_W^2$ in the W propagator (see Figure 2.2).

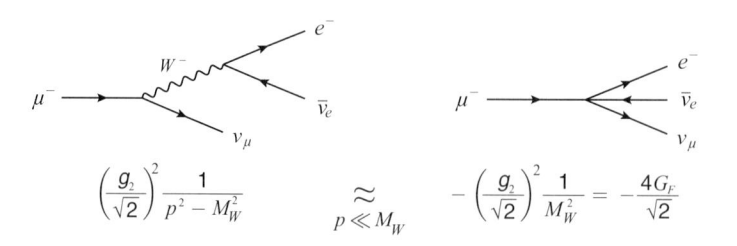

Figure 2.2 Muon decay in the Standard Model and in Fermi's effective weak theory.

An extension of Fermi's theory is the "effective electroweak theory" for scales $p \ll M_W$, which contains four-fermion couplings generated by the charged and neutral weak currents. It is in particular used to describe the weak decays of hadrons with b or c quarks (see Chapter 8). Renormalisation group equations permit the resummation to all orders in perturbation theory of logarithms in p/M_W, which arise from loop corrections to the effective vertices.

Effective theories are also widely used in QCD. Heavy-quark effective theory (HQET) implements an expansion in p/m_b or p/m_c and can be applied to hadrons containing b or c quarks. It provides not only a classification of their mass spectrum, but also yields relations between decay matrix elements and is an important ingredient in heavy-flavour physics (see Chapter 8). A different effective theory based on the heavy-quark limit is non-relativistic QCD (NRQCD), which can be used to treat $b\bar{b}$ or $c\bar{c}$ bound states, as well as the $t\bar{t}$ system close to production threshold. An effective theory for the low-energy interactions of hadrons made from light quarks is chiral perturbation theory, valid for scales $p \ll 1\,\text{GeV}$.

QCD calculations of hard-scattering processes at scale p are typically done in an effective theory where quarks with masses $m_q \gg p$ have been integrated out. The physics of these heavy quarks is contained in a distinct running coupling $\alpha_s^{(n_F)}$ for the theory with n_F active quark flavours and a series of matching conditions relating $\alpha_s^{(n_F)}$ to $\alpha_s^{(n_F-1)}$ (see Section 9.2 of [6]).

Finally, the Standard Model itself can be regarded as an effective theory for momentum scales of order v, where particles and interactions at scales $\Lambda \gg v$ have been integrated out. The Standard Model Lagrangian then corresponds to the zeroth-order term of an expansion in powers of v/Λ. Higher-order terms of this expansion are a convenient way to quantify deviations from the Standard Model due to physics at high scales, since only a finite number of new coupling terms arise at each power of v/Λ. Operators of mass dimension $4+n$ are associated with a factor $1/\Lambda^n$ in the Lagrangian. At order v^2/Λ^2 one finds, for instance, four-fermion interactions which are sometimes called contact terms. Some of these also appear in the effective electroweak theory discussed above; others violate symmetries or approximate symmetries of the Standard Model (e. g. the vertex for the transition $uu \to \bar{d}e^+$, which violates B and L and contributes to proton decay modes like $p \to \pi^0 e^+$).

In an effective field theory approach, one should match the Standard Model to a theory that explicitly contains the new physics at scale Λ. Above this scale, the strong and electroweak coupling constants change their renormalisation group running because new particles are included in the associated β functions. This is important in connection with the unification of couplings (see Chapters 7, 10 and Section 15.1.4 of [6]). With suitable new particles having suitable masses, all three gauge couplings of the Standard Model can have a common value at some high scale M_{GUT}. At that scale one can perform a further matching to a "grand unified theory". In such a theory there is a single gauge coupling, associated with a large gauge group like $SU(5)$. This group contains $SU(3)_c \times SU(2)_L \times U(1)_Y$ as a subgroup and thus unifies the three gauge interactions of the Standard Model.

References

1 Weinberg, S. (2004) The making of the Standard Model, *Eur. Phys. J. C*, **34**, 5, doi:10.1140/epjc/s2004-01761-1.
2 't Hooft, G. (2007) The making of the Standard Model, *Nature*, **448**, 271, doi:10.1038/nature06074.
3 Böhm, M., Spiesberger, H., and Hollik, W. (1986) On the 1-loop renormalization of the electroweak Standard Model and its application to leptonic processes. *Fortsch. Phys.*, **34**, 687, doi:10.1002/prop.19860341102.
4 Denner, A. (1993) Techniques for calculation of electroweak radiative corrections at the one-loop level and results for W-physics at LEP200, *Fortschr. Phys.*, **41**, 307, http://arxiv.org/abs/0709.1075.
5 Peskin, M.E. and Schroeder, D.V. (1995) *An Introduction to Quantum Field Theory*, Addison-Wesley, Reading.
6 Particle Data Group, Amsler, C. et al. (2008) Review of Particle Physics, *Phys. Lett. B*, **667**, 1, doi:10.1016/j.physletb.2008.07.018
7 Tevatron Electroweak Working Group (2009) Combination of CDF and DØ Results on the Mass of the Top Quark, http://arxiv.org/abs/0903.2503.
8 Scherer, S. (2003) Introduction to chiral perturbation theory, *Adv. Nucl. Phys.*, **27**, 277, http://arxiv.org/abs/hep-ph/0210398.
9 Peccei, R.D. (1998) Discrete and Global Symmetries in Particle Physics, http://arxiv.org/abs/hep-ph/9807516.
10 Rubakov, V.A. and Shaposhnikov, M.E. (1996) Electroweak baryon number non-conservation in the early universe and in high energy collisions, *Usp. Fiz. Nauk*, **166**, 493, http://arxiv.org/abs/hep-ph/9603208.
11 Manohar, A.V. (1996) Effective Field Theories, http://arxiv.org/abs/hep-ph/9606222.

3
Electroweak and Standard Model Tests: the Quest for Precision

Klaus Mönig and Georg Steinbrück

3.1
The Standard Model at Born Level

As discussed in the previous chapter, the unified theory of electroweak interactions is based on local gauge invariance under the combined groups of weak isospin and weak hypercharge, $SU(2)_L \otimes U(1)_Y$. The coupling constants corresponding to the two groups, g_2 and g_1, are related to the electron charge e via the Weinberg angle, θ_W: $e = g_2 \sin\theta_W = g_1 \cos\theta_W$. The electroweak interaction is intricately linked to the concept of *handedness* or *chirality*. The electromagnetic interaction depends only on the electromagnetic charge, regardless of whether the particle is left-handed or right-handed. The weak interaction, on the other hand, does distinguish between left-handed and right-handed states.

The fundamental interaction vertices of the weak force include the weak gauge bosons $W^\pm$, Z and the fermions (leptons or quarks). The *neutral-current* (NC) and *charged-current* (CC) interactions have different chirality properties. The form of the interaction Lagrangian and the associated currents have been given in (2.64) and (2.65). In terms of the mixing angle and fermion charge, the ratio of the vector and axial vector couplings, g_V^f and g_A^f is given by

$$\frac{g_V^f}{g_A^f} = 1 - 4|Q_f|\sin^2\theta_W . \tag{3.1}$$

The relative strengths of the neutral and charged couplings in electroweak theory, together with the mechanism of *electroweak symmetry breaking* (the Higgs mechanism), determines the ratio of the *gauge-boson masses*, which at Born level is simply

$$M_W = M_Z \cos\theta_W . \tag{3.2}$$

The total width of the W boson and its mass are related via

$$\Gamma_W = \frac{3 G_F M_W^3}{\sqrt{8}\pi} . \tag{3.3}$$

Physics at the Terascale, First Edition. Edited by Ian C. Brock and Thomas Schörner-Sadenius
© 2011 WILEY-VCH Verlag GmbH & Co. KGaA, Weinheim.
Published 2011 by WILEY-VCH Verlag GmbH & Co. KGaA.

In contrast to the W boson coupling, the couplings of the Z boson are non-universal, as they depend on the fermion type. In the Born approximation and assuming massless fermions, the partial width for fermion, f, is given by

$$\Gamma_{f\bar{f}} = n_F \frac{G_F M_Z^3}{6\pi\sqrt{2}} \left(\left(g_V^f\right)^2 + \left(g_A^f\right)^2 \right), \tag{3.4}$$

where n_F is the number of colours: for leptons $n_F = 1$ and for quarks $n_F = N_C = 3$.

Additional insight into the structure of the Standard Model can be gained by analysing *polar angles* and *asymmetries*. In e^+e^- collisions, the angle θ of the outgoing fermion f with respect to the incident e^- direction is described at Born level by the distribution

$$\frac{d\sigma}{d\cos\theta} \propto 1 + \cos^2\theta + \frac{8}{3} A_{FB} \cos\theta. \tag{3.5}$$

The *forward-backward asymmetry* is defined as

$$A_{FB} = \frac{\sigma_F - \sigma_B}{\sigma_F + \sigma_B}, \tag{3.6}$$

where σ_F and σ_B are the cross sections for producing negatively charged leptons in the forward hemisphere and backward hemisphere, respectively. At LEP, the hemispheres are defined with respect to the direction of the incoming e^- beam. A_{FB} depends on the centre-of-mass energy of the e^+e^- system. At the Z peak, the pole asymmetry for pure Z boson exchange is given by

$$A_{FB}^{0,f} = \frac{3}{4} \mathcal{A}_e \mathcal{A}_f, \tag{3.7}$$

where $\mathcal{A}_f$ are the polarisation parameters,

$$\mathcal{A}_f = \frac{2 g_V^f g_A^f}{\left(g_V^f\right)^2 + \left(g_A^f\right)^2} = \frac{l_f^2 - r_f^2}{l_f^2 + r_f^2}, \tag{3.8}$$

and $l_f = (g_V^f + g_A^f)/2$ and $r_f = (g_V^f - g_A^f)/2$ are the left-handed and right-handed couplings, respectively. The average longitudinal polarisation of fermion f, is given by $\mathcal{P}_f = -\mathcal{A}_f$, where for a purely left-handed fermion $\mathcal{P}_f = -1$ and for a right-handed one $\mathcal{P}_f = +1$. With the vector couplings $g_V^e = \frac{1}{2}(-1 + 4\sin^2\theta_W)$ for negatively charged leptons, $g_V^u = \frac{1}{2}\left(1 - \frac{8}{3}\sin^2\theta_W\right)$ for u-type quarks and $g_V^d = \frac{1}{2}\left(-1 + \frac{4}{3}\sin^2\theta_W\right)$ for d-type quarks (see (2.66)), the polarisation measurement provides a very sensitive way to extract $\sin\theta_W$. With polarised incident electron beams, the left-right asymmetry,

$$A_{LR} = \frac{\sigma_L - \sigma_R}{\sigma_L + \sigma_R} = \mathcal{A}_e, \tag{3.9}$$

can be measured and the polarisation parameter for electrons, $\mathcal{A}_e$, can be extracted directly. σ_L and σ_R are the total cross sections for a left-handed and right-handed polarised incident electron beam, respectively.

It is especially interesting to study the Z boson couplings to individual lepton and quark flavours separately, particularly the coupling to heavy quarks. Experimentally, the asymmetries and polarisation parameters are extracted separately for b and c quarks, denoted by $A_{FB}^{0,b(c)}$ and $\mathcal{A}_{b(c)}$. Finally, the *left-right–forward-backward asymmetry* for fermion f, defined as

$$A_{LR,FB}^f = \frac{(\sigma_L(\cos\theta > 0) - \sigma_L(\cos\theta < 0)) - (\sigma_R(\cos\theta > 0) - \sigma_R(\cos\theta < 0))}{\sigma_L + \sigma_R}$$
$$= \frac{3}{4}\mathcal{A}_f , \qquad (3.10)$$

can be measured. It is directly sensitive to $\mathcal{A}_f$, which can again be used to extract $\sin\theta_W$. The *left-right–forward-backward asymmetry* can also be defined as a function of $\cos\theta$:

$$A_{LR}(\cos\theta) = \frac{(\sigma_L(\cos\theta) - \sigma_L(-\cos\theta)) - (\sigma_R(\cos\theta) - \sigma_R(-\cos\theta))}{(\sigma_L(\cos\theta) + \sigma_L(-\cos\theta)) + (\sigma_R(\cos\theta) + \sigma_R(-\cos\theta))}$$
$$= 2|\mathcal{P}_e|\mathcal{A}_f \frac{\cos\theta}{1+\cos^2\theta} , \qquad (3.11)$$

where $\mathcal{P}_e$ is the polarisation of the electron beam. By integrating θ over the limited fiducial range of a detector, this measurement can be used to determine $\mathcal{A}_f$.

3.2
The Gain from Additional Precision

With measurements at the 10% level, the gauge structure of the Standard Model can be fixed to a large extent. However, there is a significant gain from going beyond this precision. The couplings of the gauge bosons to fermions are modified by loop effects that enter at order $\alpha \sim 1\%$. The couplings can also be modified by new Born-level effects that can come either from a small mixing of the Z with a heavier gauge boson, usually called Z', or from the heavy gauge boson directly. In the latter case, the effects are suppressed by the large mass appearing in the propagator of the heavy particle.

3.2.1
Sensitivity to New Physics in Loops

In precision calculations of electroweak quantities, loop corrections as shown in Figure 3.1 must be taken into account. The corrections to the W propagator violate the weak isospin due to the different mass of the up-type and the down-type fermions. They are of particular importance, because they are proportional to the squared mass difference of the involved particles [1]. In the Standard Model this

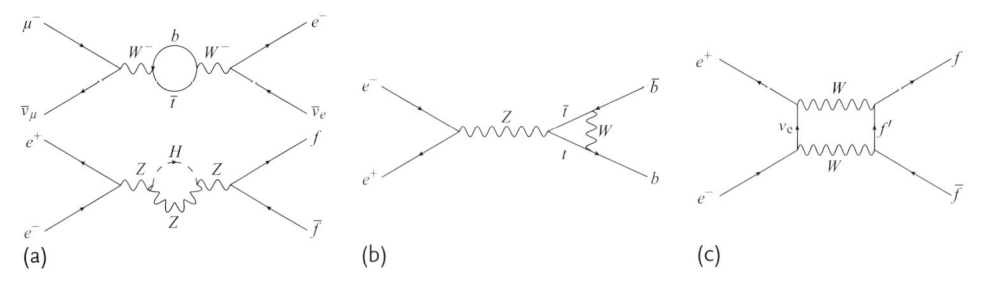

Figure 3.1 Examples of 1-loop diagrams: (a) loop corrections, (b) vertex corrections, (c) box diagrams.

leads to a large sensitivity to the top-quark mass. In extensions of the Standard Model, all isospin-violating effects like a fourth fermion generation with a large mass splitting can be tested. This diagram is also very important for pure Z boson observables measured at LEP 1 and SLC since G_F, which is measured in charged-current processes, is used for the determination of the input parameters. However, other loop corrections are also numerically relevant. The Higgs contribution to the Z propagator gives rise to a contribution proportional to $\ln \frac{M_H}{M_Z}$, which can be used to set a limit on the Higgs mass. A mass-degenerate fourth fermion generation would also cause an effect that should be clearly visible in present-day data. In a full n-loop calculation, all diagrams contributing to the considered order must be taken into account. Because of the strong resonance enhancement at LEP 1 and SLC, however, only classes of diagrams that have the same resonance structure as the tree-level graph (Figure 3.1a,b) are numerically important, while box diagrams (Figure 3.1c) can be neglected. For this reason, close to the Z resonance all loop corrections can be parametrised using effective couplings in the Born formulae:

$$g_A^f \to \sqrt{1 + \Delta \rho_f} \, g_A^f , \qquad (3.12)$$

$$\frac{g_V^f}{g_A^f} = 1 - 4|Q_f| \sin^2 \theta_{\text{eff}}^f , \quad \sin^2 \theta_{\text{eff}}^f = (1 + \Delta \kappa_f) \sin^2 \theta_W , \qquad (3.13)$$

$$M_W^2 = \frac{1}{2} M_Z^2 \left(1 + \sqrt{1 - \frac{4\pi \alpha}{\sqrt{2} G_F M_Z^2} \frac{1}{1 - \Delta r}} \right) . \qquad (3.14)$$

Because of the vertex corrections (Figure 3.1b) the couplings are flavour dependent. For $f \neq b$ the shifts are small and well known [2]. For the b quark there is a significant contribution from corrections involving the top quark in the Standard Model and additional contributions in models with an extended top sector, like Little Higgs models, or with a charged Higgs boson[13].

At lower energy, a couple of observables are sensitive to *new physics* in loops as well. The most prominent example is the *anomalous magnetic moment* of the muon,

---

13) Because of the structure of the CKM matrix, graphs coupling a top quark to a d or s quark can be neglected.

$g_\mu - 2$, which is sensitive to supersymmetry (SUSY) [3]. Here the very small size of effects from new physics is compensated by the extremely high experimental precision. This makes it, however, challenging to calculate the Standard Model contributions reliably [3]. Another class of observables is rare decays. Typically the Standard Model prediction is extremely small, so that new physics contributions can be relatively large. This feature reduces QCD corrections, which are normally difficult to handle at low energies.

3.2.2
Sensitivity to Born-Level Effects at High Mass Scales

With precision measurements one is also sensitive to effects from new physics at much higher energy scales. The propagator for the exchange of a vector boson of mass M, neglecting its width, is proportional to $1/(M^2 - s)$, where $\sqrt{s}$ is the centre-of-mass energy. For $s \ll M^2$, this can be approximated by $1/M^2$. From a pure exchange of the new particle one thus expects contributions of order $1/M^4$, while from the interference with the Standard Model amplitude, $1/M^2$ contributions are expected. At energies close to the Z mass such contributions are strongly suppressed by the resonance enhancement of the Z exchange amplitude relative to the non-resonant graphs. However, here one is sensitive to Z–Z' mixing, which modifies the weak mixing angle in a manner similar to loop corrections. Away from the resonance, propagator effects can be sizable. PEP at SLAC and PETRA at DESY were able to measure the Z couplings at the 10% level at energies much below the Z mass.

3.3
Measurements

Many precise measurements constrain the parameters of the Standard Model. The gauge sector is determined by three parameters, the couplings of the $U(1)_Y$ and the $SU(2)_L$ gauge groups and one parameter fixing the gauge-boson masses. Thus, only three measurements are needed to fix the model. For practical reasons, the measurements with the highest precision are used. These are: the electromagnetic fine-structure constant, α, at zero momentum transfer; the *Fermi constant* measured in muon decay, G_F; and the mass of the Z boson, M_Z, measured at LEP. The couplings of the Z boson are best tested by measurements at the peak of the resonance at LEP and SLD. These measurements are, together with the W boson mass, also the quantities which are most sensitive to loop effects coming from the Higgs boson. They are also sensitive to many other corrections from new physics, especially if these break the weak isospin by a large mass splitting within the doublet.

The experimentally most precise Standard Model test is the magnetic moment of the muon, $g_\mu - 2$. The weak corrections are tiny. However, it turns out that $g_\mu - 2$ is very sensitive to SUSY [3]. Also some other low energy observables, like the decay

$b \to s\gamma$, turn out to be sensitive to new physics because of their very small values in the Standard Model.

Tests at high energies in t-channel processes, where no resonances occur, or in s-channel processes away from the Z resonance can test new physics at a high scale by direct effects from the propagator of the new particles. Such tests have been performed at LEP 2 and HERA.

3.3.1
Parameters Measured on the Z Resonance

At LEP and SLC several cross sections and asymmetries in the vicinity of the Z resonance have been measured. The cross sections are sensitive to the absolute size of the couplings, while the asymmetries are sensitive to the ratio of the vector to the axial vector couplings, which are, on the Z resonance, sensitive to the weak mixing angle $\sin^2 \theta_W$. LEP 1 operated from 1989 to 1995 at centre-of-mass energies in a window of ± 3 GeV around the Z mass with luminosities up to 2×10^{31} cm^{-2} s^{-1}. A total of 17 million Z decays were recorded by the four experiments, ALEPH, DELPHI, L3 and OPAL. SLC was operated from 1989 to 1998 at the Z pole with much lower luminosity. However, SLC provided electron beam polarisation of up to 80%, and a smaller beam size allowed for much better b-tagging capabilities. In total 500 000 Z decays have been recorded at SLC, the vast majority by the SLD experiment. A complete description of all electroweak results obtained at LEP 1 and SLC can be found in [4].

The LEP and SLC experiments measured cross sections and asymmetries which are dependent on experimental cuts and other environmental parameters like the beam energy. These measurements were converted into pseudo-observables like the Z mass, partial widths and asymmetries, corrected to pure Z exchange. In further steps the pseudo-observables were combined between the different measurements and experiments and interpreted within the Standard Model.

From a scan of the Z resonance, the Z mass, M_Z, width, Γ_Z, and peak cross section, $\sigma^0_{\text{had}} = \frac{12\pi}{M_Z^2} \frac{\Gamma_e \Gamma_{\text{had}}}{\Gamma_Z^2}$, can be obtained[14]. The experiments measured the cross sections for hadronic and leptonic Z decays as a function of the centre-of-mass energy. To obtain the resonance parameters, these cross sections need to be corrected for initial-state radiation (ISR). Due to photon radiation from the incoming electrons, the centre-of-mass energy of the hard interaction is lower than twice the beam energy. The ISR correction can be calculated by folding the radiator function, which is known to second-order QED, with the cross section at the energy after radiation. To obtain the interesting resonance parameters, the experiments fitted the cross sections to a theory prediction that parametrises the Z exchange diagram in terms of effective couplings which are left free to vary in the fit. The photon exchange diagram and initial-state radiative corrections are taken from QED. The results are model-independent under the assumption that only photon and Z exchange contribute to the $e^+ e^- \to f \bar{f}$ cross section. The procedure is detailed

---

14) The superscript "0" always indicates that the observable is corrected to pure Z exchange.

in [4]. Figure 3.2 shows the scan results for hadronic Z decays as measured by the experiments. Two curves are shown: the fit to the data and the distribution after correcting for ISR. To reduce correlations between the measurements, the ratio of the hadronic to the leptonic Z cross section, $R_\ell^0 = \frac{\Gamma_{had}}{\Gamma_\ell}$, was used. To account for common errors, the leptonic forward-backward asymmetries at the different energies were fitted together with the cross sections, and the pole asymmetry, $A_{FB}^{0,\ell}$, is added as a parameter to the lineshape fit. All leptonic observables were obtained first taking e, μ and τ separately and then assuming lepton universality. The results were first produced by each experiment individually and were then combined taking common uncertainties into account. Table 3.1 shows the combined results assuming lepton universality.

The LEP experiments measured the τ polarisation from the decay modes $\tau \to e\nu\bar{\nu}, \mu\nu\bar{\nu}, \pi\nu, \rho\nu, a_1\nu$, corresponding to about 80% of the τ decays. The τ polar-

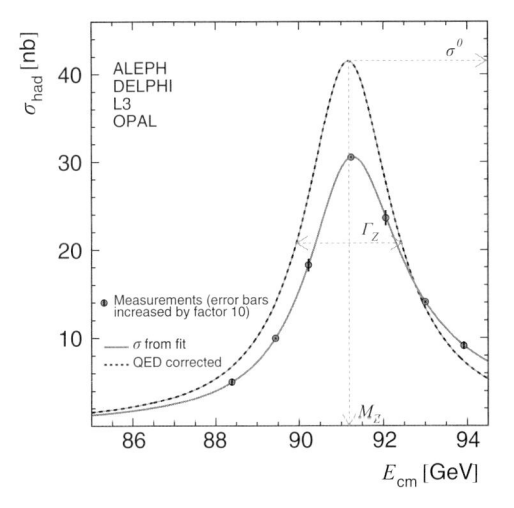

Figure 3.2 Z resonance as measured by the LEP experiments using hadronic events. The solid curve represents a fit to the data, the dashed line represents the lineshape after correcting for the effects of initial-state radiation (adapted from [4]).

Table 3.1 Results of the LEP Z scan assuming lepton universality.

			Correlations			
		M_Z	Γ_Z	σ_{had}^0	R_ℓ^0	$A_{FB}^{0,\ell}$
M_Z [GeV]	91.1875 ± 0.0021	1.000				
Γ_Z [GeV]	2.4952 ± 0.0023	−0.023	1.000			
σ_{had}^0 [nb]	41.540 ± 0.037	−0.045	−0.297	1.000		
R_ℓ^0	20.767 ± 0.025	0.033	0.004	0.183	1.000	
$A_{FB}^{0,\ell}$	0.0171 ± 0.0010	0.055	0.003	0.006	−0.056	1.000

isation was measured as a function of the production angle of the τ. From a fit to the angular dependence, $\mathcal{A}_e$ and $\mathcal{A}_\tau$ can be obtained simultaneously with almost zero correlation and similar precision. The data from the four collaborations are shown in Figure 3.3. Assuming lepton universality, they can be averaged to $\mathcal{A}_\ell = 0.1465 \pm 0.033$.

SLD measured the *left-right asymmetry* taking advantage of the polarised electron beam at the SLC. Apart from electrons, which are also produced by t-channel exchange, all Z decays can be used, leading to small statistical errors. Switching the polarisation bunch by bunch in a random way, all systematics basically cancel apart from the measurement of the beam polarisation. For this reason, A_{LR} provided the most accurate single measurement of the weak mixing angle, despite the low luminosity at the SLC. Including some lepton asymmetries, which however only give a marginal improvement, SLD obtained $\mathcal{A}_\ell = 0.1513 \pm 0.0021$.

LEP and SLD also measured several observables with b and c quarks with good precision: the ratios of the b and c partial widths of the Z to its hadronic width, R_b^0 and R_c^0; the b and c *forward-backward asymmetries*, $A_{FB}^{0,b}$ and $A_{FB}^{0,c}$; and the left-right–forward-backward asymmetries, $A_{LR,FB}^b$ and $A_{LR,FB}^c$, which are sensitive to $\mathcal{A}_b$ and $\mathcal{A}_c$ at SLC. To test the energy dependence, all measurements were combined at the centre-of-mass energies $\sqrt{s} = 89.55$, 91.26 and 92.94 GeV. Figure 3.4 shows the measured asymmetries as a function of $\sqrt{s}$. In a second step, all measurements were transported to the peak energy assuming Standard Model energy dependence and the results were converted into pole asymmetries. The corrections from the peak to the pole asymmetries are small. The results of this combination are shown in Table 3.2.

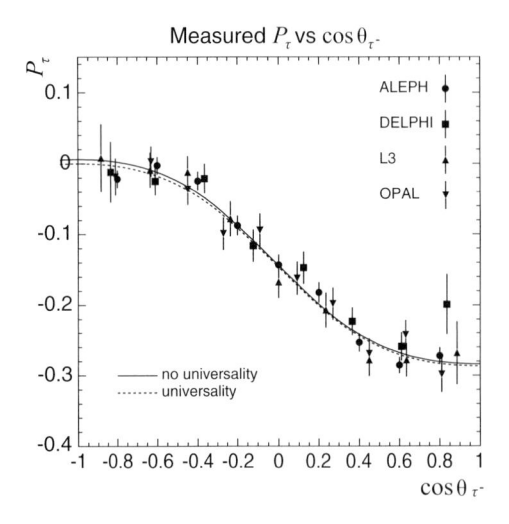

Figure 3.3 τ polarisation as a function of the τ production angle as measured by the LEP collaborations. The two curves show the fit results with and without the assumption of lepton universality (adapted from [4]).

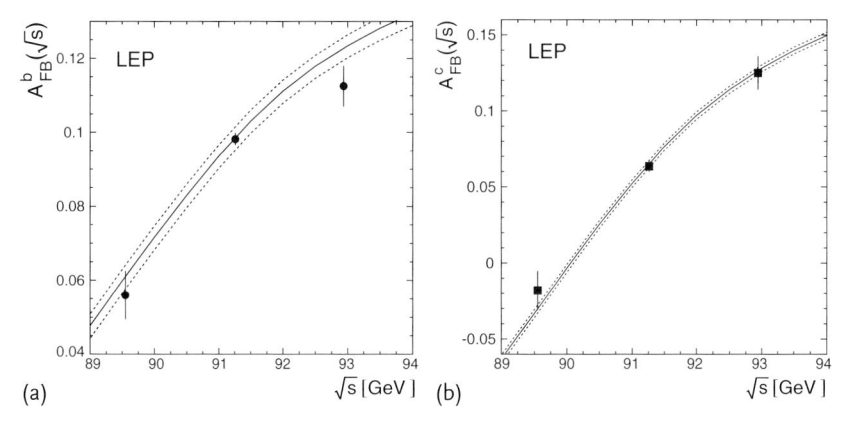

Figure 3.4 Forward-backward asymmetry for (a) b and (b) c quarks as a function of $\sqrt{s}$. The points are the data, the solid line represents the Standard Model prediction for $m_t = 172.6\,\text{GeV}$, $M_H = 300\,\text{GeV}$. The upper (lower) dashed line shows the variation for $M_H = 100\,\text{GeV}$ ($1000\,\text{GeV}$) (adapted from [4]).

Table 3.2 Results of the fit to the LEP/SLD heavy flavour data.

Observable	Result	Correlations					
R_b^0	0.21629 ± 0.00066	1.00					
R_c^0	0.1721 ± 0.0030	-0.18	1.00				
$A_{FB}^{0,b}$	0.0992 ± 0.0016	-0.10	0.04	1.00			
$A_{FB}^{0,c}$	0.0707 ± 0.0035	0.07	-0.06	0.15	1.00		
$\mathcal{A}_b$	0.923 ± 0.020	-0.08	0.04	0.06	-0.02	1.00	
$\mathcal{A}_c$	0.670 ± 0.027	0.04	-0.06	0.01	0.04	0.11	1.00

3.3.2
W Boson Mass and Couplings

3.3.2.1 Measurement of the W Boson Mass

The direct measurement of the mass of the W boson provides a stringent test of quantum corrections to the Standard Model and, together with the measurement of the top-quark mass, can be used to constrain the mass of the Higgs boson. It is, however, one of the most difficult measurements due to the need for very high precision. The first measurements of the mass of the gauge bosons were performed by the UA1 and UA2 experiments at the CERN proton–antiproton collider (Sp$\bar{\text{p}}$S), where the W and Z bosons were first directly observed. However, the first precision measurements of the W boson mass were done by the four LEP experiments during the LEP 2 phase. In this second running phase of the LEP collider at CERN, the centre-of-mass energy was increased above the threshold $2M_W$, enabling the production of W boson pairs. The Feynman diagrams for W pair production are shown

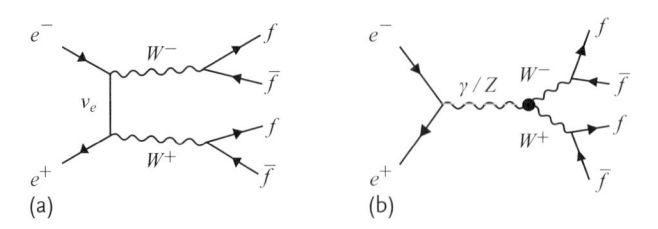

Figure 3.5 Feynman diagrams for WW production at Born level at an e^+e^- collider. (a) t-channel process via neutrino exchange; (b) s-channel diagram via photon or Z exchange.

in Figure 3.5. With a branching fraction of 68.5%, the W boson decays hadronically into a quark–antiquark pair, whereas the remaining 31.5% of the decays result in a charged lepton and a neutrino. Events with both W bosons decaying hadronically are characterised by at least four hadronic jets in the detector. One W boson decaying hadronically and the other leptonically leads to two jets, a highly energetic lepton and large missing energy due to the undetected neutrino. Both W bosons decaying leptonically results in events with two high energy leptons and large missing energy. These three classes of events are analysed separately and contribute to a combined measurement of M_W at LEP. Several methods are used to kinematically reconstruct the events and assign jets to partons. While the lowest total uncertainty comes from the semileptonic decay channel, the purely experimental precision is highest in the fully hadronic channel. Here, the largest uncertainty is from so-called "colour reconnection" effects, that is, QCD interactions between the hadronic decay products of the two W bosons which, due to the very short W lifetime, are as close as 0.5 fm. To achieve the highest possible precision, all measurements from all four LEP experiments are combined into one single result, taking all correlations into account. This leads to $M_W = 80.376 \pm 0.033$ GeV [5].

Because of the more complex initial state, the measurement of M_W is more difficult at hadron colliders. Nevertheless, it has now reached a precision comparable to measurements at lepton machines. At a hadron collider, the dominant production mechanism of W bosons is via s-channel quark–antiquark annihilation. Since the hadronic decay channel is overwhelmed by background from QCD production of multiple jets, only the leptonic decay channels are used. While for the measurements at LEP, the beam energy could be used as a constraint in the fit for the final state, at a hadron collider the situation is more complicated. Because of the unknown fractions of the proton momentum carried by the interacting quarks, the longitudinal momentum of the initial state is unknown. Also, the final state cannot be fully reconstructed due to the presence of a neutrino. The mass of the W boson can therefore only be inferred on a statistical basis from variables which are correlated with it: the transverse momentum of the neutrino, which can be obtained from the momentum imbalance of the event; the transverse momentum of the lepton; or a quantity called the transverse W boson mass, M_T, defined via $M_T^2 = 2 E_{T\nu} E_{Tl} - 2 \boldsymbol{p}_{T\nu} \cdot \boldsymbol{p}_{Tl}$. Figure 3.6 shows the M_T distribution as measured by the DØ collaboration. The current combined result for the W boson

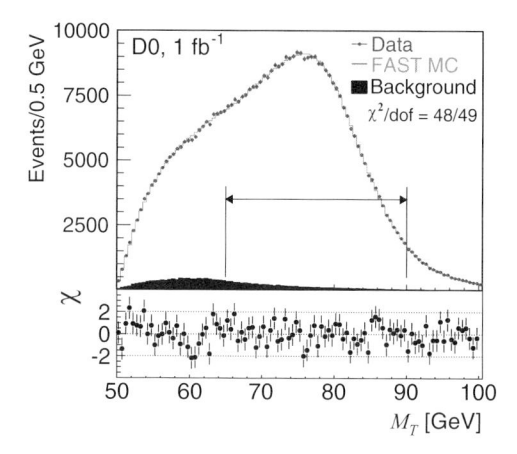

Figure 3.6 The transverse W boson mass (adapted from [6]).

mass at the Tevatron is $M_W = 80.420 \pm 0.031$ GeV [7]. Combined with the LEP results, the world average for the direct measurement of the mass of the W boson is 80.399 ± 0.023 GeV [7].

3.3.2.2 W Boson Couplings

The gauge structure of the Standard Model has some striking features concerning the couplings of its force carriers. In Quantum Electrodynamics, the photons carry no electric charge and therefore do not interact with themselves. In contrast, the carriers of the weak force carry weak charge, leading to trilinear and quartic self-couplings. This is a direct consequence of the non-Abelian nature of the underlying $SU(2)$ gauge group. Testing these gauge-boson self-interactions is an important test of the gauge structure of the Standard Model. Physics beyond the Standard Model can introduce additional couplings, changing the cross sections and the event kinematics. The most general Lorentz-invariant Lagrangian that can be written down contains seven complex couplings for each of the WWZ and $WW\gamma$ vertices. However, this large number of free parameters can be reduced by requiring C and P conservation, electromagnetic gauge invariance and further gauge constraints, leaving only three parameters: the relative coupling strength of the W boson to the neutral current, g_1^Z, and two form factors for the $WW\gamma$ vertex, κ_γ and λ_γ. At tree level in the Standard Model, these are

$$g_1^Z = 1, \quad \kappa_\gamma = 1, \quad \lambda_\gamma = 0. \tag{3.15}$$

The form factors are closely related to the magnetic dipole moment, μ_W, and the electric quadrupole moment, Q_W, of the W boson,

$$\mu_W = \frac{e}{2M_W}(1 + \kappa_\gamma + \lambda_\gamma), \quad Q_W = -\frac{e}{M_W^2}(\kappa_\gamma - \lambda_\gamma). \tag{3.16}$$

The current world averages using data collected at the LEP experiments for these three parameters are $g_1^Z = 0.984^{+0.022}_{-0.019}$, $\kappa_\gamma = 0.973^{+0.044}_{-0.045}$ and $\lambda_\gamma =$

$-0.028^{+0.020}_{-0.021}$ [8]. The results are thus all compatible with the Standard Model expectations.

3.3.3
Two-Fermion Processes Off Resonance

During the LEP 2 era, data were collected at several centre-of-mass energies above the Z pole between 130 and 209 GeV, allowing for energy-dependent measurements of the process $e^+e^- \rightarrow f\bar{f}$ [5]. These so-called two-fermion processes were studied as a test of the Standard Model and to search for signs of new physics in the interference of new processes with the off-resonance Z/γ^* exchange. Recently, corresponding measurements have also been performed at the Tevatron.

Two-fermion events often include initial-state radiation of photons, tending to reduce the effective centre-of-mass energy back to the Z boson mass. These so-called "radiative return" events are less interesting for analysis since the underlying physics has been studied in great detail during LEP 1. In order to focus on two-fermion events with true centre-of-mass energies well above M_Z, only events with $\sqrt{s'}/\sqrt{s} > 0.85$ are typically used for analysis. Here $\sqrt{s'}$ is the invariant mass of the final-state lepton pair. The two-fermion cross section versus LEP centre-of-mass energy, separately for final states with muons, taus and hadrons, is shown in Figure 3.7. Electrons are treated differently, since also t-channel diagrams contribute.

At LEP, the centre-of-mass energy-dependent forward-backward asymmetry, A_{FB}, is measured in the muon and tau channels. A_{FB} can be altered by non-Standard Model processes such as the production of additional heavy Z bosons, Z'.

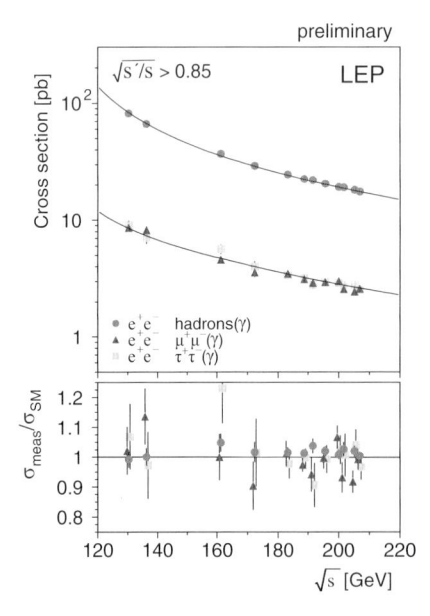

Figure 3.7 Cross-section results for two-fermion processes at LEP (adapted from [5]).

Hence any deviation in the asymmetry measurement from the Standard Model could be a sign for new physics. The combined LEP results and a comparison with the Standard Model can be seen in Figure 3.8. The agreement is very good and no deviation is found. The forward-backward asymmetry has also been measured at the Tevatron, extending the dilepton invariant mass reach beyond LEP. At a hadron collider, the asymmetry measurement is complicated by two facts. First, it is not known on an event by event basis which quarks annihilate to form the Z boson. Second, the longitudinal momenta of the initial-state partons are not known. The valence-quark and sea-quark content of the proton and antiproton and their momenta are described by parton distribution functions on a statistical basis only and with finite precision. The effect of the unknown momenta of the colliding quarks can be minimised by defining the asymmetry using the angle θ^* instead of θ. θ^* is the polar angle of the negative lepton with respect to the proton direction in the rest frame of the lepton–antilepton pair.

Unlike at LEP, where each dataset taken at a specific centre-of-mass energy contributed one data point in the asymmetry measurement, the data taken at one $p\bar{p}$ centre-of-mass energy were binned in the dilepton invariant mass. A better mass reach can be obtained by collecting more data. Therefore, a significantly improved measurement of the asymmetry can be expected with more Tevatron data.

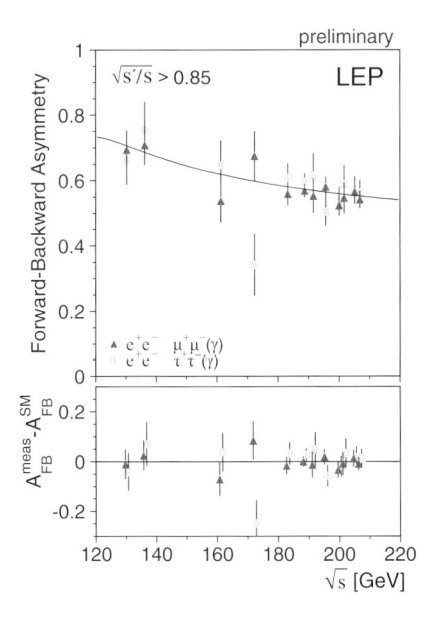

Figure 3.8 The forward-backward asymmetry for two-fermion processes at LEP (adapted from [5]).

3.3.4
Low Energy Data

Most of the measurements discussed in this chapter and in this book as a whole are done at high energy colliders, exploring the energy frontier of particle physics. However, there are a number of precision measurements performed using low energy data, which provide particularly stringent tests of the Standard Model and sensitivity to physics beyond it. These measurements should be seen as complementary to the ones done at high energies. Among the host of such sensitive low energy measurements, two are of particular interest. These are the measurement of the anomalous magnetic moment of the muon, an observable measured to better than one part per million, and the flavour changing transition $b \to s\gamma$. The latter is discussed in Chapter 8.

$g_\mu - 2$

The Dirac equation predicts the relation of the magnetic moment, M, of the muon and its spin, S, to follow

$$M = g_\mu \frac{e}{2m_\mu} S, \tag{3.17}$$

with the gyromagnetic ratio $g_\mu = 2$. However, quantum loop effects result in small but precisely calculable contributions to g_μ, commonly parametrised as $a_\mu = \frac{g_\mu - 2}{2}$. Within the Standard Model, these loop corrections are subdivided into three categories: electromagnetic (QED) corrections resulting from lepton and photon loops; electroweak corrections stemming from weak boson loops; and hadronic quark and gluon loop corrections. Whereas the electromagnetic corrections are by far the most sizable, they are also calculable to the highest precision, leaving the hadronic loop contributions to contribute the largest theoretical uncertainties to the Standard Model value of a_μ. The most precise experimental determination of a_μ was performed at the Brookhaven E821 experiment, where the precession of the spin of muons circulating in the AGS storage ring in a constant magnetic field was analysed. The world average is $a_\mu = (116\,592\,080 \pm 63) \times 10^{-11}$, which deviates from the Standard Model value of $(116\,591\,790 \pm 65) \times 10^{-11}$ by 3.2σ [3]. While not very significant, this deviation could be a hint for loop contributions by particles beyond the Standard Model such as supersymmetric particles.

3.4
Constraints from Precision Data

The precision measurements described in the last section can be used to obtain information on unknown parameters of the Standard Model, mainly the Higgs mass, or to constrain models of new physics. For this, the measurements are compared

3.4.1
Status of Theoretical Predictions

All Z pole observables and the W mass are calculated to second order in perturbation theory [9, 10]. In addition, important higher-order contributions of order $\alpha \alpha_s^2$, $G_F^2 m_t^4$, $G_F^2 m_t^2 M_Z^2$, $G_F m_t^2 \alpha_s$, $G_F m_t^2 \alpha_s^2$ are included [2]. This leads to theoretical uncertainties of $\delta \sin^2 \theta_{\text{eff}}^\ell = \pm 4.9 \times 10^{-5}$ and $\delta M_W = \pm 4$ MeV which are much below the experimental errors.

Of particular importance is the running of the fine-structure constant, α, which changes from its well-known value of $\alpha = 1/137.036$ at small momentum transfer by 10% when increasing the scale to M_Z [8]. To calculate $\alpha(M_Z^2)$ fermion loops between $s = 0$ and $s = M_Z^2$ must be taken into account. For lepton loops this is a straightforward problem. Also quark loops at higher s, where perturbative QCD can be applied, can be calculated reliably. At lower s the quark contributions can be obtained from the experimentally measured cross section $e^+e^- \to$ hadrons using the optical theorem. Several analyses of $\alpha(M_Z^2)$ exist which differ in the range and method in which perturbative QCD is applied [11]. The corrections to $\alpha(s)$ can be split into the different contributions

$$\alpha(s) = \frac{\alpha(0)}{1 - \Delta\alpha(s)} = \frac{\alpha(0)}{1 - \Delta\alpha_{e\mu\tau}(s) - \Delta\alpha_{\text{top}}(s) - \Delta\alpha_{\text{had}}^{(5)}(s)}, \qquad (3.18)$$

where $\Delta\alpha_{\text{had}}^{(5)}(s)$ represents the contribution from light quarks to which the above discussion applies. The LEP Electroweak Working Group uses conservatively

$$\Delta\alpha_{\text{had}}^{(5)}(M_Z^2) = 0.02758 \pm 0.00035, \qquad (3.19)$$

which results from an analysis that largely relies on experimental measurements [12]. The Gfitter group [13] uses

$$\Delta\alpha_{\text{had}}^{(5)}(M_Z^2) = 0.02768 \pm 0.00022, \qquad (3.20)$$

which has a significantly smaller error because the analysis relies more strongly on perturbative QCD [14]. Both numbers, however, contribute much less to the error of the electroweak fits than the measurements at higher energy, so that the choice of the analysis to use is of minor importance.

The pure QED corrections to $g_\mu - 2$ are known to complete fourth order with leading fifth-order contributions [3]. The weak corrections are strongly suppressed by the weak boson masses and known to second order. Their combined uncertainty is around 10^{-11} and is completely negligible compared to the present experimental precision. There are, however, also hadronic loop corrections which then receive QCD corrections at higher order. For the hadronic vacuum polarisation similar techniques as for $\alpha(M_Z^2)$ are used and the uncertainty is of order 5×10^{-10}. Another class of corrections comes from light-by-light scattering where a quark loop connects four photon lines. These corrections can only be calculated using mesons instead of quarks, leading to an uncertainty around 4×10^{-10}.

Observables using B-meson decays usually suffer from large QCD corrections. The uncertainties on quark masses also enter into the calculations. The most sensitive processes are thus those which are rare or forbidden in the Standard Model, so that also the corrections are small on an absolute scale. As an example, the decay $b \to s\gamma$, which is very sensitive to models with charged Higgs bosons and to SUSY, is known in second order α_s, leading to an uncertainty of 7% [15].

3.4.2
Standard Model Constraints

From the precision measurements at LEP and SLC effective couplings of the Z to the different fermions can be extracted. Figure 3.9 shows the vector and axial-vector coupling of the Z to leptons and b quarks compared to the Standard Model prediction. The lepton couplings are compatible with lepton universality and with the prediction. It can already be seen that the data prefer a light Higgs boson. The prediction for the b-quark couplings does, to a good approximation, not depend on the free parameters of the model.

All leptonic asymmetries can be expressed in terms of the *effective weak mixing angle*, $\sin^2 \theta_{\text{eff}}^{\ell}$. This is also true for the quark asymmetries, especially the b-quark asymmetry. First the $\sin^2 \theta_W$ dependence of the vector coupling is proportional to the fermion charge and thus a factor three smaller for b quarks than for leptons. In addition, the sensitivity of $\mathcal{A}$ to g_V/g_A goes to 0 for $g_V/g_A \to 1$, so that in total $A_{\text{FB}}^{0,b}$ is about 80 times more sensitive to $\sin^2 \theta_{\text{eff}}^{\ell}$ than to $\sin^2 \theta_{\text{eff}}^{b}$. Figure 3.10 shows the

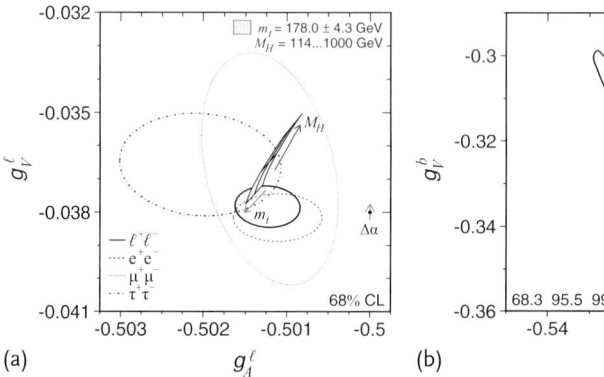

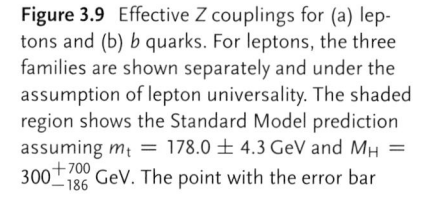

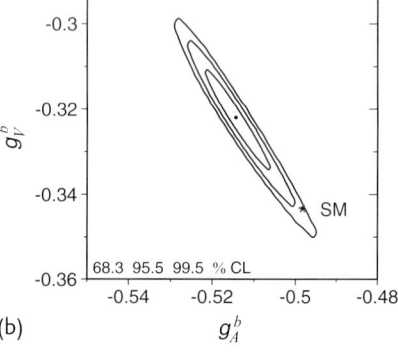

Figure 3.9 Effective Z couplings for (a) leptons and (b) b quarks. For leptons, the three families are shown separately and under the assumption of lepton universality. The shaded region shows the Standard Model prediction assuming $m_t = 178.0 \pm 4.3$ GeV and $M_H = 300_{-186}^{+700}$ GeV. The point with the error bar shows the prediction if electroweak radiative corrections are switched off with the uncertainty using $\Delta\alpha_{\text{had}}^{(5)}(M_Z^2) = 0.02758 \pm 0.00035$. For b quarks, the dependence of the prediction on the model parameters is negligible (adapted from [4]).

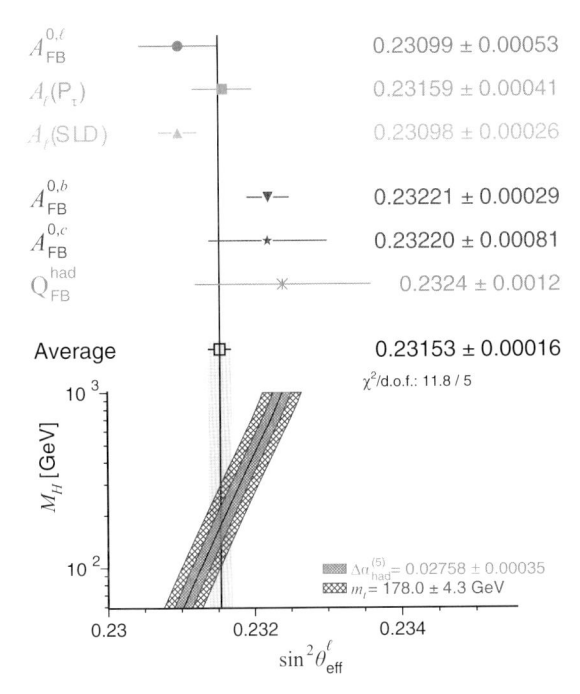

$A_{FB}^{0,\ell}$	•	0.23099 ± 0.00053
$A_\ell(P_\tau)$	■	0.23159 ± 0.00041
$A_\ell(SLD)$	▲	0.23098 ± 0.00026
$A_{FB}^{0,b}$	▼	0.23221 ± 0.00029
$A_{FB}^{0,c}$	★	0.23220 ± 0.00081
Q_{FB}^{had}	✳	0.2324 ± 0.0012
Average		0.23153 ± 0.00016
		χ^2/d.o.f.: 11.8 / 5

$\Delta\alpha_{had}^{(5)} = 0.02758 \pm 0.00035$
$m_t = 178.0 \pm 4.3$ GeV

Figure 3.10 Measurements of $\sin^2\theta_{eff}^\ell$ at LEP and SLC. Also shown is the prediction of the Standard Model (adapted from [4]).

$\sin^2\theta_{eff}^\ell$ values obtained from all asymmetries measured at LEP and SLD. Combining all numbers results in $\sin^2\theta_{eff}^\ell = 0.23153 \pm 0.00016$. The relatively bad χ^2 of the combination comes from the 3.2 standard-deviation disagreement between the two most precise measurements, the left-right asymmetry at SLD and the forward-backward asymmetries of b quarks at LEP. Although many tests have been done, no problem with any of the measurements has been found. It also turns out to be impossible to explain this difference with new physics appearing in loops. The most probable explanation therefore is a statistical fluctuation.

To check the global consistency of the data with the Standard Model, a fit of all the data to the model is performed taking all correlations into account, leaving the unknown parameters of the Standard Model free in the fit. To treat the experimental uncertainties correctly, measurements of fundamental model parameters like M_Z and m_t are simultaneously included in the dataset and left free in the fit. Several precise measurements of the strong coupling, α_s, exist, but they are all dominated by theoretical uncertainties. In the electroweak observables, α_s enters mainly as a correction to the hadronic decay width of the Z. Because of the unknown correlations between the theoretical uncertainties, no external α_s measurements are included in the fit.

Fitting the electroweak results from LEP 1 and SLC, the W mass and width from LEP 2 and the Tevatron and the top mass from the Tevatron with the fit described in [13] yields

$$M_H = 83^{+30}_{-23} \text{ GeV},\tag{3.21}$$

$$\alpha_s(M_Z) = 0.119 \pm 0.003,\tag{3.22}$$

with $\chi^2/\text{ndf} = 16.4/13$ corresponding to a probability of 23%. This shows that the data are well described by the Standard Model of electroweak interactions. The fitted value of α_s agrees well with the world average of $\alpha_s(M_Z) = 0.1176 \pm 0.0020$ [8].

Figure 3.11 shows the agreement of the individual observables with the fit. Similar fits are produced by other groups yielding consistent results [16, 17]. Figure 3.12a shows the change in χ^2 of the fit as a function of the Higgs mass. The best fit value lies in the region that is already excluded by the direct searches at LEP. The data are, however, well compatible with Higgs masses slightly above the limit. Assuming that the Higgs mechanism is as predicted by the Standard Model, an upper limit of $M_H < 158$ GeV at 95% confidence level can be obtained. The direct limits from LEP [18] and the Tevatron [19] can also be included in the fit converting the exclusion bounds into a χ^2 value. This fit yields $\chi^2/\text{ndf} = 19.9/14$ corresponding to a probability of 21%. Figure 3.12b shows the $\Delta\chi^2$ for this case. The general conclusions remain unchanged, however the mass limit is tightened to $M_H < 151$ GeV.

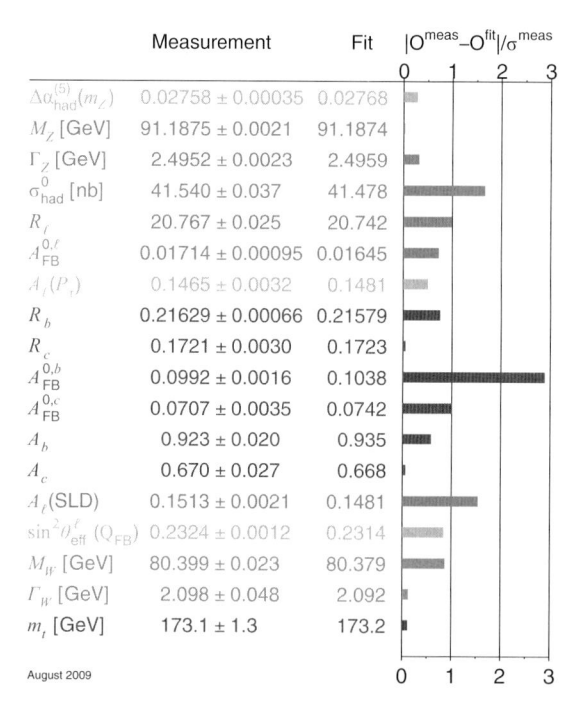

	Measurement	Fit
$\Delta\alpha^{(5)}_{had}(m_Z)$	0.02758 ± 0.00035	0.02768
M_Z [GeV]	91.1875 ± 0.0021	91.1874
Γ_Z [GeV]	2.4952 ± 0.0023	2.4959
σ^0_{had} [nb]	41.540 ± 0.037	41.478
R_l	20.767 ± 0.025	20.742
$A^{0,l}_{FB}$	0.01714 ± 0.00095	0.01645
$A_l(P_\tau)$	0.1465 ± 0.0032	0.1481
R_b	0.21629 ± 0.00066	0.21579
R_c	0.1721 ± 0.0030	0.1723
$A^{0,b}_{FB}$	0.0992 ± 0.0016	0.1038
$A^{0,c}_{FB}$	0.0707 ± 0.0035	0.0742
A_b	0.923 ± 0.020	0.935
A_c	0.670 ± 0.027	0.668
A_l(SLD)	0.1513 ± 0.0021	0.1481
$\sin^2\theta^l_{eff}(Q_{FB})$	0.2324 ± 0.0012	0.2314
M_W [GeV]	80.399 ± 0.023	80.379
Γ_W [GeV]	2.098 ± 0.048	2.092
m_t [GeV]	173.1 ± 1.3	173.2

August 2009

Figure 3.11 Electroweak observables used in the fit and their agreement with the Standard Model prediction (adapted from [16]).

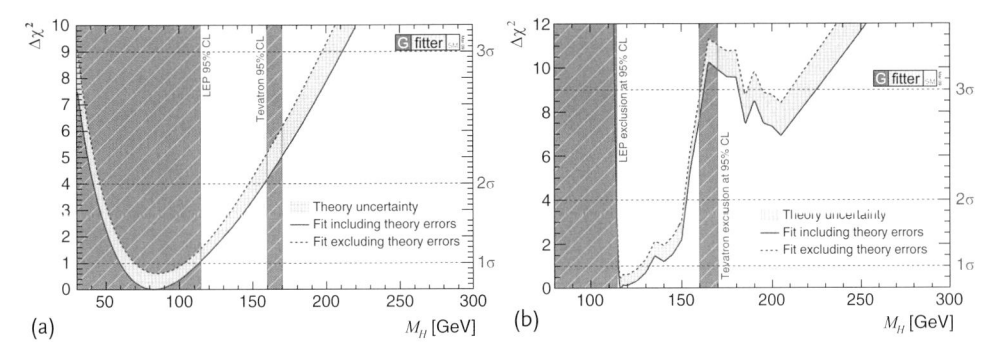

Figure 3.12 $\Delta\chi^2$ as a function of the Higgs boson mass for a fit to the electroweak precision data (a) without and (b) with the constraint from direct searches (adapted from [13]).

3.4.3
Constraints on New Physics

The precision data are sensitive to all models beyond the Standard Model that modify the predictions either directly on Born level or via loop corrections, and many analyses for a variety of models have been done.

The most popular extension of the Standard Model is SUSY which has the advantage of being fully calculable. Some particular aspects of SUSY are discussed here, and a detailed account of its general application to collider physics is given in Chapter 7. Several analyses exist that try to fit SUSY to the precision data [13, 20, 21]. SUSY is a decoupling theory which means that SUSY with heavy superpartners resembles the Standard Model with a light Higgs and is thus compatible with the data discussed in Section 3.4.2. This is shown for $\sin^2 \theta_{\text{eff}}^\ell$ and M_W in Figure 3.13 [22]. To obtain better results more information has been taken into account. WMAP has measured the anisotropy of the cosmic microwave background to a very high precision from which a dark-matter density of $\Omega h^2 = 0.1099 \pm 0.0062$ can be obtained [23]. This quantity can also be calculated for the lightest supersymmetric particle (LSP) density in SUSY. The fits assume that either the dark matter is completely given by the LSP or that the LSP density cannot be larger than the measured dark-matter density.

The anomalous magnetic moment of the muon, $g_\mu - 2$, deviates from its Standard Model prediction by 3.2σ [3] which can be easily explained by light SUSY. However, as explained in Section 3.3, there still exists the discrepancy between the Standard Model prediction using e^+e^- or τ data for the hadronic vacuum polarisation [24] and also the latest BABAR data for $e^+e^- \to \pi^+\pi^-$ show some discrepancy with the older data [25]. Assuming the constrained model with gravity mediated SUSY breaking and gauge unification at the GUT scale, the fits predict relatively light SUSY masses that can be discovered at the LHC with little luminosity If $g_\mu - 2$ is dropped from the fit the minimum is kept at the same place due to the dark matter constraint, however no useful limit exists anymore.

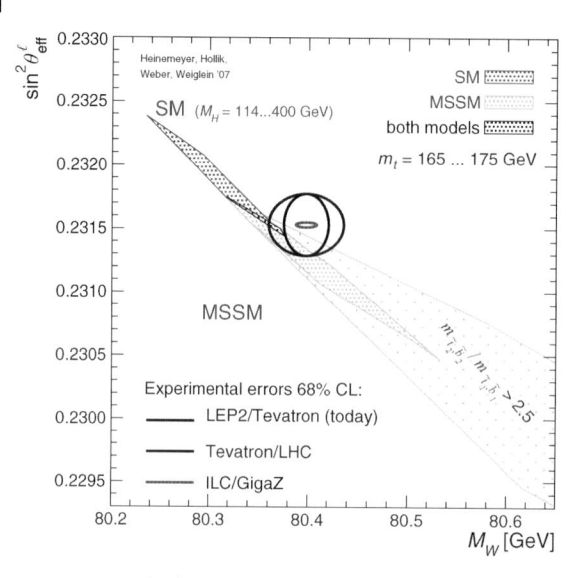

Figure 3.13 $\sin^2 \theta_{\text{eff}}^{\ell}$ versus M_W for the Standard Model and the MSSM. The projections assume that the central values remain unchanged (adapted from [22]).

Several studies exist that analyse the precision data with respect to an extended gauge sector, especially a new heavy neutral gauge boson, normally denoted as Z'. Close to the Z mass, the Z exchange diagram dominates the cross section completely, so that there is no significant sensitivity to the Z' mass; however, there is a large sensitivity to Z–Z' mixing. The Z' mass can be constrained by lower energy measurements like those of $g_\mu - 2$ and atomic parity violation [26]. Very strong constraints can also be obtained from two-fermion production at LEP 2 where the data are directly sensitive to Z' exchange [5]. In all these cases the sensitivity comes from the interference of the Z' with the Standard Model, introducing a large dependence on the Z' couplings and thus the details of the Z' model. This is complementary to Z' searches at the Tevatron [27, 28], where the Z' can be seen as a resonance in final states with $\mu^+\mu^-$ or e^+e^- pairs. The parton distribution functions decrease rapidly at high values of Bjorken-x, so that the limits have only a weak dependence on the coupling strength and are thus largely model independent. The mass limits are typically around 1 TeV, while the limit on the mixing angle, θ, is usually around $|\sin \theta| < 0.001$.

Models of new physics are often analysed in terms of the so-called ε parameters [29] or the *STU* parameters [30]. The ε parameters are defined as

$$\varepsilon_1 = \Delta\rho,$$

$$\varepsilon_2 = \cos^2 \theta_0 \Delta\rho + \frac{\sin^2 \theta_0}{\cos^2 \theta_0 - \sin^2 \theta_0} \Delta r_w - 2\sin^2 \theta_0 \Delta\kappa',$$

$$\varepsilon_3 = \cos^2 \theta_0 \Delta\rho + (\cos^2 \theta_0 - \sin^2 \theta_0) \Delta\kappa', \qquad (3.23)$$

with

$$\sin^2 \theta_{\text{eff}}^\ell = (1 + \Delta\kappa') \sin^2 \theta_0 , \qquad (3.24)$$

$$\sin^2 \theta_0 = \frac{1}{2}\left(1 - \sqrt{1 - 4\frac{\pi\alpha(M_Z^2)}{\sqrt{2}G_F M_Z^2}}\right) = 0.23098 \pm 0.00012 . \qquad (3.25)$$

The parameters are defined such that they are zero for the Standard Model on Born level. All the large isospin violating corrections, that come for example from the t–b mass difference, are absorbed in ε_1. The W mass enters only in ε_2 which keeps its Standard Model value in most of its extensions. ε_3 contains for example a logarithmic dependence on the Higgs mass.

The STU parameters can be obtained from the ε parameters by

$$S \simeq +\varepsilon_3 \frac{4\sin^2\theta_0}{\alpha(M_Z^2)} - c_S ,$$

$$T \simeq \varepsilon_1 \frac{1}{\alpha(M_Z^2)} - c_T ,$$

$$U \simeq -\varepsilon_2 \frac{4\sin^2\theta_0}{\alpha(M_Z^2)} - c_U , \qquad (3.26)$$

where the c_X are the Standard Model predictions assuming a given set of Standard Model parameters. The STU parameters are normalised such that they are zero in case of no new physics, and expected deviations are of order one. The S and T parameters assuming $U = 0$ are shown in Figure 3.14 together with the Standard Model prediction and the contours from the individual observables.

A *fourth fermion generation* would give a huge contribution to T from the mass splitting inside the doublets, but would also give a contribution of $S = \frac{N_C}{6\pi}(1 - $

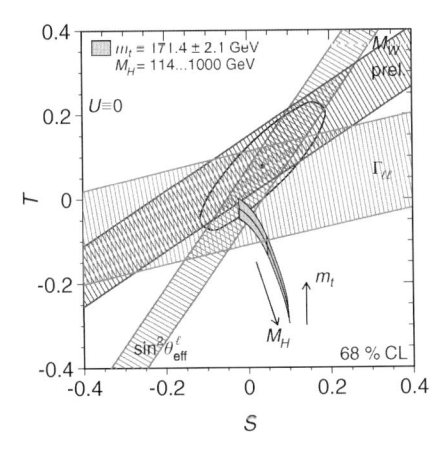

Figure 3.14 STU parameters obtained from the precision electroweak data. $S = T = U = 0$ has been fixed for $M_H = 150$ GeV. $U = 0$ has been assumed in the fit (adapted from [5]).

$2Y \ln \frac{m_u^2}{m_d^2}$) which does not vanish for non-degenerate fermions [31]. Choosing correct combinations of the mass scale and mass splitting of the fourth generation and of the Higgs boson mass can lead to a scenario which is compatible with the precision data.

To constrain such models more strongly, the major axis of the S–T ellipse must be reduced. To some extent this is possible with better W mass measurements, but a greater improvement would come from a better measurement of the Z leptonic width.

3.4.4
Expectations from LHC and ILC

The LHC has access to two of the observables entering the electroweak fits, m_t and M_W. Both quantities can be measured with similar techniques as at the Tevatron, but with much larger statistics. m_t is completely dominated by systematic uncertainties, so that at the moment $\Delta m_t = 1\,\text{GeV}$ is assumed for the LHC (see Chapter 9). For M_W, most systematic uncertainties come from the energy calibration and are of statistical nature. The large statistics at the LHC allow much more use of leptonic Z decays to be made, reducing significantly extrapolation errors. There seems to be a consensus that $\Delta M_W = 15\,\text{MeV}$ is possible and this value will be used for the estimates. However, there are also claims that $\Delta M_W = 5\,\text{MeV}$ is possible [32].

At a possible future e^+e^- linear collider the top mass can be measured in a threshold scan. The calibration systematics are replaced by the knowledge of the

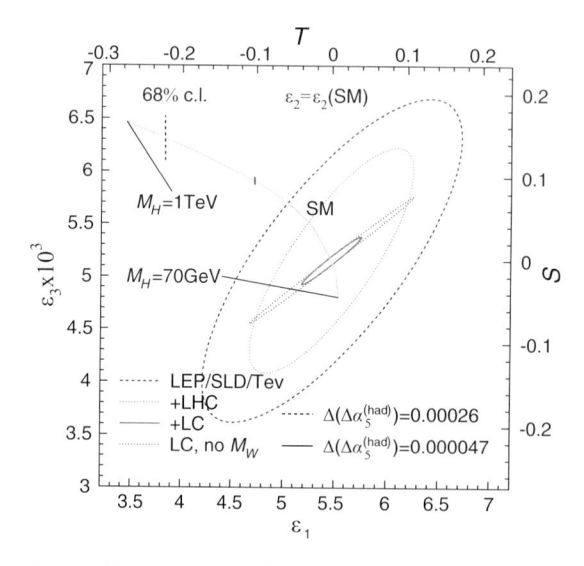

Figure 3.15 Improvement of the precision data due to LHC and ILC. Shown is the improvement in the ε and ST parameters assuming $U = 0$ (adapted from [33]).

beam energy, which should be known to a precision of 10^{-5}. Also the ambiguity in transforming the measured mass into the $\overline{\text{MS}}$ mass is much smaller for the top mass obtained from a threshold scan than for a direct reconstruction method. A final top-mass error of $\Delta m_t = 80$ MeV seems to be possible.

It is also possible to run the linear collider at lower energies, namely on the Z resonance and at the W pair production threshold. Close to the Z resonance one billion Z events can be expected in a few months of running with polarised electron and positron beams. This allows the left-right asymmetry to be measured with a precision of 10^{-4} corresponding to $\Delta \sin^2 \theta_{\text{eff}}^\ell = 1.3 \times 10^{-6}$. This measurement is only useful if at the same time $\Delta \alpha_{\text{had}}^{(5)}(M_Z^2)$ can be improved, which seems to be possible. With a scan of the Z lineshape also the measurement of its leptonic width can be improved. Because of a large influence of systematic uncertainties coming from effects such as beamstrahlung and beam energy spread, the precision is difficult to estimate; however, a factor two compared to LEP seems possible. With a scan of the W threshold, M_W may be measured with a precision of around $\Delta M_W = 6$ MeV, which is useful if the optimistic estimate of the LHC precision cannot be achieved.

Figure 3.15 shows the improvement of the ε and ST parameters due to the ILC measurements. It should be noted that the minor axis of the ellipse is given by the $\sin^2 \theta_{\text{eff}}^\ell$ measurements, which also determine the precision of M_H in the electroweak fit. The major axis is mainly given by the precision of Γ_ℓ and M_W. For unambiguous model tests like the fourth-generation analysis shown above this improvement is required.

References

1. Hollik, W.F.L. (1990) Radiative corrections in the Standard Model and their role for precision tests of the electroweak theory, *Fortschr. Phys.*, **38**, 165, doi:10.1002/prop.2190380302.
2. Bardin, D.Y. et al. (1997) Electroweak Working Group Report, http://arxiv.org/abs/hep-ph/9709229.
3. Jegerlehner, F. and Nyffeler, A. (2009) The muon g − 2, *Phys. Rep.*, **477**, 1, doi:10.1016/j.physrep.2009.04.003.
4. The ALEPH, DELPHI, L3, OPAL and SLD Collaborations, the LEP Electroweak Working Group, the SLD electroweak and heavy flavour groups (2006) Precision electroweak measurements on the Z resonance, *Phys. Rep.*, **427**, 257, doi:10.1016/j.physrep.2005.12.006.
5. The ALEPH, DELPHI, L3, OPAL, SLD Collaborations, the LEP Electroweak Working Group, the SLD Electroweak and Heavy Flavour Groups (2006) A Combination of Preliminary Electroweak Measurements and Constraints on the Standard Model, http://arxiv.org/abs/hep-ex/0612034.
6. DØ Collab., Abazov, V.M. et al. (2009) Measurement of the W boson mass, *Phys. Rev. Lett.*, **103**, 141801, doi:10.1103/PhysRevLett.103.141801.
7. The Tevatron Electroweak Working Group (2009) Updated Combination of CDF and DØ Results for the Mass of the W Boson, http://arxiv.org/abs/0908.1374.
8. Particle Data Group, Amsler, C. et al. (2008) Review of Particle Physics, *Phys. Lett. B*, **667**, 1.
9. Awramik, M., Czakon, M., Freitas, A., and Weiglein, G. (2004) Precise prediction for the W-boson mass in the Stan-

10. Awramik, M., Czakon, M., Freitas, A., and Weiglein, G. (2004) Complete two-loop electroweak fermionic corrections to the effective leptonic weak mixing angle $\sin^2 \theta_{\text{eff}}^{\text{lept}}$ and indirect determination of the Higgs boson mass, *Phys. Rev. Lett.*, **93**, 201805, doi:10.1103/PhysRevLett.93.201805.

11. Teubner, T. (2008) Hadronic contributions to the theoretical value of $(g-2)_\mu$ and $\alpha(q^2)$, *Nucl. Phys. Proc. Suppl.*, **181–182**, 20, doi:10.1016/j.nuclphysbps.2008.09.001.

12. Burkhardt, H. and Pietrzyk, B. (2005) Low energy hadronic contribution to the QED vacuum polarization, *Phys. Rev. D*, **72**, 057501, doi:10.1103/PhysRevD.72.057501.

13. Flacher, H. *et al.* (2009) Revisiting the global electroweak fit of the Standard Model and beyond with Gfitter, *Eur. Phys. J. C*, **60**, 543, doi:10.1140/epjc/s10052-009-0966-6.

14. Hagiwara, K., Martin, A.D., Nomura, D., and Teubner, T. (2007) Improved predictions for $g-2$ of the muon and $\alpha_{\text{QED}}(M_Z^2)$, *Phys. Lett. B*, **649**, 173–179, doi:10.1016/j.physletb.2007.04.012.

15. Misiak, M. *et al.* (2007) Estimate of $B(\overline{B} \to X_s \gamma)$ at $O(\alpha_s^2)$, *Phys. Rev. Lett.*, **98**, 022002, doi:10.1103/PhysRevLett.98.022002.

16. ALEPH, DELPHI, CDF, DØ, L3, OPAL, SLD Collaborations, the LEP Electroweak Working Group, the Tevatron Electroweak Working Group, the SLD Electroweak and Heavy Flavour Groups (2009) Precision Electroweak Measurements and Constraints on the Standard Model, http://arxiv.org/abs/0911.2604.

17. Erler, J. and Langacker, P. (2008) Electroweak physics, *Acta Phys. Pol. B*, **39**, 2595, http://arxiv.org/abs/0807.3023.

18. ALEPH, DELPHI, L3, OPAL collaborations and the LEP Working Group for Higgs boson searches, Barate, R. *et al.* (2003) Search for the Standard Model Higgs boson at LEP, *Phys. Lett. B*, **565**, 61, doi:10.1016/S0370-2693(03)00614-2.

19. The TEVNPH working group for the CDF and DØ Collaborations (2009) Combined CDF and DØ Upper Limits on Standard Model Higgs Boson Production with up to 4.2 fb^{-1} of Data, http://arxiv.org/abs/0903.4001.

20. Bechtle, P., Desch, K., Uhlenbrock, M., and Wienemann, P. (2009) Constraining SUSY models with Fittino using measurements before, with and beyond the LHC, *Eur. Phys. J.*, http://arxiv.org/abs/0907.2589.

21. Buchmueller, O. *et al.* (2008) Predictions for supersymmetric particle masses using indirect experimental and cosmological constraints, *JHEP*, **09**, 117, doi:10.1088/1126-6708/2008/09/117.

22. Heinemeyer, S., Hollik, W., Weber, A.M., and Weiglein, G. (2008) Z pole observables in the MSSM, *JHEP*, **04**, 039, doi:10.1088/1126-6708/2008/04/039.

23. WMAP Collab., Dunkley, J. *et al.* (2009) Five-year Wilkinson Microwave Anisotropy Probe (WMAP) observations: likelihoods and parameters from the WMAP data, *Astrophys. J. Suppl.*, **180**, 306, doi:10.1088/0067-0049/180/2/306.

24. Davier, M., Hoecker, A., Malaescu, B., Yuan, C. Z. and Zhang, Z. (2009) Reevaluation of the hadronic contribution to the muon magnetic anomaly using new $e^+ e^- \to \pi^+ \pi^-$ cross section data from BABAR, http://arxiv.org/abs/hep-ex/0908.4300.

25. BABAR Collab., Aubert, B. *et al.* (2009) Precise measurement of the $e^+ e^- \to \pi^+ \pi^-$ (gamma) cross section with the Initial State Radiation method at BABAR, *Phys. Rev. Lett.*, **103**, 231801. http://arxiv.org/abs/hep-ex/0908.3589.

26. Erler, J., Langacker, P., Munir, S., and Pena, E.R. (2009) Improved constraints on Z' bosons from electroweak precision data, *JHEP*, **08**, 017. http://arxiv.org/abs/0906.2435.

27. CDF Collab., Aaltonen, T. *et al.* (2009) Search for High-mass $e^+ e^-$ resonances in $p\overline{p}$ collisions at $\sqrt{s} = 1.96$ TeV, *Phys. Rev. Lett.*, **102**, 031801, doi:10.1103/PhysRevLett.102.031801.

28. DØ Collab (2006) Search for Heavy Z' Bosons in the Dielectron Channel with 200 pb^{-1} of Data with the DØ Detector, http://www-d0.fnal.gov/Run2Physics/

29 Altarelli, G., Barbieri, R., and Caravaglios, F. (1995) The epsilon variables for electroweak precision tests: a reappraisal, *Phys. Lett. B*, **349**, 145, doi:10.1016/0370-2693(95)00179-O.

30 Peskin, M.E. and Takeuchi, T. (1992) Estimation of oblique electroweak corrections, *Phys. Rev. D*, **46**, 381, doi:10.1103/PhysRevD.46.381.

31 Kribs, G.D., Plehn, T., Spannowsky, M., and Tait, T.M.P. (2007) Four generations and Higgs physics, *Phys. Rev. D*, **76**, 075016, doi:10.1103/PhysRevD.76.075016.

32 Besson, N., Boonekamp, M., Klinkby, E., Petersen, T., and Mehlhase, S. (2008) Re-evaluation of the LHC potential for the measurement of m_W, *Eur. Phys. J. C*, **57**, 627, doi:10.1140/epjc/s10052-008-0774-4.

33 Mönig, K. (2001) What is the case for a return to the Z pole? http://arxiv.org/abs/hep-ex/0101005.

Reference 28 continued: WWW/results/prelim/NP/N03/N03.pdf, DØ note 4375-Conf.

4
Hard QCD: Still Going Strong
Sven-Olaf Moch and Klaus Rabbertz

4.1
Introduction

About 40 years ago the gauge theory of strong interactions, which is known today under the name of Quantum Chromodynamics (QCD), started its triumphal path in particle physics. One of its roots goes back to the successful explanation of the observed spectra of hadrons (baryons and mesons) and their arrangement in multiplets by G. Zweig and M. Gell-Mann. For this purpose, Zweig and Gell-Mann invented hadron constituents which later were called *quarks* by M. Gell-Mann, borrowing the term from James Joyce's "Finnegans Wake". Quarks were postulated to come in the three flavours *up*, *down* and *strange*. All hadrons known at that time could be composed out of these three quarks and their respective antiquarks. Some apparent gaps in the observed hadrons with respect to the multiplets even led to the successful prediction of a new meson and also of the Ω^- baryon. Peculiarly, however, the postulated quarks had to carry fractional electric charges of $+2/3$, $-1/3$ and $-1/3$, a feature which was never observed in nature.

At the same time, the SLAC–MIT experiment at the Stanford Linear Accelerator Centre confirmed the "scaling" behaviour of structure functions measured in deep inelastic electron–nucleon scattering that was conjectured by J.D. Bjorken in 1969. Essentially, scaling means that the scattering cross section does not depend on the absolute energy or the momentum transfer (squared), Q^2, of the interaction but on dimensionless quantities like energy ratios or angles. This prediction is strikingly different from a cross section falling steeply with increasing momentum transfer as expected from the product of elastic scattering and structure functions representing a finite size of the nucleon charge distribution. Since the momentum transfer of the electron–nucleon scatters can also be interpreted in terms of the spatial resolution at which the nucleons are probed, the scaling behaviour translates into an independence from the resolution scale and strongly suggests that the constituents of the hadrons behave as point-like objects at high energies. In a different context such point-like constituents of hadrons had also been proposed by R.D. Feynman who had given them the name *partons*.

The dynamical description of strong interactions was advanced by employing a non-Abelian gauge theory based on the special unitary group $SU(3)$ of a new quantum number called *colour*, usually denoted as $SU(3)_c$. Every quark carries one of the three colours *red*, *green* or *blue*. By associating a different colour to each quark of a baryon, one can elegantly solve the so-called statistics problem; that is, the Δ^{++} resonance (three u quarks with total spin 3/2) would otherwise have a fully symmetric wave function, in contradiction to the Pauli principle that demands it to be fully antisymmetric. In analogy to Quantum Electrodynamics (QED), which is based on the unitary group $U(1)$, the dynamics come into play via bosonic exchange particles that mediate the strong force between the colour-charged quarks. However, owing to the more complex structure of the $SU(3)$ group and the fact that it is a non-Abelian group, there are eight colour-charged and therefore self-interacting gauge vector bosons called *gluons* (and not only an uncharged photon-like messenger particle).

The dynamics of quarks and gluons as encoded in the QCD Lagrangian (Equation 2.27 of Section 2.3) have a number of striking implications which were quickly noticed in the early 1970s. First of all, despite the additional colour degree of freedom there do not seem to be any new hadrons associated with it. The explanation comes by looking closer into the dynamics mediated by the gluons. D.J. Gross, H.D. Politzer and F. Wilczek established in 1972 that the strong force, in contrast to the electromagnetic one, becomes small at very high energies, that is, at subnuclear dimensions, and that it increases with distance. This property was called *asymptotic freedom*. As Gross, Politzer and Wilczek noted, an immediate consequence of asymptotic freedom are scaling violations in the structure functions of deep inelastic electron–nucleon scattering, which had been observed in the SLAC–MIT experiment at the same time. Thus, the scaling of structure functions, as explained by interactions with point-like constituents which in the limit of very small distances react like free particles, along with logarithmic violations as predicted by QCD helped to demonstrate that Feynman's partons represent nothing else than the coloured quarks, antiquarks and gluons of QCD.

As a consequence of asymptotic freedom, only entities which are colour singlets without any net colour charge can be observed as free particles. These are not subject to strong interactions any more. This is in complete accordance with experiment where only colourless baryons or mesons with integer multiples of the electric charge are detected. Coloured quarks with fractional electric charges or other coloured objects like gluons are confined to subnuclear dimensions. This property of QCD is called *confinement*. Nevertheless the fractional charges have a direct influence on measurable quantities, notably the ratio R of the total cross section of the reaction $e^+e^- \to$ hadrons compared to $e^+e^- \to \mu^+\mu^-$. At low centre-of-mass energies one obtains $R = n_c \cdot \sum_q e_q^2 \cdot (1 + \text{corrections})$ where $n_c = 3$ is the number of colours and e_q is the charge of each quark type that can be produced. Each time the collision energy is increased such that the limit for the pairwise ($q\bar{q}$) production of heavier quarks like *charm* or *bottom* is surpassed, a step in this ratio is observed.

Today, one believes that QCD is the correct gauge theory to describe strong interactions, one of the four fundamental forces of nature. It has a broad range of applications from the hard interactions between coloured quarks and gluons at Terascale energies down to the low energy formation of hadrons and mesons. Numerous excellent text books and review articles have appeared in the course of time and provide extensive coverage. References [1–5] are recommended specifically in the context of QCD at Terascale colliders. Further useful information, data, programs, and so on can be found online[15].

4.2
The Strong Coupling

The key property of QCD which is of relevance at Terascale colliders is the fact that (unlike in QED) the strong coupling $\alpha_s \ll 1$ at high energies. This is a genuine quantum effect due to anti-screening of the colour charges caused by the self-interaction of gluons. A perturbative expansion in the strong coupling, α_s, can thus be applied and provides the most powerful tool for theoretical predictions. This includes the running of the strong coupling which is governed by the renormalisation equation,

$$\frac{d}{d \ln \mu^2} \alpha_s(\mu) = \beta(\alpha_s) = -\sum_i \beta_i \left(\frac{\alpha_s}{4\pi}\right)^{i+1}, \qquad (4.1)$$

where the expansion coefficients of the β function are known to four-loop order [6] and start off with $\beta_0 = \frac{11}{3} C_A - \frac{4}{3} T_R n_F$ (note the "−" sign!). Here n_F denotes the number of flavours, and the QCD structure constants (or colour charges) take the values $C_F = 4/3$ and $C_A = 3$. $T_R = 1/2$ by conventional normalisation of the $SU(3)$ matrices.

To date numerous measurements have been conducted and employed to extract α_s at various momentum scales Q. Figure 4.1 demonstrates the consistency of measurements from e^+e^- annihilation, deep inelastic scattering and heavy quarkonia. The curve corresponds to the QCD prediction for a combined world average of $\alpha_s(M_Z) = 0.1184 \pm 0.0007$ at the scale set by the mass of the Z boson.

The alternative approach of lattice gauge theory, which applies QCD to a world discretised in space and time and which is not restricted to expansions around $\alpha_s \ll 1$, has over the years continuously improved its techniques to determine the strong coupling in a completely non-perturbative manner. The results are in excellent agreement with the standard methods. In addition, lattice QCD brought theory closer to the goal for which QCD was originally contrived, namely to explain the measured hadron spectra [8]. For a general overview on the status of QCD predictions in comparison to experiments and in particular with respect to α_s see [7] and [9]

15) PDG: http://pdg.web.cern.ch/pdg,
    HEPDATA: http://durpdg.dur.ac.uk/HEPDATA,
    HepForge: www.hepforge.org.

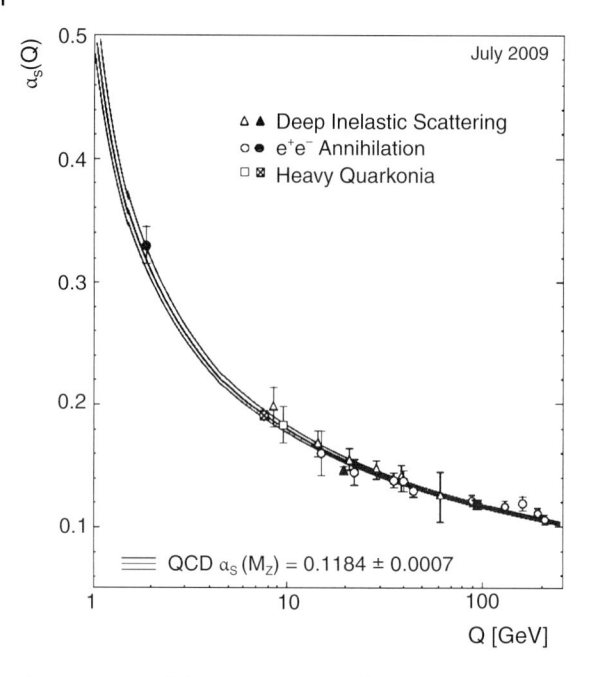

Figure 4.1 One of the great successes of QCD: Consistent determinations of the running coupling, $\alpha_s(Q)$, from various processes at different momentum scales Q (adapted from [7]).

4.3
Perturbative QCD at Colliders

Perturbative QCD has celebrated great successes in the description of hard scattering processes, leading to an ever-increasing accuracy of the theoretical predictions. This includes a precise derivation of the violation of an exact scaling of the structure functions as well as the discovery of the gluon at the e^+e^- storage ring PETRA at DESY. The latter evidently could not be accomplished by measuring a real gluon, which is a coloured object and therefore confined to subnuclear dimensions, but by measuring the corresponding jet rates. Likewise, at hadron colliders one needs to separate the long-distance effects, for example details of the dynamics of coloured partons inside the colourless initial hadrons, from the hard scattering reaction of individual partons at large momentum transfer Q.

This basic property of QCD is *factorisation* which rests on the fact that one can separate the dynamics from different scales [10]. For hard hadron–hadron scattering this property implies that the constituent partons from each incoming hadron interact at short distance, that is, in a situation where $\alpha_s(Q) \ll 1$, while the strongly interacting system of partons inside the hadron at nuclear distances is clearly nonperturbative. The coupling strength α_s inevitably becomes too large for perturbation theory to be applicable. Schematically QCD factorisation can be depicted as in Figure 4.2.

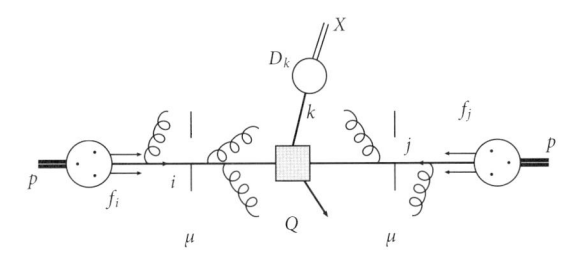

Figure 4.2 Factorisation of the hard scattering cross section in proton–proton collisions in the centre-of-mass frame for the QCD-improved parton model, (4.2).

Thus, for a cross section $\sigma_{pp \to X}$ of some hadronic final state X in, say, proton–proton scattering one can write

$$\sigma_{pp \to X} = \sum_{ijk} \int dx_1 dx_2 dz\, f_i(x_1, \mu) f_j(x_2, \mu)$$
$$\times \hat{\sigma}_{ij \to k}\left(x_1, x_2, z, Q^2, \alpha_s(\mu), \mu\right) D_{k \to X}(z, \mu), \quad (4.2)$$

where all functions have a clear physical interpretation. The parton distribution functions (PDFs) in the proton f_i ($i = q, \bar{q}, g$) describe the fraction x_i of the hadron momentum carried by the quark or gluon. The convolution of f_i and f_j determines, for a given collider, the so-called parton luminosity that will be discussed in Section 4.5. The details of the integration range in the convolution in (4.2) are controlled by the kinematics of the hard scattering process.

Since the proton is a very complicated bound state, the PDFs cannot be calculated in perturbation theory. Rather they have to be obtained from global fits to experimental data. A collection of such fits can be found online[16]. The (hard) parton cross section $\hat{\sigma}_{ij \to k}$ that depends on the parton types i, j and k is calculable perturbatively in QCD in powers of the strong coupling α_s. It describes how the constituent partons from incoming protons interact at short distances, $\mathcal{O}(1/Q)$. The final state X may denote hadrons, mesons, jets, and so on and needs another transition from the perturbative hard partons in the final state to the observed particles. The necessary function $D_{k \to X}$ can be a fragmentation function (FF) or a jet algorithm. Here the interface with showering algorithms (based on a Monte Carlo approach) becomes crucial, see Chapter 5.

All quantities in (4.2) are functions of the renormalisation and factorisation scale which are usually taken to be the same: $\mu \equiv \mu_r = \mu_f$. Physical observables like the cross section $\sigma_{pp \to X}$ in (4.2) cannot depend on these scales. In the perturbative approach, this implies that any residual dependence on μ in $\sigma_{pp \to X}$ has to be of a higher order compared to the accuracy of the cross-section calculation. This property can be cast in the form

$$\frac{d}{d \ln \mu^2} \sigma_{pp \to X} = \mathcal{O}(\alpha_s^{l+1}). \quad (4.3)$$

for a prediction to $\mathcal{O}(\alpha_s^l)$.

16) LHAPDF: http://projects.hepforge.org/lhapdf.

This equation motivates the commonly adopted approach to quantify uncertainties in theoretical predictions based on a variation of the renormalisation and factorisation scale by conventional factors of 1/2 and 2.

4.4
Hard Parton Scattering

Of central interest for any theory prediction of cross sections at hadron colliders is the description of the hard parton scattering process, that is, the calculation of the quantity $\hat{\sigma}_{ij}$ in (4.2). Important reactions for measurements at Terascale colliders like the Tevatron at Fermilab or the LHC at CERN are the production of jets with high transverse momenta and of W and Z bosons, possibly in association with jets (see Section 4.8). In addition, *top* quarks created either in pair- or single-*top* production mode will play a prominent role (see Chapter 9). The projected luminosity at the LHC promises large production rates for jets as well as for these Standard Model (SM) particles. The discovery of the SM Higgs boson (see Chapter 6) and new massive particles as postulated in theory extensions beyond the SM (BSM) (see Chapters 7 and 10) are the key elements of the physics case for the LHC experiments. The production rates for all these processes rely on the hard parton cross sections $\hat{\sigma}_{ij}$, and various approaches to the calculation of this quantity exist ranging from inclusive to fully differential in terms of kinematic variables. Building on exact matrix elements, the leading-order (LO) approximation of $\hat{\sigma}_{ij}$ in the perturbative approach provides a first estimate, although often with large theoretical uncertainties. Tree-level calculations in the SM, in its supersymmetric extensions or in other BSM models are automated to a large extent, and hard scattering cross sections are typically obtained through numerical phase-space integration with flexible kinematic cuts. The limitation of approaches based on exact matrix elements is the multiplicity of final-state particles. More than eight jets (particles) in the final state are currently at the edge of computational capabilities.

Whenever a good precision for theoretical predictions is required, at least the next-to-leading order (NLO) radiative corrections need to be considered. Due to (4.3) this is the first instance where a meaningful theoretical uncertainty can be quoted. Currently, the NLO QCD corrections to many important hard scattering processes with final-state multiplicities of 2–5 particles are known either numerically, yielding (differential) parton-level Monte Carlo events, or analytically for (inclusive) cross sections.

Unfortunately, the computation of radiative corrections in QCD is still involved, and at higher orders automation is less straightforward. Nevertheless, recently, with inspiration from string theory, significant progress based on on-shell methods and generalised unitarity has been made. NLO computations require renormalisation (see Chapter 2) and, in QCD, also collinear and infrared safety, that is, the cancellation of collinear and soft singularities. Beyond NLO, for example at next-to-next-to-leading order (NNLO) in QCD perturbation theory, some selected results are known mostly for inclusive kinematics. This level of precision is typ-

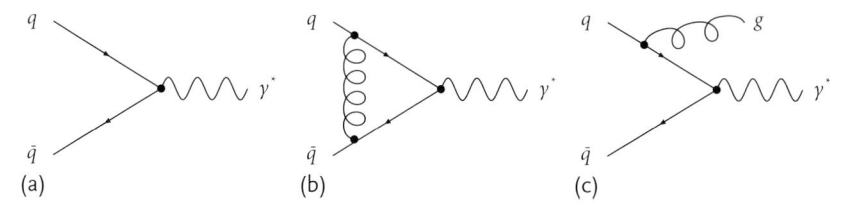

Figure 4.3 Feynman diagrams for the Drell–Yan process ($q\bar{q} \to \gamma^* \to \ell^+\ell^-$) at LO and NLO in QCD. (a) Born amplitude; (b) NLO virtual correction; (c) real radiation.

ically needed whenever predictions to better than $\mathcal{O}(10\%)$ accuracy are required. Beyond NNLO only very few results are known in QCD, for example the four-loop prediction for the running coupling discussed above [6].

To illustrate the discussion above it is instructive to consider the QCD radiative corrections for the Drell–Yan process, that is, dilepton production through the process $q\bar{q} \to \gamma^* \to \ell^+\ell^-$. The corresponding Feynman diagrams are given in Figure 4.3. Focusing, for simplicity, on quark–antiquark annihilation ($q\bar{q} \to \gamma^*$) and taking $n_F = 1$, the LO cross section reads

$$\hat{\sigma}^{(0)}_{q\bar{q}} = e_q^2 \delta(1-x)\sigma^{(0)}, \tag{4.4}$$

where $\sigma^{(0)}$ is the Born cross section for $e^+e^- \to \gamma^*$ in QED.

At NLO two classes of corrections enter which are separately divergent in the infrared limit. Thus, a regularisation is required which is typically carried out by working in $D = 4 - 2\epsilon$ dimensions of space-time. The virtual corrections are proportional to the Born cross section which factorises:

$$\hat{\sigma}^{(1),v}_{q\bar{q}} = e_q^2 \sigma^{(0)} C_F \frac{\alpha_s}{4\pi} \delta(1-x) \left(\frac{\mu^2}{Q^2}\right)^\epsilon \left(-\frac{4}{\epsilon^2} - \frac{6}{\epsilon} - 16 + 14\zeta_2 + \mathcal{O}(\epsilon)\right). \tag{4.5}$$

The divergences of the exchanged gluon in Figure 4.3b becoming soft and collinear manifest themselves in double poles $1/\epsilon^2$ in (4.5). Upon adding the real and virtual corrections for the squared amplitudes in Figure 4.3 and integrating over the final-state phase space, one obtains for the sum of $\hat{\sigma}^{(1),r}_{q\bar{q}}$ and $\hat{\sigma}^{(1),v}_{q\bar{q}}$

$$\hat{\sigma}^{(1)}_{q\bar{q}} = e_q^2 \sigma^{(0)} C_F \frac{\alpha_s}{4\pi} \left(\frac{\mu^2}{Q^2}\right)^\epsilon \left\{-\frac{1}{\epsilon}\left(\frac{8}{1-x} - 4 - 4x + 6\delta(1-x)\right)\right.$$
$$+ 16 \frac{\ln(1-x)}{1-x} - (16 + 8\zeta_2)\delta(1-x)$$
$$\left. -4(1+x)(2\ln(1-x) - \ln(x)) - 8\frac{\ln(x)}{1-x} + \mathcal{O}(\epsilon)\right\}. \tag{4.6}$$

Equation 4.6 illustrates the general structure of NLO QCD corrections with initial-state partons, where a collinear divergence proportional to the so-called splitting function $P_{qq}^{(0)}$ remains. Introducing the coefficient function $c_{q\bar{q}}^{(1)}$ of the hard scattering, (4.6) can be expressed as

$$\hat{\sigma}_{q\bar{q}}^{(1)} = e_q^2 \sigma^{(0)} \frac{\alpha_s}{4\pi} \left(\frac{\mu^2}{Q^2}\right)^\epsilon \left\{\frac{1}{\epsilon} 2 P_{qq}^{(0)}(x) + c_{q\bar{q}}^{(1)}(x) + \mathcal{O}(\epsilon)\right\} . \tag{4.7}$$

In order to arrive at finite partonic (physical) cross sections at a factorisation scale μ one needs to absorb the collinear divergence proportional to $P_{qq}^{(0)}$ in renormalised parton distributions,

$$f_q^{\text{ren}}(\mu^2) = f_q^{\text{bare}} - e_q^2 \frac{\alpha_s}{4\pi} \frac{1}{\epsilon} \left(\frac{\mu^2}{Q^2}\right)^\epsilon P_{qq}^{(0)}(x) , \tag{4.8}$$

where the subtraction on the right-hand side may be understood as a pure counter-term in an operator definition of the corresponding PDF in a minimal subtraction scheme. Thus, one arrives at

$$\hat{\sigma}_{q\bar{q}} = e_q^2 \sigma^{(0)} \left(\delta(1-x) + \frac{\alpha_s}{4\pi} \left\{c_{q\bar{q}}^{(1)}(x) - \ln\left(\frac{Q^2}{\mu^2}\right) 2 P_{qq}^{(0)}(x)\right\}\right) . \tag{4.9}$$

Here, a couple of remarks should be made: The real gluon emission in the soft and collinear limit gives rise to terms (so-called "+"-distributions) in the NLO coefficient function $c_{q\bar{q}}^{(1)}$ which are of the type $\alpha_s \ln(1-x)/(1-x)$, see (4.6) and (4.7). These "+"-distributions are responsible for the fact that the higher-order corrections (colloquially called K factors) are numerically large, an observation that puzzled people in the early days of QCD. The universal nature of the large corrections appears at l-loop precision as $(\alpha_s)^l \ln^{2l-1}(1-x)/(1-x)$ and their relation to the Sudakov logarithms is understood. Sudakov logarithms typically arise from edges of the phase space, and their appearance in hard cross sections requires resummation – a field where a lot of progress has been made in the last two decades. If one were to retain only those terms $(\alpha_s)^l \ln^{2l-1}(1-x)/(1-x)$ at each order this would define the so-called leading logarithmic accuracy and resummation in closed form would give rise to the Sudakov exponential. The Sudakov exponential defines emission probabilities of real radiation (to leading logarithmic accuracy) and is an essential ingredient in simulations of hard cross sections with parton-shower Monte Carlos. The latter are very important tools for understanding multi-parton scattering and the underlying event (see Chapter 5), and their improved merging with exact NLO calculations is currently an active field of technical developments (see Section 5.6).

4.5
Parton Luminosity

At hadron colliders one has wide-band beams of quarks and gluons, and the necessary proton PDFs, f_i ($i = q, \bar{q}, g$), entering in the master equation (4.2) are not directly accessible in QCD perturbation theory. However, their scale dependence (evolution) is. The independence of any physical observable on the scale μ

(see (4.3)) immediately gives rise to evolution equations for the PDFs, resulting in a system of coupled integro-differential equations governed by the complete set of splitting functions, P_{ij}:

$$\frac{d}{d\ln\mu^2}\begin{pmatrix} f_{q_i}(x,\mu) \\ f_g(x,\mu) \end{pmatrix} = \sum_j \int_x^1 \frac{dz}{z} \begin{pmatrix} P_{q_i q_j}(z) & P_{q_i g}(z) \\ P_{g q_j}(z) & P_{gg}(z) \end{pmatrix} \begin{pmatrix} f_{q_j}(x/z,\mu) \\ f_g(x/z,\mu) \end{pmatrix}. \tag{4.10}$$

The splitting functions, that is, the kernels of these differential equations, correspond to the different possible parton splittings. They are universal quantities and can be calculated in perturbation theory from the collinear singularity of any hard scattering process, see (4.7) for the specific example of $P_{qq}^{(0)}$ to first order. Physically, the evolution equation (4.10) states that one becomes sensitive to lower momentum partons as the resolution of the scattering reaction is increased, that is, as the scale μ becomes larger. The universality allows the determination of sets of PDFs in global fits to experimental data. Given an input distribution at a low scale, say $Q = 3\,\text{GeV}$, which has to be determined from comparison to data, (4.10) can be solved to predict the PDFs at a high scale. Thus, upon evolution this information from fits to reference processes can be used to provide cross-section predictions at LHC energies.

Generally, the parton distributions in hadrons are distinguished by the flavour quantum numbers, which are additive. The valence distribution originates from differences of quarks and antiquarks $q - \bar{q}$. The proton is composed of the sea distribution (i.e. the sum over all flavours $q + \bar{q}$) and of the gluon, g. Modern parametrisations of parton distribution from global fits account in particular for the effects of experimental measurement limitations and come with the according uncertainties on the parameters of the fit. Much of the experimental information originates from deep inelastic scattering data on structure functions (e.g. the electron–proton collider HERA at DESY), but also fixed-target experiments for muon–proton (deuterium) or (anti)neutrino–proton scattering as well as information on gauge-boson production at the Tevatron collider contribute. With these experimental data, the flavour separation of PDFs and the determination of the gluon PDF becomes feasible. At the LHC the production of W and Z bosons will be of particular importance, since these processes are expected to play a prominent role in understanding and calibrating the parton flux at the LHC itself (see Chapter 17.4.3).

At hadron colliders the necessary PDF information about the parton kinematics and the available partonic phase space for a given hard scattering reaction can be summarised in the parton luminosity L_{ij},

$$\frac{d^2 L_{ij}}{d\hat{s}dy} = \frac{1}{s}\{f_{i/p}(x_1,\mu) f_{j/p}(x_2,\mu) + (1 \leftrightarrow 2)\}. \tag{4.11}$$

Here $\hat{s}(s)$ are the partonic (hadronic) centre-of-mass energies squared with $\hat{s} = x_1 x_2 s$. The inclusion of s into the definition of L_{ij} allows a comparison between different colliders, for example LHC and Tevatron. The rapidity, y, of a final-state particle of mass M is given by $y = \ln(x_1/x_2)/2$, that is, the parton momentum fractions read $x_1 = (Me^y)/\sqrt{s}$ and $x_2 = (Me^{-y})/\sqrt{s}$ in the centre-of-mass system of the incoming hadron momenta, p_1 and p_2, with $p_1 = \sqrt{s}(x_1, 0, 0, x_1)/2$ and $p_2 = \sqrt{s}(x_2, 0, 0, -x_2)/2$.

The luminosity $dL_{ij}/d\hat{s}$ integrated over y is shown in Figure 4.4a. At the Tevatron $L_{q\bar{q}} > L_{qg} > L_{gg}$ whereas one can clearly see that at LHC the situation is reversed. The parton luminosity $L_{q\bar{q}}$ ranks third at LHC, and due to the large gluon component in the PDFs we have $L_{qg} > L_{gg} > L_{q\bar{q}}$. This feature is responsible for example for the dominance of SM Higgs production in the gluon fusion channel at the LHC.

The dependence of a final state of mass M at a given rapidity, y, on the parton kinematics in x and Q^2 is illustrated in Figure 4.4b, which also emphasises the impact of deep inelastic scattering data, for example from the HERA collider, on LHC predictions. Considering the allowed region for parton kinematics at LHC in Figure 4.4, it is clear that there is a large overlap in x with the range covered by deep inelastic scattering experiments. However, the relevant hard scale, Q, at the LHC is typically two to three orders higher due to the increased centre-of-mass energy, $\sqrt{s}$, where the dominant values of the momentum fractions $x_{1,2} \sim (Me^{\pm y})/\sqrt{s}$ depend on M and y.

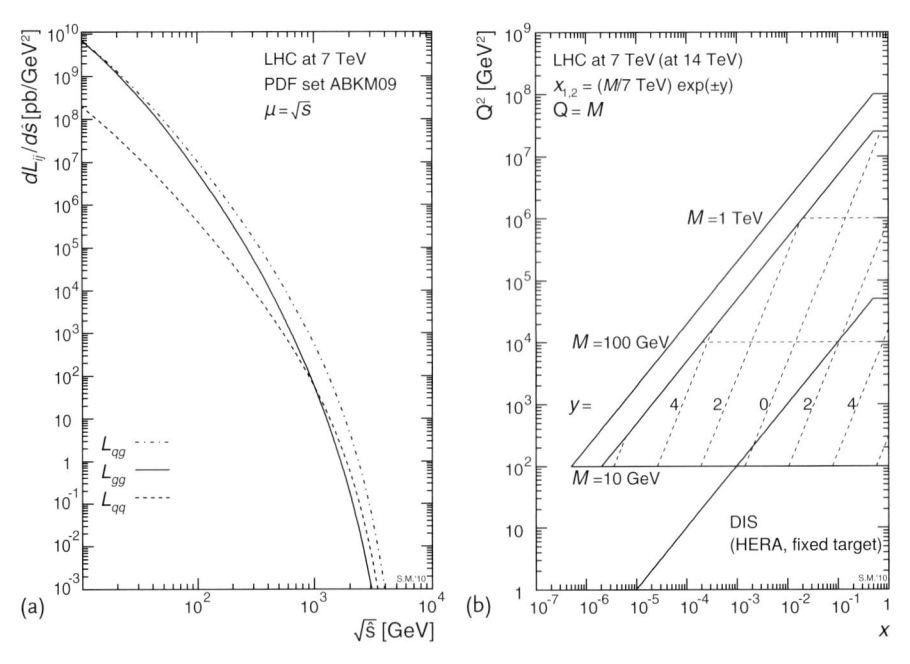

Figure 4.4 (a) The parton luminosity as a function of the partonic centre-of-mass energy $\sqrt{\hat{s}}$ for the LHC with $\sqrt{s} = 7$ TeV using the PDF set of [11]; (b) the parton kinematics at LHC with $\sqrt{s} = 7$ TeV in comparison to deep inelastic scattering at HERA and at fixed-target experiments.

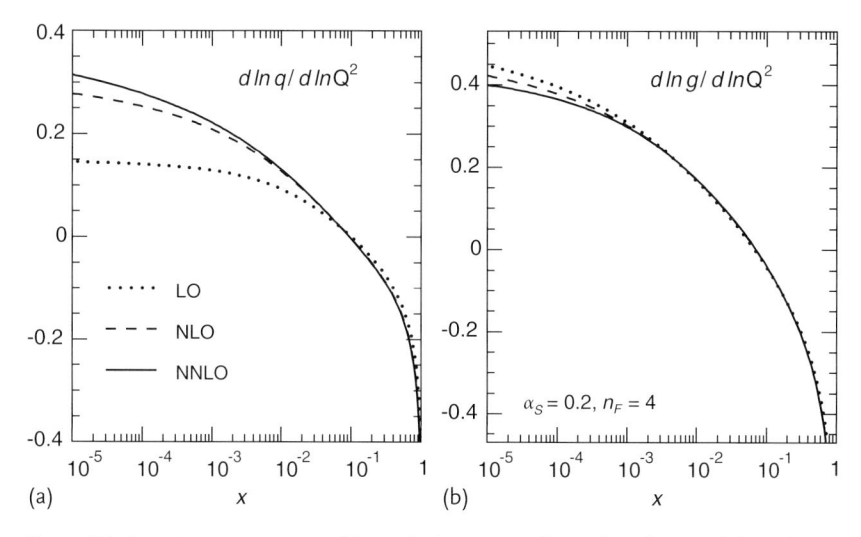

Figure 4.5 Perturbative expansion of the scale derivatives of typical quark (a) and gluon (b) distributions at $\mu \approx 5.5$ GeV (adapted from [12]).

The large difference in the hard momentum scale, Q, between HERA and LHC requires the parton evolution based on (4.10) to be sufficiently accurate in perturbative QCD. The necessary perturbative precision for quantitative predictions is approached at NNLO [12]. The stability of the evolution is shown in Figure 4.5 where the scale derivatives of the quark and gluon distributions at $\mu \approx 5.5$ GeV are presented. Obviously, the expansion is very stable except for very small momentum fractions $x \lesssim 10^{-4}$, demonstrating that the perturbative evolution equation (4.10) is applicable down to very small x. In terms of LHC parton kinematics, this corresponds to perturbative stability for central rapidities $|y| \lesssim 2$, while modifications are at most expected in the very forward (backward) regions $|y| \gtrsim 4$.

4.6
Fragmentation Functions and Event Shapes

The QCD description of scattering processes with final-state mesons or hadrons also raises the question how precisely the coloured partons produced in the hard scattering reaction transform into colourless objects. Assuming again that large-distance effects of QCD are decoupled from the original hard reaction and do not distort the measurable picture too much, one can predict that the initial partons will show up in the detectors as collimated streams of hadrons. For an individual hadron with for example measured charge and energy this transition is called *fragmentation*, see (4.2). The necessary fragmentation functions (FF), $D_{k \to X}$, parametrise the probability of a parton to fragment into a hadron, h, carrying a fraction, z, of the parton's momentum. Thus, the FFs are the final-state analogue of the

PDFs mentioned above. They also obey an evolution equation similar to (4.10), although with time-like kinematics and the matrix of splitting functions in (4.10) transposed. Most of the experimental data on FFs have been accumulated in the past at the LEP collider at CERN taking advantage of the experimentally favourable situation at e^+e^- machines. At hadron colliders, the rates and spectra of (charged) hadron multiplicities are typically early measurements during the commissioning phase, and also the LHC experiments have reported first results. The current status on FFs is summarised in [9].

Another possibility to relate the observed final-state hadrons to the confined initiating partons is by defining characteristic quantities which are not sensitive to the details of soft non-perturbative effects of QCD. In early studies of strong interactions in particle collisions, the main interest was to differentiate between a 2-jet-like structure favoured by QCD and the expectations from other models. For a 2-to-2 scattering like $e^+e^- \to q\bar{q} \to$ hadrons, a pencil-like momentum flow is predicted. Consequently, one strategy consists in the search for a principal axis along which the momenta of all produced hadrons are orientated. Then it is possible to derive a continuous dimensionless number, generically called *event shape* and denoted by F here, that characterises the extent to which the particles are bundled around this axis.

The first evidence of jet production was established in 1975 with the event shape *sphericity* which is, however, not collinear and infrared safe and therefore cannot be dealt with in perturbative QCD. A safe event measure, *thrust*, is defined as the normalised sum of the projections of all momenta onto the event axis, the thrust axis $\boldsymbol{n}_T$, that maximises this value:

$$T := \max_{\boldsymbol{n},\boldsymbol{n}^2=1} \frac{\sum_i |\boldsymbol{p}_i \cdot \boldsymbol{n}|}{\sum_i |\boldsymbol{p}_i|} = \frac{\sum_i |\boldsymbol{p}_i \cdot \boldsymbol{n}_T|}{\sum_i |\boldsymbol{p}_i|} . \tag{4.12}$$

The thrust value varies between $T = 1$ for a strictly linear orientation of all momenta where the thrust axis coincides with this direction (e. g. the $q\bar{q}$ pair produced back-to-back), and $T = T_{\min}$ with $T_{\min} = 0.5$ in case of a completely spherically symmetric distribution of the produced hadrons in e^+e^- annihilation. For simplicity, one redefines event shapes, F, such that they take on a value of $F = 0$ for the simplest reaction. Hence, thrust becomes $1 -$ thrust, $\tau := 1 - T$. Many other event shapes are in use like jet mass, jet broadening, the C and D parameters or Fox–Wolfram moments, some of which go beyond the characterisation of the 2-jet likeness of an event.

While in e^+e^- collisions the relation between the thrust axis $\boldsymbol{n}_T$ and the outgoing back-to-back $q\bar{q}$ pair is straightforward, this is not so simple anymore for lepton–hadron or hadron–hadron scattering. In the first case the solution is to examine an event in the Breit frame of reference where the incoming parton is back-scattered by the purely space-like photon exchanged with the electron probe. For hadron–hadron collisions the centre-of-mass system cannot be determined because of the unknown fractions of the hadron momenta carried by the interacting partons. Therefore, the events are analysed in the plane perpendicular to the beam directions where the vector sum of all transverse momenta should give zero. Definitions for some possible event shapes are given in [13]. However, due to experimental constraints the range in polar angle θ or equivalently in pseudorapidity, $\eta = -\ln\tan\frac{\theta}{2}$, with respect to the beam directions is limited. The CMS collaboration at the LHC investigated [14] the normalised event shape distribution of central transverse thrust,

$$\tau_{\perp,C} = 1 - T_{\perp,C} = \max_{n_C, n_C^2 = 1} \frac{\sum_{i \in C} p_{\perp,i} \cdot n_C}{\sum_{i \in C} |p_{\perp,i}|}, \qquad (4.13)$$

where the limit $|\eta| < \eta_C = 1.3$ is determined by the geometry of the central barrel calorimeter of the CMS detector [15]. Although in principle all energy depositions in the calorimeter can be used to evaluate the transverse thrust of an event, for reasons of the energy calibration strategy a jet algorithm (see next section) is used for pre-clustering. In addition to characterising an event in terms of its *jetiness*, the normalised event-shape distributions exhibit a reduced sensitivity to the jet energy calibration, which is usually the by far dominant source of experimental systematic uncertainty. Figure 4.6 shows a preliminary measurement of the central transverse thrust distribution including statistical and systematic uncertainties for 78 nb^{-1} of integrated luminosity at 7 TeV centre-of-mass energy together with predictions of several Monte Carlo (MC) generator programs. Leading-order dijet configurations can clearly be distinguished from multi-jet events, as can be seen from a comparison of the thrust distributions for different jet multiplicities in Figure 4.6 (2-jet: Figure 4.6a; and multi-jet: Figure 4.6b), where the error bars represent the statistical uncertainty on the data, and the shaded bands represent the sum of statistical and systematic uncertainties. Early measurements of event shapes allow differences in the modelling of QCD multi-jet production to be studied and are a valuable input to MC generator tuning.

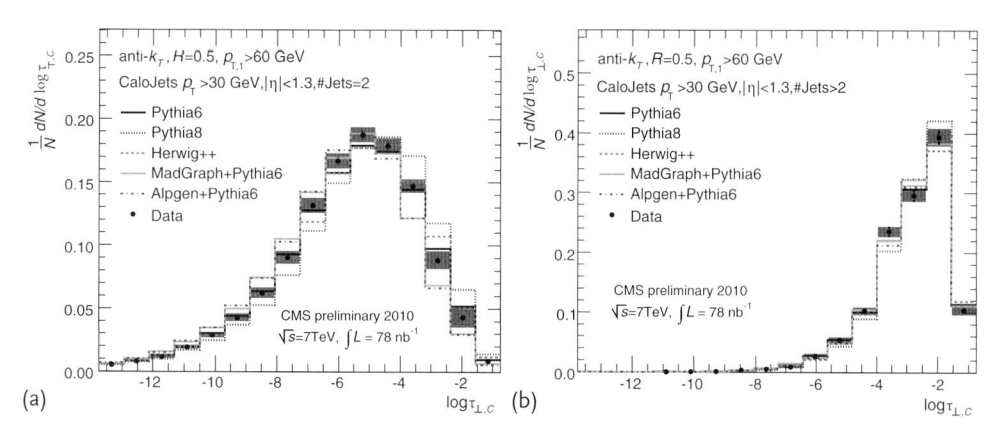

Figure 4.6 Preliminary measurement of central transverse thrust distribution ($\tau_{\perp,C}$, on a logarithmic scale) (a) for 2-jet events and (b) multi-jet events together with predictions of several Monte Carlo generator programs. (adapted from [14]).

The collinear and infrared safe event shapes are calculable to any order in perturbative QCD, are actually known to NNLO for e^+e^- collisions and can be combined with resummed predictions which improve the theoretical description of the differential event shape distributions close to the 2-jet limit where multiple radiations lead to large logarithmic corrections. In general, event shapes are somewhat more sensitive to details of the transition from partons to the hadronic final state than jet quantities addressed in the next section. This becomes apparent in additional power-suppressed corrections to the theory predictions which are proportional to $1/Q^p$, with Q being the relevant energy scale of the reaction and $p = 1$ for most event shapes. Historical overviews on event shape and jet measurements can be found in [16, 17].

4.7
Jet Production

The notion of *jets* was already used earlier in this chapter to denote a collimated stream of hadrons oriented in the direction of the initiating parton, but no exact definition was given up to now. Even though we talked of jets, most comparisons between experiment and theory were performed in terms of quark fragmentation [18] or event shapes and axes as presented before. The continuous measure of an event shape was sufficient to discover the gluon, as the main interest was to look for deviations from the primary expectation of a 2-jet-like structure in a model with only quarks.

The first prescription of a jet algorithm where particles (or energy depositions) are grouped together depending on whether they are inside an angular cone around a particular direction was given by G. Sterman and S. Weinberg in 1977 [19]. This prescription was extended in order to analyse hadron–hadron collisions in terms of a number of cone-shaped jets of a chosen jet size or radius, R, which are localised around the highest concentrations of energy in an event. The basic difference between event shapes and jet measures is that, instead of associating an inclusive continuous measure to each event (the event shape value), an integer number, the number of jets, is used for event characterisation. The separation between these two classes, however, is not strict. One can categorise, for example, all events into the class *2-jet-like* if they fulfil $\tau < 0.1$, and *multi-jet-like* otherwise. Similarly, the distance value where an event flips from having n to $n-1$ jets has a characteristic value and hence is a kind of event shape.

Cone jet algorithms were in widespread use for hadron–hadron collisions because they are efficient in finding a configuration of jet cones even for large numbers of particles or energy depositions. However, it proved to be difficult to be sure whether a configuration is optimal and to deal with overlapping cones. Diverging choices to characterise jets in the different experiments lead to a proposal for standardisation of cone jet algorithms, the Snowmass Accord [20].

Another approach to define jets was chosen in e^+e^- collisions which have less complicated final states. The general procedure here is to calculate all pairwise distances according to a specific distance measure, to combine the two closest objects into one (following a recombination prescription) and to iterate this procedure until all distances exceed a minimal value. The leftover set of objects are then called jets. Modern versions of these sequential recombination algorithms are the k_T and anti-k_T jet algorithms, applicable in e^+e^-, lepton–hadron as well as hadron–hadron collisions. An extensive overview of jet definitions in QCD is presented in [21].

Since QCD is supposed to be the theory of all strong interactions, it is desirable to be able to compare jets and jet properties between all kinds of collisions. As a consequence one should try to use cone jet algorithms also in e^+e^- reactions and k_T-type algorithms in the higher-multiplicity environment of hadron–hadron collisions. In preparation for Run 2 of the Tevatron experiments CDF and DØ, a strategy was developed [22]. The problem of efficiency of the k_T algorithm, however, whose time consumption supposedly rises proportional to N^3 (with N being the number of input objects) could be remedied only in [23] where it was shown that the complexity can be reduced to $N \ln N$. This major breakthrough opened up the possibility to apply k_T-type sequential recombination algorithms in high-multiplicity environments like at the LHC or even in heavy ion collisions.

Even after explicitly addressing the problem of collinear and infrared safety of jet algorithms in the Tevatron Run 2 workshop [22], cone jets still tend to suffer either from instabilities due to collinear splittings or the addition of soft particles. This makes comparisons between experiment and perturbative QCD results very difficult. An excellent discussion of these issues is given in [21].

In order to converge on common procedures for the future between the different experiments and the theory community it was proposed [24] that the following elements should always be specified:

- The jet definition which specifies all details of the procedure by which an arbitrary set of four-momenta from physical objects is mapped into a set of jets. The jet definition is composed of a jet algorithm (e. g. the inclusive longitudinally boost-invariant k_T algorithm), together with all its parameters (e. g. the jet-radius parameter, R, the split-merge overlap threshold f, the seed-threshold p_T cut, etc.) and the recombination scheme (e. g. the four-vector recombination scheme or "E scheme") according to which the four-momenta are recombined during the clustering procedure. Ideally, a reference to a full specification of the jet algorithm is given. If this is not available, the jet algorithm should be described in detail. Two elaborate jet software libraries including source code exist in the form of the FastJet or SpartyJet packages which are available on the web[17].
- The final state ("truth-level") specification. Consistent comparisons between experimental results, or between experimental results and Monte Carlo simulations, are only possible if the jet definition is supplemented with an exact specification of the set of the physical objects to which it was applied, or to which a quoted jet measurement has been corrected. This could for example be the

---

17) SpartyJet: http://projects.hepforge.org/spartyjet; FastJet: http://fastjet.fr.

set of momenta of all hadrons with a lifetime above some threshold. A popular choice for this threshold is 10 ps.

Concentrating on the measurement of inclusive jets, the first result clearly demonstrating the occurrence of jets in hadron–hadron collisions was published by the UA2 collaboration in 1982 [25]. In addition to their measurement of inclusive jets, the CDF publication [26] contains also a comprehensive historical overview. In total one finds a very good agreement between the predictions of perturbative QCD and corresponding measurements in all experiments to date. Figure 4.7 gives an overview of how data compare to theory for a multitude of different experiments including also lepton–hadron scattering. New preliminary results using the anti-k_T jet algorithm with 60 nb^{-1} of integrated luminosity at 7 TeV centre-of-mass energy

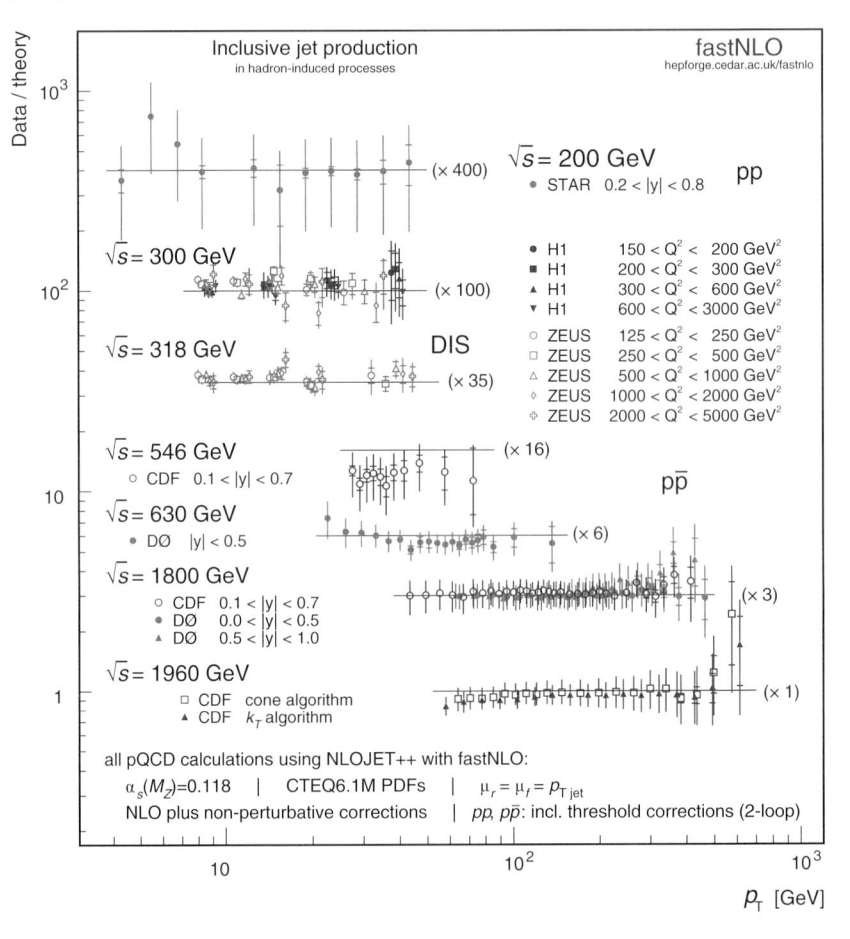

Figure 4.7 Summary of comparisons of jet data and theory for a multitude of different experiments in hadron–hadron and lepton–hadron scattering at different centre-of-mass energies (adapted from [27]).

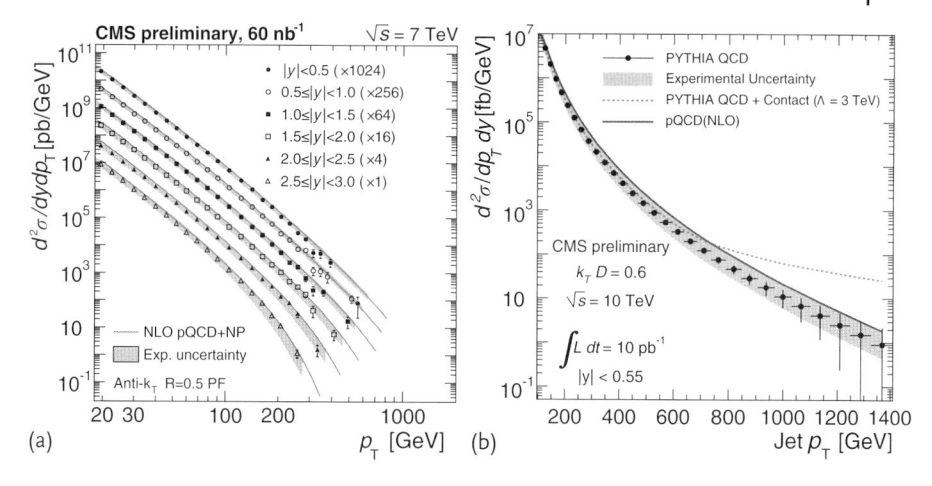

Figure 4.8 Preliminary inclusive jet measurement for the anti-k_T jet algorithm with 60 nb^{-1} of integrated luminosity at 7 TeV (a) and expectations for 10 TeV with an assumed luminosity of 10 pb^{-1} with k_T jets (b), both from CMS (adapted from [28, 29]).

are shown in Figure 4.8a and are contrasted with expectations for 10 TeV centre-of-mass energy with an assumed luminosity of 10 pb^{-1} with k_T jets. Possible deviations due to new physics, for example in the form of contact interactions, are pointed out.

4.8
Gauge-Boson Production

One of the most prominent hard scattering processes at the Terascale is the production of $W^\pm$ and Z bosons which gives rise to very clean experimental signals through their subsequent decay into leptons ($l^+\nu_l, l^-\bar{\nu}_l$ or l^+l^-). With the large rates expected for the production of $W^\pm$ and Z bosons especially at the LHC and the very good experimental identification of electrons and muons, these measurements will be dominated entirely by systematic uncertainties.

On the theory side, the radiative QCD corrections to the hard parton cross sections $\hat{\sigma}_{ij \to W^\pm/Z}$ are sizable. Their computation has been sketched in Section 4.4. Currently, they are known to NNLO accuracy for fully differential predictions, including cuts on transverse momenta or the rapidity of real radiation.

With the higher-order quantum corrections included, the theory predictions exhibit two nice features: first, the perturbative stability of the results in Figure 4.9 nicely demonstrates the apparent convergence of the perturbative expansion; second, the theoretical uncertainties according to (4.3) obtained by varying the renormalisation and factorisation scale μ by the conventional (although arbitrary) factor of two around the gauge-boson masses $M_{W,Z}$ are greatly reduced at NNLO, making for example the $W^\pm$ and Z bosons' rapidity distribution a very precisely predicted

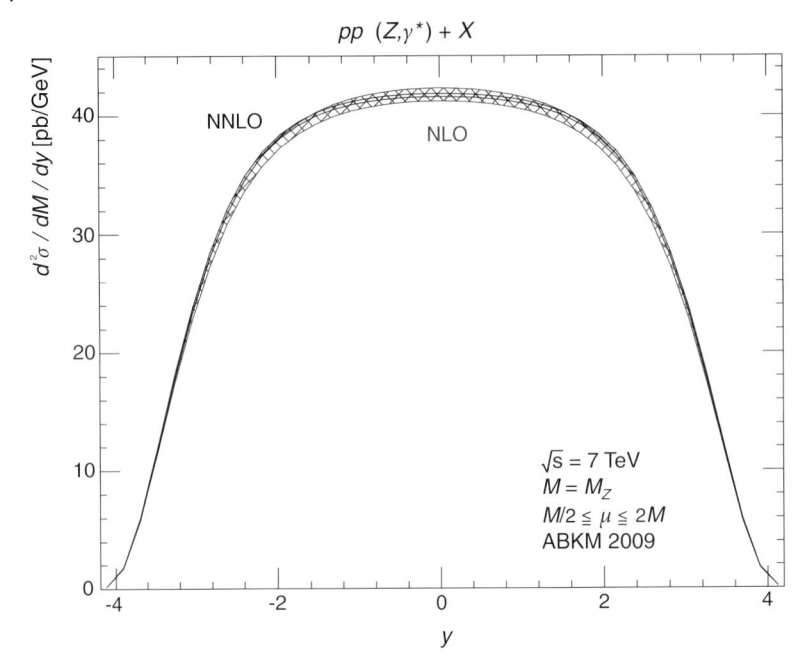

Figure 4.9 The rapidity-dependent cross sections for photon and Z-boson production at the LHC with estimates of the theoretical uncertainty from variations of the scale μ [30] (courtesy of L.J. Dixon).

observable (see Figure 4.9). The largest residual uncertainty on the theory side then rests in the parton luminosity L_{ij}, equation (4.11), with different global fits of PDFs giving slightly different predictions. Because of this fact, $W^{\pm}$ and Z boson production are often considered *standard candle* processes for the parton luminosity at the LHC.

Also of great importance is the production of $W^{\pm}$ and Z bosons with additional jets. These processes are an important background for example for Higgs production in the decay channel $H \rightarrow WW$ and subsequent semileptonic decays of the W bosons. Additional jets arise if the real-emission partons have sufficiently high transverse momentum. For instance the Feynman diagram in Figure 4.3c would contribute at LO in QCD to $\gamma^* + 1$ jet. The production rates for all these processes, $W^{\pm}/Z + n$ jets, are also sizable and, due to the large theoretical uncertainty coming from scale variations at LO, the computation of NLO QCD corrections is currently a field of major activity aiming at more precise predictions for the LHC.

4.9
Jet Shapes

Jet shapes look into the internal structure of jets. They are intermediate observables which depend on the perturbatively described parent parton and its first hard radiations, the perturbatively motivated parton showering and finally the phenomenological models of hadronisation. On the theory side (see for example [1]), one expects asymptotically that the average multiplicity in a gluon jet is enhanced by the ratio $C_A/C_F = 9/4$ of the colour factors in comparison to jets initiated by the light (anti)quarks u, d or s. At the same time the power with which the particle density in a jet diminishes with increasing angle can be estimated to be smaller by C_F/C_A. Experimentally, centre-of-mass energies of at least 50 GeV are required to have sufficiently collimated jets that then can be investigated for differences in quark and gluon fragmentation. In 1989, the AMY experiment at the TRISTAN e^+e^- storage ring reported measurements performed on 2- and 3-jet events that confirm at least qualitatively the expectations from QCD. At the even higher energies of the LEP collider the agreement with theory became even more evident. The presence of heavy quarks (c, b) and their weak decays leads to a slightly different internal jet structure which will not be discussed here. In total, the differences can also be described in terms of different fragmentation functions for gluons, light and heavy quarks.

In hadron–hadron collisions at the Tevatron two techniques have been employed to differentiate between gluon and light-quark jets. In the first method jets clustered with the k_T algorithm are subdivided again into subjets whose multiplicity can then be related statistically to gluon or quark jet parents [31]. The second approach looks into the fractional transverse momentum $1 - \psi(r/R)$ of a jet with size R outside the jet core for a particular radius, $r < R$ [32]. Figure 4.10a shows the measurement by the CDF collaboration for $r = 0.3$, which interpolates between the values for pure gluon or quark jets depending on their respective percentages at a specific jet p_T from Monte Carlo simulations with the generator. Figure 4.10b gives the simulation-based expectations of the CMS collaboration for a similar analysis ($r = 0.2$) at 14 TeV centre-of-mass energy at the LHC [33].

In addition to the abundant production of jets at the LHC allowing more detailed investigations of their internal structure than ever before, new types of jets can be envisioned. On the one hand, the high jet transverse momenta reaching far beyond the current Tevatron limit of around 700 GeV lead to even more collimated streams of hadrons. On the other hand, high-p_T jets might also be created by highly boosted massive particles such as top quarks or W, Z or Higgs bosons. Since the origin of these massive particles could possibly again be traced back to the decay of new objects, it becomes extremely important to find criteria to distinguish these jets from normal QCD jets. In the context of identifying the production of heavy Higgs bosons or $t\bar{t}$ pairs, this has already been discussed in terms of subjet reconstruction [34]. Newer jet-shape observables like planar flow or angularities were investigated in [35]. In particular, differences between a normal jet substructure and that due to 2-body or 3-body decays of heavy resonances have been pointed

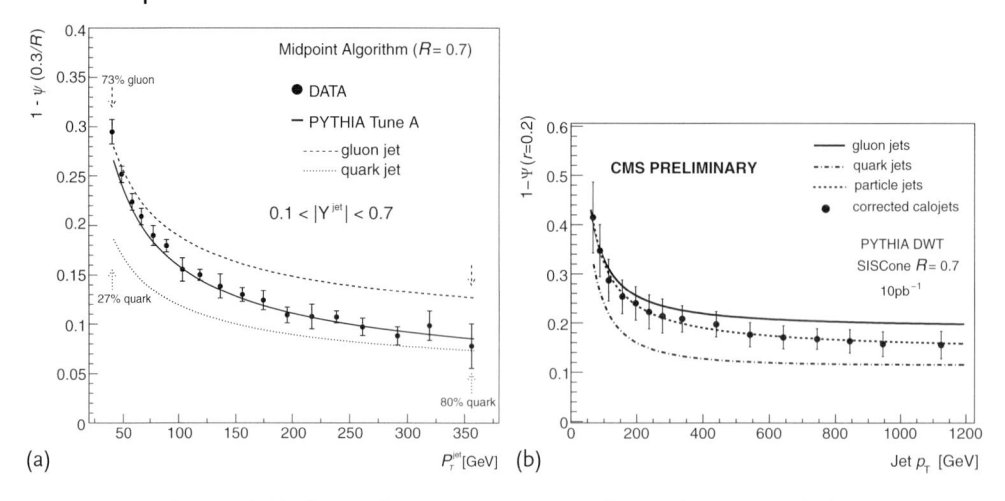

Figure 4.10 The fractional transverse momentum of a jet with size R outside the jet core, $1 - \psi(r/R)$, is presented for jets measured in (a) $p\bar{p}$ collisions at 1.96 TeV by CDF or (b) simulated in the CMS calorimeters for pp collisions at 14 TeV (adapted from [33] and [32]).

out. Apart from SM particles, these jets could also be initiated by new objects such as 2-prong decays of a Z' or 3-prong decays of a fourth quark generation (t'). Figure 4.11 gives two examples for differences in properties of normal QCD jets in contrast to jets from decayed massive particles. In summary, the substructure of jets at the LHC promises to become a very interesting field for studies of QCD as well as of new physics; the challenge will be to find means to differentiate especially between "normal" jets and jets of boosted massive particles.

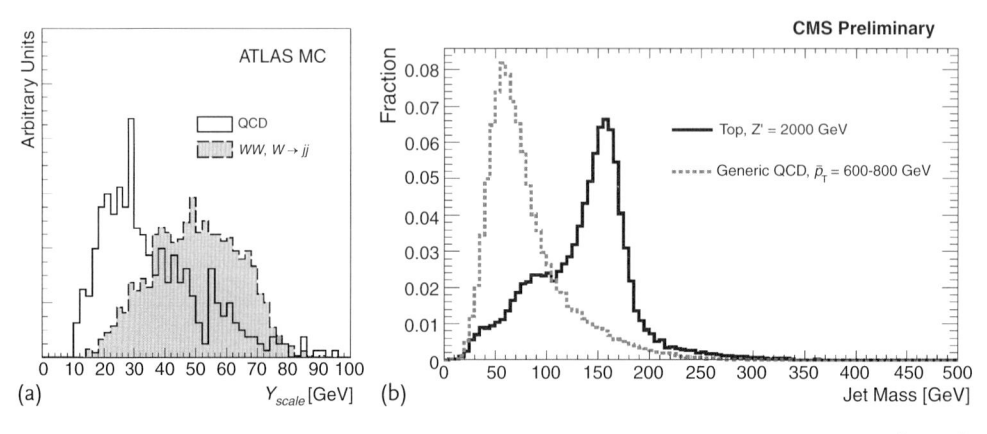

Figure 4.11 (a) Distribution of the scale variable y_{scale} indicating the threshold for splitting a given jet with mass $M_{jet} > 40$ GeV into two subjets for simulated QCD dijet processes and hadronically decaying boosted W bosons. (b) Jet mass for top-quark jets from $Z' \to tt'$ with $M_{Z'} = 2$ TeV compared to non-top-quark jets from generic QCD background with similar jet transverse momenta (adapted from [36] and [37]).

4.10
Tests of the QCD Gauge Structure

Experiments in the 1960s and 1970s demonstrated that the fundamental properties of QCD like asymptotic freedom and confinement give a good description of the data in terms of hadron spectra or the scaling behaviour of structure functions. More detailed observations on scaling violations, event shapes and 3-jet events, as required due to the presence of gluons, further affirmed the belief in QCD as the theory of strong interactions. There is only one free parameter in QCD calculations for massless partons, the strong coupling, $\alpha_s(Q)$, which has to be determined experimentally. Since its dependence on the energy scale, Q, is predicted as well, measurements of α_s performed in a multitude of QCD-related processes over a wide range of Q can all be related to each other.

In hadron collisions, the additional complication of the less well-known gluon density of the initial hadrons comes into play. Special care has to be taken to avoid circular reasoning by exploiting the same data twice to determine α_s and the gluon density independently of each other, while they usually appear as a product in the calculations. A recent result from inclusive jet production by the DØ collaboration [38] taking this into account by restricting the range in jet transverse momentum is shown in Figure 4.12a. At the LHC, a much wider range in jet transverse momenta will be at disposal for further tests of QCD.

It has been argued [1] that although experimental results from event shapes and 3-jet cross sections prefer a non-Abelian $SU(3)_c$ gauge theory, also other possibilities like an "Abelian QCD" with $[U(1)]^3$ as the gauge group are, in principle, imaginable. In that case, there would be no gluon self-coupling as in the triple-gluon vertex and the colour charges $C_F = 4/3$ and $C_A = 3$ of QCD would be altered. As a consequence the relative weights of higher-order corrections to 3-jet quantities or, more directly, the predictions for 4-jet final states would change which can be

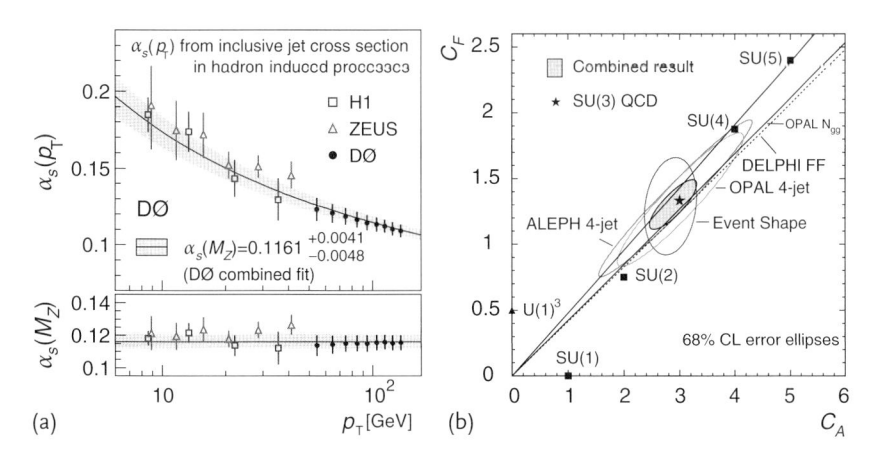

Figure 4.12 The strong coupling $\alpha_s(p_T)$ and $\alpha_s(M_Z)$ as measured by DØ (a) and a summary of the QCD structure constants C_F and C_A as measured at LEP (b) (adapted from [38] and [39]).

tested experimentally. For the latter, the $q\bar{q}q\bar{q}$ final state of e^+e^- collisions, associated with a $g \to q\bar{q}$ vertex, exhibits a factor of $C_F T_R n_F$, again with $T_R = 1/2$ and $n_F = 5$ for the number of active quark flavours, assuming SM particles only. The $q\bar{q}gg$ final state contributes two summands to the cross section with factors C_F^2 and $C_F C_A$, where only the latter includes triple-gluon vertices $g \to gg$ and is absent in the Abelian case. Corresponding NLO corrections have been known since 1997, allowing the simultaneous measurement of α_s and the colour charges C_F and C_A. Figure 4.12 [39] summarises the different possible measurements. At LHC energies, the high-p_T jet production via $gg \to gg$, involving also the quartic-gluon vertex, becomes important and gives new opportunities for exciting QCD observations.

4.11
Outlook

First LHC collision data have been recorded up to a centre-of-mass energy of 7 TeV, and the LHC experiments are rapidly analysing the data at this new energy frontier. Some measurements have been published and even higher energies will be reached in the future, opening a window to an unprecedented multitude of new physics analyses involving multiple-jet and multiple-boson production. This rich programme of measurements not only re-establishes the SM, but also sets the scene for searches for new phenomena. 2010 marks the beginning of a new era of particle physics.

References

1. Ellis, R.K., Stirling, W.J., and Webber, B.R. (1996) *QCD and Collider Physics*, Cambridge University Press.
2. Ynduráin, F.J. (1999) *The Theory of Quark and Gluon Interactions*, Springer, Berlin.
3. Dissertori, G., Knowles, I.G., and Schmelling, M. (2002) *Quantum Chromodynamics: High Energy Experiments and Theory*, Oxford University Press.
4. Campbell, J.M., Huston, J.W., and Stirling, W.J. (2007) Hard interactions of quarks and gluons: a primer for LHC physics, *Rept. Prog. Phys.*, **70**, 89, doi:10.1088/0034-4885/70/1/R02.
5. Moch, S. (2008) Hard QCD at hadron colliders, *J. Phys. G*, **35**, 073001, doi:10.1088/0954-3899/35/7/073001.
6. van Ritbergen, T., Vermaseren, J.A.M., and Larin, S.A. (1997) The four-loop beta function in quantum chromodynamics, *Phys. Lett. B*, **400**, 379, doi:10.1016/S0370-2693(97)00370-5.
7. Bethke, S. (2009) The 2009 world average of $\alpha_s(M_Z)$, *Eur. Phys. J. C*, **64**, 689, doi:10.1140/epjc/s10052-009-1173-1.
8. Dürr, S. *et al.* (2008) Ab-initio determination of light hadron masses, *Science*, **322**, 1224, doi:10.1126/science.1163233.
9. Particle Data Group, Amsler, C. *et al.* (2008) Review of Particle Physics, *Phys. Lett. B*, **667**, 1, doi:10.1016/j.physletb.2008.07.018.
10. Collins, J.C., Soper, D.E., and Sterman, G.F. (1988) Factorization of hard processes in QCD, *Adv. Ser. Direct. High Energy Phys.*, **5**, 1. http://arxiv.org/abs/hep-ph/0409313.
11. Alekhin, S., Blümlein, J., Klein, S., and Moch, S. (2010) The 3-, 4-, and 5-flavor NNLO parton from deep-

inelastic-scattering data and at hadron colliders, *Phys. Rev. D*, **81**, 014032, doi:10.1103/PhysRevD.81.014032.
12. Vogt, A., Moch, S., and Vermaseren, J.A.M. (2004) The three-loop splitting functions in QCD: the singlet case, *Nucl. Phys. B*, **691**, 129, doi:10.1016/j.nuclphysb.2004.04.024.
13. Banfi, A., Salam, G.P., and Zanderighi, G. (2004) Resummed event shapes at hadron–hadron colliders, *JHEP*, **08**, 062, doi:10.1088/1126-6708/2004/08/062.
14. CMS Collab. (2010) Hadronic event shapes in pp collisions at 7 TeV, http://cdsweb.cern.ch/record/1280682.
15. CMS Collab. (2008) The CMS experiment at the CERN LHC, *JINST*, **3**, S08004, doi:10.1088/1748-0221/3/08/S08004.
16. Cahn, R. and Goldhaber, G. (2009) *The Experimental Foundations of Particle Physics*, 2nd edn., Cambridge University Press, http://www.slac.stanford.edu/spires/find/hep/www?irn=2223830.
17. Kastrup, H.A. and Zerwas, P.M. (eds) (1993) *QCD – 20 Years Later*, vol. 1–2, World Scientific, Singapore.
18. Feynman, R.P., Field, R.D., and Fox, G.C. (1978) Quantum-chromodynamic approach for the large-transverse-momentum production of particles and jets, *Phys. Rev. D*, **18**, 3320, doi:10.1103/PhysRevD.18.3320.
19. Sterman, G.F. and Weinberg, S. (1977) Jets from quantum chromodynamics, *Phys. Rev. Lett.*, **39**, 1436, doi:10.1103/PhysRevLett.39.1436.
20. Huth, J.E. *et al.* (1992) Towards a standardization of jet definitions, in: *5th DPF Summer Study on High-energy Physics* (ed. E.L. Berger), Chap. 7, p. 134, World Scientific, Singapore, http://cdsweb.cern.ch/record/238477.
21. Salam, G.P. (2010) Towards jetography, *Eur. Phys. J. C*, **67**, 637, doi:10.1140/epjc/s10052-010-1314-6.
22. Blazey, G.C. *et al.* (2000) Run II Jet Physics: Proceedings of the Run II QCD and Weak Boson Physics Workshop, http://arxiv.org/abs/hep-ex/0005012.
23. Cacciari, M. and Salam, G.P. (2006) Dispelling the N^3 myth for the k_t jet-finder, *Phys. Lett. B*, **641**, 57, doi:10.1016/j.physletb.2006.08.037.
24. Buttar, C. *et al.* (2008) Standard Model Handles and Candles Working Group: Tools and Jets Summary Report, http://arxiv.org/abs/0803.0678.
25. UA2 Collab., Banner, M. *et al.* (1982) Observation of Very Large Transverse Momentum Jets at the CERN anti-p p Collider, *Phys. Lett.*, **B118**, 203, doi:10.1103/10.1016/0370-2693(82)90629-3.
26. CDF Collab., Affolder, A.A. *et al.* (2001) Measurement of the inclusive jet cross section in $\bar{p}p$ collisions at $\sqrt{s} = 1.8$ TeV, *Phys. Rev. D*, **64**, 032001, doi:10.1103/PhysRevD.64.032001.
27. Kluge, T., Rabbertz, K., and Wobisch, M. (2006) Fast pQCD calculations for PDF fits, http://arxiv.org/abs/hep-ph/0609285.
28. CMS Collab. (2010) Measurement of the Inclusive Jet Cross Section in pp Collisions at 7 TeV, http://cdsweb.cern.ch/record/1280682.
29. CMS Collab. (2009) Initial Measurement of the Inclusive Jet Cross Section at 10 TeV with CMS, http://cdsweb.cern.ch/record/1195745.
30. Anastasiou, C., Dixon, L.J., Melnikov, K., and Petriello, F. (2004) High precision QCD at hadron colliders: electroweak gauge boson rapidity distributions at NNLO, *Phys. Rev. D*, **69**, 094008, doi:10.1103/PhysRevD.69.094008.
31. DØ Collab., Abazov, V.M. *et al.* (2002) Subjet multiplicity of gluon and quark jets reconstructed with the k_T algorithm in $p\bar{p}$ collisions, *Phys. Rev. D*, **65**, 052008, doi:10.1103/PhysRevD.65.052008.
32. CDF Collab., Acosta, D.E. *et al.* (2005) Study of jet shapes in inclusive jet production in $p\bar{p}$ collisions at $\sqrt{s} = 1.96$ TeV. *Phys. Rev. D*, **71**, 112002, doi:10.1103/PhysRevD.71.112002.
33. CMS Collab. (2008) Transverse Momentum Distribution within Jets in pp Collisions at 14 TeV, CMS Physics Analysis Summary QCD-08-005, http://cdsweb.cern.ch/record/1281585.
34. Seymour, M.H. (1994) Searches for new particles using cone and cluster jet algo-

rithms: a comparative study, *Z. Phys. C*, **62**, 127, doi:10.1007/BF01559532.

35 Almeida, L.G. *et al.* (2009) Substructure of high-p_T Jets at the LHC, *Phys. Rev. D*, **79**, 074017, doi:10.1103/PhysRevD.79.074017.

36 ATLAS Collab. (2008) Expected Performance of the ATLAS Experiment – Detector, Trigger and Physics, *Tech. Rep. CERN-OPEN-2008-020*, CERN, http://arxiv.org/abs/0901.0512.

37 CMS Collab. (2009) A Cambridge-Aachen (C-A) based Jet Algorithm for boosted top-jet tagging, http://cdsweb.cern.ch/record/1194489.

38 DØ Collab., Abazov, V.M. *et al.* (2009) Determination of the strong coupling constant from the inclusive jet cross section in $p\overline{p}$ collisions at $\sqrt{s} = 1.96$ TeV, *Phys. Rev. D*, **80**, 111107, doi:10.1103/PhysRevD.80.111107.

39 Kluth, S. (2004) Jet physics in e^+e^- annihilation from 14-GeV to 209-GeV, *Nucl. Phys. Proc. Suppl.*, **133**, 36, doi:10.1016/j.nuclphysbps.2004.04.134.

5
Monte Carlo Generators and Fixed-order Calculations: Predicting the (Un)Expected

Stefan Gieseke and Zoltán Nagy

In this chapter the basic concepts underlying the transition from partonic degrees of freedom towards hadronic ones are outlined. First, the perturbative calculation of cross sections and their evolution from large to small scales in parton showers is discussed. This is supplemented with a short description of hadronisation models and the so-called underlying event, both of which are non-perturbative phenomena.

5.1
Fixed-Order Born-Level Calculations

The simplest calculations to be carried out are *Born-level fixed-order calculations*. These involve a *phase-space integral* over the *tree-level matrix element* squared and a jet measurement function. The structure of the Born-level cross section is

$$\sigma[F_J] = \int_m d\Gamma^{(m)}(\{p\}_m) |\mathcal{M}(\{p\}_m)|^2 F_J(\{p\}_m) , \qquad (5.1)$$

where $d\Gamma^{(m)}(\{p\}_m)$ is the phase space-integral measure, $\mathcal{M}(\{p\}_m)$ represents the m-parton tree-level matrix element and $F_J(\{p\}_m)$ is the jet measurement function that defines the physical observable.

The calculation of the cross section in (5.1) is relatively simple, and the integral is free of *infrared and ultraviolet singularities*. The matrix element is typically a complicated expression, but can often be generated in an automated way; several implementations of such *automated calculation* can be found in the literature: ALPGEN [1], GRACE [2], HELAC [3], MADGRAPH [4], SHERPA [5].

Tree-level cross sections can predict the shapes of distributions, but typically have several defects: (i) Since the result is only of leading order in the strong coupling expansion, it strongly depends on the unphysical renormalisation and factorisation scheme. (ii) Predictions for exclusive physical quantities suffer from large logarithms. In the phase-space regions where these logarithms are dominant the predictions are unreliable. (iii) In the Born-level calculation, every jet is represented by a single parton and there is no information about the inner jet structure. (iv) In

Physics at the Terascale, First Edition. Edited by Ian C. Brock and Thomas Schörner-Sadenius.
© 2011 WILEY-VCH Verlag GmbH & Co. KGaA, Weinheim.
Published 2011 by WILEY-VCH Verlag GmbH & Co. KGaA.

a real measurement, hadrons are observed in the detector and jets typically consist of many of them. However, hadronisation effects cannot be taken into account in Born-level calculations.

5.2
Next-to-Leading Order Calculations

One can improve the precision of Quantum Chromodynamics (QCD) predictions by calculating the next term in the perturbative expansion, the next-to-leading order correction (NLO). However, although this is only one order more than the Born cross section, the complexity of the calculations increases substantially. One has to face enormous algebraic and analytic complexity.

The naive structure of an NLO QCD calculation is

$$\sigma_{\text{NLO}} = \int_N d\sigma^B + \int_{N+1} d\sigma^R + \int_N d\sigma^V . \quad (5.2)$$

Here σ^B, σ^R and σ^V correspond to the Born-level term and the *real and virtual corrections*, respectively. Since both the real and virtual terms are singular in $d = 4$ dimensions, the expression (5.2) is well defined only in $d = 4 - 2\epsilon$ dimensions. However, the sum of the two contributions is finite. One therefore first has to regularise the integrals. In the real part, the singularities come from regions in the phase space integral where a gluon becomes soft or two partons become collinear; the integral over these degenerate phase-space regions leads to contributions proportional to $1/\epsilon$ and $1/\epsilon^2$. The infrared singularity structure of the virtual contributions is exactly the same but with opposite sign, thus ensuring the cancellation with the real part.

In order to achieve the cancellation in the calculations, one has to reorganise it in such a way that they can be carried out in $d = 4$ dimensions:

$$\sigma_{\text{NLO}} = \int_N d\sigma^B + \int_{N+1} \left[d\sigma^R - d\sigma^A\right]_{\epsilon=0} + \int_N \left[d\sigma^V + \int_1 d\sigma^A\right]_{\epsilon=0} . \quad (5.3)$$

In (5.3), the approximated version of the real contribution is first subtracted and then added back in a different form. In the second term on the right-hand side, $d\sigma^A$ cancels the singularities of $d\sigma^R$ such that it is safe to perform the integral in $d = 4$ dimensions. In the third term, the explicit singularities of $d\sigma^V$ are cancelled by $\int_1 d\sigma^A$, where the integral over the unresolved phase space is performed analytically. It is important that the approximated real contribution, σ^A, has a universal structure. This term is based on the soft and collinear factorisation property of the QCD matrix elements. A general subtraction scheme was defined by Catani and Seymour [6], and an extension of this method for massive fermions is also available [7].

With the NLO corrections, the dependence on the renormalisation and factorisation scales is significantly reduced. Furthermore, at NLO one of the final-state jets is represented by two partons, thus giving some information about the inner jet structure. However, this information is still very crude. In addition, NLO calculations of exclusive quantities still suffer from large logarithms, and hadronisation effects still cannot be considered. Therefore, for some applications it is necessary to go even one order higher in the perturbative expansion, namely to the next-to-next-to-leading order (NNLO).

5.3
Next-to-Next-to-Leading Order Calculations

For some processes and/or jet observables it is important to know the cross sections at next-to-next-to-leading order level, for example when the NLO K factor (defined as the ratio of the NLO to the LO cross section) is larger than 2, indicating that the NLO correction doesn't reduce the scale dependence sufficiently. Recently some simple but important processes have been calculated at NNLO [8, 9], and there are ongoing efforts to define a general scheme for NNLO calculations [10–12].

5.4
Leading-Order Parton Showers

Fixed-order calculations are systematically defined order by order and usually give a good description of the data over the regions of phase space where large-p_T events are dominant. However, at any given order one still has to deal with the presence of the large logarithms, and the hadronisation effect still cannot be considered.

There is, however, another way to calculate cross sections in the perturbative framework, parton-shower calculations. In parton showers, the aim is to calculate cross sections *to all orders*, and for this purpose real and virtual QCD matrix elements are needed. These matrix elements, however, are very complicated for high multiplicities and/or multi-loop diagrams. It is impossible to treat them in an exact way beyond a certain order and a certain number of external legs. Partons showers are therefore based on approximations, namely: (i) The real and virtual QCD amplitudes are approximated by their universal soft and collinear factorisation properties. (ii) Only strongly ordered emissions are considered. This ordering is ensured by an evolution parameter which might be called *shower time*. (iii) The shower evolution leaves the total cross section invariant. This is the so-called *unitarity condition*.

5.4.1
Shower Evolution

A typical parton-shower algorithm for hadron–hadron collisions works with states with two initial-state partons, a and b, and some number m of final-state partons that can be labelled with integers $1, 2, \ldots, m$. The momenta of these partons can then be specified by giving $\{p\}_m = \{p_a, p_b, p_1, \ldots, p_m\}$. Each parton also carries a flavour $f \in \{g, u, \bar{u}, d, \bar{d}, \ldots\}$, so that the momenta and flavours can be specified with $\{p, f\}_m$.

In the most general case one has to keep track of the colour flow and spin information. Since the physical states are based on the QCD density operator (in quantum space), two colour and two spin indices have to be associated to each initial-state and final-state parton. The complete set of $m + 2$ partons is denoted by $\{p, f, s', c', s, c\}_m$, where $\{c', c\}_m$ denotes the colour connections and $\{s', s\}_m$ denotes the spin of the partons. One can now consider the states $|\{p, f, s', c', s, c\}_m)$ to form a basis for a vector space in the sense of statistical mechanics[18]. After some amount of shower evolution starting from a base state $|\{p, f, s', c', s, c\}_2)$ with two final-state partons, a state $|\rho)$ is reached that is a linear combination of base states, so that $(\{p, f, s', c', s, c\}_m|\rho)$ represents the probability, in the shower model, for the state $|\rho)$ to consist of $m + 2$ partons with momenta, flavours, spins and colours $\{p, f, s', c', s, c\}_m$. Any measurement function, $F(\{p\}_m)$, can be also represented as a vector in the statistical space, $F(\{p\}_m) = (\{p, f, s', c', s, c\}_m|F)$. The statistical space is a suitable representation for the QCD density operator that is useful to describe and understand the partonic dynamics in hadronic collisions.

Every measurement in partonic systems has a typical resolution scale, μ. This means that the measurement is sensitive to radiation that is harder than the resolution of the observable and that all softer or collinear interactions are invisible. It is therefore a reasonable approximation to introduce the density operator, ρ, as a function of the resolution scale squared, μ^2. Interactions with squared scales greater than μ^2 are included in the density operator, $\rho = |M\rangle\langle M|$, as resolved visible radiation, while interactions with scales smaller than μ^2 are integrated out or included in the parton distribution functions (PDFs). In fact, the logarithm of the resolution scale, μ^2, is a more useful variable than μ^2 itself. Therefore, the shower time is defined as $t = \log(Q_0^2/\mu^2)$. The shower time helps to separate resolvable and unresolvable radiation in partonic systems.

Two elementary operators can be introduced in the statistical space. The operator $\mathcal{H}_I(t)$ describes a single resolvable emission. This operator always increases the number of partons and changes the flavour, spin and colour structure of the event. Virtual and unresolvable radiation are described by the operator $\mathcal{V}_I(t)$. This operator does not change the number of partons and the flavour and spin structure of the event, but it can still modify the colour structure. Both operators can be derived

---

18) Note that the statistical space is not identical to the quantum space and that the vectors in the statistical space are denoted by $|\ldots)$ while the vectors in quantum space are denoted by the standard bra-ket notation, $|\ldots\rangle$.

from the soft and collinear factorisation properties of the Born and one-loop-level QCD amplitudes.

The parton shower produces an approximate version of the QCD density operator $|\rho\rangle$ as a function of the resolution scale. One can introduce the shower evolution operator that evolves a physical state from time t_1 to time t_2,

$$|\rho(t_2)\rangle = \mathcal{U}(t_2, t_1)|\rho(t_1)\rangle . \tag{5.4}$$

The shower evolution operator $\mathcal{U}(t_2, t_1)$ can be constructed from the elementary splitting operators $\mathcal{H}_I(t)$ and $\mathcal{V}_I(t)$ by considering all the possible emissions in a strongly ordered way that can take the system from t_1 to t_2. This leads to the following shower evolution operator:

$$\mathcal{U}(t_2, t_1) = \mathbb{T} \exp\left\{\int_{t_1}^{t_2} d\tau [\mathcal{H}_I(\tau) - \mathcal{V}_I(\tau)]\right\} . \tag{5.5}$$

The requirement that shower evolution leaves the total cross section invariant relates operator $\mathcal{V}_I(t)$ to operator $\mathcal{H}_I(t)$. The observable that measures the total cross section is $F_1(\{p\}_m) = 1$, and the corresponding vector in the statistical space is denoted by $|1\rangle$. Thus, from the unitary condition one can find

$$(1|\mathcal{U}(t_2, t_1) = (1| \implies (1|\mathcal{V}_I(t) = (1|\mathcal{H}_I(t) . \tag{5.6}$$

Roughly speaking this means that the operator of the unresolvable radiation can be obtained directly from the resolvable splitting operator. In the next section, the operator $\mathcal{H}_I(t)$ will be defined.

One can see that the shower evolution operator obeys an integral equation,

$$\mathcal{U}(t_f, t_2) = \mathcal{N}(t_f, t_2) + \int_{t_2}^{t_f} dt_3 \mathcal{U}(t_f, t_3) \mathcal{H}_I(t_3) \mathcal{N}(t_3, t_2) . \tag{5.7}$$

The shower evolution starts from the hard scattering at shower time t_2. The evolution operator is the sum of two terms. The first term in (5.7) represents parton evolution without splitting. This operator is the generalisation of the Sudakov exponent and gives the probability that nothing happens between time t_2 and t_f,

$$\mathcal{N}(t_2, t_1) = \mathbb{T} \exp\left\{-\int_{t_1}^{t_2} d\tau \mathcal{V}_I(\tau)\right\} . \tag{5.8}$$

Remember that $\mathcal{V}_I(\tau)$ is a non-trivial operator in colour space which makes the implementation hard. Further approximations are needed to be able to deal with this non-trivial colour structure in the Monte Carlo implementations.

The second term in (5.7) represents the splitting. The partonic state is evolved without splitting to an intermediate time, t_3, and splitting happens as given by the splitting operator $\mathcal{H}_I(t_3)$. The system is then evolved with possible further splittings from t_3 to t_f. This evolution equation is depicted in Figure 5.1.

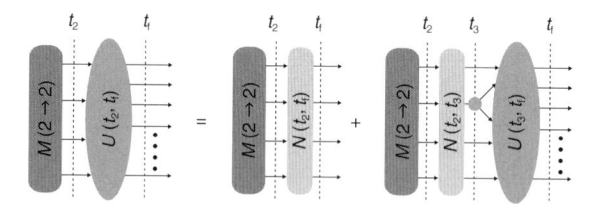

Figure 5.1 Visualisation of the evolution equation. The shower (*medium grey oval*) starts from the hard matrix element (*dark grey rounded rectangle*). The partons are evolved to the final scale without splitting (*light grey rounded rectangle*) or with splitting at an intermediate time (*small grey circle*) and evolved to the final scale with possible splittings.

5.4.2
The Splitting Operator

The splitting operator of the leading-order shower is derived from the factorisation property of the QCD matrix elements in the soft and collinear limits. This factorisation property is universal. When two partons become collinear, or when a gluon becomes soft, the QCD matrix element becomes singular in the relevant phase-space regions. The matrix element factorises into a singular factor and the hard matrix element. This singular factor helps to construct the splitting operator of the shower algorithm. The most general structure of the splitting operator is

$$(\{\hat{p}, \hat{f}, \hat{s}', \hat{c}', \hat{s}, \hat{c}\}_{m+1} | \mathcal{H}(t) | \{p, f, s', c', s, c\}_m)$$
$$= \sum_{l=a,b,1,\ldots,m} \delta\left(t - T_l(\{\hat{p}, \hat{f}\}_{m+1})\right) (\{\hat{p}, \hat{f}\}_{m+1} | \mathcal{P}_l | \{p, f\}_m) \frac{m+1}{2}$$
$$\times \frac{n_c(a) n_c(b) \eta_a \eta_b}{n_c(\hat{a}) n_c(\hat{b}) \hat{\eta}_a \hat{\eta}_b} \frac{f_{\hat{a}/A}(\hat{\eta}_a, \mu_F^2) f_{\hat{b}/B}(\hat{\eta}_b, \mu_F^2)}{f_{a/A}(\eta_a, \mu_F^2) f_{b/B}(\eta_b, \mu_F^2)} \sum_k \Psi_{lk}(\{\hat{f}, \hat{p}\}_{m+1})$$
$$\times \sum_{\beta=L,R} (-1)^{1+\delta_{lk}} (\{\hat{c}', \hat{c}\}_{m+1} | \mathcal{G}_\beta(l, k) | \{c', c\}_m)$$
$$\times (\{\hat{s}', \hat{s}\}_{m+1} | \mathcal{Y}_\beta(l, k; \{\hat{f}, \hat{p}\}_{m+1}) | \{s', s\}_m) . \qquad (5.9)$$

This splitting operator describes the most general case when all the colour and spin correlations and heavy flavour effects are considered during the shower evolution.

The splitting operator describes how the state changes when a real emission happens. Equation 5.9 contains a sum over all the possible emitters ($\sum_l$) that takes into account all emission possibilities (a parton radiating a gluon, a gluon splitting to a $q\bar{q}$ pair). Next comes a delta function that defines the evolution variable. There is some freedom in this definition, although it is not completely arbitrary. The operator $\mathcal{P}_l$ represents the momentum and flavour mapping. The flavour mapping is determined by the QCD vertices; however, for the momentum mapping there are fewer constraints.

In the second line of (5.9), the ratios of the flux factors and of the PDFs of the incoming hadrons can be seen. In the case of an initial-state splitting, these ratios differ from one. The spin-averaged splitting function $\Psi_{lk}(\{\hat{f},\hat{p}\}_{m+1})$ is the function that is usually related to the Altarelli–Parisi splitting functions. They have different forms depending on whether $l \neq k$, which corresponds to the emitted parton $m+1$ being a gluon, or $l = k$, which corresponds to the new parton being a quark or an antiquark. Thus, we have

$$\Psi_{lk} = \frac{\alpha_s}{2\pi} \frac{1}{\hat{p}_l \cdot \hat{p}_{m+1}} \left[2A_{lk} \frac{\hat{p}_l \cdot \hat{p}_k}{\hat{p}_k \cdot \hat{p}_{m+1}} \theta(l \neq k) + H_{ll}^{\text{coll}}(\{\hat{f},\hat{p}\}_{m+1}) \right]. \tag{5.10}$$

The partitioning function $A_{lk}(\{\hat{p}\}_{m+1})$ has to obey the condition that $A_{lk} + A_{kl} = 1$, but otherwise is arbitrary. Its task is to distribute the soft gluon contributions along the collinear directions.

The first term in brackets in (5.10) is singular both in the soft and collinear limit, while the second term ($H_{ll}^{\text{coll}}(\{\hat{f},\hat{p}\}_{m+1})$) leads only to collinear singularities. The expression next to the partitioning function is the well-known eikonal factor and represents a gluon emission from the $l - k$ colour dipole. It is singular in the collinear limit when $\hat{p}_l \parallel \hat{p}_{m+1}$ or $\hat{p}_k \parallel \hat{p}_{m+1}$. In the implementation it is important to define the A_{lk} function such that all the $\hat{p}_k \parallel \hat{p}_{m+1}$ collinear singularities are shuffled from the Ψ_{lk} splitting function to the Ψ_{kl} splitting function. With such a partitioning function the spin-averaged splitting kernel, Ψ_{lk}, can be considered as a half-dipole antenna. The parton shower based on this splitting kernel is usually called the *partitioned dipole shower*.

In the last line of (5.9), the $\mathcal{G}_\beta(l,k)$ operator represents the insertion of the emitted parton in the colour space. For gluon emission the definition is

$$(\{\hat{c}',\hat{c}\}_{m+1}|\mathcal{G}_R(l,k)|\{c',c\}_m) = {}_D\langle\{\hat{c}\}_{m+1}|t_l^\dagger|\{c\}_m\rangle\langle\{c'\}_m|t_k|\{\hat{c}'\}_{m+1}\rangle_D. \tag{5.11}$$

This operator inserts a gluon on the leg l in the $|\{c\}_m\rangle$ state and on the leg k in the $\langle\{c'\}_m|$ state. The other colour operator is $\mathcal{G}_L(l,k) = \mathcal{G}_R(k,l)$. When $l \neq k$, this operator is usually not diagonal and can introduce negative contributions which makes the implementation of the full splitting operator impossible. In order to deal with this one has to use further approximations.

The $\mathcal{Y}_\beta$ operator represents the non-trivial spin correlations which come from the pure collinear part of the $g \to g+g$ and $g \to q+\bar{q}$ vertices. In many cases the spin correlations are not important and their effects are negligible. One can consider spin averaging and in this approximation the $\mathcal{Y}_\beta$ operator becomes diagonal and the dependence on the spins is trivial, so that one can describe the evolution of the states without referring to spin at all. Thus, the prescription is vastly simplified and one can work with basis states $|\{p,f,c',c\}_m\rangle$.

More details on the formalism and the general structure of parton-shower algorithms can be found in [13, 14].

5.5
Implementations and Shower Schemes

In this section the similarities and differences between the existing parton-shower implementations are briefly reviewed. In the previous sections, a very general framework has been set up for leading-order parton-shower algorithms. There are basically three points where the essential differences lie, namely the *momentum mapping*, the *evolution parameter* and the choice of the *soft partitioning function*.

5.5.1
Angular-Ordered Shower

What would happen if one used angular ordering for successive parton emissions? Then ordering in $t_\angle$ is ordering in splitting angle, starting from large angles and progressing to smaller angles as $t_\angle$ increases. Since the emissions are not ordered by the hardness of the emission, one has to also introduce a veto in order to avoid splittings with too little transverse momentum. Thus, the shower time is given by

$$t_\angle = T_l\left(\{\hat{p},\hat{f}\}_{m+1}\right) = \log 2 - \log \frac{\hat{p}_l \cdot \hat{p}_{m+1}\hat{Q}^2}{\hat{p}_l \cdot \hat{Q}\hat{p}_{m+1}\cdot \hat{Q}} = \log \frac{2}{1-\cos\vartheta_{l,m+1}}, \quad (5.12)$$

where $\hat{Q}$ is the total momentum of the incoming partons after the splitting. In the rest frame of $\hat{Q}$, one can see that the expression under the logarithm is the emission angle.

For the soft gluon partitioning function one can use

$$A'_{lk} = \theta\left(\vartheta_{l,m+1} < \vartheta_{l,k}\right)\frac{1-\cos\vartheta_{m+1,k}}{1-\cos\vartheta_{l,k}}. \quad (5.13)$$

These functions are positive. They do not sum to 1, but one gets the right result after suitable averaging over azimuthal angles. Since the emissions are ordered in angle, the theta function in (5.13) is always equal to 1. This means that the spin-averaged splitting function becomes independent of the momentum of the colour-connected parton,

$$\Psi_l^{(a.o.)} = \frac{\alpha_s}{2\pi}\frac{2}{\hat{p}_l \cdot \hat{p}_{m+1}}\left[\frac{\hat{p}_l \cdot \hat{Q}}{\hat{p}_{m+1}\cdot\hat{Q}}\theta(l\neq k) + H_{ll}^{\text{coll}}(\{\hat{f},\hat{p}\}_{m+1})\right]. \quad (5.14)$$

Note that the expression under the brackets gives the standard Altarelli–Parisi splitting kernels in the strict collinear limit.

Now, in the spinless splitting operator one can perform the sum over the colour-connected partons ($\sum_k$). This simplifies the colour structure enormously, and the splitting operator becomes

$$
\begin{aligned}
&(\{\hat{p}, \hat{f}, \hat{c}', \hat{c}\}_{m+1} | \mathcal{H}(t) | \{p, f, c', c\}_m) \\
&= \sum_{l=a,b,1,\ldots,m} \delta\left(t - T_l(\{\hat{p}, \hat{f}\}_{m+1})\right) (\{\hat{p}, \hat{f}\}_{m+1} | \mathcal{P}_l | \{p, f\}_m) \\
&\quad \times \frac{n_c(a) n_c(b) \eta_a \eta_b}{n_c(\hat{a}) n_c(\hat{b}) \hat{\eta}_a \hat{\eta}_b} \frac{f_{\bar{a}/A}(\hat{\eta}_a, \mu_F^2) f_{\hat{b}/B}(\hat{\eta}_b, \mu_F^2)}{f_{a/A}(\eta_a, \mu_F^2) f_{b/B}(\eta_b, \mu_F^2)} \Psi_l^{(a.o.)}(\{\hat{f}, \hat{p}\}_{m+1}) \\
&\quad \times (m+1)(\{\hat{c}', \hat{c}\}_{m+1} | \mathcal{G}_R(l, l) | \{c', c\}_m) \,.
\end{aligned}
$$
(5.15)

This operator is diagonal in colour space and can easily be implemented in a general-purpose Monte Carlo program.

A very similar algorithm is implemented in HERWIG [15, 16]. HERWIG uses the emission angle weighted with the energy of the mother parton as the evolution parameter. The major difference between HERWIG and the procedure described above, however, is in the momentum mapping: in the described scheme, the momentum configuration is reconstructed in every step of the shower evolution, whereas in HERWIG the splitting variables are kept track of during the evolution, but the momentum reconstruction is done only at the end of the shower. It is relatively easy to implement the HERWIG algorithm since there are no complicated colour correlations and the non-trivial colour correlations coming from the wide-angle soft regions are completely dropped by the angular-ordering conditions.

One of the purposes of parton-shower algorithms is to sum up large logarithms. It has been shown that the algorithm described above can accurately sum up the leading (LL) and next-to-leading logarithms (NLL) of a certain class of exclusive observables, for example the k_T jet algorithm and the event-shape observables thrust and C parameter in e^+e^- annihilation [17–19]. However, it certainly fails for observables which are sensitive to wide-angle soft radiation, like the JADE algorithm or some non-global observables.

The other main purpose of parton-shower algorithms is to provide reasonable approximations for the hard cross section over the whole phase space. The angular-ordered shower is not really good at this point. One of the sources of error is the soft partitioning function in (5.13). This obeys $A_{lk} + A_{kl} = 1$ only approximately, and the approximation may be crude. The basic approximation in the shower program is the soft and collinear factorisation of the QCD matrix elements. One of the main purposes of the evolution variable is to control the goodness of these approximations in the parton shower when they are applied successively to the fully exclusive matrix elements. In general one should use some kind of hardness variable (like virtuality of transverse momentum) that is sensitive to both soft and collinear radiations. The emission angle is sensitive only to the collinear radiation.

5.5.2
Partitioned Dipole Shower with Leading-Colour Approximation

The partitioned dipole shower with leading-colour approximation is one of the most popular schemes and is implemented in various codes. It chooses as the evolution variable the transverse momentum of the splitting with respect to the momentum of the mother parton and an auxiliary momentum that is usually chosen to be the momentum of the colour-connected parton. Thus, the shower time is given by

$$t_\perp = T_l(\{\hat{p}, \hat{f}\}_{m+1}) = \log \frac{\hat{Q}^2}{-k_\perp^2} \,. \tag{5.16}$$

Here $\hat{Q}$ is the total incoming momentum after the splitting and $k_\perp$ is the transverse momentum of the emitted partons. Actually the transverse momentum ordering does not play any important role in this type of shower algorithm. One can also use the virtuality or any other sensible variable to order the evolution.

There are several possibilities for the soft partitioning function. Usually, the simplest choice is made:

$$A_{lk} = \frac{\hat{p}_k \cdot \hat{p}_{m+1}}{\hat{p}_k \cdot \hat{p}_{m+1} + \hat{p}_l \cdot \hat{p}_{m+1}} \,. \tag{5.17}$$

This function obeys the condition $A_{lk} + A_{kl} = 1$ exactly. This might help to cover the phase space away from the soft and collinear region, but the complicated colour correlations make the implementation very hard. Using the leading-colour approximation, each gluon is treated as carrying colour $3 \otimes \bar{3}$ instead of 8. The allowed parton pairs forming a dipole are then colour-connected partners. Furthermore, leading-colour contributions come only from diagonal colour configurations when $\{c\}_m = \{c'\}_m$. Thus, the description of the states is greatly simplified and one can work with simpler base states, $|\{p, f, c\}_m\rangle$:

$$(\{\hat{p}, \hat{f}, \hat{c}\}_{m+1} | \mathcal{H}(t) | \{p, f, c\}_m)$$
$$= \sum_{l=a,b,1,\ldots,m} \delta(t - T_l(\{\hat{p}, \hat{f}\}_{m+1})) (\{\hat{p}, \hat{f}\}_{m+1} | \mathcal{P}_l | \{p, f\}_m)$$
$$\times \frac{n_c(a) n_c(b) \eta_a \eta_b}{n_c(\hat{a}) n_c(\hat{b}) \hat{\eta}_a \hat{\eta}_b} \frac{f_{\hat{a}/A}(\hat{\eta}_a, \mu_F^2) f_{\hat{b}/B}(\hat{\eta}_b, \mu_F^2)}{f_{a/A}(\eta_a, \mu_F^2) f_{b/B}(\eta_b, \mu_F^2)}$$
$$\times (m+1) \sum_k \Psi_{lk}(\{\hat{f}, \hat{p}\}_{m+1}) \langle\{\hat{c}\}_{m+1} | a_{lk}^\dagger | \{c\}_m\rangle \,. \tag{5.18}$$

Here $a_{lk}^\dagger$ is an operator in colour space that acts in quantum space. It describes the colour part of the splitting in the leading-colour approximation. For example when $k \neq l$ this operator inserts a gluon between parton l and k on a colour string.

One can see the main difference between the angular-ordered shower and the leading-colour approximation. In the angular-ordered shower a further approximation has been made in the splitting function Ψ_{lk} by having special choices for the partitioning function and for the evolution variable. In the leading-colour approximation, in contrast, there is an extra approximation only in the colour space and colour operator.

Such an algorithm is implemented in PYTHIA [20, 21] and in the parton shower based on the Catani–Seymour dipole factorisation formulae [22–24]. The precision of these algorithms depends on the choice of the evolution variable and the soft partition functions. Recently it has been shown that a shower with transverse momentum ordering and with a soft partitioning function as given in (5.17) fails to sum up the large logarithm of the Z boson transverse momentum in the Drell–Yan process at NLL level [25].

Usually the shower algorithms are less sensitive to the choice of the momentum mapping, but there is an example when the shower fails to sum up large logarithms even at LL level. The shower based on the Catani–Seymour dipole factorisation formulae doesn't reproduce the correct series of the large logarithms of the Z boson transverse momentum in the Drell–Yan process [25]. This is because of the bad choice of the momentum mapping in the initial-state radiations.

5.5.3
Antenna Dipole Shower with Leading-Colour Approximation

The antenna dipole shower is very much a reorganisation of the leading-colour partitioned dipole shower. The total emission probability has been partitioned into a fraction, A_{lk}, associated with the splitting of parton l and a fraction, $A_{kl} = 1 - A_{lk}$, associated with the splitting of parton k. From (5.9) one can see that there is a separate momentum mapping, $\mathcal{P}_l$, for each parton, l, that splits. Thus, in a highly compressed notation, one can write that the emission from the $l-k$ dipole is partitioned into two terms,

$$\mathcal{H}_{lk}^{\text{part}}(t) \propto [\mathcal{P}_l A_{lk} + \mathcal{P}_k A_{kl}] \frac{\hat{p}_l \cdot \hat{p}_k}{\hat{p}_{m+1} \cdot \hat{p}_l \; \hat{p}_{m+1} \cdot \hat{p}_k} . \tag{5.19}$$

In the leading-colour approximation these two terms come with the same colour structure. One might be tempted to get rid of this ambiguity of the partitioning function, particularly in the leading-colour approximation where one can use pairs of colour-connected partons as the dipoles. The idea would then be to consider each $l-k$ dipole as a unit that can emit a gluon. Thus, the basic building blocks are $2 \to 3$ parton splittings. In a dipole antenna shower, there is no A_{lk}; instead, there is a separate momentum mapping $\mathcal{P}_{lk}$ for each dipole $l-k$. Thus, (5.19) is replaced by

$$\mathcal{H}_{lk}^{\text{ant}}(t) \propto \mathcal{P}_{lk} \frac{\hat{p}_l \cdot \hat{p}_k}{\hat{p}_{m+1} \cdot \hat{p}_l \; \hat{p}_{m+1} \cdot \hat{p}_k} . \tag{5.20}$$

The freedom to choose A_{lk} now resides in the freedom to choose $\mathcal{P}_{lk}$. This function must be symmetric under the interchange of partons l and k. It is usually defined in such a way that momentum is conserved locally in the $2 \to 3$ splitting, without taking momentum from any of the other partons. For a final-state dipole, this means $p_l + p_k = \hat{p}_l + \hat{p}_k + \hat{p}_{m+1}$. Note that it is possible to have a shower that is simultaneously a partitioned dipole shower and an antenna dipole shower. One can get this case by defining

$$\mathcal{P}_{lk} = \theta(\vartheta_{l,m+1} < \vartheta_{k,m+1})\mathcal{P}_l + \theta(\vartheta_{k,m+1} < \vartheta_{l,m+1})\mathcal{P}_k, \tag{5.21}$$

where the ϑ_{ij} is the angle between momenta i and j and $\mathcal{P}_l$ represents the momentum mapping that is used in (5.9). Then $A_{lk} = \theta(\vartheta_{l,m+1} < \vartheta_{k,m+1})$.

A shower based on this approach may be called a *dipole antenna shower*. The pioneering development along these lines is the final-state shower of ARIADNE [26]. More recent examples include those in VINCIA [27].

5.6
Matching Parton Showers to Fixed-Order Calculations

Parton-shower algorithms are usually very good in phase-space regions where radiation with small transverse momentum dominates. Away from these regions, however, the performance can be very poor, and the uncertainty of the shower predictions can be very large. Fortunately, fixed-order calculations perform much better in these critical regions. Thus, it would be an obvious step to combine the strengths of these two approaches and to combine ("match") fixed-order calculations to parton-shower algorithms.

5.6.1
Born-Level Matching

The standard parton showers have a deficiency which is illustrated in Figure 5.2. Figure 5.2a depicts a term contributing to the standard shower. In this term, there are Sudakov factors and $1 \to 2$ parton splitting functions. If we omit the Sudakov factors, one has the $1 \to 2$ parton splittings as depicted in Figure 5.2b. These splittings are approximations based on the splitting angles being small or one of the daughter partons having small momentum. Thus, the shower splitting probability with two splittings approximates the exact squared matrix element for $2 \to 4$ scat-

Standard shower contribution   Small pT approximation $|M(2 \to 4)|^2$
(a)   (b)   (c)

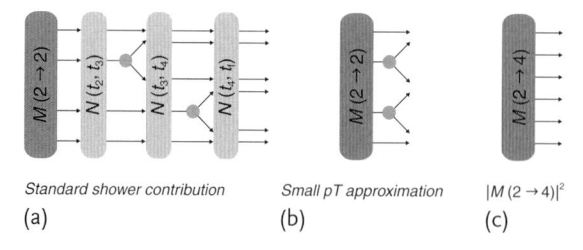

Figure 5.2 (a) the $2 \to 4$ cross section in shower approximation; (b) The shower approximation omitting the Sudakov factors; (c) the exact tree level $2 \to 4$ cross section. the cross section based on splitting functions (b) is a collinear/soft approximation to this.

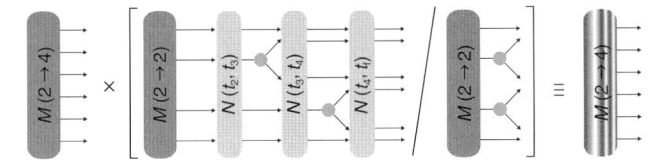

Figure 5.3 An improved version of the $2 \to 4$ cross section. First, the 4-parton configuration is generated according to the exact matrix element. Then the shower approximation (with Sudakov factors) is taken, divided by the approximate collinear squared matrix element, and multiplied by the exact tree-level squared matrix element. The graphical symbol on the right-hand side represents this Sudakov reweighted cross section.

tering. This approximation is good in parts of the final-state phase space, but not in all of it. Thus, one might want to replace the approximate squared matrix element of Figure 5.2b with the exact squared matrix element of Figure 5.2c. However, the exact squared matrix elements lack the Sudakov factors.

One can improve the approximation as illustrated in Figure 5.3. The exact matrix element is reweighted by the ratio of the shower approximation with Sudakov factors to the shower approximation without Sudakov factors, thus indirectly inserting the Sudakov factors into the exact squared matrix element (CKKW algorithm) [28, 29]. A similar algorithm was proposed in [30] (MLM algorithm). The CKKW algorithm uses the k_T jet algorithm to define the ratio needed to calculate the Sudakov reweighting factor while the *MLM method* uses the cone jet-finding algorithm to define a unique emission history for the reweighting procedure. The advantage of MLM method over the CKKW method is that the algorithm uses the native Sudakov factors of the underlying parton shower. There is a further step in implementing this idea. CKKW divides the shower evolution into two stages, $0 < t < t_\mathrm{ini}$ and $t_\mathrm{ini} < t < t_\mathrm{f}$, where t_ini is a parameter that represents a mod-

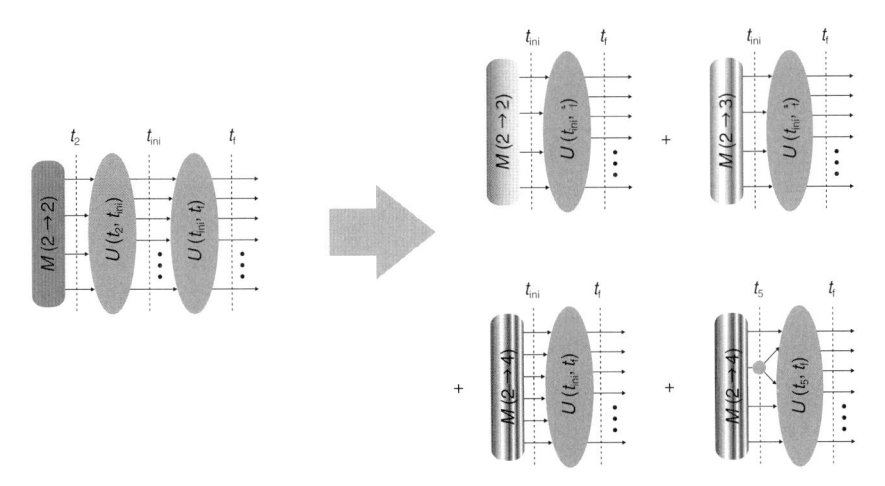

Figure 5.4 Shower with CKKW jet-number matching. The calculation for *n* jets at scale t_ini is based on the Sudakov reweighted tree-level cross section for the production of *n* partons.

erate p_T scale and t_f represents the very small p_T scale at which the shower stops and hadronisation is simulated. With this division, the Sudakov reweighting can be performed for the part of the shower at scale harder than t_{ini}, as depicted in Figure 5.4. Evolution from t_{ini} to t_f is done via the ordinary shower algorithm. This method can be extended for NLO-level matrix elements, and it has been worked out for electron–positron annihilation [31, 32].

5.6.2
Next-to-Leading Order Matching

The Born-level matching discussed in the previous section can improve the shower prediction and predict the shape of the distributions. However, the uncertainties are still large and no information about the normalisation of the cross sections can be provided. This is not surprising because all parts of the calculation are defined at leading-order level. The obvious step to get more information about the distributions and to fix the normalisation would be to define the parton shower at NLO level. Unfortunately this project is very complicated and so far there is no such algorithm available in the literature. One can, however, try to find some intermediate solution, for example by matching a LO parton shower to fixed-order calculations at NLO.

Matching parton showers with NLO fixed-order calculations is a very active field of parton-shower research. There are two basic approaches. The first one is the MC@NLO project [33, 34]. The main idea here is to avoid the double counting of the NLO contributions by introducing extra counter-terms. This method has been applied to several $2 \to 0 + X$ processes without colour in the final state and also to some $2 \to 1 + X, 2 \to 2 + X$ processes with heavy QCD particles in the final state [33, 34].

The other approach was originally proposed by Krämer and Soper [35, 36] and was implemented for $e^+ e^- \to 3$ jets. The idea here is to include the first step of the shower in the NLO calculation and then start the parton showering from this configuration. Several matching algorithms based on this concept have been proposed in the last few years [22, 31, 32]. One of the most popular algorithms nowadays is the method [37–39].

The structure of the NLO fixed-order calculations after applying the subtraction scheme to remove the infrared singularities is

$$\sigma_{\text{NLO}} = \int_m \left[d\sigma^B + d\sigma^V + d\sigma^C + \int_1 d\sigma^A \right] F_J^{(m)}$$
$$+ \int_{m+1} \left[d\sigma^R F_J^{(m+1)} - d\sigma^A F_J^{(m)} \right]. \quad (5.22)$$

Here $d\sigma^B$, $d\sigma^R$, $d\sigma^V$, $d\sigma^C$ and $d\sigma^A$ are the Born-level term, the real and virtual contributions, the collinear counterterm and the subtraction term of the NLO scheme, respectively. The physical quantity is defined by the functions $F_J^{(m)}$ and $F_J^{(m+1)}$. Note that the $d\sigma^A$ term is subtraction-scheme dependent.

The naive way to add parton-shower corrections is to replace the jet functions F_J with the shower interface function. The shower interface function starts the subsequent parton shower from the corresponding m or $m + 1$ parton configuration. This approach is, however, not good because it leads to double counting when the shower starts from the Born term, $d\sigma^B$, in the first step of the evolution that generates part of the NLO contribution in an approximated way. On the other hand these NLO contributions are already considered exactly by the NLO terms $d\sigma^R$ and $d\sigma^V$.

To avoid double counting Frixione and Webber [33, 34] organised the calculation in the following way:

$$\sigma_{MC} = \int_m \left[d\sigma^B + d\sigma^V + d\sigma^C + \int_1 d\sigma^A \right] I_{MC}^{(2 \to m)}$$
$$+ \int_{m+1} \left[d\sigma_{m+1}^R - d\sigma_{m+1}^{MC} \right] I_{MC}^{(2 \to m+1)}$$
$$+ \int_{m+1} \left[d\sigma_{m+1}^{MC} - d\sigma_{m+1}^A \right] I_{MC}^{(2 \to m)} . \quad (5.23)$$

Here the contribution $d\sigma_{m+1}^{MC}$ is extracted from the underlying parton-shower algorithm. The functions $I_{MC}^{(2 \to m)}$ and $I_{MC}^{(2 \to m+1)}$ are the interface functions to the shower. There are different choices for the m and $m + 1$ parton interface functions, resulting in

$$I_{MC}^{(2 \to m)} \sim \mathcal{U}(t_f, t_m) \quad \text{and} \quad I_{MC}^{(2 \to m+1)} \sim \mathcal{U}(t_f, t_{m+1}) \mathcal{N}(t_{m+1}, t_m) . \quad (5.24)$$

In the m-parton case the shower starts from the m-parton configuration. In the $m + 1$ parton case one first inserts a Sudakov factor representing the probability of nothing happening between the m-parton and $(m+1)$-parton states and then starts the shower from the $(m + 1)$-parton configuration.

5.7 Hadronisation

In the previous sections it was shown how to evolve the final state of a hard short-distance process down to a cutoff scale Q_0, also dubbed *resolution scale μ* previously. The partonic final state that has been reached at this stage has to be transformed into a hadronic final state, as it can be measured in a real experiment. This is one of the main purposes of Monte Carlo event generator programs. It is clear that the partonic final state will quite heavily depend on Q_0. There will be an increasing number of soft particles emitted if this cutoff is lowered.

A measurement should not depend on the cutoff scale which lies somewhat between the short-distance and long-distance physics. Therefore, in an ideal model the hadronisation should compensate for the dependence on Q_0. In practice, we do not have an understanding of the hadronisation from first principles. This understanding would imply a sound understanding of the confinement mechanism, which is still an open problem in QCD. Therefore, we have to model the hadronisation of partonic final states.

The scale Q_0 can be understood as the smallest scale at which one trusts the perturbative expansion in α_s. It should be of the order of the typical distance scale of hadrons, which is clearly in the domain of non-perturbative physics. The value of Q_0 has to be adjusted to data. All models that will be considered in the following lead to Q_0 of the order of 1 GeV. This implies that currently, the understanding of the perturbative part of the simulation gives a better description of the data, even though the applicability of perturbation theory at scales of 1 GeV is already debatable.

From the point of view of the perturbative phase it is very sensible that a parton-shower model has served the purpose of evolving the coloured final state towards small scales. In this way, potentially large logarithms have already been taken into account, as the parton shower resums the leading large logarithms. Hence, a possible additional dependence on large scales will be suppressed. Therefore, a fairly clear distinction between perturbative and non-perturbative physics can be made. Historically, hadronisation models had been developed first without parton showers. In the following, several different hadronisation models will be discussed, starting with *independent fragmentation* [40–42]. Then the *Lund string model* [43–46] and finally the *cluster hadronisation model* [47–49] (that already started off with parton showers) will be discussed.

The important features of hadronisation models are best understood for a simple $q\bar{q}$ system, that is, two colour charges that rapidly move in opposite directions. Once isolated, the strong colour field that emerges around the particles allows for the creation of new particles out of the vacuum. In the next three subsections the three aforementioned models are discussed in some detail.

5.7.1
Independent Fragmentation

The first attempt to describe all particles within hadronic jets exclusively was made with the so-called *independent fragmentation* model. In this model, a single quark is paired with an antiquark from an additional $q\bar{q}$ pair that emerges out of the vacuum and forms a meson, carrying a fraction z of the quark momentum. The mechanism repeats such that the quark with momentum $(1-z)$ that was left behind will again pair up with a new antiquark from the vacuum and so forth. The fraction, z, of longitudinal momentum that is taken away from the free quark, q, and given to the hadron, h, is determined randomly according to the so-called *fragmentation function* $D_h^q(z)$. Additional transverse momentum is given to the mesons according to a Gaussian. If gluons are present instead of initial quarks, they are first longitudinally split into $q\bar{q}$ pairs. This can be done following the extreme that either the quark or the antiquark will carry all of the gluon's momentum or for example following the $g \to q\bar{q}$ splitting function.

This model gives a reasonable description of the structure of jets in e^+e^- annihilations from low to medium energy, for example their clear limits in transverse momentum, resulting from the Gaussian. On the other hand the model has some shortcomings. There is always a last quark left after iterating the fragmentation process until there is only some momentum left that is smaller than a given cutoff. This means that there is always some residual colour charge and flavour left which must somehow be neutralised with the remainder of the event. Furthermore, there is no smooth transition from two jets that have emerged from two rather collinear hard partons into one jet, when applying a jet algorithm with a low resolution parameter. The two model jets can always be distinguished. Hence, the requirement that the transition between parton shower and hadronisation is smoothly regulated by the parameter Q_0 cannot hold. Whenever a fairly hard, collinear gluon has been radiated it will be visible as an extra jet, giving a different structure to the overall event.

5.7.2
Lund String Fragmentation

The idea of hadrons being created in the strong colour field between coloured particles that rapidly move apart has itself most literally manifested in the *Lund string fragmentation model*. The "Lund string" that is formed between the two colour charges is nothing but a thin flux tube that forms non-perturbatively with a constant energy per unit length between the charges. This energy density is called the "string tension", κ, and turns out to be of order 1 GeV/fm. Therefore, increasing the distance between the charges by one length unit requires always the same, constant amount of energy. This reflects the well-established linear static potential between quark–antiquark pairs that is known to describe their bound states (quarkonia, e. g. the J/ψ meson) very accurately. The physical idea now is that this increasing energy will be picked up by short-range $q\bar{q}$ fluctuations of the vacuum, to put the emerging quarks on their mass shell. These newly created colour charges can cut the string into two smaller strings with much lower total energy. If repeated long enough, there will be many $q\bar{q}$ pairs left with quite small relative movement. These will finally form the hadrons.

The original strings can be described on the basis of a classical action in only one space and one time dimension, reflecting that they only have a longitudinal degree of freedom. Then, the free classical strings perform "yo-yo" like motions with constant speed and sudden turnarounds. In space-time such a string is covering a chain of diamond-like shapes. When quantum mechanics is added, the strings can break up and one finds the so-called area law

$$\frac{dP}{dA} = P_0 e^{-P_0 A}, \qquad (5.25)$$

describing the probability P that a string breaks up. A is the area of the backward light-cone that has been swept by the classical string. Rather than working in the space-time picture, it is possible to transform the picture into momentum space. Then, one can rewrite the area law in terms of transverse mass, $m_\perp^2 = m^2 + p_\perp^2$, and the light-cone momentum fraction, z, as

$$\frac{d^2 P}{dz dm_\perp^2} = \frac{b}{z} \exp\left(-b\frac{m_\perp^2}{z}\right). \tag{5.26}$$

When applied to the case of the Lund model, hadrons are directly produced from the string excitations. The longitudinal momentum is then chosen according to the *Lund symmetric fragmentation function* [46]

$$f(z) = \frac{N}{z}(1-z)^a \exp\left(-b\frac{m_\perp^2}{z}\right), \tag{5.27}$$

that is compatible with the requirement that on average one obtains the same results, regardless of whether one is starting the fragmentation on the left or the right end of the string. Furthermore, a transverse mass of the hadron pair is chosen according to an exponential,

$$\exp\left(\frac{-\pi m_\perp^2}{\kappa}\right). \tag{5.28}$$

The hadron flavour is then selected according to phase-space weights and other considerations. One should note that this hadronisation model has been very successfully applied to describe a large variety of data. At the same time, one might criticise that the inherent dynamics of this hadronisation may outweigh the dynamics of the perturbative sector.

5.7.3
Cluster Hadronisation

The *cluster hadronisation* model is based on the assumption that the dynamics of the hadronisation should be as simple as possible while the information of the perturbative parton evolution should be carried forward as far as possible. Most notably, the model uses the effect of *preconfinement of colour* [50–52], meaning that partons that come out of the perturbative phase close to each other in momentum space also tend to form colour singlets. This is used quite literally in this model.

After the parton evolution has finished, all partons are put onto their constituent mass shells, including the gluon. The gluons are then split into non-perturbative $q\bar{q}$ pairs, leaving a coloured final state of only quarks and antiquarks behind that only contains colour **3** and $\bar{\mathbf{3}}$ states. The history of these colour states is followed up throughout the parton evolution, allowing one to pair up colour-connected particles that will then form a colourless cluster.

The resulting mass spectrum of these clusters has some remarkable features. It is centred around the parton-shower cutoff and has a power-like tail of higher masses. Interestingly, this spectrum turns out to be independent of the hard scale from which the final-state parton shower has been initiated, meaning that the dynamics at this point in the simulation can be regarded as truly non-perturbative. Hence, this should be a good starting point for a hadronisation model.

To proceed, one first has to split the low number of heavy clusters into pairs of lighter clusters. In this step an additional $q\bar{q}$ pair is created. The kinematics is chosen such that in the original cluster's rest frame, the constituents keep moving in the original direction and the newly created quarks don't acquire any transverse momentum. Therefore, one regards this splitting as being "string-like". In the last step, the remaining clusters which are all considered light enough, will decay into pairs of hadrons. For this, an additional quark–antiquark (diquark–antidiquark) pair is created in order to form two mesons (baryons). The decay of the cluster into the hadron pair is isotropic in the cluster's rest frame. The species of hadrons is chosen according to a combination of weights from the available momentum phase space, angular momentum, flavour multiplets and so on. In the exceptional case where the cluster is too light to allow a decay into the two lightest possible hadrons, a single light hadron is formed and some remaining energy is shared with a neighbouring cluster in order to allow this hadron to be put on its mass shell.

Also this model has been shown to describe a large variety of experimental data, on the other hand, it might allow for less flexibility as there are less parameters than in the Lund string model.

In summary, the model of independent fragmentation has been an important step towards the development of modern event generators but is only rarely used today [53]. The Lund string fragmentation is used in all Monte Carlo event generators from Lund, most notably PYTHIA 6.4 [54] and its successor PYTHIA 8.1 [21] and is implicitly used by many other programs that link to PYTHIA for the hadronisation phase of the simulation. The cluster hadronisation model has always been used in HERWIG [55] and its successor HERWIG++ [16]. In a slight modification [56], the cluster model is also used in newer versions of SHERPA [57].

5.8
The Underlying Event

An additional important aspect in modelling final states at hadron colliders is the description of the *underlying event*. The underlying event may be defined as "everything except the hard process of interest". From the modelling point of view this includes parton-shower emissions from the initial and final states as well as jets from multiple partonic interactions.

5.8.1
Relevance of the Underlying Event at Hadron Colliders

The default perspective to understand hadronic collisions at the LHC is that of collinear factorisation, where we have some *hard* final state and can calculate this signal inclusively with respect to all softer radiation in the same collision. In this description, the hard particles are initiated by point-like interactions of two partons, each from one of the colliding hadrons. This is fine as long as the signal can be measured uniquely. However, many interesting signals at the LHC are characterised by the presence of jets and their properties. A jet is measured in turn as a collection of particles where no distinction with respect to the particle's origin can be made.

At hadron colliders, particles will not only be produced by interactions of a single pair of partons, but also via additional partonic interactions and/or even by additional soft interactions of the beam remnants, of which a modelling is strictly beyond the limit of perturbation theory and hence beyond partonic degrees of freedom. The additional partonic interactions are important to render the total cross section for hadronic interactions finite and smaller than the well-measured experimental cross sections. Otherwise, these interactions would violate unitarity. Therefore, in order to measure, for example, the energy of a jet accurately, all soft particles, also those from softer interactions or additional hard partonic interactions, will be present and have to be modelled correctly and as accurately as possible. Only then one is able to reduce systematic errors from the so-called jet energy scale measurement.

5.8.2
Measuring the Underlying Event

There are currently two main strategies to measure the underlying event at hadron colliders. One is based on the fact that the physics that is dominating the underlying event is mostly of soft, or lower, transverse momentum, such that most of the events that are recorded with only a minimal trigger will be dominated by underlying event physics. These measurements are called *minimum bias* measurements. The other strategy is to trigger some hard events with a reasonably large rate, for example dijet events or Drell–Yan lepton pair production, and to try to subdivide regions in phase space that are probably dominated by the hard process or the underlying event, respectively.

The latter case assumes that, to some extent, the underlying event is uncorrelated with the hard process. The best example for such an analysis is the dijet analysis [58]. After finding the two hardest jets of, say, at least 20 GeV in a dijet event in the central detector ($|\eta| < 1$), one calls the ϕ region around the hardest (or trigger) jet the "towards region". The region in the opposite direction, where one typically has a second hard, recoiling jet, is called the "away region" and the remaining region, mostly transverse to these two, the "transverse region".

One then assumes that the dijet event itself produces most of its activity in a narrow cone around the two leading jets – further hard jets being suppressed by the smallness of the strong coupling, α_s. Additional activity will mostly stem from collinear emissions around the leading two jets. However, the underlying event, that is assumed to be totally uncorrelated to the two jets, will, at least statistically, populate the whole region in ϕ space evenly. Then, the obvious step to extract the underlying event activity is to measure only jets in the region transversely to the jets (the "transverse region") where this is the only activity.

When saying "activity" one usually thinks of transverse energy (or transverse momentum) or multiplicity. Commonly used observables for the underlying event are then either the scalar sum of transverse momenta, p_T^{sum}, or the multiplicity, N_{ch}, of all (charged) particles in the transverse region. Usually, one considers the average of these as a function of the transverse momentum of the leading jet, $p_\perp^{lead}$. Then one finds, after overcoming a trigger effect, that this activity is independent of $p_\perp^{lead}$ and therefore forms a plateau. The height of these plateaus and their ratio for $\langle p_\perp^{sum} \rangle$ and $\langle N_{ch} \rangle$ are the most discussed observables of the underlying event at hadron colliders.

Other very important measurements are of events with four hard objects, either four jets [59] or one photon and three jets [60]. These measurements have shown that, to a large extent, the origin of the four objects are two simultaneous hard processes rather than a single hard process with four hard objects.

5.8.3
Modelling the Underlying Event in Monte Carlo Event Generators

The measurements of events with four hard objects in the final state discussed above give preference to models that employ the mechanism of multiple hard partonic scatters as a model for the underlying event [61–65]. Extending these scatters into the soft, non-perturbative regime then allows minimum bias interactions to be described. The crucial assumption for this sort of model is that there are uncorrelated hard events happening at the same time. This assumption can be justified as follows: the hadrons during their interaction are understood as clouds of partons, extending to distances ~ 1 fm. The individual parton–parton interactions, however, are, when considered hard, happening on much smaller distance scales. Therefore, it is very unlikely that there are additional partons exchanged between these multiple scatters.

On the basis of this assumption it is relatively straightforward to calculate the average multiplicity of hard scatters during one event. At the same time, one is able to unitarise the total cross section and also fix parameters when considering further properties of hadronic interactions. Given this multiplicity one can then sample the additional hard scatters and employ parton showering, hadronisation and hadronic decays exactly as in the modelling of the hard process anyway.

The details of the implementation are the starting point for serious modelling, for example when one is concerned with the parton distribution function for the additional hard scatters. There is no clear theoretical justification for using the

PDFs in the same (or some modified) way as for the hard scatter, and therefore the existing event generators use different strategies to model these "remnant PDFs".

References

1. Mangano, M.L., Moretti, M., Piccinini, F., Pittau, R., and Polosa, A.D. (2003) ALPGEN, a generator for hard multiparton processes in hadronic collisions, *JHEP*, **07**, 001, doi:10.1088/1126-6708/2003/07/001.
2. Yuasa, F. et al. (2000) Automatic computation of cross sections in HEP, *Prog. Theor. Phys. Suppl.*, **138**, 18, doi:10.1143/PTPS.138.18.
3. Kanaki, A. and Papadopoulos, C.G. (2000) HELAC: A package to compute electroweak helicity amplitudes, *Comput. Phys. Commun.*, **132**, 306, doi:10.1016/S0010-4655(00)00151-X.
4. Maltoni, F. and Stelzer, T. (2003) MadEvent: Automatic event generation with MadGraph, *JHEP*, **02**, 027, doi:10.1088/1126-6708/2003/02/027.
5. Krauss, F., Kuhn, R., and Soff, G. (2002) AMEGIC++ 1.0, a matrix element generator in C++, *JHEP*, **02**, 044, doi:10.1088/1126-6708/2002/02/044.
6. Catani, S. and Seymour, M.H. (1997) A general algorithm for calculating jet cross sections in NLO QCD, *Nucl. Phys. B*, **485**, 291, doi:10.1016/S0550-3213(96)00589-5.
7. Catani, S., Dittmaier, S., Seymour, M.H., and Trocsanyi, Z. (2002) The dipole formalism for next-to-leading order QCD calculations with massive partons, *Nucl. Phys. B*, **627**, 189, doi:10.1016/S0550-3213(02)00098-6.
8. Binoth, T. and Heinrich, G. (2004) Numerical evaluation of phase space integrals by sector decomposition, *Nucl. Phys. B*, **693**, 134, doi:10.1016/j.nuclphysb.2004.06.005.
9. Anastasiou, C., Melnikov, K., and Petriello, F. (2004) Higgs boson production at hadron colliders: differential cross sections through next-to-next-to-leading order, *Phys. Rev. Lett.*, **93**, 262002, doi:10.1103/PhysRevLett.93.262002.
10. Gehrmann-De Ridder, A., Gehrmann, T., and Glover, E.W.N. (2005) Antenna subtraction at NNLO, *JHEP*, **09**, 056, doi:10.1088/1126-6708/2005/09/056.
11. Weinzierl, S. (2009) The infrared structure of $e^+e^- \to$ 3 jets at NNLO reloaded, *JHEP*, **07**, 009, doi:10.1088/1126-6708/2009/07/009.
12. Somogyi, G. and Trocsanyi, Z. (2008) A subtraction scheme for computing QCD jet cross sections at NNLO: integrating the subtraction terms I, *JHEP*, **08**, 042, doi:10.1088/1126-6708/2008/08/042.
13. Nagy, Z. and Soper, D.E. (2007) Parton showers with quantum interference, *JHEP*, **09**, 114, doi:10.1088/1126-6708/2007/09/114.
14. Nagy, Z. and Soper, D.E. (2008) Parton showers with quantum interference: leading color, with spin, *JHEP*, **07**, 025, doi:10.1088/1126-6708/2008/07/025.
15. Marchesini, G. et al. (1992) HERWIG 5.1 – a Monte Carlo event generator for simulating hadron emission reactions with interfering gluons, *Comput. Phys. Commun.*, **67**, 465, doi:10.1016/0010-4655(92)90055-4.
16. Bähr, M. et al. (2008) HERWIG++ physics and manual, *Eur. Phys. J. C*, **58**, 639, doi:10.1140/epjc/s10052-008-0798-9.
17. Catani, S., Dokshitzer, Y.L., Olsson, M., Turnock, G., and Webber, B.R. (1991) New clustering algorithm for multijet cross sections in e^+e^- annihilation, *Phys. Lett. B*, **269**, 432, doi:10.1016/0370-2693(91)90196-W.
18. Catani, S., Trentadue, L., Turnock, G., and Webber, B.R. (1993) Resummation of large logarithms in e^+e^- event shape distributions, *Nucl. Phys. B*, **407**, 3, doi:10.1016/0550-3213(93)90271-P.
19. Catani, S., Webber, B.R., and Marchesini, G. (1991) QCD coherent branching and semi-inclusive processes

at large x, *Nucl. Phys. B*, **349**, 635, doi:10.1016/0550-3213(91)90390-J.

20 Sjöstrand, T. and Skands, P.Z. (2005) Transverse-momentum-ordered showers and interleaved multiple interactions, *Eur. Phys. J. C*, **39**, 129, doi:10.1140/epjc/s2004-02084-y.

21 Sjöstrand, T., Mrenna, S., and Skands, P.Z. (2008) A brief introduction to PYTHIA 8.1, *Comput. Phys. Commun.*, **178**, 852, doi:10.1016/j.cpc.2008.01.036.

22 Nagy, Z. and Soper, D.E. (2006) A new parton shower algorithm: Shower evolution, matching at leading and next-to-leading order level, http://arxiv.org/abs/hep-ph/0601021.

23 Dinsdale, M., Ternick, M., and Weinzierl, S. (2007) Parton showers from the dipole formalism, *Phys. Rev. D*, **76**, 094003, doi:10.1103/PhysRevD.76.094003.

24 Schumann, S. and Krauss, F. (2008) A parton shower algorithm based on Catani-Seymour dipole factorisation, *JHEP*, **03**, 038, doi:10.1088/1126-6708/2008/03/038.

25 Nagy, Z. and Soper, D.E. (2010) On the transverse momentum in Z-boson production in a virtuality ordered parton shower, *JHEP*, **03**, 097, doi:10.1007/JHEP03(2010)097.

26 Lönnblad, L. (1992) Ariadne version 4: A Program for simulation of QCD cascades implementing the color dipole model, *Comput. Phys. Commun.*, **71**, 15, doi:10.1016/0010-4655(92)90068-A.

27 Giele, W.T., Kosower, D.A., and Skands, P.Z. (2008) A simple shower and matching algorithm, *Phys. Rev. D*, **78**, 014 026, doi:10.1103/PhysRevD.78.014026.

28 Catani, S., Krauss, F., Kuhn, R., and Webber, B.R. (2001) QCD matrix elements + parton showers, *JHEP*, **11**, 063, doi:10.1088/1126-6708/2001/11/063.

29 Lönnblad, L. (2002) Correcting the colour-dipole cascade model with fixed order matrix elements, *JHEP*, **05**, 046, doi:10.1088/1126-6708/2002/05/046.

30 Mangano, M.L., Moretti, M., Piccinini, F., and Treccani, M. (2007) Matching matrix elements and shower evolution for top-quark production in hadronic collisions, *JHEP*, **01**, 013, doi:10.1088/1126-6708/2007/01/013.

31 Nagy, Z. and Soper, D.E. (2005) Matching parton showers to NLO computations, *JHEP*, **10**, 024, http://arxiv.org/abs/hep-ph/0503053.

32 Lavesson, N. and Lönnblad, L. (2008) Extending CKKW-merging to one-loop matrix elements, *JHEP*, **12**, 070, doi:10.1088/1126-6708/2008/12/070.

33 Frixione, S. and Webber, B.R. (2002) Matching NLO QCD computations and parton shower simulations, *JHEP*, **06**, 029, doi:10.1088/1126-6708/2002/06/029.

34 Frixione, S., Nason, P., and Webber, B.R. (2003) Matching NLO QCD and parton showers in heavy flavour production, *JHEP*, **08**, 007, doi:10.1088/1126-6708/2003/08/007.

35 Krämer, M. and Soper, D.E. (2004) Next-to-leading order QCD calculations with parton showers, I. Collinear singularities, *Phys. Rev. D*, **69**, 054019, doi:10.1103/PhysRevD.69.054019.

36 Krämer, M., Mrenna, S., and Soper, D.E. (2006) Next-to-leading order QCD jet production with parton showers and hadronization, *Phys. Rev. D*, **73**, 014022, doi:10.1103/PhysRevD.73.014022.

37 Nason, P. (2004) A new method for combining NLO QCD with shower Monte Carlo algorithms, *JHEP*, **11**, 040, doi:10.1088/1126-6708/2004/11/040.

38 Frixione, S., Nason, P., and Oleari, C. (2007) Matching NLO QCD computations with Parton Shower simulations: the POWHEG method. *JHEP*, **11**, 070, doi:10.1088/1126-6708/2007/11/070.

39 Alioli, S., Nason, P., Oleari, C., and Re, E. (2010) A general framework for implementing NLO calculations in shower Monte Carlo programs: the POWHEG BOX, *JHEP*, **06**, 043, doi:10.1007/JHEP06(2010)043.

40 Field, R.D. and Feynman, R.P. (1977) Quark elastic scattering as a source of high transverse momentum mesons, *Phys. Rev. D*, **15**, 2590–2616, doi:10.1103/PhysRevD.15.2590.

41 Field, R.D. and Feynman, R.P. (1978) A parametrization of the properties

41 of quark jets, *Nucl. Phys. B*, **136**, 1, doi:10.1016/0550-3213(78)90015-9.
42 Feynman, R.P., Field, R.D., and Fox, G.C. (1978) Quantum-chromodynamic approach for the large-transverse-momentum production of particles and jets, *Phys. Rev. D*, **18**, 3320, doi:10.1103/PhysRevD.18.3320.
43 Artru, X. and Mennessier, G. (1974) String model and multiproduction, *Nucl. Phys. B*, **70**, 93, doi:10.1016/0550-3213(74)90360-5.
44 Bowler, M.G. (1981) e^+e^- production of heavy quarks in the string model, *Z. Phys. C*, **11**, 169, doi:10.1007/BF01574001.
45 Andersson, B., Gustafson, G., and Söderberg, B. (1983) A general model for jet fragmentation, *Z. Phys. C*, **20**, 317, doi:10.1007/BF01407824.
46 Andersson, B., Gustafson, G., and Söderberg, B. (1986) A probability measure on parton and string states, *Nucl. Phys. B*, **264**, 29, doi:10.1016/0550-3213(86)90471-2.
47 Field, R.D. and Wolfram, S. (1983) A QCD model for e^+e^- annihilation, *Nucl. Phys. B*, **213**, 65, doi:10.1016/0550-3213(83)90175-X.
48 Gottschalk, T.D. (1983) A realistic model for e^+e^- annihilation including parton bremsstrahlung effects, *Nucl. Phys. B*, **214**, 201, doi:10.1016/0550-3213(83)90658-2.
49 Webber, B.R. (1984) A QCD model for jet fragmentation including soft gluon interference, *Nucl. Phys. B*, **238**, 492, doi:10.1016/0550-3213(84)90333-X.
50 Amati, D. and Veneziano, G. (1979) Preconfinement as a property of perturbative QCD, *Phys. Lett. B*, **83**, 87, doi:10.1016/0370-2693(79)90896-7.
51 Bassetto, A., Ciafaloni, M., and Marchesini, G. (1979) Color singlet distributions and mass damping in perturbative QCD, *Phys. Lett. B*, **83**, 207, doi:10.1016/0370-2693(79)90687-7.
52 Marchesini, G., Trentadue, L., and Veneziano, G. (1981) Space-time description of colour screening via jet calculus techniques, *Nucl. Phys. B*, **181**, 335, doi:10.1016/0550-3213(81)90357-6.
53 Paige, F.E., Protopopescu, S.D., Baer, H., and Tata, X. (2003) ISAJET 7.69: A Monte Carlo event generator for pp, $\bar{p}p$, and e^+e^- reactions, http://arxiv.org/abs/hep-ph/0312045.
54 Sjöstrand, T., Mrenna, S., and Skands, P.Z. (2006) PYTHIA 6.4 physics and manual, *JHEP*, **05**, 026, doi:10.1088/1126-6708/2006/05/026.
55 Corcella, G. et al. (2001) HERWIG 6.5: an event generator for Hadron Emission Reactions With Interfering Gluons (including supersymmetric processes), *JHEP*, **01**, 010, doi:10.1088/1126-6708/2001/01/010.
56 Winter, J.C., Krauss, F., and Soff, G. (2004) A modified cluster-hadronization model, *Eur. Phys. J. C*, **36**, 381, doi:10.1140/epjc/s2004-01960-8.
57 Gleisberg, T. et al. (2009) Event generation with SHERPA 1.1, *JHEP*, **02**, 007, doi:10.1088/1126-6708/2009/02/007.
58 CDF Collab., Affolder, A.A. et al. (2002) Charged jet evolution and the underlying event in $p\bar{p}$ collisions at 1.8 TeV, *Phys. Rev. D*, **65**, 092002, doi:10.1103/PhysRevD.65.092002.
59 Axial Field Spectrometer Collab., Akesson, T. et al. (1987) Double parton scattering in pp collisions at $\sqrt{s} = 63$ GeV, *Z. Phys. C*, **34**, 163, doi:10.1007/BF01566757.
60 CDF Collab., Abe, F. et al. (1997) Double parton scattering in $\bar{p}p$ collisions at $\sqrt{s} = 1.8$ TeV, *Phys. Rev. D*, **56**, 3811, doi:10.1103/PhysRevD.56.3811.
61 Sjöstrand, T. and van Zijl, M. (1987) A multiple-interaction model for the event structure in hadron collisions, *Phys. Rev. D*, **36**, 2019, doi:10.1103/PhysRevD.36.2019.
62 Sjöstrand, T. and Skands, P.Z. (2004) Multiple interactions and the structure of beam remnants, *JHEP*, **03**, 053, doi:10.1088/1126-6708/2004/03/053.
63 Butterworth, J.M., Forshaw, J.R., and Seymour, M.H. (1996) Multiparton interactions in photoproduction at HERA, *Z. Phys. C*, **72**, 637, doi:10.1007/s002880050286.
64 Bähr, M., Gieseke, S., and Seymour, M.H. (2008) Simulation of multiple partonic interactions in Herwig++,

JHEP, **07**, 076, doi:10.1088/1126-6708/2008/07/076.

65 Bähr, M., Butterworth, J.M., Gieseke, S., and Seymour, M.H. (2009) Soft interactions in Herwig++, http://arxiv.org/abs/0905.4671.

6
The Higgs Boson: Still Elusive After 40 Years

Markus Schumacher and Michael Spira

The Higgs mechanism is a cornerstone of the Standard Model (SM). The introduction of the fundamental Higgs field [1–4] renders the standard electroweak theory weakly interacting up to high energy scales without violating the unitarity bounds of scattering amplitudes. Because of spontaneous symmetry breaking in the Higgs sector, the electroweak gauge bosons Z and W as well as the fermions acquire masses through the interaction with the Higgs fields (see Chapter 2). Since the gauge symmetry, though hidden, is still preserved, the theory of electroweak interactions is renormalisable [5–9]. In the Standard Model, one weak isospin Higgs doublet is introduced and leads to the existence of one elementary Higgs particle after electroweak symmetry breaking. The Higgs boson couplings to the electroweak gauge bosons and all fermions grow with their masses. The only unknown parameter of the Higgs boson is the value of its mass, M_H. Once this is known, all production and decay properties of the SM Higgs boson are fixed [10, 11]. The search for the Higgs boson is a crucial endeavour at present and future experiments for establishing the standard formulation of the electroweak theory.

6.1
The Higgs Boson Mass

Even though the mass of the Higgs boson cannot be predicted in the Standard Model, stringent upper and lower bounds can nevertheless be derived from internal consistency conditions and extrapolations of the model to high energies.

The Higgs boson has been introduced as a fundamental particle to render scattering amplitudes of two particles into two particles involving longitudinally polarised W bosons, W_L, compatible with unitarity. On the basis of the general principle of time–energy uncertainty, particles must decouple from a physical system if their mass grows indefinitely. The mass of the Higgs particle must therefore be bounded to restore unitarity in the perturbative regime: from the asymptotic expansion of the elastic $W_L W_L$ S-wave scattering amplitude including W and Higgs boson exchanges, $A(W_L W_L \to W_L W_L) \to -G_F M_H^2/4\sqrt{2}\pi$, it follows that

$M_H^2 \leq 2\sqrt{2}\pi/G_F \sim (850\,\text{GeV})^2$. Within the canonical formulation of the Standard Model, consistency conditions therefore require a Higgs boson mass below 1 TeV.

Restrictive bounds on the value of the SM Higgs boson mass follow from hypotheses of the energy scale Λ up to which the Standard Model can be extended before new physical phenomena emerge. These would be associated with new strong interactions between the fundamental particles. The key to these bounds is the evolution of the quartic Higgs coupling, λ, (see Chapter 2) with the energy due to quantum fluctuations. The basic contributions are depicted in Figure 6.1. The Higgs boson loop itself gives rise to an indefinite increase of the coupling while the fermionic top-quark loop, with increasing top mass, drives the coupling to smaller values, finally even to values below zero. The variation of λ and of the top-Yukawa coupling g_t with energy, parametrised by $t = \log \mu^2/v^2$, may be written as

$$\begin{aligned} \frac{d\lambda}{dt} &= \frac{3}{8\pi^2}\left[\lambda^2 + \lambda g_t^2 - g_t^4\right] &&: \lambda(v^2) = M_H^2/v^2\,, \\ \frac{dg_t}{dt} &= \frac{1}{32\pi^2}\left[\frac{9}{2}g_t^3 - 8g_t g_s^2\right] &&: g_t(v^2) = \sqrt{2}m_t/v\,, \end{aligned} \quad (6.1)$$

where g_s denotes the strong coupling of QCD. Only the leading contributions from Higgs, top and QCD loops are taken into account.

For moderate top masses, λ rises indefinitely, $\partial\lambda/\partial t \sim +\lambda^2$, and the coupling becomes strong shortly before reaching the Landau pole,

$$\lambda(\mu^2) = \frac{\lambda(v^2)}{1 - \frac{3\lambda(v^2)}{8\pi^2}\log\frac{\mu^2}{v^2}}\,. \quad (6.2)$$

Re-expressing the initial value $\lambda(v^2)$ by the Higgs boson mass, the condition $\lambda(\Lambda) < \infty$ can be translated into an upper bound on the Higgs boson mass,

$$M_H^2 \leq \frac{8\pi^2 v^2}{3\log\frac{\Lambda^2}{v^2}}\,. \quad (6.3)$$

This mass bound is related logarithmically to the energy, Λ, up to which the Standard Model is assumed to be valid. The maximum value of M_H for the minimum cut-off $\Lambda \sim 1\,\text{TeV}$ is given by $\sim 750\,\text{GeV}$. This bound is close to the estimate of $\sim 700\,\text{GeV}$ in lattice calculations for $\Lambda \sim 1\,\text{TeV}$, which allow proper control of non-perturbative effects near the boundary.

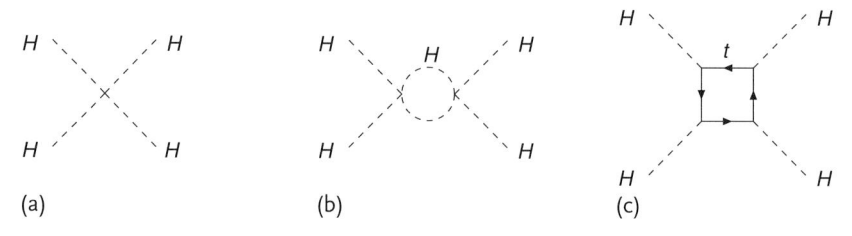

Figure 6.1 Diagrams generating the evolution of the quartic Higgs coupling, λ. (a) tree level, (b) Higgs boson loop, (c) top-quark loop.

Table 6.1 Higgs boson mass bounds for two values of the cut-off Λ.

Λ	M_H
1 TeV	55 GeV $\lesssim M_H \lesssim$ 700 GeV
10^{16} GeV	130 GeV $\lesssim M_H \lesssim$ 190 GeV

A lower bound on the Higgs boson mass can be derived from the requirement of vacuum stability. Since top-loop corrections reduce λ when the top-Yukawa coupling increases, λ becomes negative if the top mass becomes too large. In such a case, the self-energy potential would become deeply negative and the electroweak ground state would no longer be stable. To avoid this instability, the Higgs boson mass must exceed a minimum value for a given top mass. This lower bound depends on the cut-off Λ.

For any given Λ, the allowed values of M_H are shown in Figure 6.2. For a central top mass $m_t = 175$ GeV, the allowed Higgs boson mass values are collected in Table 6.1 for two specific cut-off values Λ. If the Standard Model is assumed to be valid up to the scale of grand unification (GUT), the Higgs boson mass is restricted to a narrow window between 130 and 190 GeV. The observation of a Higgs boson mass above or below this window would demand a new physics scale below the GUT scale.

The Higgs boson contributes logarithmically to the electroweak precision observables at higher orders as discussed in Chapter 3. Thus, a global fit to the electroweak data provides an indirect determination of the Higgs boson mass range that is consistent with the precision data of LEP, SLC and the Tevatron. These indirect fits

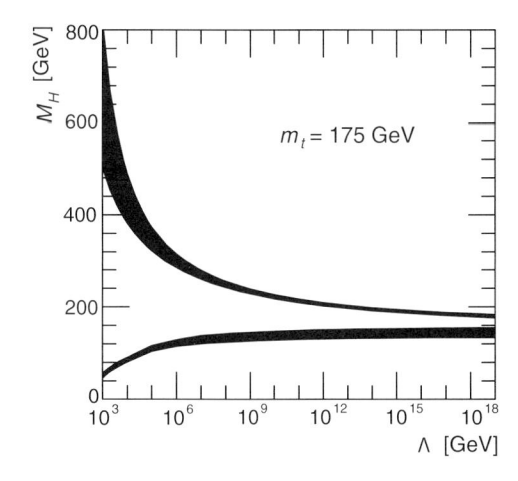

Figure 6.2 Bounds on the mass of the Higgs boson in the SM. Here, Λ denotes the energy scale at which the Higgs boson system of the SM would become strongly interacting (upper bound); the lower bound follows from the requirement of vacuum stability. The size of the bands represents the theoretical uncertainties (adapted from [12]).

impose an upper bound on the Higgs boson mass of about 158 GeV at 95% C.L., clearly favouring a light Higgs particle.

6.2
Higgs Boson Decays

The partial width of Higgs boson decays to lepton and quark pairs is given at leading order (LO) by

$$\Gamma(H \to f\bar{f}) = N_C \frac{G_F}{4\sqrt{2}\pi} m_f^2(M_H^2) M_H \beta_f^3 , \qquad (6.4)$$

$N_C = 1$ or 3 being the colour factor, m_f the fermion mass and $\beta_f = \sqrt{1 - 4m_f^2/M_H^2}$ the fermion velocity. Asymptotically, the fermionic width grows only linearly with the Higgs boson mass. The bulk of QCD radiative corrections can be mapped into the scale dependence of the quark mass, evaluated at the Higgs boson mass.

Above the WW and ZZ decay thresholds, the partial widths for the decays into WW and ZZ may be written as

$$\Gamma(H \to VV) = \delta_V \frac{G_F}{16\sqrt{2}\pi} M_H^3 (1 - 4x + 12x^2) \beta_V , \qquad (6.5)$$

where $x = M_V^2/M_H^2$ and $\delta_V = 2$ and 1 for $V = W$ and Z, respectively. The factor of 2 difference between the decays into WW and ZZ is due to the fact that the two identical Z bosons cannot be distinguished. For large Higgs boson masses, the vector bosons are longitudinally polarised. Since the wave functions of these states are linear in the energy, the widths grow as the third power of the Higgs boson mass. Below the threshold for two real bosons, the Higgs particle can decay into $V^{(*)}V^{(*)}$ pairs, one or both of the vector bosons being virtual. The electroweak corrections to these decay widths are of moderate size.

In the Standard Model, gluonic Higgs boson decays are mediated by top-quark and bottom-quark loops, photonic decays in addition by W loops. Since these decay modes are significant only far below the top and W thresholds, they are described by the approximate expressions [10, 11] which are valid in the limit $M_H^2 \ll 4M_W^2, 4m_t^2$,

$$\Gamma(H \to gg) = \frac{G_F \alpha_s^2(M_H^2)}{36\sqrt{2}\pi^3} M_H^3 \left[1 + \left(\frac{95}{4} - \frac{7N_F}{6}\right)\frac{\alpha_s}{\pi}\right] , \qquad (6.6)$$

$$\Gamma(H \to \gamma\gamma) = \frac{G_F \alpha^2}{128\sqrt{2}\pi^3} M_H^3 \left[\frac{4}{3} N_C e_t^2 - 7\right]^2 , \qquad (6.7)$$

where N_F is the number of quark flavours and e_t is the electric charge of the top quark. The QCD radiative corrections, which include the ggg and $gq\bar{q}$ final states in (6.6), are very important; they increase the partial width by about 65%. Even though photonic Higgs boson decays are very rare, they nevertheless open an attractive resonance-type channel for the search for Higgs particles. The higher-order

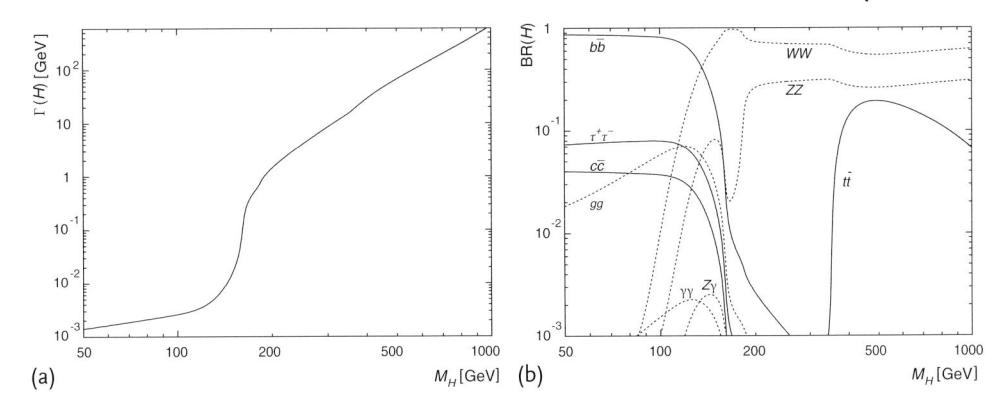

Figure 6.3 (a) Total decay width (in GeV) of the SM Higgs boson as a function of its mass; (b) branching ratios of the dominant decay modes of the SM Higgs particle. All relevant higher-order corrections are taken into account (adapted from [13]).

corrections to the photonic Higgs boson decay mode are small in the Higgs boson mass range relevant for the LHC.

The Higgs boson decay modes can be divided into two different mass ranges. For $M_H \lesssim 135$ GeV the Higgs boson mainly decays into $b\bar{b}$ and $\tau^+\tau^-$ pairs with branching ratios of about 85% and 8%, respectively (see Figure 6.3b). The decay modes into $c\bar{c}$ and gluon pairs, with the latter mediated by top-quark and bottom-quark loops, accumulate a branching ratio of up to about 7%, but do not play a relevant role at the LHC. Since the dominant Higgs boson decays will be swamped by the QCD background, one of the most important Higgs boson decays in this mass range at the LHC is the decay into photon pairs, which is mediated by W, top-quark and bottom-quark loops. It has a branching fraction of up to 2×10^{-3}. For Higgs boson masses above 135 GeV, the main decay modes are those into WW and ZZ pairs, where below the corresponding kinematic thresholds the vector bosons are off-shell. These decay modes dominate over the decay into $t\bar{t}$ pairs, the branching ratio of which does not exceed ∼20% as can be inferred from Figure 6.3b. The total decay width of the Higgs boson, shown in Figure 6.3a, does not exceed about 1 GeV below the WW threshold. For very large Higgs boson masses, the total decay width grows up to the order of the Higgs boson mass itself, so that the interpretation of the Higgs boson as a resonance becomes questionable. This Higgs boson mass range coincides with the upper bound of the Higgs boson mass from triviality, see Table 6.1.

6.3
Higgs Boson Production at the LEP Collider

The main production mechanisms for Higgs bosons in e^+e^- collisions are *Higgs-strahlung*, $e^+e^- \rightarrow Z^* \rightarrow ZH$ (see Figure 6.4a) and *WW fusion*, $e^+e^- \rightarrow \bar{\nu}_e \nu_e (WW) \rightarrow \bar{\nu}_e \nu_e H$ (see Figure 6.4b). In Higgs-strahlung the Higgs boson is

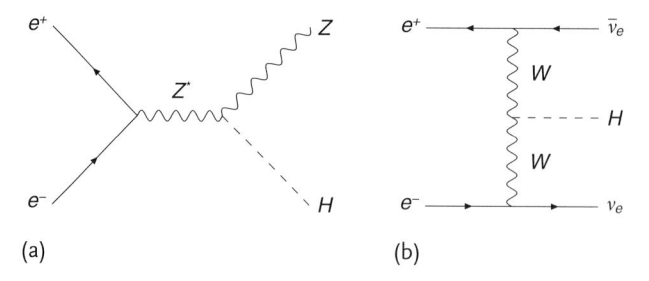

Figure 6.4 Typical diagrams for all relevant Higgs boson production mechanisms at e^+e^- colliders at leading order: (a) Higgs-strahlung; (b) WW fusion.

emitted from the Z-boson line, while WW fusion is a formation process of Higgs bosons in the collision of two quasi-real W bosons radiated off the electron and positron beams.

The cross section for Higgs-strahlung can be written in a compact form as

$$\sigma(e^+e^- \to ZH) = \frac{G_F^2 M_Z^4}{24\pi s}\left[(g_V^e)^2 + (g_A^e)^2\right]\lambda^{1/2}\frac{\lambda + 12M_Z^2/s}{\left[1 - M_Z^2/s\right]^2}, \quad (6.8)$$

where $g_V^e = -1/2 + 2\sin^2\theta_W$ and $g_A^e = -1/2$ are the vector and axial-vector Z charges of the electron and $\lambda = [1-(M_H+M_Z)^2/s][1-(M_H-M_Z)^2/s]$ is the usual two-particle phase-space function. The cross section behaves as $\sigma \sim (G_F M_Z^2)^2/s$, i.e. it is of second order in the weak coupling and is inversely proportional to s. Higher-order contributions to the cross section amount to $\mathcal{O}(10\%)$ and are under theoretical control.

Since the cross section vanishes for high energies, the Higgs-strahlung process is most useful when searching for Higgs bosons in the range where the collider energy is of the same order as the Higgs boson mass, $\sqrt{s} \gtrsim \mathcal{O}(M_H)$. Since the recoiling Z mass in the two-body reaction $e^+e^- \to ZH$ is mono-energetic, the mass of the Higgs boson can be reconstructed from the energy of the Z boson, $M_H^2 = s - 2\sqrt{s}E_Z + M_Z^2$, without any need to analyse the decay products of the Higgs boson. For leptonic Z decays, missing-mass techniques provide a very clear signal, since the Higgs mass can be reconstructed from the Z decay alone, i.e. independent of the Higgs decay mode.

Also the cross section for the WW fusion process can be cast implicitly into a compact form:

$$\sigma(e^+e^- \to \bar{\nu}_e\nu_e H) = \frac{G_F^3 M_W^4}{4\sqrt{2}\pi^3}\int_{\kappa_H}^1\int_x^1\frac{dxdy}{[1+(y-x)/\kappa_W]^2}f(x,y),$$

$$f(x,y) = \left(\frac{2x}{y^3} - \frac{1+3x}{y^2} + \frac{2+x}{y} - 1\right)\left[\frac{z}{1+z} - \log(1+z)\right]$$

$$+ \frac{x}{y^3}\frac{z^2(1-y)}{1+z}, \quad (6.9)$$

with $\kappa_H = M_H^2/s$, $\kappa_W = M_W^2/s$ and $z = y(x-\kappa_H)/(\kappa_W x)$. The electroweak higher-order corrections of $\mathcal{O}(10\%)$ are under control. At LEP 2 energies the Higgs-strahlung process would have been the dominant Higgs boson production mode. The fusion process becomes relevant only at the highest energies, reaching a 5% contribution to the total production rate at $\sqrt{s} = 209\,\text{GeV}$.

6.4 Higgs Boson Production at Hadron Colliders

The main Higgs boson production mechanism at the Tevatron and the LHC will be the *gluon fusion process* $gg \to H$. This process is mediated by top-quark and bottom-quark loops (see Figure 6.5a). Because of the large size of the top-Yukawa couplings and the gluon densities in the relevant Bjorken-x range, gluon fusion comprises the dominant Higgs boson production mechanism for the whole Higgs boson mass range.

The QCD corrections to the top-quark and bottom-quark loops have been known for a long time and include the full Higgs boson and quark-mass dependencies [10, 11]. They increase the total cross section by 50–100%. Setting the top-quark mass to a very large value provides an approximation that is accurate to $\sim 10\%$ for all Higgs boson masses. In this limit the next-to-next-to-leading order (NNLO) QCD corrections have been calculated; they increase the total cross section by a further ~ 20–30%. A full massive NNLO calculation is only partially available [14, 15], so that the NNLO results can only be trusted for intermediate Higgs boson masses, $M_H \lesssim 300\,\text{GeV}$. The approximate NNLO results have been improved by a soft-gluon resummation at the next-to-next-to-leading log (NNLL) level, which yields

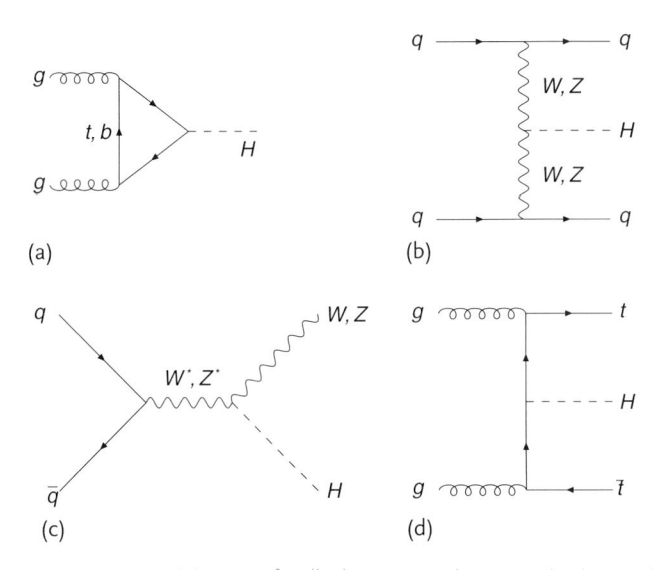

Figure 6.5 Typical diagrams for all relevant Higgs boson production mechanisms at hadron colliders at leading order: (a) gluon fusion; (b) vector boson fusion; (c) Higgs-strahlung; (d) Higgs bremsstrahlung off top quarks.

another increase of the total cross section by ~10%. Electroweak corrections have been computed as well and turn out to be small. The theoretical uncertainties of the total cross section can be estimated to be ~15%.

For large Higgs boson masses, the *W and Z boson fusion processes* $qq \to qq + WW/ZZ \to qqH$ (see Figure 6.5b) become competitive at the LHC. These processes do not play a role at the Tevatron. They are, however, relevant at the LHC in the intermediate Higgs boson mass range, since the additional forward jets offer the opportunity to reduce the background processes significantly. Since there is no colour exchange between the two quark lines at next-to-leading order (NLO), the NLO QCD corrections can be derived from the NLO corrections to deep inelastic lepton-nucleon scattering. They turn out to be $\mathcal{O}(10\%)$ for the total cross section. The full electroweak corrections are of similar size. Quite recently the NLO QCD and electroweak corrections to the differential cross sections have also been computed, resulting in modifications of the relevant distributions by up to ~30%. The residual uncertainties are of $\mathcal{O}(5\%)$.

In the intermediate mass range $M_H \lesssim 2M_Z$, Higgs-strahlung off W bosons and Z bosons, $q\bar{q} \to Z^*/W^* \to H + Z/W$ (see Figure 6.5c), provides alternative signatures for the Higgs boson searches at the Tevatron and the LHC. Since only the initial-state quarks are strongly interacting at LO, the NLO QCD corrections can be inferred from the Drell–Yan process. They increase the total cross section by $\mathcal{O}(30\%)$. Recently this calculation has been extended up to NNLO. The NNLO corrections of ~5–10% are smaller and demonstrate better theoretical control. Moreover, the full electroweak corrections result in a decrease of the total cross section by 5–10%. The total theoretical uncertainty is of $\mathcal{O}(5\%)$.

Higgs boson *radiation off top quarks* $pp \to q\bar{q}/gg \to Ht\bar{t}$ (see Figure 6.5d) plays a role at the LHC for smaller Higgs boson masses below ~150 GeV, while it is much less important at the Tevatron due to the kinematic suppression. The LO cross section has been computed a long time ago. During the last years the full NLO QCD corrections have been calculated, resulting in a moderate increase of the total cross section by ~20% at the LHC. The effects on the relevant parts of final-state particle distributions are also of moderate size, i.e. $\mathcal{O}(10\%)$, so that earlier experimental analyses are not expected to alter much as a result of these corrections.

All SM Higgs boson production cross sections at the Tevatron and the LHC, including NLO QCD corrections, are shown in Figure 6.6. The gluon fusion and vector boson fusion cross sections are about two orders of magnitude larger at the LHC than at the Tevatron. This is due to the higher energy and the smaller Bjorken-x values which contribute to the gluon fusion cross section at the LHC. The vector boson fusion process is further suppressed at the Tevatron with respect to the LHC due to the additional jets in the final state which form a large invariant mass and thus reduce the available phase space of the final-state particles more strongly at the Tevatron. The Higgs-strahlung cross sections $q\bar{q} \to W/Z + H$ differ by only one order of magnitude between the Tevatron and the LHC. Higgs boson radiation off top quarks on the other hand is strongly suppressed at the Tevatron due to the large masses of the final-state top quarks and Higgs bosons. At the Tevatron, Higgs-

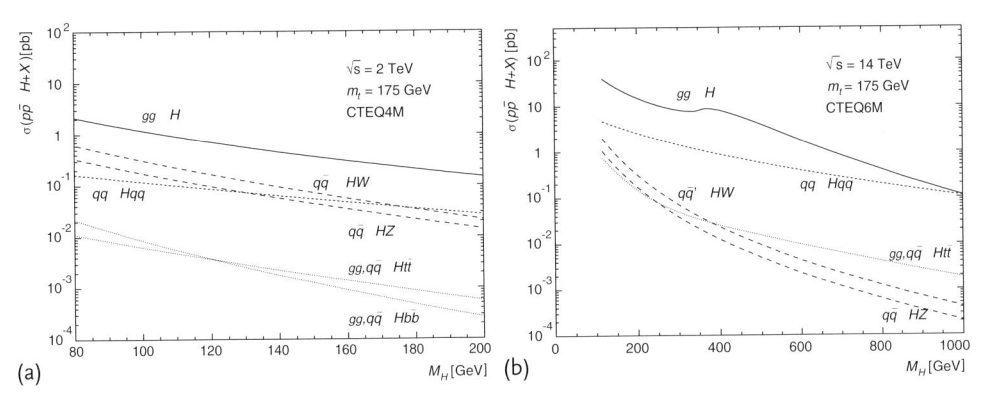

Figure 6.6 Higgs boson production cross sections (a) at the Tevatron and (b) the LHC for the various production mechanisms as a function of the Higgs boson mass. The full QCD-corrected results for gluon fusion, $gg \rightarrow H$, vector boson fusion, $qq \rightarrow VVqq \rightarrow Hqq$, vector boson bremsstrahlung, $q\bar{q} \rightarrow V^* \rightarrow HV$, and associated production, $gg, q\bar{q} \rightarrow Ht\bar{t}$, are shown (adapted from [10, 16–18]).

strahlung and gluon fusion play the dominant role for the experimental searches, while the Higgs boson search at the LHC is mainly based on the gluon fusion and the vector boson fusion processes.

6.5
Past and Present Searches at LEP and Tevatron

The challenge of searches for the Higgs boson is to develop a selection which allows the a priori very small signal-to-background ratio to be increased to a level of the order 1/1 to 1/10. In order to reach this goal, sophisticated selection strategies have been developed using either simple rectangular cuts or making use of multivariate techniques such as likelihood ratios, artificial neural networks or boosted decision trees. The combination of production mechanisms and decay modes of the Higgs boson and the particles, which are eventually produced in association with the Higgs particle, yields a plethora of final-state topologies which can be used to search for Higgs boson production. The following considerations have to be taken into account when choosing the most promising search channels: sufficient a priori signal rate determined by cross sections and branching ratios; methods to trigger on signal events with high efficiency; methods to suppress the background to an acceptable level; and means to estimate the uncertainties on the background prediction and, in the case of exclusion, also the uncertainties on the signal prediction. For the discovery of a new resonance (which means that the probability to observe the actual number of events is less than 2.85×10^{-7} under the background-only hypothesis) only the background prediction and its uncertainty are needed. Wrong assumptions on the shape of discriminating observables for the signal process will lead to a reduced sensitivity. Hypothetically, one observed event can lead to a claim of discovery, given the search is basically background-free ($\approx 3 \times 10^{-7}$ expected background events). In order to exclude the signal hypothesis at the 95% confidence level, a minimum signal yield of 3 expected events after application of the full selection chain is needed. In this case, also the uncertainties on the signal rate and shape of the discriminating distributions will influence the derived exclusion limits – e.g. a too optimistic assumption for the mass resolution for Higgs boson candidates will yield a too stringent exclusion.

Although the Higgs mechanism was incorporated in the electroweak theory four decades ago, significant parts of the mass range of the SM Higgs boson could only be investigated after the start of the Large Electron Positron Collider (LEP) in 1989. According to the 1988 Particle Data Book [19], "the only cast-iron constraint on the mass of the Higgs boson was $M_H > 14$ MeV at 95% confidence level. A combination of theoretical arguments and bounds from K, B and Υ decays probably excluded the range below 4 GeV".

6.5.1
Searches at LEP

The most stringent exclusion until recently was provided by searches at the e^+e^- collider LEP. Until 2000, a data sample of approximately 0.5(2.5) fb^{-1} was collected during the LEP 2 programme by the four experiments ALEPH, DELPHI, L3 and OPAL at centre-of-mass energies in excess of 206 (189) GeV [21]. Searches have been performed for the Higgs-strahlung process $e^+e^- \rightarrow Z^* \rightarrow ZH$ in various combinations of Z and Higgs boson decay modes: $ZH \rightarrow q\bar{q}b\bar{b}$, $\nu\bar{\nu}b\bar{b}$, $e^+e^-b\bar{b}$, $\mu^+\mu^-b\bar{b}$, $\tau^+\tau^-q\bar{q}$, $q\bar{q}\tau^+\tau^-$. These final states cover approximately 80% of the total production rate for a Higgs boson with a mass of 115 GeV. The signal cross section in comparison to the dominant background processes is shown in Figure 6.7a. The strong dependence of the cross section on the Higgs boson mass at a given centre-of-mass energy is clearly visible. The a priori signal-to-background ratio is of the order of 10^{-4} at the edge of the LEP sensitivity. Given the long time intervals between bunch crossings of 22 µs, the instantaneous luminosity of 10^{32} cm^{-2} s^{-1} and the magnitude of the total cross section, triggering on all signal topologies was not a problem. The selection made use of the good identification performance for *b*-flavoured jets, tau leptons, electrons and muons. The Higgs boson candidate mass could be reconstructed in all final states including the decay $Z \rightarrow \nu\bar{\nu}$ by exploiting energy-momentum conservation and constraining the mass of two final-state particles to the Z-boson mass. After all selection cuts, a signal-to-background ratio of the order of 2 to 1 could be achieved at the edge of the LEP sensitivity for masses of 115 GeV, with a 50% signal efficiency. The background was dominated by W^+W^- and ZZ pair production decaying to four-fermion final states. The mass distribution obtained from the combination of the four LEP experiments is shown in Figure 6.8a. No significant hints for the production of a Higgs boson were observed which allowed all Higgs boson mass hypotheses below 114.4 GeV to be excluded with at least 95% confidence level (see Figure 6.8b) [21]. The expected exclusion limit, assuming only background processes are present in nature,

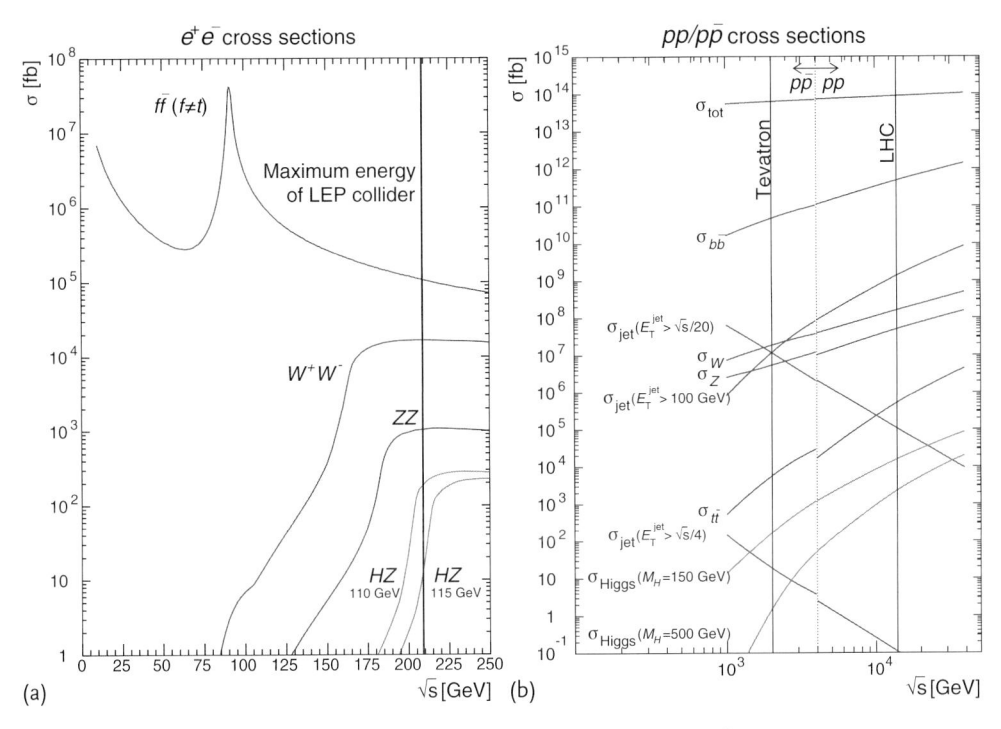

Figure 6.7 Comparison of leading-order cross sections for Higgs boson production and various background processes at (a) e^+e^- colliders and (b) hadron colliders calculated with [20]. The discontinuities in the cross-section curves at hadron colliders are due to the change from $p\bar{p}$ to pp colliders.

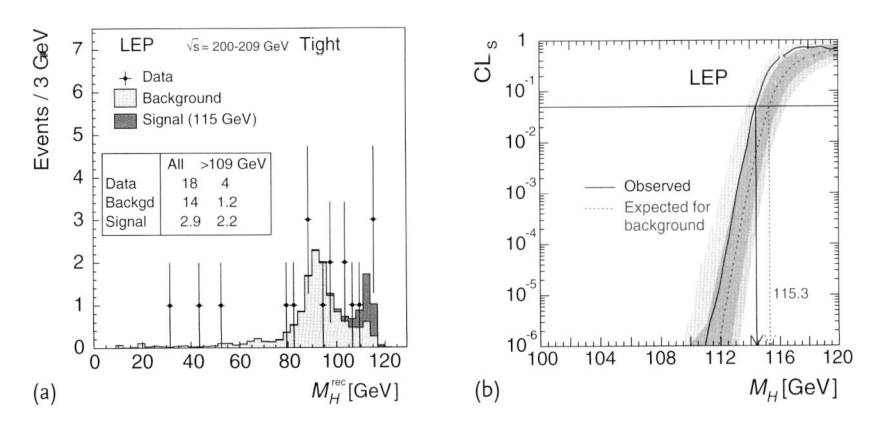

Figure 6.8 Mass distribution of the SM Higgs boson search at LEP (a) and exclusion at 95% confidence level of Higgs boson mass hypotheses at LEP (b) (adapted from [21]).

is 115.3 GeV. The observed limit is lower due to a slight deviation of the observed mass spectrum from the background expectation at the level of 1.7 standard deviations.

6.5.2
Searches at the Tevatron

The search for the Higgs boson has been continued at the Tevatron $p\bar{p}$ collider, that, since the year 2001, has been operating at a centre-of-mass energy of 1.96 TeV. Until January 2010 a dataset corresponding to an integrated luminosity of up to 5.4 fb^{-1} has been analysed by each of the two experiments CDF and DØ. In contrast to the search at LEP, the tiny a priori signal-to-background ratio (see Figure 6.7b), huge background cross section and small time interval between bunch crossings (ca. 400 ns) put severe requirements on the trigger capabilities. As a consequence, purely hadronic final states which have the largest signal rate, for example production in gluon fusion or vector boson fusion (VBF) with decays $H \to b\bar{b}$, $H \to W^+W^- \to q\bar{q}'q''\bar{q}'''$, cannot be exploited for the Higgs boson search. At least one lepton, photon or large transverse missing energy is already required at the trigger level. In total, roughly 90 different selections have been performed by CDF and DØ in order to look for Higgs boson production in an optimal way. In the lower mass range, from the LEP exclusion to approximately 150 GeV, searches have been performed for the following combination of production and decay modes (and their charge conjugates): $W^+H \to l^+\nu b\bar{b}$; $ZH \to l^+l^-(\nu\bar{\nu})b\bar{b}$; VBF, ZH, $W^+H \to q\bar{q}^{(\prime)}\tau^+\tau^-$; $t\bar{t}H$, $H \to b\bar{b}$; and $H \to \gamma\gamma$. The overall sensitivity is dominated by the associated production with weak gauge bosons and the decay $H \to b\bar{b}$. These channels cover less than 1% of the total production rate corresponding to only $\mathcal{O}(10\,\text{fb})$ taking into account the branching ratios of W and Z bosons to leptons. The other channels contribute only marginally, due to their very small production rates in the SM at the level of 1 fb or less.

Besides exploiting kinematic differences, the main tools for the suppression of backgrounds are good identification of b-flavoured jets and tau leptons and a precise reconstruction of the mass of Higgs boson candidates from dijet or ditau lepton topologies. With the currently analysed datasets, no hints for Higgs boson production in the low mass range have been observed, and cross sections corresponding to 2 to 3 times the SM prediction can be excluded at the 95% confidence level [22].

In the mass range from 140 to 200 GeV, the dominant decay mode of the Higgs boson is to a pair of W bosons. Only decays into $e^+\nu_e$ or $\mu^+\nu_\mu$ and their charge conjugates are considered as search topologies. All dominant production modes (gluon fusion, VBF, associated production with a W boson) have been considered. Again only $\approx 5\%$ of the total signal production rate can be exploited corresponding to roughly 20–25 fb for Higgs boson masses around 160 GeV. The experiments have followed two different approaches: firstly, inclusive searches where only requirements on the Higgs boson decay products are applied; secondly, topologies have been split up according to the jet multiplicity in the final state in order to enhance the sensitivity to gluon fusion, VBF and W associated production. Because of the presence of two neutrinos stemming from the decays of W bosons with small transverse momenta, the mass of the Higgs boson candidate cannot be reconstructed. However, the scalar nature of the Higgs boson (spin = 0) and the parity-violating structure of the W decays results in small azimuthal opening angle between the two charged leptons from the Higgs boson decay. This information is exploited in order to discriminate the signal process from background processes. The background-subtracted distribution of the signal-to-background ratio of the final discriminant, which is input to the statistical interpretation of the results, is shown in Figure 6.9a. The majority of signal events are in the phase-space region where the signal-to-background ratio is in the range of 1 : 100 to 1 : 1. The latest combination of the Tevatron results in spring 2010 allowed the mass exclusion obtained at LEP to be extended for the first time. The Higgs boson mass range from 162 to 166 GeV has been excluded at 95% confidence level, with an expected exclusion sensitivity between 159 and 169 GeV [23] (see Figure 6.9b).

The Tevatron will probably collect at least $10\,\text{fb}^{-1}$ before its shutdown, which would double the analysed dataset compared to the results discussed here. With an improved analysis performance, it is expected that all SM mass hypotheses below 185 GeV can be excluded with 95% confidence level, if no deviation from the background expectation is observed. Evidence for Higgs boson production corresponding to a three standard-deviation discrepancy from the background expectation can be expected in the mass range from about 150 to 175 GeV and a few GeV above the LEP exclusion [24]. In the past the envisaged improvements of the analysis performance have been achieved during reanalysis of the data. The estimation of prospects do not take into account the current findings from the already analysed data sets, i.e. the derivation of the expected ranges for evidence do not include the current exclusion limits.

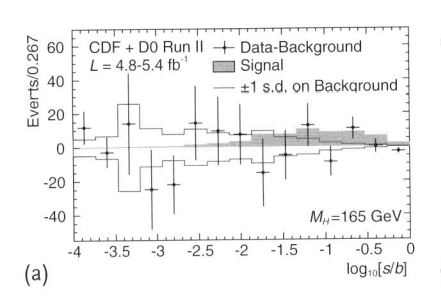

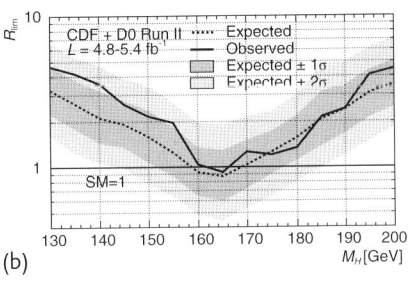

Figure 6.9 (a) Background-subtracted data distributions for the discriminant histograms, summed for bins with similar signal-to-background ratio. The signal expectation for a Higgs boson with a mass of 165 GeV is shown as a shaded area, the uncertainties on the background are shown as solid lines. (b) Observed and expected 95% confidence level upper limits on the SM Higgs boson production with respect to the SM expectation obtained from $H \to W^+W^-$ searches (adapted from [23]).

6.6
Prospects for Higgs Boson Searches at the LHC

In March 2010 the LHC started collecting data at a centre-of-mass energy significantly beyond that of the Tevatron. Firstly approximately 1 fb^{-1} of data are foreseen to be collected at a centre-of-mass energy of 7 TeV. After a shutdown the LHC will then reach its design centre-of-mass energy of 14 TeV and several years later also its design luminosity of 10^{34} cm^{-2}s^{-1}. The production rates at 7 TeV are reduced by a factor of 2 to 4 with respect to those at 14 TeV, depending on the production mechanism and the Higgs boson mass. The simulation studies from the ATLAS and CMS experiments summarised briefly below were performed for a centre-of-mass energy of 14 TeV and the so-called low-luminosity scenario of 2×10^{33} cm^{-2} s^{-1} [18, 25].

The same general considerations apply as for the searches at the Tevatron. However, the increased luminosity, the increased total cross section (see Figure 6.7b) and the reduced time between bunch crossing of 25 ns put even more stringent requirements on the trigger. Again, only a small portion of the produced signal yield can be searched for. For Higgs boson masses close to the LEP bound, the inclusive searches for the $H \to \gamma\gamma$ and the search for $H \to \tau^+\tau^-$ in VBF are the most promising. For intermediate and large Higgs boson masses, the search will be performed for $H \to W^+W^- \to l^+\nu l^{-\prime}\bar{\nu}$ in the inclusive mode and in the VBF topology, and for $H \to ZZ \to l^+l^-l^{+\prime}l^{-\prime}$ inclusively. The signal rate at LHC in these channels is two orders of magnitude higher than at the Tevatron, so that the decay modes $H \to \gamma\gamma, \tau^+\tau^-, ZZ \to l^+l^-l^{+\prime}l^{-\prime}$ yield a sufficiently large signal rate to be accessible at the LHC.

For the search in the $H \to \gamma\gamma$ decay mode, excellent identification of photons and their discrimination from jets and an excellent mass resolution for diphoton invariant masses are needed. The mass spectra expected for representative Higgs boson masses at CMS are shown in Figure 6.10a.

In the decay $H \to \tau^+\tau^-$, due to the presence of three or four neutrinos, the mass can only be reconstructed in the collinear approximation which assumes that the tau decay products follow the flight direction of the tau lepton. This is a good approximation, as, due to its small mass, the tau lepton receives a large Lorentz boost. In order to achieve a reasonable mass resolution, the Higgs boson needs to be produced with a significant transverse momentum. At Born level, this excludes production in gluon fusion, where the Higgs boson is produced with vanishing transverse momentum. In VBF production, the Higgs boson receives a large transverse momentum due to the presence of two additional jets in the final state. These so-called *tagging jets* are produced with a large rapidity difference and a large invariant dijet mass, and there is no hadronic activity between them. These characteristics of the signal topology are exploited in the event selection. The expected ditau invariant mass spectrum at ATLAS is shown in Figure 6.10b.

For the associated production with W or Z bosons, the cross section is only enhanced by one order of magnitude when comparing LHC to the Tevatron. In contrast to the Tevatron, those topologies are not considered to contribute significantly to the discovery potential at LHC, because the signal-to-background ratios are small

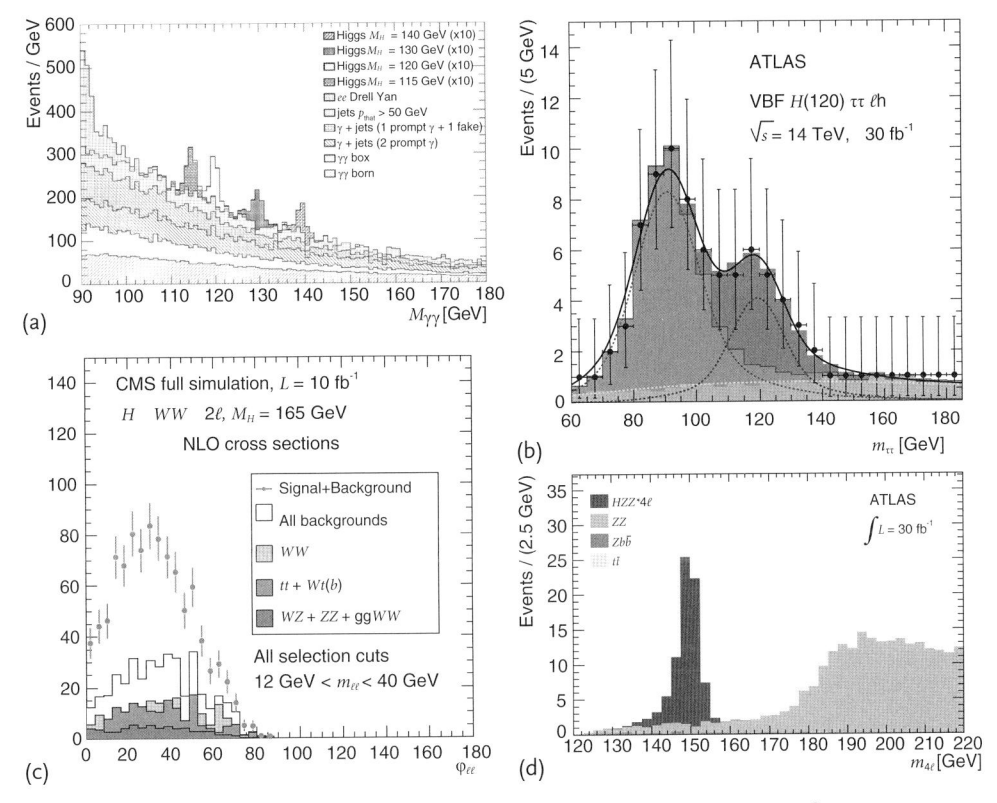

6.6 Prospects for Higgs Boson Searches at the LHC

Figure 6.10 (a) expected mass distribution after collecting $10\,\text{fb}^{-1}$ in the $H \to \gamma\gamma$ decay mode at CMS (adapted from [18]); (b) expected mass distribution after collecting $30\,\text{fb}^{-1}$ in the $H \to \tau^+\tau^-$ decay mode produced in VBF at ATLAS (adapted from [25]); (c) expected distribution of the azimuthal opening angle between the leptons after collecting $10\,\text{fb}^{-1}$ in the $H \to W^+W^- \to l^+\nu l^{-\prime}\overline{\nu}$ decay mode at CMS (adapted from [18]); (d) expected mass distribution after collecting $30\,\text{fb}^{-1}$ in the $H \to ZZ \to l^+l^-l^{+\prime}l^{-\prime}$ decay mode at ATLAS (adapted from [25]).

and the background estimates are not well under control. The same statements also hold for the search for the associated production with a pair of top quarks $t\bar{t}H$ with subsequent decay $H \to b\bar{b}$ [18, 25]. Recently, the search for $W(Z)H$, $H \to b\bar{b}$ in the so-called boosted topology has been proposed [26], and first experimental studies look promising [27].

For masses above 130 GeV, the decay modes into massive weak gauge bosons dominate the discovery potential. The decay mode $H \to ZZ \to l^+l^-l^{+\prime}l^{-\prime}$ is considered to be the golden channel at the LHC. An excellent identification of isolated leptons and very good resolution for the four-lepton invariant mass yield a good signal-to-background ratio (see Figure 6.10d).

At $M_H \approx 2M_W$, the branching ratio into a pair of Z bosons is suppressed; hence, the search for $H \to W^+W^- \to l^+\nu l^{-\prime}\overline{\nu}$ complements the search. The selection strategy is the same as at the Tevatron. The discriminating observable, the

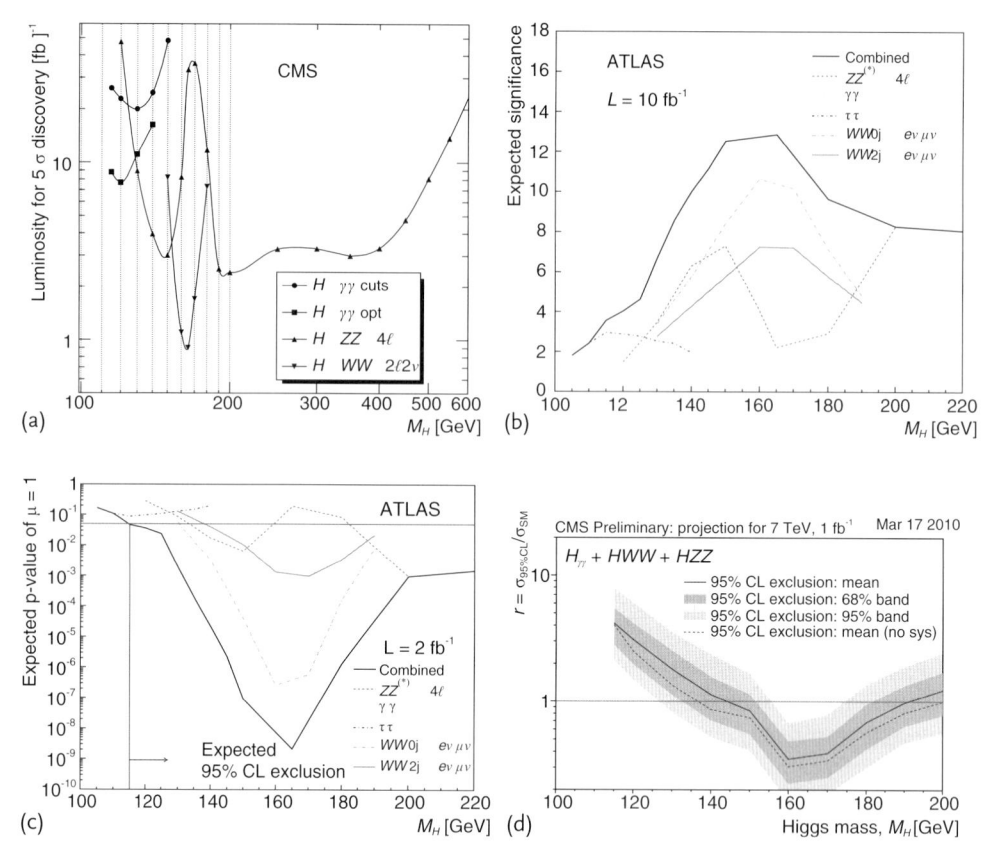

Figure 6.11 (a) luminosity required for discovery at 14 TeV with CMS (adapted from [18]); (b) expected significance with 10 fb^{-1} at 14 TeV with ATLAS (adapted from [25]); (c) expected exclusion with 2 fb^{-1} at 14 TeV with ATLAS (adapted from [25]); (d) expected exclusion with 1 fb^{-1} at 7 TeV with CMS (adapted from [28]).

azimuthal opening angle between the final-state leptons, is shown in Figure 6.10c for CMS.

The overall sensitivity for the discovery at 14 TeV with CMS and ATLAS is summarised in Figure 6.11a,b. With a dataset of 30 fb^{-1}, the whole allowed mass range in the SM can be covered. The sensitivity for discovery starts after collecting $\approx$ 1 fb^{-1} at a centre-of-mass energy of 14 TeV. The most challenging mass range is the one adjacent to the LEP exclusion, where a large integrated luminosity and the combination of several channels are needed.

With a dataset of 2 fb^{-1} ATLAS expects to exclude all Higgs boson mass hypotheses from the LEP limit to 460 GeV (see Figure 6.11c).

CMS has also evaluated the sensitivity for Higgs boson searches at a centre-of-mass energy of 7 TeV for an integrated luminosity of 1 fb^{-1}, by rescaling the findings at 14 TeV with the ratio of the cross sections at the two collision energies [28]. In the $H \to W^+W^- \to l^+\nu l^{-\prime}\bar{\nu}$ decay mode, a sensitivity for discovery in the mass range between 160 and 170 GeV is expected. Exclusion at 95% confidence level is expected in the mass range between 150 and 185 GeV. A combination with the decay modes $H \to \gamma\gamma$ and $H \to ZZ \to l^+l^-l'^+l'^-$ extends the mass range for exclusion of the SM Higgs boson hypothesis to the interval from 145 to 190 GeV (see Figure 6.11d).

6.7
Implications of Observation or Exclusion

If a Standard Model-like Higgs boson is realised in nature, it will be discovered at the LHC over the whole theoretically allowed mass range. At least one Higgs boson will be observable in the case of the CP-conserving Minimal Supersymmetric Standard Model (MSSM) [18, 25]. However, there are also extensions of the SM, for example the CP-violating MSSM, Next-to-MSSM (NMSSM), HEIDI-Models (see e.g. [29] for those models and references therein), where current LHC Monte Carlo studies are not sufficient to establish a "no lose" situation [29, 30]. Difficult scenarios arise when large branching ratios of a Higgs boson to invisible particles, to gluons or light flavoured quarks occur or when the resonance gets very broad [29]. Hence the situation may occur that the whole SM mass range is excluded, but that nevertheless Higgs bosons with different properties are realised in nature.

If a new particle is observed, the next questions will be whether this particle is the messenger of the mechanism responsible for electroweak symmetry breaking, whether it is realised in the way the Standard Model predicts, or whether it belongs to an extended Higgs boson sector.

In the Standard Model, the mass of the Higgs boson fixes its properties completely. At the LHC, the mass can best be measured in the decays $H \to \gamma\gamma$ and $H \to ZZ^{(*)} \to l^+l^-l'^+l'^-$. The expected statistical uncertainty in CMS after collecting 30 fb^{-1} corresponds to 0.1% for masses below 200 GeV and degrades to 1% for a mass of 500 GeV [18]. Ultimately, the precision of the mass determination will be limited by the knowledge of the lepton and photon energy scales. After collecting 300 fb^{-1}, a precision of 0.1% for Higgs boson masses below 400 GeV is expected in ATLAS, which degrades to 1% at a mass of 600 GeV [31].

Under the assumption that the observed particle is a Higgs boson, the measured value of the mass can be compared to the range of theoretically allowed values in the SM and the indirect determination from electroweak precision observables. If the SM is valid to the Planck scale, the mass of the Higgs boson should be in the range 130 to 190 GeV. The observation of a value significantly outside this corridor implies that the SM is not valid up to the Planck scale. The measured mass determines the energy scale at which the Standard Model needs to be extended by new physics.

A comparison of the direct measurement of the Higgs boson mass with the indirect determination in the SM electroweak fit is a crucial test of the consistency of the model. For a Higgs mass of 120 GeV and assuming an improvement in the accuracy of the determinations of the masses of the top quark and the W boson from currently 1.2 GeV and 24 MeV to 1.0 GeV and 15 MeV, respectively, a precision of $\pm^{45}_{35}$ GeV at 68% confidence level can be achieved. This can be improved by a few GeV if the uncertainty on $\Delta\alpha^5_{had}(M_Z^2)$ can be decreased by a factor of 3.

An important step to establish the Higgs mechanism will be to measure the couplings of the observed Higgs boson candidate to other particles and finally the Higgs self-coupling. At the LHC only products of production cross section times branching ratios in ≈ 10 final states can be extracted from the data directly. As the total decay width cannot be determined for masses below 200 GeV, no determination of absolute couplings is possible without further theory assumptions. Under the assumption that only one resonance is produced, the ratio of partial decay widths can be extracted. On the basis of older ATLAS MC studies using leading-order cross sections, it is expected that the ratios Γ_τ/Γ_W, Γ_γ/Γ_W and Γ_Z/Γ_W can be extracted with a precision of 50 to 10% depending on which ratio, mass and integrated luminosity are considered (see [32] and references therein). This will probably already allow the discrimination between some extended Higgs sectors and the one of the SM. At a future e^+e^- linear collider, the complete profile of the Higgs boson, including absolute values of the couplings, can be determined with high precision, and hopefully also the Higgs self-coupling can be measured (see [32] and references therein). This will allow more stringent tests of the SM and its discrimination from SM extensions. The key advantage of a lepton collider for the investigation of the Higgs boson profile is that the Higgs boson can be observed in a model-independent way by analysing the recoil mass spectrum of the two final-state leptons in the process $e^+e^- \to ZH \to l^+l^-H$.

If no new particle compatible with the expected characteristics of a Higgs boson is observed and also no Higgs bosons are realised in nature, another mechanism is needed to restore unitarity at high energies. Signatures of these models are either new resonances or anomalous couplings in the scattering of longitudinally polarised massive weak gauge bosons. Such effects will be observable at the LHC and a future e^+e^- linear collider, but might require larger integrated luminosities.

References

1 Higgs, P.W. (1964) Broken symmetries, massless particles and gauge fields, *Phys. Lett.*, **12**, 132, doi:10.1016/0031-9163(64)91136-9.

2 Higgs, P.W. (1964) Broken symmetries and the masses of gauge bosons, *Phys. Rev. Lett.*, **13**, 508, doi:10.1103/PhysRevLett.13.508.

3 Englert, F. and Brout, R. (1964) Broken Symmetry and the Mass of Gauge Vector Mesons, *Phys. Rev. Lett.*, **13**, 321, doi:10.1103/PhysRevLett.13.321.

4 Guralnik, G.S., Hagen, C.R., and Kibble, T.W.B. (1964) Global conservation laws and massless particles, *Phys. Rev. Lett.*, **13**, 585, doi:10.1103/PhysRevLett.13.585.

5 Veltman, M.J.G. (1968) Perturbation theory of massive Yang-Mills fields, *Nucl. Phys. B*, **7**, 637, doi:10.1016/0550-3213(68)90197-1.

6 't Hooft, G. (1971) Renormalization of massless Yang-Mills fields, *Nucl. Phys. B*, **33**, 173, doi:10.1016/0550-3213(71)90395-6.

7. 't Hooft, G. (1971) Renormalizable Lagrangians for massive Yang-Mills fields, *Nucl. Phys. B*, **35**, 167, doi:10.1016/0550-3213(71)90139-8.
8. 't Hooft, G. and Veltman, M.J.G. (1972) Regularization and renormalization of gauge fields, *Nucl. Phys. B*, **44**, 189, doi:10.1016/0550-3213(72)90279-9.
9. 't Hooft, G. and Veltman, M.J.G. (1972) Combinatorics of gauge fields, *Nucl. Phys. B*, **50**, 318, doi:10.1016/S0550-3213(72)80021-X.
10. Spira, M. (1998) QCD effects in Higgs physics, *Fortsch. Phys.*, **46**, 203, http://arxiv.org/abs/hep-ph/9705337.
11. Djouadi, A. (2008) The anatomy of electroweak symmetry breaking. Tome I: The Higgs boson in the Standard Model, *Phys. Rept.*, **457**, 1, doi:10.1016/j.physrep.2007.10.004.
12. Hambye, T. and Riesselmann, K. (1997) Matching conditions and Higgs boson mass upper bounds re-examined, *Phys. Rev. D*, **55**, 7255, doi:10.1103/PhysRevD.55.7255.
13. Djouadi, A., Kalinowski, J., and Spira, M. (1998) HDECAY: A program for Higgs boson decays in the standard model and its supersymmetric extension, *Comput. Phys. Commun.*, **108**, 56, doi:10.1016/S0010-4655(97)00123-9.
14. Pak, A., Rogal, M., and Steinhauser, M. (2010) Finite top quark mass effects in NNLO Higgs boson production at LHC, *JHEP*, **02**, 025, doi:10.1007/JHEP02(2010)025.
15. Harlander, R.V., Mantler, H., Marzani, S., and Ozeren, K.J. (2010) Higgs production in gluon fusion at next-to-next-to-leading order QCD for finite top mass, *Eur. Phys. J. C*, **66**, 359–372, doi:10.1140/epjc/s10052-010-1258-x.
16. Spira, M. (1998) Higgs Boson Production and Decay at the Tevatron, http://arxiv.org/abs/hep-ph/9810289.
17. Higgs Working Group Collab., Carena, M.S. *et al.* (2000) Report of the Tevatron Higgs Working Group of the Tevatron Run 2 SUSY/Higgs Workshop, http://arxiv.org/abs/hep-ph/0010338.
18. CMS Collab., Bayatian, G.L. *et al.* (2007) CMS Technical Design Report, Volume II: Physics Performance, *J. Phys. G*, **34**, 995, doi:10.1088/0954-3899/34/6/S01.
19. Yost, G.P. *et al.* (1988) Review of particle properties: Particle Data Group, *Phys. Lett. B*, **204**, 1, doi:10.1016/0370-2693(88)90505-9.
20. Sjostrand, T., Mrenna, S., and Skands, P.Z. (2006) PYTHIA 6.4 physics and manual, *JHEP*, **05**, 026, http://arxiv.org/abs/hep-ph/0603175.
21. LEP Working Group for Higgs boson searches and ALEPH, DELPHI, L3 and OPAL Collab., Barate, R. *et al.* (2003) Search for the Standard Model Higgs boson at LEP, *Phys. Lett. B*, **565**, 61, doi:10.1016/S0370-2693(03)00614-2.
22. TEVNPH Working Group for the CDF and DØ Collab. (2009) Combined CDF and DØ Upper Limits on Standard Model Higgs Boson Production with 2.1–5.4 fb^{-1} of Data, http://arxiv.org/abs/0911.3930.
23. CDF and DØ Collab., Aaltonen, T. *et al.* (2010) Combination of Tevatron searches for the Standard Model Higgs boson in the W^+W^- decay mode, *Phys. Rev. Lett.*, **104**, 061802, doi:10.1103/PhysRevLett.104.061802.
24. Peters, K. (2009) Higgs searches, http://arxiv.org/abs/0911.1469.
25. ATLAS Collab., Aad, G. *et al.* (2009) Expected Performance of the ATLAS Experiment – Detector, Trigger and Physics, http://arxiv.org/abs/0901.0512, CERN-OPEN-2008-020.
26. Butterworth, J.M., Davison, A.R., Rubin, M., and Salam, G.P. (2008) Jet Substructure as a new Higgs-search channel at the Large Hadron Collider, *Phys. Rev. Lett.*, **100**, 242001, doi:10.1103/PhysRevLett.100.242001.
27. ATLAS Collab. (2009), ATLAS Sensitivity to the Standard Model Higgs in the HW and HZ Channels at High Transverse Momenta, http://cdsweb.cern.ch/record/1201444, ATL-PHYS-PUB-2009-088.
28. CMS Collab. (2010), The CMS physics reach for searches at 7 TeV, http://cdsweb.cern.ch/record/1264099, CMS-NOTE-2010-008.
29. Accomando, E. *et al.* (2006) Workshop on CP Studies and Non-Standard Higgs

Physics, http://arxiv.org/abs/hep-ph/0608079.

30 Adam, N.E. *et al.* (2008) Higgs Working Group Summary Report, http://arxiv.org/abs/0803.1154.

31 ATLAS Collab. (1999) ATLAS Detector and Physics Performance. Technical Design Report, Vol. 2, http://atlas.web.cern.ch/Atlas/GROUPS/PHYSICS/TDR/access.html, CERN-LHCC-99-15.

32 De Roeck, A. *et al.* (2010) From the LHC to future colliders, *Eur. Phys. J. C*, **66**, 525, doi:10.1140/epjc/s10052-010-1244-3.

7
Supersymmetry

Herbert Dreiner and Peter Wienemann

7.1
Introduction

The Standard Model (SM) of particle physics has a particle content of three families of quarks and leptons, one complex Higgs doublet and the gauge symmetry

$$G_{SM} = SU(3)_c \times SU(2)_L \times U(1)_Y . \qquad (7.1)$$

This is an internal symmetry which does not affect the space-time properties of the particles. Furthermore, the SM is based on the *external* Poincaré symmetry. Given the particle content and the symmetries the interactions are fixed. In particular the Yukawa interactions, which give masses to the quarks and leptons, are given by

$$\mathcal{L}_{\text{Yuk}} = (h_e)_{ij} \bar{\ell}_{Li} \phi_H e_{Rj} + (h_d)_{ij} \bar{q}_{Li} \phi_H d_{Rj} + (h_u)_{ij} \bar{q}_{Li} \tilde{\phi}_H u_{Rj} , \qquad (7.2)$$

where $i, j, k = 1, 2, 3$ are generation indices (cf. 2.59) which uses a different notation). These (renormalisable!) interactions automatically satisfy lepton and baryon number conservation.

The basic idea of the supersymmetric Standard Model (SSM) is to extend the SM external symmetry to the *super Poincaré algebra* [1]. In order to have an action which is invariant under this extended set of symmetry transformations, the particle content must be expanded to include new, supersymmetric (SUSY) particles. Furthermore, new interactions arise. It is the purpose of this chapter to first introduce the new particle content and its possible interactions in Sections 7.2–7.5. Then, in Section 7.6, we discuss in detail how SUSY searches are performed and how SUSY parameters could be measured at the LHC and ILC.

7.2
Supersymmetry Transformations and Fields

The basic idea of supersymmetry is that the action is invariant under transformations relating fermions and bosons by a spinorial generator, Q, which has spin

Physics at the Terascale, First Edition. Edited by Ian C. Brock and Thomas Schörner-Sadenius.
© 2011 WILEY-VCH Verlag GmbH & Co. KGaA, Weinheim.
Published 2011 by WILEY-VCH Verlag GmbH & Co. KGaA.

1/2 [2],

$$|\text{FERMION}\rangle \xleftrightarrow{Q} |\text{BOSON}\rangle . \tag{7.3}$$

These field pairs are called *superpartners*. They coincide in all quantum numbers, except spin. The infinitesimal field transformations are given by

$$\text{Scalar (spin 0):} \quad \phi \longrightarrow \phi' = \phi + \Delta\phi , \tag{7.4}$$

$$\text{Spinor (spin 1/2):} \quad \psi \longrightarrow \psi' = \psi + \Delta\psi , \tag{7.5}$$

where the shifts in the fields are given by

$$\Delta\phi = \epsilon \cdot |\text{FERMION}\rangle , \quad \Delta\psi = \epsilon \cdot |\text{BOSON}\rangle . \tag{7.6}$$

Here, ϵ is a constant (space-time independent) spinor in global supersymmetry. In order to obtain a SUSY invariant action, each SM particle must be given a superpartner, hence doubling the particle content. Representative examples are given in Table 7.1. Overall, for each chirality of a SM fermion there is a separate superpartner, for example for the left-handed electron a complex scalar $\tilde{e}_L$ exists, which is called the left-handed selectron:

$$\left(e_L^-, e_R^-\right) \longleftrightarrow \left(\tilde{e}_L^-, \tilde{e}_R^-\right) . \tag{7.7}$$

Note that the superpartner of the electron can not be spin 1 instead of spin 0. A superfield which contains a spin 1 and a spin 1/2 is necessarily in a real representation. The electron in the SM is, however, necessarily in a complex representation, due to its $SU(2)$ chiral properties.

Since the quantum numbers (except spin) are identical, superpartners must also have equal masses. Since no 511 keV scalar partner of the electron has been observed, supersymmetry must be broken. From a low energy point of view this

Table 7.1 Examples for superpartners. The SSM requires at least two Higgs doublets. The other charged and neutral leptons, as well as the quarks, have analogous superpartners which are called sleptons and squarks.

SM field		Superpartner	Name
$e_{L,R}^-$ (spin $= \frac{1}{2}$)	$\longleftrightarrow$	$\tilde{e}_{L,R}^-(s=0)$	scalar electron (selectron)
$t_{L,R}(s=\frac{1}{2})$	$\longleftrightarrow$	$\tilde{t}_{L,R}(s=0)$	scalar top (stop)
$W^\pm(s=1)$	$\longleftrightarrow$	$\tilde{W}^\pm(s=\frac{1}{2})$	wino
$H^\pm(s=0)$	$\longleftrightarrow$	$\tilde{H}^\pm(s=\frac{1}{2})$	higgsino
$\gamma, Z^0(s=1)$	$\longleftrightarrow$	$\tilde{\gamma}, \tilde{Z}^0(s=\frac{1}{2})$	photino, zino
$H^0, h^0(s=0)$	$\longleftrightarrow$	$\tilde{H}^0, \tilde{h}^0(s=\frac{1}{2})$	higgsino
$g_{a=1,\ldots,8}(s=1)$	$\longleftrightarrow$	$\tilde{g}_a(s=\frac{1}{2})$	gluino

corresponds to adding mass terms for the SUSY particles in the Lagrangian. Collider experiments have hitherto set lower bounds on these masses of the order of 100 GeV.

However, the fields listed in Table 7.1 are not necessarily the mass eigenstates of the theory which would be observed in an experiment. After electroweak symmetry breaking the only good gauge symmetries are $SU(3)_c$ and $U(1)_{em}$. Now, for example, the SUSY partner of the $W^\pm$ boson, the wino, $\tilde{W}^\pm$, and the SUSY partner of the scalar charged Higgs boson, $H^\pm$, have the same good quantum numbers, they are electrically charged colour singlets and have spin 1/2. Thus, they can mix and in fact they do. The new mass eigenstates are called charginos and denoted $\tilde{\chi}^\pm_{i=1,2}$. Note that the electron also has the same gauge quantum numbers. We shall return to its possible mixing later. Similarly the photino, zino and two neutral higgsinos mix to form the neutralinos: $\tilde{\chi}^0_{1,2,3,4}$, ordered according to mass. The case of the neutrino will be discussed later.

For an introductory motivation on supersymmetry, in particular the gauge hierarchy problem, see for example [3].

7.3
Superfields and Superpotential

In the mathematical construction of supersymmetric invariant Lagrangians it is convenient to introduce *superfields*. Loosely speaking these are combined functions. For example for the right-handed $SU(2)$ singlet electron we have

$$\Phi_{e_R} \equiv E^c \sim \phi^*_{e_R} + \epsilon \psi^c_{e_R} . \tag{7.8}$$

All non-gauge interactions can then be described by the holomorphic function $W(\Phi_i)$ (not $\Phi^\dagger$!), which is called the *superpotential*. The index i runs over all the possible chiral superfields consisting of (spin 0, spin 1/2) pairs. The other chiral superfields are also denoted by upper-case letters: the lepton-doublet and quark-doublet superfields, $L = (N, E)_L$, $Q = (U, D)_L$, and the quark singlet superfields, D^c, U^c. In addition there are two Higgs doublet superfields, $H_{1,2}$ [19].

Employing the symmetries of the model, G_{SM} and supersymmetry, one can write down the most general renormalizable superpotential interactions,

$$\begin{aligned} W_{TOT} &= W_{P_6} + W_{\not{L}} + W_{\not{B}}, \\ W_{P_6} &= (h_e)_{ij} L_i H_1 E^c_j + (h_d)_{ij} Q_i H_1 D^c_j + (h_u)_{ij} Q_i H_2 U^c_j + \mu H_1 H_2 , \\ W_{\not{L}} &= \lambda_{ijk} L_i L_j E^c_k + \lambda'_{ijk} L_i Q_j D^c_k + \kappa_i L_i H_2 , \\ W_{\not{B}} &= \lambda''_{ijk} U^c_i D^c_j D^c_k . \end{aligned} \tag{7.9}$$

19) There are at least two Higgs doublets required in supersymmetry, since a single chiral higgsino doublet cannot be cancelled in the anomaly. Furthermore, due to the holomorphic nature of the superpotential, one needs two Higgs doublets to give masses to both down-type and up-type quarks.

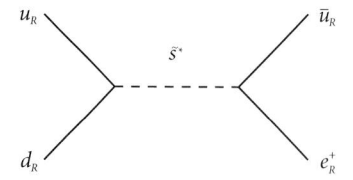

Figure 7.1 Parton-level diagram for the proton decay $p \to e^+ \pi^0$. The spectator u quark is not shown. The vertex on the left arises from the operator $\lambda''_{112}\overline{UDS}$, the vertex on the right from the operator $\lambda'_{112}L_1Q_1\overline{S}$. The exchanged particle is a strange squark.

Here $i, j, k \in 1, 2, 3$ are generation indices; h and λ denote dimensionless couplings. μ, κ_i have mass dimension 1. The terms in W_{P_6} are the supersymmetric analogues of the SM Yukawa interactions in (7.2). The terms in $W_{\not L}$ and $W_{\not B}$ are new. The $W_{\not L}$ terms violate lepton number, the $W_{\not B}$ terms violate baryon number. Together they lead to an unacceptable level of proton decay, as depicted in Figure 7.1. This process is only suppressed by $M_{\tilde{d}_j}^4 = \mathcal{O}[(100\,\text{GeV})^4]$, where $M_{\tilde{d}_j}$ denotes the mass of the d squark. In order to be consistent with the strict experimental bounds it is required to have

$$\lambda'_{i1j} \cdot \lambda''_{11j} < 2 \times 10^{-27} \left(\frac{M_{\tilde{d}_j}}{100\,\text{GeV}}\right)^2, \quad i = 1, 2, \quad j \neq 1. \tag{7.10}$$

Clearly this can only be satisfied if either $\lambda'_{i1j} = 0$ or $\lambda''_{11j} = 0$. Various symmetries or theories have been considered to achieve this: grand unified symmetries, string-theory models and global continuous symmetries. We shall focus here on discrete symmetries.

7.4
Discrete Symmetries

The most widely used symmetry is *R-parity*, which is a discrete multiplicative symmetry: $R_p = (-1)^{L+3B+2S}$. Here L is the lepton number, B is the baryon number and S is the spin of the particle. Thus, all SM particles have $R_p = +1$ and their SUSY partners have $R_p = -1$. R_p leads to the superpotential $W_{\text{TOT}} = W_{P_6}$ for which the proton is stable. This is equivalently achieved by *matter parity*,

$$\begin{aligned}(L, E^c, Q, U^c, D^c) &\longrightarrow -(L, E^c, Q, U^c, D^c), \\ (H_1, H_2) &\longrightarrow +(H_1, H_2).\end{aligned} \tag{7.11}$$

A stable proton can also be achieved by *baryon parity*

$$\begin{aligned}(Q, U^c, D^c) &\longrightarrow -(Q, U^c, D^c), \\ (L, E^c, H_1, H_2) &\longrightarrow +(L, E^c, H_1, H_2),\end{aligned} \tag{7.12}$$

leading to $W_{\text{TOT}} = W_{P_6} + W_{\not{L}}$, or *lepton parity*,

$$(L, E^c) \longrightarrow -(L, E^c),$$
$$(Q, U^c, D^c, H_1, H_2) \longrightarrow +(Q, U^c, D^c, H_1, H_2), \quad (7.13)$$

leading to $W_{\text{TOT}} = W_{P_6} + W_{\not{B}}$, or the $\mathbb{Z}_3$-symmetry *baryon triality*, B_3,

$$\Psi_j \to e^{i a_j 2\pi/3} \Psi_j, \quad \begin{array}{c|cccccc} & Q & U^c & D^c & L & E^c & H_d & H_u \\ \hline a_j & 0 & 2 & 1 & 2 & 2 & 2 & 1 \end{array},$$

which also leads to $W_{\text{TOT}} = W_{P_6} + W_{\not{L}}$ and a stable proton.

From a theoretical point of view, how can these symmetries be distinguished? It is expected that discrete symmetries are violated by quantum gravity effects, unless they are the remnant of a spontaneously broken gauge symmetry [4]. Such special remnant symmetries are called *discrete gauge symmetries*. If the original $U(1)$ gauge symmetry is anomaly-free, then the remnant discrete symmetry must satisfy certain conditions [5, 6]. If it does, it is called *discrete gauge anomaly-free*. It was found that with the supersymmetric SM particle content, only two $\mathbb{Z}_2$ or $\mathbb{Z}_3$ symmetries are discrete gauge anomaly-free: R-parity and baryon triality. Baryon triality has the advantage of prohibiting dangerous dimension-five (non-renormalisable) proton-decay operators [5–7]. There is one further higher discrete gauge anomaly-free symmetry: *proton hexality*, P_6, which gives $W = W_{P_6}$ without allowing for the dangerous dimension-five operators [8]. We thus find P_6 and B_3 equally well motivated from a theory point of view. The P_6 model, often also called the R-parity conserving model, is usually denoted the *MSSM*. The B_3 model is also often denoted the R-parity violating model. The resulting phenomenology of these two models is very different.

7.5
R-Parity Conservation (P_6 Model) vs. R-Parity Violation (B_3 Model)

We summarised the model building and phenomenological differences of the P_6 and the B_3 models in Table 7.2. Before discussing the details of these models, let us briefly introduce the *minimal supergravity* (mSUGRA) model of SUSY. If we promote SUSY to a local symmetry, then we must have for our transformation parameter $\epsilon \to \epsilon(x, t)$, cf. (7.6). When constructing the full theory on this basis, it turns out that SUSY then automatically includes local coordinate transformations as in general relativity. The theory contains gravity! There is thus an intimate connection between SUSY and gravity. This theory is called *supergravity* (SUGRA). Furthermore, as mentioned in Section 7.2, SUSY must be broken, if it exists. Phenomenologically this is resolved by adding heavy mass terms for all the newly introduced particles shown in Table 7.1. These are then new free parameters in the theory. In the mSUGRA model these parameters are universal at the unification scale of about 10^{16} GeV. That is at this scale all the new heavy squarks and all the

Table 7.2 Summary of the main features of the two supersymmetric Standard Models.

P_6 SSM	B_3 SSM
+ Supersymmetric SM fields − Extra discrete symmetry: P_6	+ Supersymmetric SM fields − Extra discrete symmetry: B_3
Baryon and lepton number conserved + Natural dark-matter candidate: $\tilde{\chi}_1^0$ − Must add field(s) ν_R and mass scale $M_{\text{Maj.}}$	Lepton number violated − Must add dark-matter candidate, e.g. axino + Naturally light neutrino masses
Produce SUSY particles only in pairs LSP is lightest neutralino LSP is stable Typically missing E_T signature	Can singly produce SUSY particles (Almost) Any SUSY particle can be the LSP LSP decays, typically in detector Several charged leptons signature

new heavy sleptons have one *universal mass* which is usually denoted M_0. Similarly all the SUSY partners of the gauge bosons have a universal mass at this scale, denoted $M_{1/2}$. In addition there is a *trilinear scalar interaction* which must be introduced and the universal parameter is A_0. In all SUSY theories the Higgs sector is parametrised by $\tan\beta$ (the ratio of the two Higgs vacuum expectation values) and the Higgs mixing parameter μ. In mSUGRA models only the sign of μ enters. We thus have the five parameters

$$M_0, M_{1/2}, A_0, \tan\beta, \text{sgn}(\mu) . \tag{7.14}$$

We now discuss our two models, P_6 and B_3 in some detail.

7.5.1
Model Building

Both the P_6 and B_3 models have the minimal SSM field content (Table 7.1) and an additional symmetry to guarantee proton stability.

In the P_6 model, lepton and baryon number are conserved. Because of R-parity, the *lightest SUSY particle* (LSP) is stable. Cosmologically it must then be electrically neutral [9]. The sneutrino is experimentally excluded [10], leaving the lightest neutralino, $\tilde{\chi}_1^0$. In large portions of the mSUGRA parameter space this is in fact the case. A typical SUSY spectrum in mSUGRA models is shown in Table 7.3. Note that the strongly interacting particles are typically heavier than the only weakly interacting particles.

It is well known that at least two neutrinos of the SM must be massive to explain the observed oscillation data [11]. In the P_6 SSM this requires the implementation of the see-saw mechanism [12, 13]. Thus, the model must be extended by one or more singlet fields (right-handed neutrinos) as well as a new very high Majorana neutrino mass scale, $M_{\text{Maj}} = \mathcal{O}(10^{10}\text{ GeV})$.

In the B_3 model, lepton number is violated through the additional operators in $W_{\not{L}}$. These directly lead to massive neutrinos [14], which are naturally light [15]. No new fields or energy scales are needed. However, due to these operators the LSP is

Table 7.3 Masses of the supersymmetric particles for the SPS1a spectrum. The masses increase from left to right and from top to bottom. Note that the sleptons are lighter than the squarks and the gluinos. The lightest neutralino is the LSP. At the end, the masses of the five Higgs bosons are given.

Particle	Mass (GeV)	Particle	Mass (GeV)	Particle	Mass (GeV)
$\tilde{\chi}_1^0$	97	$\tilde{\tau}_1$	137	$\tilde{e}_R^-, \tilde{\mu}_R^-$	146
$\tilde{\chi}_2^0$	181	$\tilde{\chi}_1^\pm$	182	$\tilde{\nu}_{e,\mu,\tau}$	188
$\tilde{e}_L^-, \tilde{\mu}_L^-$	206	$\tilde{\tau}_2$	210	$\tilde{\chi}_3^0$	363
$\tilde{\chi}_2^\pm$	378	$\tilde{\chi}_4^0$	380	$\tilde{t}_1$	400
$\tilde{b}_{1,2}$	518; 550	$\tilde{u}_1, \tilde{d}_1$	552	$\tilde{c}_1, \tilde{s}_1$	552
$\tilde{u}_2, \tilde{c}_2$	567	$\tilde{d}_2, \tilde{s}_2$	575	$\tilde{t}_2$	591
$\tilde{g}$	610	h^0, H^0	110; 397	$A^0, H^\pm$	397; 405

unstable and is not suitable as a dark-matter candidate. Another particle, such as the *axino*, must be introduced [16]. However, the axion/axino might be required in both models anyway to solve the strong CP problem [17].

Since the LSP is no longer constrained to be the $\tilde{\chi}_1^0$, in principle any SUSY particle can be the LSP. In B_3 mSUGRA models the following LSPs have been found [18]:

$$\tilde{\chi}_1^0, \tilde{\tau}_1, \tilde{e}_R, \tilde{\mu}_R, \tilde{\nu}_{i=1,2,3}, \tilde{b}_1, \tilde{t}_1, \tilde{s}_R, \tilde{d}_R , \tag{7.15}$$

which all lead to very distinctive phenomenology.

Furthermore, since lepton number is violated, the charged leptons ($e^\pm, \mu^\pm, \tau^\pm$) mix with the previous charginos, to form five mass eigenstates. Similarly the neutrinos mix with the neutralinos. The latter is actually the origin of one of the neutrino masses. The others arise from loop contributions. This lepton-gaugino mixing can also lead to very distinctive signatures.

7.5.2
Phenomenology

If kinematically accessible, the dominant mechanism of SUSY particle production at the LHC in P_6 models involves pairs of squarks and gluinos via the strong $SU(3)_c$ coupling. At the parton level we can have

$$g + g \to \tilde{g} + \tilde{g}, \ \tilde{q}_{L,R} + \tilde{q}_{L,R}^* , \tag{7.16}$$

$$q + \bar{q} \to \tilde{q}_{L,R} + \tilde{q}_{L,R}^*, \tilde{g} + \tilde{g} , \tag{7.17}$$

$$g + q \to \tilde{q}_{L,R} + \tilde{g} . \tag{7.18}$$

Here, superpartners of antiparticles are denoted by an asterisk. Representative tree-level Feynman diagrams are shown in Figure 7.2.

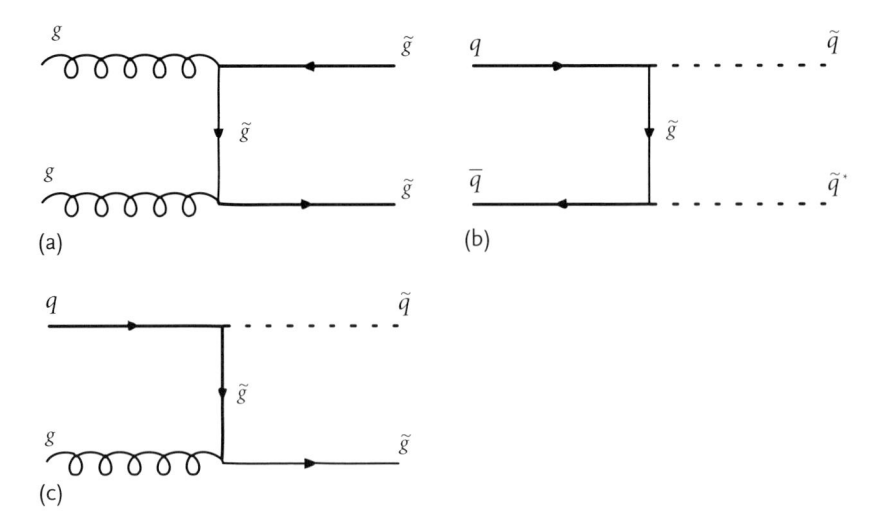

Figure 7.2 Representative *t*-channel diagrams for the parton-level processes: (a) $gg \to \tilde{g}\tilde{g}$; (b) $q\bar{q} \to \tilde{q}\tilde{q}^*$; and (c) $qg \to \tilde{q}\tilde{g}$. There are further *s*-channel diagrams as well.

The squarks and gluinos have short lifetimes and cascade decay to the $\tilde{\chi}_1^0$ LSP within the detector,

$$\tilde{g} \to q\tilde{q}_L \to qq\tilde{\chi}_2^0 \to qq\ell^\pm \tilde{\ell}_R^\mp \to qq\ell^\pm \ell^\mp \tilde{\chi}_1^0 \,. \tag{7.19}$$

The final state $\tilde{\chi}_1^0$ then results in a significant amount of missing transverse energy ($\not{E}_T$ or MET), which is a typical P_6 SUSY signature.

In the B_3 models, since *R*-parity is violated, SUSY particles can be produced singly and on resonance, for example [19],

$$\text{LQD}^c: \quad u + \bar{d} \to \tilde{\mu}^+ \to \mu^+ + \tilde{\chi}_1^0 \,, \tag{7.20}$$

cf. (7.9). The cross section is proportional to the square of a small coupling ($(\lambda'_{211})^2$ in this case), but the kinematic reach is much higher. Via the same rotated Feynman diagram a neutralino LSP can decay:

$$\text{LQD}^c: \quad \tilde{\chi}_1^0 \to u + \bar{d} + \mu^-, \quad d + \bar{d} + \nu_\mu \,, \tag{7.21}$$

where an additional decay to a neutrino via the same operator is indicated. Note also that due to the Majorana nature of the neutralino the same neutralino can decay to the charge-conjugate final states. The lifetime is approximately given by

$$\tau_{\tilde{\chi}_1^0} \sim \frac{128 \sin^2\theta_W \pi^2}{3\alpha(\lambda'_{211})^2} \frac{M_{\tilde{q}}^4}{M_{\tilde{\chi}_1^0}^5}$$

$$= 2.5 \times 10^{-18}\,\text{s} \left(\frac{0.01}{\lambda'_{211}}\right)^2 \left(\frac{M_{\tilde{q}}}{100\,\text{GeV}}\right)^4 \left(\frac{50\,\text{GeV}}{M_{\tilde{\chi}_1^0}}\right)^5, \tag{7.22}$$

which will typically occur within the detector. Here θ_w is the electroweak mixing angle. However, one should keep in mind that for example the stau or other SUSY particles can be the LSP, see (7.15). These also typically decay in the detector, however with other signatures. In all cases, we no longer have the missing transverse energy signature (or it is diluted in the case of the neutrino decay mode).

7.6 Measuring Supersymmetry

Looking for supersymmetry is one of the key objectives of the present experimental high energy physics programme. Establishing SUSY experimentally proceeds in several distinct steps:

1. look for physics beyond the SM (BSM);
2. test the supersymmetric origin of the observed new physics phenomena;
3. determine the underlying SUSY model and constrain its parameters as precisely as possible.

Because of the rich phenomenology of SUSY, the experimental techniques to look for and test supersymmetry experimentally can vary significantly. In the following we will focus on the *R*-parity conserving or P_6 model discussed above. We also restrict ourselves to mSUGRA-like spectra as shown in Table 7.3. We provide a selection of methods which can be applied at the Large Hadron Collider and a future e^+e^- linear collider (LC). A more detailed overview can be found for example in [20–22].

7.6.1 Searches at the LHC

The LHC has excellent chances to discover BSM physics. In order to be sensitive to a large variety of models, initial search strategies often exploit generic features. With conserved *R*-parity, this involves $\not{E}_T$ due to the escaping $\tilde{\chi}_1^0$ LSP, as discussed above. In addition SUSY is often accompanied by hard jets and leptons, since the production is mostly dominated by strong interactions, see (7.16)–(7.18). For mSUGRA-like mass hierarchies, the squarks and gluinos are heavy and cascade decay to the lighter particles ending in the $\tilde{\chi}_1^0$ LSP, see (7.19).

The hard jets typically originate from the decays of the strongly interacting SUSY particles and leptons mostly come from gaugino or gauge-boson decays. In summary, SUSY events are often characterised by several hard jets, a lot of missing transverse energy and optionally leptons. Search channels are usually categorised by the number of leptons they require. Important cases are no leptons, one lepton, two opposite-sign (OS) leptons, two same-sign (SS) leptons, and tri-leptons. The discovery potential of the different search channels is displayed in Figure 7.3 for the CMS experiment assuming an integrated luminosity of $10\,\text{fb}^{-1}$ at $\sqrt{s} =$

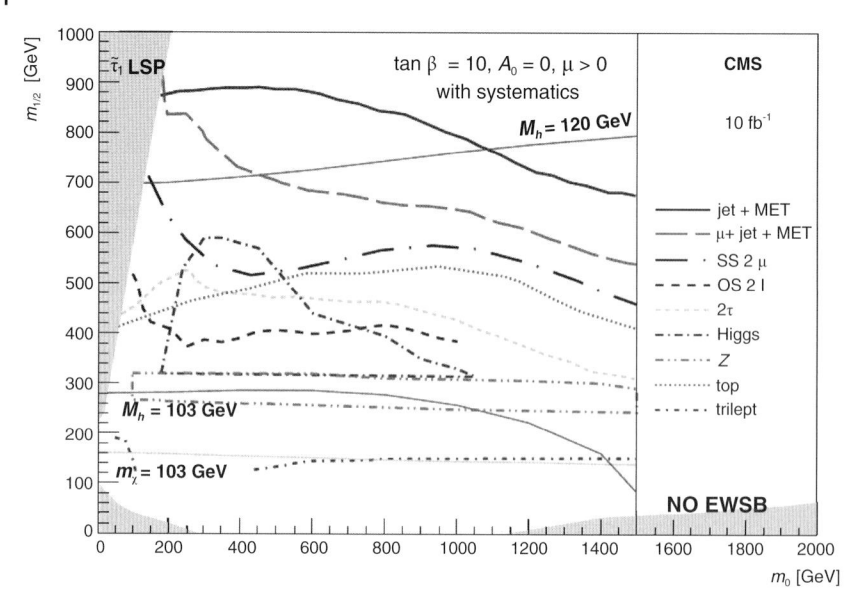

Figure 7.3 Discovery potential of the CMS experiment as a function of the mSUGRA parameters m_0 and $m_{1/2}$. The other free parameters are fixed to $\tan\beta = 10$, $A_0 = 0$ GeV, $\text{sign}(\mu) = +1$. The assumed integrated luminosity is $10\,\text{fb}^{-1}$ at $\sqrt{s} = 14$ TeV (adapted from [21]).

14 TeV [21]. This study has been carried out within the framework of mSUGRA for fixed values of $\tan\beta = 10$, $A_0 = 0$ GeV and $\text{sign}(\mu) = +1$. Parameter points below the shown curves can be discovered with at least 5σ significance. The covered discovery regions correspond to a mass reach of approximately 2 TeV for squarks and gluinos. For $1\,\text{fb}^{-1}$ at $\sqrt{s} = 14$ TeV this reduces to roughly 1.5 TeV [21].

7.6.2
Measuring SUSY Properties at the LHC

After evidence for BSM physics has been found, the objective will be to collect as much information as possible on the properties of the discovered new physics. One important source of information is the underlying mass spectrum.

7.6.2.1 Mass Reconstruction

Because of the escaping $\tilde{\chi}_1^0$ LSPs in events, it is usually impossible to reconstruct mass peaks. This has triggered a lot of work on mass reconstruction techniques during the last 15 years. A comprehensive overview of the field is impossible here. Instead, we focus on two exemplary methods (see [23] for a more complete review).

One possibility to overcome the problem imposed by two missing particles is to consider sufficiently long decay cascades [24–26]. Then there are enough constraints to solve for the unknown masses. An example is again the decay chain (7.19). If we consider for example the dilepton invariant-mass spectrum of the leptons from the $\tilde{\chi}_2^0$ decay, it has a triangular shape with a sharp drop-off at the position

$$m_{\ell\ell}^{\max} = m_{\tilde{\chi}_2^0}\sqrt{1 - m_{\tilde{\ell}}^2/m_{\tilde{\chi}_2^0}^2}\sqrt{1 - m_{\tilde{\chi}_1^0}^2/m_{\tilde{\ell}}^2}\,. \tag{7.23}$$

By including the quark from the squark decay, additional observable invariant mass spectra can be constructed: $m_{q\ell(\text{low})}^{\max}$, $m_{q\ell(\text{high})}^{\max}$, $m_{q\ell\ell}^{\min}$ for events with $m_{\ell\ell}^{\max}/\sqrt{2} \leq m_{\ell\ell} \leq m_{\ell\ell}^{\max}$, $m_{q\ell\ell}^{\max}$. Here max (min) refers to an upper (lower) endpoint of a spectrum. Low (high) means that the lepton is chosen which provides the lower (higher) invariant mass value. The positions of each of these edges depend again on the masses of the involved SUSY masses – analogous to (7.23). If there are at least three sequential two-body decays (as in the decay chain given in (7.19)), measurements of edge positions provide sufficient constraints to solve the system for the relevant SUSY masses.

Even for shorter decay cascades, mass measurements are feasible. Sufficient constraints are obtained by considering pairs of decay chains in an event. The three-body decays of a pair of gluinos are a simple example,

$$pp \to \tilde{g}\tilde{g} \to q\bar{q}\tilde{\chi}_1^0 q\bar{q}\tilde{\chi}_1^0\,. \tag{7.24}$$

Here we have two identical pairs of semi-invisible decays. The preferred quantity to reconstruct masses in this case is the so-called m_{T2} variable (sometimes also referred to as stransverse mass) [27, 28],

$$m_{\text{T2}}\left(m_{q\bar{q}T}^{(1)}, m_{q\bar{q}T}^{(2)}, \boldsymbol{p}_T^{q\bar{q}(1)}, \boldsymbol{p}_T^{q\bar{q}(2)}, \boldsymbol{p}_T; m_\chi\right) = \min_{\boldsymbol{p}_T^{\chi(1)} + \boldsymbol{p}_T^{\chi(2)} = \boldsymbol{p}_T} \left\{\max\left(m_T^{(1)}, m_T^{(2)}\right)\right\}, \tag{7.25}$$

where

$$m_T^2(m_\chi) = m_{q\bar{q}T}^2 + m_\chi^2 + 2\left(E_T^{q\bar{q}} E_T^\chi - \boldsymbol{p}_T^{q\bar{q}} \cdot \boldsymbol{p}_T^\chi\right) \tag{7.26}$$

denotes the square of the transverse mass. The superscripts (1), (2) are used to identify the two different decay branches. The minimisation in (7.25) runs over all possible pairs of $\tilde{\chi}_1^0$ transverse momenta which are consistent with the total missing transverse momentum measured in the event. Strictly speaking m_{T2} is not a variable but a function. Its parameter m_χ represents a mass hypothesis for the a priori unknown $\tilde{\chi}_1^0$ mass. Setting the trial mass m_χ to be equal to $m_{\tilde{\chi}_1^0}$, one obtains the inequality

$$m_{\text{T2}}(m_\chi = m_{\tilde{\chi}_1^0}) \leq m_{\tilde{g}}\,, \tag{7.27}$$

that is, the gluino mass can be determined from the endpoint of the m_{T2} spectrum, if $m_{\tilde{\chi}_1^0}$ is known from other measurements. In the case considered here it is even possible to simultaneously measure $m_{\tilde{g}}$ and $m_{\tilde{\chi}_1^0}$.

Each event of the type given in (7.24) provides a function $m_{\text{T2}}(m_\chi)$. Under certain conditions (for more information see for example [23]) the function max$_\text{events}$ $m_\text{T2}(m_\chi)$ is non-differentiable at the position $m_\chi = m_{\tilde{\chi}_1^0}$, whereas it is smooth elsewhere. Since max$_\text{events}$ $m_\text{T2}(m_{\tilde{\chi}_1^0}) = m_{\tilde{g}}$, it is possible to measure both $m_{\tilde{g}}$ and $m_{\tilde{\chi}_1^0}$ from the position of the "kink". Typical precisions for SUSY mass measurements at the LHC range from 1% to 5%.

7.6.2.2 Spin Measurements

Apart from supersymmetry there are other new physics models which provide very similar LHC signatures, for example Kaluza–Klein excitations in models with extra dimensions [29] and little Higgs models [30]. A discrimination of these models is essential for our understanding. A key difference is the spin of the new particles. If SUSY is the correct model, superpartners must differ by 1/2 in spin.

The conventional way to determine the spin of a particle is to check the angular distribution of decay products in the rest frame of the mother particle. However, in R-parity conserving models, the rest frame of the SUSY particles cannot be reconstructed. Nevertheless, there are several different techniques to check particle spins.

If the mass spectrum of the new particles is known, production cross sections can be indicators of their spins [31]. Even if masses are assumed to be equal, particles with different spins have very different production rates due to differences in the number of degrees of freedom, couplings, spin-statistics and angular-momentum conservation. Unfortunately this approach is accompanied by ambiguities. A fermion can for example be faked by two almost mass-degenerate scalars. An additional complication and model dependence is that only topological cross sections, that is, products of production rates and branching fractions, can be measured. Branching fractions can vary significantly with changing model parameters.

A second method is to measure angular correlations of decay products. Because of unknown boosts, the correlation information must be expressed in terms of Lorentz-invariant quantities, for example invariant-mass distributions. Historically the first spin analysis of SUSY particles at the LHC [32] considered the quantity

$$\mathcal{A}(m_{q\ell\pm}) = \left(\frac{d\Gamma(m_{q\ell+})}{dm_{q\ell+}} - \frac{d\Gamma(m_{q\ell-})}{dm_{q\ell-}}\right) \bigg/ \left(\frac{d\Gamma(m_{q\ell+})}{dm_{q\ell+}} + \frac{d\Gamma(m_{q\ell-})}{dm_{q\ell-}}\right),$$

(7.28)

where $m_{q\ell\pm}$ is the jet-lepton invariant mass from the decay cascade

$$\tilde{q}_\text{L} \to q\tilde{\chi}_2^0 \to q\ell_\text{n}^\pm \tilde{\ell}_\text{R}^\mp \to q\ell_\text{n}^\pm \ell_\text{f}^\mp \tilde{\chi}_1^0 .$$

(7.29)

The indices n and f are used to distinguish the lepton from the $\tilde{\chi}_2^0$ decay (so-called "near" lepton) from the lepton from the $\tilde{\ell}_\text{R}$ decay (so-called "far" lepton). The definition of the asymmetry $\mathcal{A}$ is motivated by the fact that in many cases the intermediate $\tilde{\chi}_2^0$ is polarised. Angular-momentum conservation then implies spin correlations between the quark and the near lepton and leads to different shapes of the jet–lepton mass spectrum for $q\ell_\text{n}^-$ ($\bar{q}\ell_\text{n}^+$) and $q\ell_\text{n}^+$ ($\bar{q}\ell_\text{n}^+$). Since one cannot distinguish experimentally between near and far lepton and between quark and antiquark (at least for light quarks), one has to combine quarks and antiquarks as well as near and far leptons. But since a proton–proton collider (as opposed to a proton–antiproton collider) produces more squarks than anti-squarks due to differences in quark and antiquark parton distribution functions, a certain charge asymmetry $\mathcal{A}$ as a function of $m_{q\ell^\pm}$ remains, provided the $\tilde{\chi}_2^0$ is actually a spin-1/2 particle.

This spin analysis has triggered a lot of activity in the field, considering more decay chains to check the spin of other SUSY particles. A comprehensive review of spin-determination techniques at the LHC is given for example in [33].

7.6.2.3 Measurements of Other SUSY Properties

In addition to the mass and spin determination, the LHC allows the measurement of various other SUSY properties – at least in certain regions of parameter space. In SUSY, the quark–squark–gluino Yukawa coupling $\hat{g}_s(q\tilde{q}\tilde{g})$ is equal to the quark–quark–gluon gauge coupling $g_s(qqg)$. To test this fundamental SUSY relation, the two production processes

$$\text{squark-pair production:} \quad qq \to \tilde{q}\tilde{q}(g)\,, \tag{7.30}$$

$$\text{Compton process:} \quad qg \to \tilde{q}\tilde{g}(g) \quad \text{with} \quad \tilde{g} \to \tilde{q}q \tag{7.31}$$

are considered [34]. The squark-pair production cross section is proportional to $\hat{g}_s^4$ whereas the cross section for the Compton process scales with $g_s^2 \hat{g}_s^2$. Since on average the Compton process produces more jets, the ratio of $\hat{g}_s/g_s$ can be determined from the distribution of the number of jets for events with same-sign dileptons and missing transverse energy using only mild theoretical assumptions. The precision of the coupling ratio measurement can be improved by including the transverse momentum spectrum of the third and fourth hardest jet which also allows a distinction between the two processes. For 300 fb^{-1} of integrated luminosity the achievable precision is a few percent.

To distinguish Majorana from Dirac gluinos one can measure the ratio $N(\ell^\pm \ell^\pm)/N(\ell^+ \ell^-)$, the number of events with same-sign leptons divided by the number of events with opposite-sign leptons [35]. Majorana neutralinos can be distinguished from Dirac neutralinos by investigating the shape of the $q\ell^\pm$ invariant-mass spectra in the decay chain of (7.29) [35].

Finally, several analyses showed that it is possible to look for CP-violating effects in the SUSY sector at the LHC [36–40].

7.6.3
Measuring SUSY Properties at the ILC

Provided that a sufficient number of SUSY particles is within the energy reach of the International Linear Collider (ILC), such a machine will augment our knowledge significantly. The advantages have their seeds in the electroweakly interacting initial state and the fact that electrons, unlike protons, are elementary particles. As a consequence, the signal-to-background ratio is much more favourable, the comparably low event rates allow detectors to be run without a trigger, the centre-of-mass energy is known and tunable, and beam particles can be polarised.

7.6.3.1 Mass Measurements

At the ILC, masses of accessible SUSY particles can be measured in two different ways. The first option is *continuum production*, where masses are reconstructed from mass and energy distributions of the final states. The second possibility is a *scan* of the production cross section close to the threshold.

Sleptons can be pair-produced in the process

$$e^+e^- \to \tilde{\ell}_i^\pm \tilde{\ell}_j^\mp, \tilde{\nu}_\ell \tilde{\nu}_\ell^* \quad \text{with} \quad \ell = e, \mu, \tau \quad \text{and} \quad i, j = R, L \quad \text{or} \quad 1, 2. \tag{7.32}$$

To exemplify the mass extraction method, let us consider the reaction $e^+e^- \to \tilde{\ell}_R^+ \tilde{\ell}_R^- \to \ell^+ \tilde{\chi}_1^0 \ell^- \tilde{\chi}_1^0$ [41–43]. Because of the scalar nature of the slepton, the energy spectrum of the leptons produced in the slepton decay is flat with endpoints given by E_- and E_+. These endpoint positions are functions of the slepton mass, the $\tilde{\chi}_1^0$ mass and the centre-of-mass energy, $\sqrt{s}$. For known $\sqrt{s}$, one can thus determine the masses from a measurement of the endpoints,

$$m_{\tilde{\ell}} = \frac{\sqrt{s}}{E_- + E_+} \sqrt{E_- E_+}, \quad m_{\tilde{\chi}_1^0} = m_{\tilde{\ell}} \sqrt{1 - \frac{E_- + E_+}{\sqrt{s}/2}}. \tag{7.33}$$

Alternatively one can perform a threshold scan of the slepton production cross section to measure the slepton masses. In the case of production of selectrons with the same chiral quantum number ($\tilde{e}_R \tilde{e}_R$, $\tilde{e}_L \tilde{e}_L$) it is advantageous to perform the threshold scan with e^-e^- collisions instead of e^+e^-. In the e^-e^- case, the threshold has a steep β dependence contrary to the β^3 dependence for e^+e^- collisions, where β is the velocity of the slepton in the lab frame [44]. Achievable precisions for slepton mass measurements at the ILC are typically $\mathcal{O}(100 \text{ MeV})$.

Neutralinos and charginos are pair-produced:

$$e^+e^- \to \tilde{\chi}_i^0 \tilde{\chi}_j^0 \quad i, j = 1, 2, 3, 4, \tag{7.34}$$

$$e^+e^- \to \tilde{\chi}_i^\pm \tilde{\chi}_j^\mp \quad i, j = 1, 2. \tag{7.35}$$

The decay channels of neutralinos and charginos depend on the mass hierarchy, cf. the example spectrum in Table 7.3. The second-lightest neutralino mostly decays according to $\tilde{\chi}_2^0 \to \ell^+\ell^- \tilde{\chi}_1^0$, either via three-body decay (if $m_{\tilde{\chi}_2^0} - m_{\tilde{\chi}_1^0} < M_Z$, $m_{\tilde{\chi}_2^0} - m_{\tilde{\chi}_1^0} < M_h$ and $m_{\tilde{\chi}_2^0} < m_{\tilde{\ell}}$) or via two subsequent two-body decays with an intermediate Z, h or $\tilde{\ell}$. If $m_{\tilde{\nu}} < m_{\tilde{\chi}_2^0}$, also completely invisible decays are possible. The lightest chargino decays according to $\tilde{\chi}_1^\pm \to \ell^\pm \nu_\ell \tilde{\chi}_1^0$, either as a three-body decay (if $m_{\tilde{\ell}^\pm} > m_{\tilde{\chi}_1^\pm}$) or via two consecutive two-body decays. As in the slepton case, the masses of the neutralinos and charginos can be measured from energy and mass spectra of the leptons. In addition threshold scans can be performed.

The masses of neutralinos and charginos can be measured with a precision of $\mathcal{O}(50 \text{ MeV})$ ($\mathcal{O}(1 \text{ GeV})$) in the case of $\tilde{\chi}_1^0$ ($\tilde{\chi}_2^0$), around 500 MeV for the lightest chargino, and a few GeV for the heavy neutralinos/charginos [45].

Squarks are likely to be beyond the energy reach of the ILC. If nevertheless a squark flavour is light enough, it is probably the lightest scalar top, $\tilde{t}_1$. If the masses fulfil the conditions $m_{\tilde{\chi}_1^\pm} < m_{\tilde{t}_1}$ and $m_{\tilde{t}_1} - m_{\tilde{\chi}_1^0} < m_t$, the $\tilde{t}_1$ mass can be measured using the process $e^+e^- \to \tilde{t}_1 \tilde{t}_1^* \to b\tilde{\chi}_1^+ \bar{b}\tilde{\chi}_1^- \to b\tau^+\nu\tilde{\chi}_1^0 \bar{b}\tau^-\nu\tilde{\chi}_1^0$. Provided that the $\tilde{\chi}_1^\pm$ and $\tilde{\chi}_1^0$ masses are known, one can reconstruct the $\tilde{t}_1$ mass from the energy spectrum of the b jets [46]. In addition one can also infer the $\tilde{t}_1$ mass from a threshold scan. If $m_{\tilde{\chi}_1^\pm} > m_{\tilde{t}_1}$ such that the decay $\tilde{t}_1 \to b\tilde{\chi}_1^+$ is kinematically forbidden, the process $e^+e^- \to \tilde{t}_1 \tilde{t}_1^* \to c\tilde{\chi}_1^0 \bar{c}\tilde{\chi}_1^0$ can also be exploited. Both the mass and mixing angle can be determined from cross-section measurements with opposite beam polarisations [47]. A mass precision of order $\mathcal{O}(1\text{ GeV})$ can be reached.

7.6.3.2 Spin Measurements

Spin measurements at the ILC are much simpler than at the LHC. Thanks to kinematic constraints one can exploit at the ILC, it is possible to determine the spin from the production-angle distribution.

Considering slepton production, the slepton direction can be determined up to a two-fold ambiguity if the involved masses are known. In the case of smuons, the polar production angle is distributed according to $\sin^2\theta$, as they are scalars. The false solutions have a flat distribution in $\cos\theta$ and can easily be subtracted [48]. For selectrons, the angular distribution is more complicated due to the additional t-channel contribution which entails a forward-backward asymmetry.

The same technique can also be applied to extract the spin of neutralinos and charginos. Then the t-channel contribution and the mixed $U(1)$ and $SU(2)$ couplings lead to more complicated asymmetric polar-angle distributions.

7.6.3.3 Measurements of Other SUSY Properties

The production of selectron pairs with different chiral quantum numbers via the processes $e^+e^- \to \tilde{e}_R^+ \tilde{e}_R^-$, $e^+e^- \to \tilde{e}_L^+ \tilde{e}_L^-$, $e^+e^- \to \tilde{e}_L^+ \tilde{e}_R^-$ and $e^+e^- \to \tilde{e}_R^+ \tilde{e}_L^-$ can be disentangled by exploiting the different dependence of their production cross sections on the electron and positron beam polarisation. This requires both electron and positron polarisation [49].

The t channel of selectron production is sensitive to the Yukawa coupling $\hat{g}(e\tilde{e}\tilde{\chi}_i^0)$ which is directly related to the SM gauge couplings via SUSY [44]. CP phases of the $U(1)$ mass parameter, M_1, and the Higgsino mass parameter, μ, can be measured via angular correlations among the decay products of neutralinos [50]. CP phases of the tri-linear couplings A_τ, A_b and A_t can be determined in a global fit of production cross sections, masses and branching fractions of third-generation sfermions [51].

7.6.4
SUSY Parameter Determination

The measurements discussed so far represent an intermediate step towards the final goal of uncovering the underlying new physics model and determining its parameters as precisely as possible. This can be accomplished by comparing all available measurements, M, with predictions of a given theory for the corresponding observables $O(P)$. These theory predictions are functions of parameters P. The aim is to vary the parameters P until the best possible agreement between M and $O(P)$ has been found. This optimisation can be repeated for all available theories. On the basis of the goodness of fit, it is possible to rank different models according to their agreement with data.

Such analyses have been carried out with available low energy (LE) precision data and with anticipated LHC and ILC measurements [52–58]. As an example, the outcome of such an analysis for the SPS1a benchmark point of Table 7.3 is shown in Figure 7.4. It is the SUSY mass spectrum corresponding to the param-

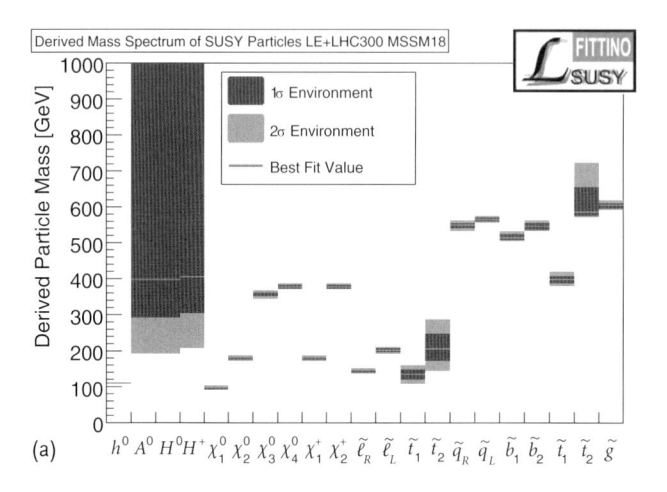

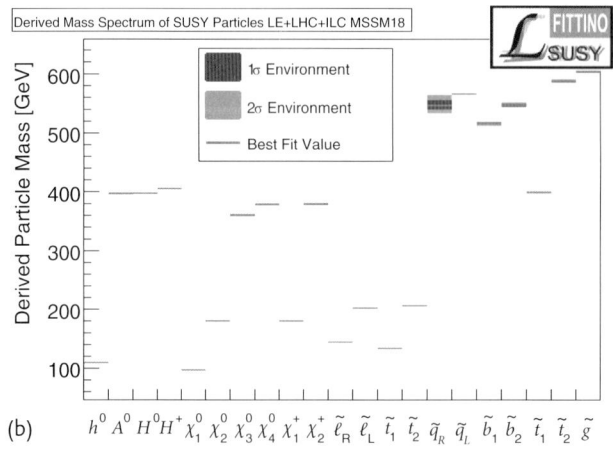

Figure 7.4 SUSY mass spectrum corresponding to the parameter constraints derived for the MSSM18 model. The study is performed for the SPS1a benchmark point. (a) Using LE and anticipated LHC data assuming an integrated luminosity of 300 fb^{-1}; (b) using ILC measurements in addition (adapted from [58], with kind permission of Springer Science+Business Media).

eter constraints derived for an MSSM18 fit [58]. MSSM18 is a simplified MSSM model which assumes that all CP phases vanish, that there is no inter-generation mixing for sfermions, that there is no L–R mixing in the first two sfermion generations and that same-type sfermion mass parameters are equal. Under these assumptions, 18 free SUSY parameters remain. As inputs to the fit, LE and LHC measurements as they are expected from simulation studies for the SUSY benchmark point SPS1a are used (Figure 7.4a). For the LHC, an integrated luminosity of $300\,\text{fb}^{-1}$ at a centre-of-mass energy of 14 TeV is assumed. Figure 7.4b shows the corresponding results if ILC measurements are used in addition. The benchmark point considered in this analysis is in rather good agreement with all presently available experimental constraints, with the exception of a somewhat too large cold dark-matter relic density.

Although the LHC will be able to constrain the masses of many SUSY particles already rather precisely within the framework of the MSSM18, third generation sfermions and heavy Higgs bosons remain only weakly constrained. The main reason is that the heavy Higgs bosons are not directly measurable at the LHC for the considered benchmark point. As a result, μ and $\tan\beta$ are only loosely bound which directly affects the Higgs sector and, via the off-diagonal entries in the sfermion mixing matrix, also the third-generation sfermions. Adding ILC measurements increases the precision by one to two orders of magnitude (except for $\tilde{q}_R$) [58].

7.7
Summary and Conclusions

We have presented a possible extension of the SM of particle physics; the external symmetry supersymmetry. Supersymmetry changes the spins of particles by 1/2 and necessitates a doubling of the particle spectrum. We have shown that there exist in principle two viable versions for this extension: R-parity conservation (here also called the P_6 model) and the R-parity violating model B_3, which violates lepton number. The former is more widely considered, as it is more economical in the number of couplings. The latter has the advantage of naturally implementing light neutrino masses.

As an example case, we have then proceeded to investigate how SUSY is being looked for at the LHC in the case of R-parity conservation. LHC has good prospects to discover (sub-)TeV-scale supersymmetry. If the SUSY search is successful, the LHC can perform rather precise mass measurements of a number of SUSY particles and might allow to check various SUSY relations like spins and couplings to be checked. The LHC results can finally be complemented by very precise measurements at the ILC.

References

1. Bailin, D. and Love, A. (2004) *Supersymmetric gauge field theory and string theory*, IOP (Graduate student series in physics), Bristol, UK.
2. Wess, J. and Zumino, B. (1974) Supergauge transformations in four dimensions, *Nucl. Phys. B*, **70**, 39, doi:10.1016/0550-3213(74)90355-1.
3. Drees, M. (1996) An Introduction to Supersymmetry, http://arxiv.org/abs/hep-ph/9611409.
4. Krauss, L.M. and Wilczek, F. (1989) Discrete gauge symmetry in continuum theories, *Phys. Rev. Lett.*, **62**, 1221, doi:10.1103/PhysRevLett.62.1221.
5. Ibanez, L.E. and Ross, G.G. (1991) Discrete gauge symmetry anomalies, *Phys. Lett. B*, **260**, 291, doi:10.1016/0370-2693(91)91614-2.
6. Ibanez, L.E. and Ross, G.G. (1992) Discrete gauge symmetries and the origin of baryon and lepton number conservation in supersymmetric versions of the Standard Model, *Nucl. Phys. B*, **368**, 3, doi:10.1016/0550-3213(92)90195-H.
7. Banks, T. and Dine, M. (1992) Note on discrete gauge anomalies, *Phys. Rev. D*, **45**, 1424, doi:10.1103/PhysRevD.45.1424.
8. Dreiner, H.K., Luhn, C., and Thormeier, M. (2006) What is the discrete gauge symmetry of the minimal supersymmetric Standard Model? *Phys. Rev. D*, **73**, 075007, doi:10.1103/PhysRevD.73.075007.
9. Ellis, J.R., Hagelin, J.S., Nanopoulos, D.V., Olive, K.A., and Srednicki, M. (1984) Supersymmetric relics from the big bang, *Nucl. Phys. B*, **238**, 453, doi:10.1016/0550-3213(84)90461-9.
10. Hebbeker, T. (1999) Can the sneutrino be the lightest supersymmetric particle? *Phys. Lett. B*, **470**, 259, doi:10.1016/S0370-2693(99)01313-1.
11. Gonzalez-Garcia, M.C. and Maltoni, M. (2008) Phenomenology with massive neutrinos, *Phys. Rept.*, **460**, 1, doi:10.1016/j.physrep.2007.12.004.
12. Minkowski, P. (1977) $\mu \to e\gamma$ at a rate of one out of 1-billion muon decays? *Phys. Lett. B*, **67**, 421, doi:10.1016/0370-2693(77)90435-X.
13. Mohapatra, R.N. and Senjanovic, G. (1980) Neutrino mass and spontaneous parity nonconservation, *Phys. Rev. Lett.*, **44**, 912, doi:10.1103/PhysRevLett.44.912.
14. Hall, L.J. and Suzuki, M. (1984) Explicit R-parity breaking in supersymmetric models, *Nucl. Phys. B*, **231**, 419, doi:10.1016/0550-3213(84)90513-3.
15. Allanach, B.C., Dedes, A., and Dreiner, H.K. (2004) *R*-parity violating minimal supergravity model, *Phys. Rev. D*, **69**, 115002, doi:10.1103/PhysRevD.69.115002.
16. Rajagopal, K., Turner, M.S., and Wilczek, F. (1991) Cosmological implications of axinos, *Nucl. Phys. B*, **358**, 447, doi:10.1016/0550-3213(91)90355-2.
17. Peccei, R.D. and Quinn, H.R. (1977) CP conservation in the presence of pseudoparticles, *Phys. Rev. Lett.*, **38**, 1440, doi:10.1103/PhysRevLett.38.1440.
18. Dreiner, H.K. and Grab, S. (2009) All possible lightest supersymmetric particles in R-parity violating minimal supergravity models, *Phys. Lett. B*, **679**, 45, doi:10.1016/j.physletb.2009.06.059.
19. Butterworth, J. and Dreiner, H.K. (1993) *R*-parity violation at HERA, *Nucl. Phys. B*, **397**, 3, doi:10.1016/0550-3213(93)90334-L.
20. ATLAS Collab., Aad, G. et al. (2009) Expected Performance of the ATLAS Experiment – Detector, Trigger and Physics, http://arxiv.org/abs/0901.0512.
21. CMS Collab., Bayatian, G.L. et al. (2007) CMS physics technical design report, Vol. II: Physics performance, *J. Phys. G*, **34**(6), 995, doi:10.1088/0954-3899/34/6/S01, IOP Publishing.
22. ECFA/DESY LC Physics Working Group, Aguilar-Saavedra, J.A, et al. (2001) TESLA Technical Design Report Part III: Physics at an e^+e^- Linear Collider, http://arxiv.org/abs/hep-ph/0106315.
23. Barr, A.J. and Lester, C.G. (2010) A Review of the Mass Measurement Tech-

niques proposed for the Large Hadron Collider, http://arxiv.org/abs/1004.2732.

24 Hinchliffe, I., Paige, F.E., Shapiro, M.D., Soderqvist, J., and Yao, W. (1997) Precision SUSY measurements at CERN LHC, *Phys. Rev. D*, **55**, 5520, doi:10.1103/PhysRevD.55.5520.

25 Allanach, B.C., Lester, C.G., Parker, M.A., and Webber, B.R. (2000) Measuring sparticle masses in non-universal string inspired models at the LHC, *JHEP*, **09**, 004, doi:10.1088/1126-6708/2000/09/004.

26 Gjelsten, B.K., Miller, 2, D.J., and Osland, P. (2004) Measurement of SUSY masses via cascade decays for SPS 1a, *JHEP*, **12**, 003, doi:10.1088/1126-6708/2004/12/003.

27 Lester, C.G. and Summers, D.J. (1999) Measuring masses of semi-invisibly decaying particles pair produced at hadron colliders, *Phys. Lett. B*, **463**, 99, doi:10.1016/S0370-2693(99)00945-4.

28 Cho, W.S., Choi, K., Kim, Y.G., and Park, C.B. (2008) Gluino Stransverse Mass, *Phys. Rev. Lett.*, **100**, 171801, doi:10.1103/PhysRevLett.100.171801.

29 Appelquist, T., Cheng, H.C., and Dobrescu, B.A. (2001) Bounds on universal extra dimensions, *Phys. Rev. D*, **64**, 035002, doi:10.1103/PhysRevD.64.035002.

30 Arkani-Hamed, N., Cohen, A.G., and Georgi, H. (2001) Electroweak symmetry breaking from dimensional deconstruction, *Phys. Lett. B*, **513**, 232, doi:10.1016/S0370-2693(01)00741-9.

31 Datta, A., Kane, G.L., and Toharia, M. (2005) Is it SUSY? http://arxiv.org/abs/hep-ph/0510204.

32 Barr, A.J. (2004) Using lepton charge asymmetry to investigate the spin of supersymmetric particles at the LHC, *Phys. Lett. B*, **596**, 205–212, doi:10.1016/j.physletb.2004.06.074.

33 Wang, L.T. and Yavin, I. (2008) A review of spin determination at the LHC, *Int. J. Mod. Phys. A*, **23**, 4647, doi:10.1142/S0217751X08042778.

34 Freitas, A., Skands, P.Z., Spira, M., and Zerwas, P.M. (2007) Examining the identity of Yukawa with gauge couplings in supersymmetric QCD at LHC, *JHEP*, **07**, 025, doi:10.1088/1126-6708/2007/07/025.

35 Choi, S.Y., Drees, M., Freitas, A., and Zerwas, P.M. (2008) Testing the Majorana nature of gluinos and neutralinos, *Phys. Rev. D*, **78**, 095007, doi:10.1103/PhysRevD.78.095007.

36 Langacker, P., Paz, G., Wang, L.T., and Yavin, I. (2007) A T-odd observable sensitive to CP violating phases in squark decay, *JHEP*, **07**, 055, doi:10.1088/1126-6708/2007/07/055.

37 Ellis, J., Moortgat, F., Moortgat-Pick, G., Smillie, J.M., and Tattersall, J. (2009) Measurement of CP violation in stop cascade decays at the LHC, *Eur. Phys. J. C*, **60**, 633, doi:10.1140/epjc/s10052-009-0964-8.

38 Deppisch, F. and Kittel, O. (2009) Probing SUSY CP violation in two-body stop decays at the LHC, *JHEP*, **09**, 110, doi:10.1088/1126-6708/2009/09/110.

39 Moortgat-Pick, G., Rolbiecki, K., Tattersall, J., and Wienemann, P. (2010) Probing CP violation with and without momentum reconstruction at the LHC, *JHEP*, **01**, 004, doi:10.1007/JHEP01(2010)004.

40 Deppisch, F.F. and Kittel, O. (2010) CP violation in sbottom decays, http://arxiv.org/abs/1003.5186.

41 Freitas, A., Martyn, H.U., Nauenberg, U., and Zerwas, P.M. (2004) Sleptons: Masses, Mixings, Couplings, http://arxiv.org/abs/hep-ph/0409129.

42 Martyn, H.U. (2004) Study of Sleptons at a Linear Collider – Supersymmetry Scenario SPS 1a, http://arxiv.org/abs/hep-ph/0406123.

43 Martyn, H.U. (2004) Detection of sleptons at a linear collider in models with small slepton neutralino mass differences, http://arxiv.org/abs/hep-ph/0408226.

44 Freitas, A., von Manteuffel, A., and Zerwas, P.M. (2004) Slepton production at e^+e^- and e^-e^- linear colliders, *Eur. Phys. J. C*, **34**, 487, doi:10.1140/epjc/s2004-01744-2.

45 Desch, K. (2005) Supersymmetry at LHC and ILC, http://arxiv.org/abs/hep-ph/0501096.

46 Feng, J.L. and Finnell, D.E. (1994) Squark mass determination at the next generation of linear e^+e^- colliders, *Phys. Rev. D*, **49**(5), 2369, doi:10.1103/PhysRevD.49.2369.

47 Finch, A., Sopczak, A., and Kluge, H. (2005) Determination of the scalar top mass at a linear e^+e^- collider, http://www.slac.stanford.edu/spires/find/hep/www?key=7317565, Prepared for International Conference on Linear Colliders (LCWS 04), Paris, France, 19–24 Apr 2004.

48 Abe, K. *et al.* (2001) Particle Physics Experiments at JLC, http://arxiv.org/abs/hep-ph/0109166.

49 Moortgat-Pick, G.A. (2004) Motivation for polarised e^- and e^+ Beams, http://arxiv.org/abs/hep-ph/0410118.

50 Bartl, A., Fraas, H., Kittel, O., and Majerotto, W. (2004) CP asymmetries in neutralino production in e^+e^- collisions, *Phys. Rev. D*, **69**, 035007, doi:10.1103/PhysRevD.69.035007.

51 Bartl, A., Hesselbach, S., Hidaka, K., Kernreiter, T., and Porod, W. (2004) Top squarks and bottom squarks in the minimal supersymmetric standard model with complex parameters, *Phys. Rev. D*, **70**, 035003, doi:10.1103/PhysRevD.70.035003.

52 LHC/LC Study Group, Weiglein, G. *et al.* (2006) Physics interplay of the LHC and the ILC, *Phys. Rept.*, **426**, 47, doi:10.1016/j.physrep.2005.12.003.

53 Allanach, B.C. and Lester, C.G. (2006) Multidimensional mSUGRA likelihood maps, *Phys. Rev. D*, **73**, 015013, doi:10.1103/PhysRevD.73.015013.

54 Bechtle, P., Desch, K., Porod, W., and Wienemann, P. (2006) Determination of MSSM parameters from LHC and ILC observables in a global fit, *Eur. Phys. J. C*, **46**, 533, doi:10.1140/epjc/s2006-02485-x.

55 Lafaye, R., Plehn, T., Rauch, M., and Zerwas, D. (2008) Measuring supersymmetry, *Eur. Phys. J. C*, **54**, 617, doi:10.1140/epjc/s10052-008-0548-z.

56 Buchmueller, O. *et al.* (2008) Predictions for supersymmetric particle masses using indirect experimental and cosmological constraints, *JHEP*, **09**, 117, doi:10.1088/1126-6708/2008/09/117.

57 AbdusSalam, S.S., Allanach, B.C., Quevedo, F., Feroz, F., and Hobson, M. (2010) Fitting the Phenomenological MSSM, *Phys. Rev. D*, **81**, 095012, doi:10.1103/PhysRevD.81.095012.

53 Bechtle, P., Desch, K., Uhlenbrock, M., and Wienemann, P. (2010) Constraining SUSY models with Fittino using measurements before, with and beyond the LHC, *Eur. Phys. J. C*, **66**, 215, doi:10.1140/epjc/s10052-009-1228-3.

8
Quark Flavour Physics

Gudrun Hiller and Ulrich Uwer

When talking about "flavour" what we mean is that known fundamental matter comes in three generations, carrying the same charges under the Standard Model gauge group $SU(3)_c \times SU(2)_L \times U(1)_Y$, and flavour is the feature that distinguishes the generations.

As a result of this flavour feature we observe a complex phenomenology, such as a spectrum of elementary matter particles spanning many orders of magnitude in mass from the electron with 0.5 MeV or the neutrinos in the sub-eV range to the top quark as heavy as 173 GeV, or the breakdown of CP symmetry, a phenomenon that, to date, has been observed in generation-changing hadronic processes only.

Flavour physics has accompanied the making of the Standard Model as it is today. Its description in terms of consistency and precision has greatly progressed over the past decade. The detailed studies in b physics played an important role here. In Section 8.1 we review our current understanding of flavour, within the Standard Model, and explain briefly how we found out what we know today from key heavy-quark measurements.

There are also new questions to be asked in relation to flavour when it comes to physics beyond the Standard Model. First, in almost all models with TeV-scale new physics, partner particles of the Standard Model ones are involved. It is conceivable that some of the partner particles, for instance quark partners, come in different flavours as well. One has then to ask about the flavour properties of these new particles, for example: does their spectrum exhibit a large spread between the different generations and do the flavours mix as the quarks do? Next, the quest for the origin of flavour and generation structure. The latter is in the Standard Model merely parametrised, but it is not known how this puzzling pattern of hierarchical quark masses and mixings arises. It may well be, that the new dynamics can give insights into the dynamics of flavour. Vice versa, studies in flavour physics can help to find deviations from the Standard Model at higher energies. We address the important connection between flavour and new physics in Section 8.2.

Physics at the Terascale, First Edition. Edited by Ian C. Brock and Thomas Schörner-Sadenius
© 2011 WILEY-VCH Verlag GmbH & Co. KGaA, Weinheim.
Published 2011 by WILEY-VCH Verlag GmbH & Co. KGaA.

8.1
Flavour Within the Standard Model

The fermionic constituents of hadronic matter, the quarks, come in generations: three copies of up-type quarks, u_i, of electric charge $Q_u = +2/3e$, called up ($i = 1$, or u), charm ($i = 2$, or c) and top ($i = 3$, or t), and three generations of down-type quarks, d_i, with $Q_d = -1/3e$, termed down ($i = 1$, or d), strange ($i = 2$, or s) and bottom ($i = 3$, or b). The ordering from the first to the third generation is with increasing mass.

All quarks differ by their mass and flavour mixing properties. Such a generation structure is caused by the Yukawa interactions, which couple left-handed quarks, q_L, to right-handed ones, q_R, $q = u, d$, together with the Higgs boson in a gauge invariant way. After spontaneous breaking of electroweak symmetry the quarks' Yukawa terms give rise to masses and mixings as

$$\mathcal{L}_Y^{\text{quarks}} = -\frac{v}{\sqrt{2}} (\bar{d}_L Y_d d_R + \bar{u}_L Y_u u_R) + \text{h.c.} \ . \tag{8.1}$$

The Yukawa matrices Y_d and Y_u, which are three by three, complex matrices in generation space are solely responsible for flavour in the Standard Model. Note that here and in most formulae in this chapter we give flavour formulae in matrix form, suppressing generation indices.

8.1.1
The CKM Quark Mixing Matrix

The Yukawa matrices, in general, do not need to be diagonal in generation space in the representation of quark fields where all gauge bosons couple universally to the generations. In fact, in the Standard Model they are not. We obtain mass eigenstates of the quarks, $\tilde{q}_A$, from the above "gauge basis", q_A, by unitary transformations $\tilde{q}_A = V_{A,q} q_A$ for $q = u, d$ and $A = L, R$, where $V_{A,q} V_{A,q}^\dagger = 1$. The unitary matrices $V_{A,q}$ are determined by requiring that they diagonalise the Yukawa matrices as

$$M_u = \begin{pmatrix} m_u & 0 & 0 \\ 0 & m_c & 0 \\ 0 & 0 & m_t \end{pmatrix} = \text{diag}(m_u, m_c, m_t) = \frac{v}{\sqrt{2}} V_{L,u} Y_u V_{R,u}^\dagger \ , \tag{8.2}$$

$$M_d = \begin{pmatrix} m_d & 0 & 0 \\ 0 & m_s & 0 \\ 0 & 0 & m_b \end{pmatrix} = \text{diag}(m_d, m_s, m_b) = \frac{v}{\sqrt{2}} V_{L,d} Y_d V_{R,d}^\dagger \ . \tag{8.3}$$

The quark masses m_q appear as usual Dirac mass terms in

$$\mathcal{L}_Y^{\text{quarks}} = -\bar{\tilde{d}}_L M_d \tilde{d}_R - \bar{\tilde{u}}_L M_u \tilde{u}_R + \text{h.c.} \ , \tag{8.4}$$

which follows immediately from (8.1) after changing from gauge fields to mass eigenstates.

8.1 Flavour Within the Standard Model

If the up-type and down-type Yukawa matrices cannot be diagonalised simultaneously, that is, by the same unitary transformations $V_{A,d} \neq V_{A,u}$, there is one important net effect related to flavour from the basis change besides having quark fields as mass eigenstates: the charged current interaction gets a flavour structure, encoded in the Cabibbo–Kobayashi–Maskawa (CKM) [1, 2] quark mixing matrix $V_{\text{CKM}} = V_{\text{L},u} V_{\text{L},d}^\dagger$. Specifically, the CKM matrix appears in the Lagrangian as

$$\mathcal{L}_{\text{CC}} = -\frac{g_2}{\sqrt{2}} \left(\bar{\tilde{u}}_\text{L} \gamma^\mu W_\mu^+ V_{\text{CKM}} \tilde{d}_\text{L} + \bar{\tilde{d}}_\text{L} \gamma^\mu W_\mu^- V_{\text{CKM}}^\dagger \tilde{u}_\text{L} \right) . \tag{8.5}$$

The element $(V_{\text{CKM}})_{ij}$ connects a left-handed up-type quark of the ith generation to a left-handed down-type quark of the jth generation. Labelling the matrix elements by quark flavours rather than generational indices, the CKM matrix reads in an intuitive way

$$V_{\text{CKM}} = \begin{pmatrix} V_{ud} & V_{us} & V_{ub} \\ V_{cd} & V_{cs} & V_{cb} \\ V_{td} & V_{ts} & V_{tb} \end{pmatrix} . \tag{8.6}$$

For a non-trivial mixing matrix $(V_{\text{CKM}})_{ij} \neq \delta_{ij}$, as realised in nature, the weak interaction allows for intergenerational changes in the charged current interactions. In the Standard Model, this is the only way to change flavour.

Since it is the product of two unitary matrices, V_{CKM} itself is unitary, $V_{\text{CKM}} V_{\text{CKM}}^\dagger = 1$, and its entries are in general complex. The presence of phases in the CKM matrix breaks the CP symmetry, which can be seen from CP transforming the charged current Lagrangian (8.5):

$$\mathcal{L}_{\text{CC}}^{\text{CP}} = -\frac{g_2}{\sqrt{2}} \left(\bar{\tilde{d}}_\text{L} \gamma^\mu W_\mu^- V_{\text{CKM}}^T \tilde{u}_\text{L} + \bar{\tilde{u}}_\text{L} \gamma^\mu W_\mu^+ V_{\text{CKM}}^* \tilde{d}_\text{L} \right) . \tag{8.7}$$

Comparing the original Lagrangian (8.5) to its CP copy (8.7) one sees that they would agree for $V_{\text{CKM}} = V_{\text{CKM}}^*$, that is, a real-valued quark mixing matrix. However, this is not the case in nature, as one has observed CP violation consistent with non-vanishing CKM phases.

The number of parameters of a general 3 by 3 unitary matrix is nine with three rotation angles, and a total of six complex phases. Why does the CKM matrix contain a single physical phase only? The answer is that we can perform global, generation-dependent re-phasings of the mass eigenstates $(\tilde{q}_\text{L})_k \to e^{i\alpha_{q_k}} (\tilde{q}_\text{L})_k$, $k = 1, 2, 3$ to remove five of the phases. (The right-handed fields need to be rotated simultaneously $(\tilde{q}_\text{R})_k \to e^{i\alpha_{q_k}} (\tilde{q}_\text{R})_k$ to keep the quark masses real.) We can remove only five phases in this way, because V_{CKM} is sandwiched between three u_L and three d_L fields, and out of six different fields one can get at most five independent relative phases.

The CKM matrix can be parametrised as

$$V_{\text{CKM}} = \begin{pmatrix} c_{12} c_{13} & s_{12} c_{13} & s_{13} e^{-i\delta} \\ -s_{12} c_{23} - c_{12} s_{23} s_{13} e^{i\delta} & c_{12} c_{23} - s_{12} s_{23} s_{13} e^{i\delta} & s_{23} c_{13} \\ s_{12} s_{23} - c_{12} c_{23} s_{13} e^{i\delta} & -c_{12} s_{23} - s_{12} c_{23} s_{13} e^{i\delta} & c_{23} c_{13} \end{pmatrix} , \tag{8.8}$$

where $s_{ij} \equiv \sin\theta_{ij}$ and $c_{ij} \equiv \cos\theta_{ij}$. This parametrisation follows closely the one describing rotations in 3D space using the Euler angles θ_{12}, θ_{13} and θ_{23}, and has a CP-violating phase δ [2]. The number of physical real-valued parameters of the CKM matrix is four. The parametrisation (8.8) is exact and completely general.

Phenomenologically, the CKM matrix is hierarchical, $\theta_{13} \ll \theta_{23} \ll \theta_{12} \ll 1$, and has an order one phase, $\delta \sim \mathcal{O}(1)$. A useful parametrisation making the hierarchy transparent is due to Wolfenstein [3]. It reads

$$V_{\text{CKM}} = \begin{pmatrix} 1 - \lambda^2/2 & +\lambda & A\lambda^3(\rho - i\eta) \\ -\lambda & 1 - \lambda^2/2 & +A\lambda^2 \\ A\lambda^3(1 - \rho - i\eta) & -A\lambda^2 & 1 \end{pmatrix} + \mathcal{O}(\lambda^4), \qquad (8.9)$$

which is an excellent approximation. The expansion is in the small parameter λ. CP violation arises for a non-zero η, and is connected to the third generation. The Wolfenstein parameters λ, A, ρ and η are determined to be [4]

$$\lambda = 0.225, \quad A = 0.81, \quad \bar{\rho} = 0.14, \quad \bar{\eta} = 0.34. \qquad (8.10)$$

Here, we give the parameters beyond the lowest order in the λ-expansion [5] $\bar{\rho} = \rho(1 - \lambda^2/2)$ and $\bar{\eta} = \eta(1 - \lambda^2/2)$, which are commonly used in fits.

It follows from the observed structure of the CKM matrix that quark transitions within the same generation are preferred, and the further one goes away from the diagonal, the stronger is the suppression. For example, the third generation quarks are decoupled from the first two at order λ^2, which is at the level of a few percent. The CP phase $\sim \arctan \eta/\rho$ is large, about $68°$.

In total, there are ten parameters in the quark flavour sector, six masses and four related to V_{CKM}. The flavour violation in the hadronic sector, including decays, productions rates at colliders and meson mixing effects, should be described by these parameters alone, if the Standard Model is correct. Since all parameters are known, this statement is very predictive and subject to numerous tests. The level to which we can test the Standard Model is linked to the precision to which we know its parameters. We are therefore interested in improving the knowledge of the quark flavour parameters.

8.1.2
The Unitarity Triangle

The unitarity of the CKM matrix, $V_{\text{CKM}} V_{\text{CKM}}^\dagger = 1$, leads to experimentally testable constraints for the matrix elements. Three of the nine complex equations are of the type $|V_{xd}|^2 + |V_{xs}|^2 + |V_{xb}|^2 = 1$ and express the fact that the probability of an up-type quark, x, to decay into any down-type quark via W-boson emission is one. Constraints on the phases of matrix elements result from the orthogonality of the matrix columns and rows. Taking the first and the third column of V_{CKM} one finds:

$$V_{ud} V_{ub}^* + V_{cd} V_{cb}^* + V_{td} V_{tb}^* = 0. \qquad (8.11)$$

Equation 8.11 is the sum of three complex numbers and defines a triangle relation in the complex plane which is graphically depicted in Figure 8.1a.

8.1 Flavour Within the Standard Model

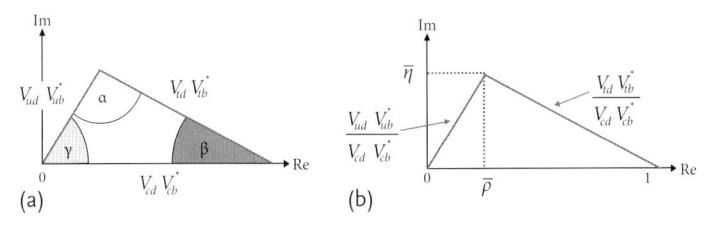

Figure 8.1 (a) unitarity condition depicted as a triangle in the complex plane; (b) rescaled unitarity triangle. The apex has the coordinates $(\bar{\rho}, \bar{\eta})$ as introduced in (8.10).

The lengths of the triangle sides are given by $|V_{xd}V_{xb}^*|$ while the three angles are defined as,

$$\alpha = \arg\left[-\frac{V_{td}V_{tb}^*}{V_{ud}V_{ub}^*}\right], \quad \beta = \arg\left[-\frac{V_{cb}V_{cd}^*}{V_{td}V_{tb}^*}\right], \quad \gamma = \arg\left[-\frac{V_{ud}V_{ub}^*}{V_{cd}V_{cb}^*}\right]. \tag{8.12}$$

By construction the three angles must sum up to 180 degrees. The six possible triangles defined by the unitarity relations all have the same area but differ significantly in their shape. Using the Wolfenstein parameters one easily sees that the triangles can be grouped into three pairs: The "db" triangle defined by (8.11) as well as the "tu" triangle[20] have sides with similar length, all of the order $A\lambda^3$.

Conventionally the "db" triangle is referred to as the "unitarity triangle". The experimental determination of its shape was the primary aim of the B-factory experiments BABAR and BELLE. The "tu" triangle on the other hand is experimentally not accessible as the top quark decays before it forms a meson, which would be necessary in order to measure the "tu" angles. The remaining four triangles are rather slim and their experimental validation is more difficult and requires high precision measurements of rare K, B_s^0 and D decays.

It is customary to rescale the unitarity triangle by normalising the triangle basis to one, as shown in Figure 8.1b. For the rescaled triangle the position of the apex is given by the complex number $\bar{\rho} + i\bar{\eta}$ defined in (8.10).

Figure 8.2 reflects the current experimental knowledge of the unitarity triangle. The figure summarises several independent measurements which constrain the shape of the triangle:

- Measurements of branching fractions of semileptonic B-meson decays constrain the length of the left side: semileptonic decays such as $B \to X_c \ell \nu$ and $B \to X_u \ell \nu$, where the X_c and X_u are hadronic states containing a c quark or a u quark respectively, involve $b \to c$ or $b \to u$ quark transitions. Their branching fractions are proportional to $|V_{cb}|^2$ or $|V_{ub}|^2$. Hadronic uncertainties limit the determination of the CKM elements from the branching fractions.
  The helicity-suppressed decay $B^+ \to \tau^+ \nu$ also involves a $b \to u$ quark transition and thus provides additional information on $|V_{ub}|$.

20) At order $\mathcal{O}(\lambda^3)$ the "db" and the "tu" triangles are equal.

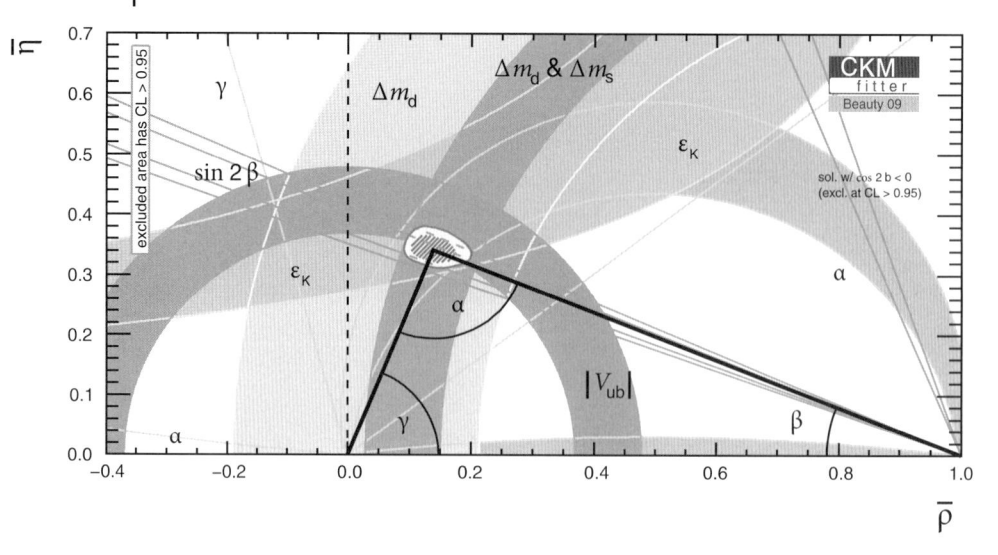

Figure 8.2 Experimental status of the "db" unitarity triangle. There is excellent agreement between constraints from CP-conserving observables measuring the triangle sides and from CP violation measurements determining the angles (adapted from [4]).

- $B_d^0 \overline{B}_d^0$ and $B_s^0 \overline{B}_s^0$ mixing constrain the length of the right side of the triangle – the mixing phenomenon of neutral mesons is the subject of the next section.
- Precise measurements of CP violation in B-meson decays offer the possibility to probe the CP violating phase in the CKM matrix and to determine the angles of the unitarity triangle. The strongest constraint currently comes from the measurement of the time-dependent CP asymmetry in the decay $B_d^0 \to J/\psi\, K_S^0$. The measurement determines the angle β to $\beta = (21.1 \pm 0.9)°$ [6]. How to measure CP violation in B decays is discussed in Section 8.3.
- Neutral kaon physics measurements put additional constraints on the position of the apex of the unitarity triangle denoted by ϵ_K in Figure 8.2.

The different independent measurements show an impressive agreement and limit the apex of the triangle to the small dashed area in Figure 8.2. The consistency of measurements validate the CKM paradigm of the Standard Model. The measurements identify quark mixing as the primary source of CP violation in the quark sector and put stringent limits on possible new physics phenomena contributing to the quantum corrections.

8.2 Flavour and New Physics

The presence of new physics brings in new questions to flavour physics: How can we test the Standard Model and find new physics with flavour physics? What can we learn about the TeV scale from present-day flavour physics data? In the following, we address both questions, along with heavy flavour key observables. The most important tools are flavour changing neutral currents (FCNCs).

8.2.1 Flavour Changing Neutral Currents

In the Standard Model neutral current interactions conserve flavour. This is, however, a result of the field content and not enforced by a symmetry. Charged currents can induce FCNCs through quantum loops. Two types of FCNC processes are known, those with flavour number change by one unit, $\Delta f = 1$, as in decays, and those with $\Delta f = 2$, as in meson mixing.

Several FCNC transitions for quarks are possible, $b \to d$, $b \to s$, $s \to d$, $c \to u$, $t \to u$ and $t \to c$. Except for the ones involving top quarks, all of them have been seen experimentally, in neutral meson mixing and in rare decays (the latter not in $c \to u$) [7]. Famous examples are rare radiative and semileptonic b decays, which have been observed at the level of the Standard Model $\mathcal{B}(b \to s\gamma) = 3 \times 10^{-4}$ (for photon energies above 1.6 GeV) and $\mathcal{B}(b \to s\ell^+\ell^-) = 4 \times 10^{-6}$ (for $m_{\ell\ell}^2 > 0.04\,\text{GeV}^2$). Rare top decays are too small to be observed if there is no new physics. In the Standard Model, the largest decay rate is into a gluon, $\mathcal{B}(t \to cg) \sim 10^{-10}$.

Let us discuss more closely the $\Delta b = 1$, $\Delta s = 1$ decay of a b quark changing into an s quark by W-up-type quark loops as shown in Figure 8.3. The Standard Model $b \to s$ amplitude can be written with its flavour structure manifest as

$$\mathcal{A}(b \to s)_{\text{SM}} = V_{ub} V_{us}^* A_u + V_{cb} V_{cs}^* A_c + V_{tb} V_{ts}^* A_t . \tag{8.13}$$

Here, A_q, $q = u, c, t$ denote the subamplitudes from each of the up-type quark flavours in the loop with the corresponding CKM elements factored out. The subamplitudes depend, as far as flavour is concerned, on the corresponding quark mass only, and for dimensional reasons on m_q^2/M_W^2. The functions $A_q = A(m_q^2/M_W^2)$ are obtained from an explicit loop calculation and depend on the full process [8].

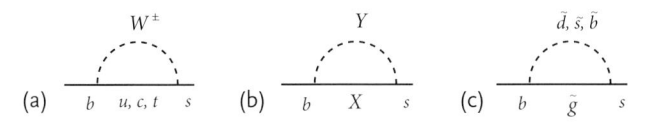

Figure 8.3 (a) the skeleton $b \to s$ FCNC transition in the Standard Model; (b) with generic new physics; (c) a possible amplitude in the MSSM with gluinos and squarks.

Using the unitarity of the CKM matrix, specifically $\sum_i V_{ib} V_{is}^* = 0$, one can rewrite (8.13) as

$$\mathcal{A}(b \to s)_{SM} = V_{tb} V_{ts}^* (A_t - A_c) + V_{ub} V_{us}^* (A_u - A_c) . \tag{8.14}$$

We see that the Standard Model $b \to s$ amplitude would vanish unless the CKM matrix is non-trivial, that is, it allows for changes between different generations. The amplitude would also vanish for identical up-type quark masses. From these two suppression mechanisms one can infer that the first term in (8.14) containing the top contribution is the dominant one. Firstly, it is of second order in the small Wolfenstein parameter, λ, entering the CKM matrix (8.9), whereas the second term is $\mathcal{O}(\lambda^4)$. Secondly, the higher level of degeneracy, that is, the *Glashow–Iliopoulos–Maiani* (GIM) mechanism, suppresses the second term even further since $(m_c^2 - m_u^2)/M_W^2 \ll 1$, but is inactive for the first term due to the presence of the top quark, with $(m_t^2 - m_c^2)/M_W^2$ of order one. With rare b decays we can therefore measure properties of the top quark, despite it being much heavier than the b quark.

The analysis of the $b \to s$ amplitude (8.14) is quite general, and it is straightforward to see that every FCNC amplitude has the following features in the Standard Model: (i) FCNCs are induced by the weak interaction at loop level; (ii) FCNCs require a non-trivial mixing matrix. Quark FCNCs are suppressed by off-diagonal CKM elements; (iii) FCNCs vanish if the intermediate quarks are degenerate in mass. Since the splitting in the up-type quark spectrum is larger than the one in the down-type one, the resulting suppression is larger for FCNCs with external up-type quarks than for down-type ones.

What makes FCNC processes powerful is their suppression in the Standard Model due to mixing (CKM), degeneracy (GIM) and absence at tree level. New physics, which does not need to have the same mechanisms built in, hence competes in FCNC processes with only a small Standard Model background. Equally important is that FCNCs feel the physics in the loops from energies much higher than the ones actually involved in the real process. We saw already that b-FCNCs probe the top quark. Some examples with Beyond-the-Standard-Model (BSM) physics are shown in Figure 8.3.

It is these features of new physics and high energy sensitivity that allow us to draw strong conclusions on the physics at the TeV scale from existing flavour data. We illustrate this by considering neutral meson mixing. In a generalised Fermi theory, the $\Delta b = 2$ mixing amplitude can be described by effective 4-Fermi interactions. For B_d mesons in the Standard Model (see Figure 8.4), one has

$$\mathcal{A}(\overline{B}_d^0 \to B_d^0)_{SM} \propto \frac{g_2^4}{16\pi^2} (V_{tb} V_{td}^*)^2 \frac{\langle B_d^0 | (\bar{d}_L \gamma_\mu b_L)(\bar{d}_L \gamma^\mu b_L) | \overline{B}_d^0 \rangle}{M_W^2} S(x_t) , \tag{8.15}$$

$$S(x_t) = x_t \left(\frac{1}{4} + \frac{9}{4} \frac{1}{(1-x_t)} - \frac{3}{2} \frac{1}{(1-x_t)^2} \right) + \frac{3}{2} \left(\frac{x_t}{x_t - 1} \right)^3 \log x_t , \tag{8.16}$$

where $x_t = m_t^2/M_W^2$ and S is the resulting loop function of the box diagrams. The up and charm quark contributions with $m_{u,c}^2/M_W^2 \simeq 0$ have been included by means of unitarity. The relevant scale is the weak scale, here the mass of the W

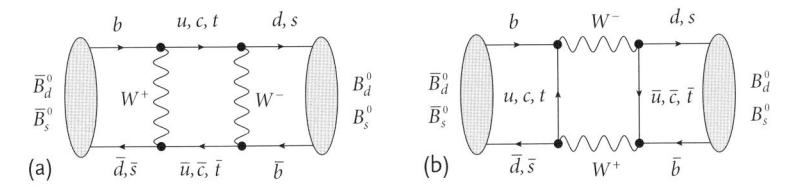

Figure 8.4 (a, b) The lowest order Standard Model contribution (box diagrams) to $B^0\overline{B}^0$ mixing.

boson. The amplitude (8.15) is strongly CKM suppressed, at the order λ^6. The 4-quark operator exhibits the Standard Model $V - A$ structure of the weak interaction.

The $B_d^0\overline{B}_d^0$ mixing amplitude in a new physics model with scale $\Lambda_{\rm NP}$ can generically be written as

$$\mathcal{A}\left(\overline{B}_d^0 \to B_d^0\right)_{\rm NP} \propto \frac{\langle B_d^0|(\bar{d}\Gamma b)(\bar{d}\Gamma' b)|\overline{B}_d^0\rangle}{\Lambda_{\rm NP}^2} \quad \text{(tree level)}, \tag{8.17}$$

$$\mathcal{A}\left(\overline{B}_d^0 \to B_d^0\right)_{\rm NP} \propto \frac{g_2^4}{16\pi^2}\frac{\langle B_d^0|(\bar{d}\Gamma b)(\bar{d}\Gamma' b)|\overline{B}_d^0\rangle}{\Lambda_{\rm NP}^2} \quad \text{(weak loop)}, \tag{8.18}$$

with arbitrary Dirac structures Γ, Γ' and assuming no specific flavour structure. Tree-level induced FCNCs (8.17) or loop suppression (8.18) are allowed for, the latter of which brings in an additional factor of $1/16\pi^2$.

Since the B_d^0–$\overline{B}_d^0$ oscillation data are consistent with the predictions of the Standard Model, the mixing amplitude cannot be dominated by the new physics contribution. Comparing (8.15) to (8.17) and (8.18) we therefore conclude the following: to be in agreement with observation, the scale of the new physics must be very high, $\Lambda_{\rm NP} \gtrsim 4\pi M_W/(g_2^2|V_{td}|) \sim \mathcal{O}(250\,{\rm TeV})$, assuming tree level, or $\Lambda_{\rm NP} \gtrsim M_W/|V_{td}| \sim \mathcal{O}(10\,{\rm TeV})$ for a weakly coupled loop induced amplitude. In either case, the connection with the electroweak scale is lost!

Had we looked at $B_s^0\overline{B}_s^0$ mixing instead of $B_d^0\overline{B}_d^0$, the bounds would weaken because the CKM suppression of the latter Standard Model amplitude is only V_{ts}^2 instead of V_{td}^2. On the other hand, the strongest constraints come from CP violation in neutral kaon mixing. A summary of all existing $\Delta f = 2$ bounds after carefully taking into account the theoretical and experimental uncertainties is given in Table 8.1 [9]. If the new physics enters the mixing amplitudes through loops, the bounds weaken by a factor of $\sim 4\pi$.

Table 8.1 Lower bounds at 95% C.L. on the scale of new physics from FCNC mixing data in TeV for arbitrary new physics.

	$K^0\overline{K}^0$	$D^0\overline{D}^0$	$B_d^0\overline{B}_d^0$	$B_s^0\overline{B}_s^0$
$\Lambda_{\rm NP}$ [TeV]	2×10^5	5×10^3	2×10^3	3×10^2

So, insisting on a low-scale new physics model with $\Lambda_{\rm NP}$ of the order TeV, such as the Minimal Supersymmetric Standard Model (MSSM), we must suppress the otherwise overwhelming new physics amplitude by flavour mechanisms such as mixing or degeneracy. In other words, the physics at the TeV scale needs a specific flavour structure, which has similar suppression power to the one present in the Standard Model.

8.2.2
Null Tests

The sensitivity to physics beyond the Standard Model is enhanced in observables which are very small or essentially vanish in the Standard Model. Such observables are termed null tests. Null tests in flavour physics typically involve CP asymmetries or the helicity structure of the decays at hand. The reason is that both CP violation[21] and the left-handedness of the currents participating in the weak interaction are firm predictions of the Standard Model.

An important example is CP violation in $b \to s$ transitions: inspection of (8.14) shows that the magnitudes of the amplitudes for $b \to s$ and its CP conjugate $\bar{b} \to \bar{s}$ processes differ at order $\text{Im}\left[V_{ub}V_{us}^*/(V_{tb}V_{ts}^*)\right] \simeq \bar{\eta}\lambda^2$ which is of order 10^{-2}. Similarly, the CP phases entering $B_s^0 \bar{B}_s^0$ mixing in the Standard Model,

$$\beta_s \equiv \arg\left[-\frac{V_{ts}V_{tb}^*}{V_{cs}V_{cb}^*}\right] \simeq \bar{\eta}\lambda^2 \tag{8.19}$$

are predicted to be suppressed down to this level as well.

8.3
B-Meson Key Measurements

In the recent past many measurements of *B*-meson properties have been performed. They provided important experimental input for the understanding of the CKM mechanism and for the search for physics beyond the description of today's Standard Model. In the context of this book we cannot present all the key measurements adequately. Instead we limit the discussion to a few conceptually important observables, their determination and their physics potential:

- mixing of neutral *B* mesons;
- measurement of the CP violating mixing phases;
- search for very rare decays.

---

21) We do not consider here the θ term from the QCD Lagrangian.

8.3.1
Production of B Mesons

Large samples of B-meson decays are required to experimentally study loop-suppressed mixing, CP violation or strongly suppressed flavour changing neutral current processes. An efficient and copious source of B mesons are electron–positron storage rings operated at a centre-of-mass energy equivalent to the mass of the $\Upsilon(4S)$ resonance. The $\Upsilon(4S)$ is produced with a cross section of about 1 nb and decays nearly entirely to a $B^0\bar{B}^0$ or a B^+B^- pair. The first dedicated machines for the production of B mesons (DORIS II with the Crystal Ball and the ARGUS experiment, and CESR with the CUSB and the CLEO experiment) were built and operated already at the end of the 1970s, shortly after the discovery of the Υ resonances and the b quark.

The $\Upsilon(4S)$ mass of 10.58 GeV is only about 20 MeV higher than the kinematic threshold for the production of two B mesons, each with mass $m_B = 5.279$ GeV. As a consequence the mesons are produced essentially at rest. A determination of the B-meson decay time (of order picoseconds) using the flight-length and thus the time-resolved oscillation measurement is not possible. To overcome this experimental limitation, so-called "asymmetric B factories" were proposed for the second generation of such experiments. For these machines the energy of the colliding electron and positron are different and the Υ is boosted with respect to the laboratory system. With an electron energy of 9 GeV and a positron energy of 3.1 GeV (PEP-II storage ring at Stanford) a boost of the Υ along the beam direction of $\beta = 0.5$ is achieved, sufficient to resolve the flight distances of the two mesons.

The two asymmetric e^+e^- B factories, PEP-II at SLAC with the BABAR experiment and KEK-B in Japan with the BELLE experiment, were operated with typical beam currents between 2 and 3 A per stored beam. The instantaneous luminosities reached values up to 2×10^{34} cm^{-2} s^{-1}, corresponding to a rate of 20 Hz of $B\bar{B}$ pairs being produced. Over their lifetime the two experiments have collected together around 1.2×10^9 $B\bar{B}$ events.

An alternative and even more copious source of B hadrons are proton–(anti)proton collisions at the highest energies. The production cross sections for a $b\bar{b}$ quark pair are 100 µb at Tevatron ($p\bar{p}$ collisions at $\sqrt{s} = 2$ TeV) and more than 500 µb at LHC (pp collisions at $\sqrt{s} = 14$ TeV). A luminosity of 2×10^{32} cm^{-2} s^{-1} as proposed for the LHCb experiment at the LHC results in a $b\bar{b}$ production rate of 100 kHz. Physicists will have access to a produced b-hadron sample of the order of 10^{12} events per year. In addition to the huge production rate, proton–proton machines have a further advantage: all b-hadron species are produced, in particular B_s^0 mesons. A clear disadvantage with respect to e^+e^- B factories is the large particle multiplicity of proton–proton interactions. The largest fraction of the observable particles do not result from B-hadron decays but from fragmentation processes and the proton remnant. The full reconstruction of B decays is therefore more difficult than at the e^+e^- B factories.

8.3.2
Mixing of Neutral B Mesons

Mixing of neutral B mesons has been observed for the first time by the ARGUS collaboration in 1987 in the system of B_d^0 mesons [10]. The observation put stringent limits on the mass of the top quark: Only top masses larger than $\sim 50\,\text{GeV}$ could cause the necessary "deactivation" of the GIM suppression to observe mixing in B_d^0 mesons. Mixing in B_s^0 mesons is about 35 times faster and it took until 2006 before B_s^0 mixing was observed at the Tevatron [11, 12].

8.3.2.1 Mixing Phenomenology

Neutral B mesons, B_q^0 with q being either a d or an s quark, exhibit the phenomenon of $B^0\overline{B}^0$ mixing[22], that is, a $\Delta b = 2$ transition between the two flavour eigenstates B^0 and $\overline{B}^0$ of the meson. By convention the B_q^0 ($\overline{B}_q^0$) is a $\bar{b}q$ ($b\bar{q}$) quark combination.

In the Standard Model $B^0\overline{B}^0$ mixing is caused by flavour changing weak processes described at lowest order by the box diagrams in Figure 8.4. Because of the possibility of the $\Delta b = 2$ transition, an initially produced B^0 or $\overline{B}^0$ evolves in time into a superposition of B^0 and $\overline{B}^0$. The time evolution $B^0(t)$ ($\overline{B}^0(t)$) of an initially produced $B^0(\overline{B}^0)$, can be described by an effective 2-dimensional Schrödinger equation:

$$i\frac{d}{dt}\begin{pmatrix}|B(t)\rangle\\|\overline{B}(t)\rangle\end{pmatrix} = \left(M - i\frac{\Gamma}{2}\right)\begin{pmatrix}|B(t)\rangle\\|\overline{B}(t)\rangle\end{pmatrix}, \tag{8.20}$$

where M and Γ are 2×2 Hermitian matrices. The CPT theorem requires that the diagonal elements are equal. In case of mixing, the off-diagonal elements M_{12} and Γ_{12} are non-zero, and as a consequence, the flavour states B^0 and $\overline{B}^0$ in which the B mesons are produced are not mass eigenstates, defined as the eigenvectors of the Hamiltonian $M - i\Gamma/2$. The mass eigenstates are obtained by diagonalising the matrix and can be expressed as linear combinations of the flavour states,

$$|B_L\rangle = p|B^0\rangle + q|\overline{B}^0\rangle \tag{8.21}$$

$$|B_H\rangle = p|B^0\rangle - q|\overline{B}^0\rangle. \tag{8.22}$$

The generally complex coefficients q and p obey the normalisation condition $|p|^2 + |q|^2 = 1$. The indices L and H refer to the lighter (heavier) mass eigenstate. The time evolution of the mass state is given by the eigenvalues,

$$|B_{H,L}(t)\rangle = e^{-i(M_{H,L} - i\Gamma_{H,L}/2)t}|B_{H,L}\rangle \tag{8.23}$$

where $|B_{H,L}\rangle = |B_{H,L}(t=0)\rangle$ denotes the mass states at $t = 0$ and $M_{H,L}$, $\Gamma_{H,L}$ are the mass and the decay width of the heavy and light B-meson state.

---

22) While in the following B^0 represents both neutral B_q species, the mixing formalism presented here can also be applied to neutral K^0 or D^0 mesons.

The time evolution of a B meson produced in a pure flavour state

$$|B^0(t)\rangle = \frac{1}{2p}(|B_L(t)\rangle + |B_H(t)\rangle) \tag{8.24}$$

$$|\overline{B}^0(t)\rangle = \frac{1}{2q}(|B_L(t)\rangle - |B_H(t)\rangle) \tag{8.25}$$

can be easily derived from (8.23) by replacing the mass eigenstates by the flavour eigenstates according to (8.21) and (8.22):

$$|B^0(t)\rangle = g_+(t)|B^0\rangle + \frac{q}{p}g_-(t)|\overline{B}^0\rangle \tag{8.26}$$

$$|\overline{B}^0(t)\rangle = g_-(t)\frac{p}{q}|B^0\rangle + g_+(t)|\overline{B}^0\rangle \tag{8.27}$$

where the coefficients are given by

$$g_+(t) = e^{-i(m-i\frac{\Gamma}{2})t}\left[+\cosh\frac{\Delta\Gamma t}{4}\cos\frac{\Delta m t}{2} - i\sinh\frac{\Delta\Gamma t}{4}\sin\frac{\Delta m t}{2}\right] \tag{8.28}$$

$$g_-(t) = e^{-i(m-i\frac{\Gamma}{2})t}\left[-\sinh\frac{\Delta\Gamma t}{4}\cos\frac{\Delta m t}{2} + i\cosh\frac{\Delta\Gamma t}{4}\sin\frac{\Delta m t}{2}\right]. \tag{8.29}$$

In the last equations the heavy and light meson parameters have been replaced by the average mass and width, m, Γ, and the mass and width difference, $\Delta m, \Delta\Gamma$:

$$m = \frac{M_H + M_L}{2}, \quad \Gamma = \frac{\Gamma_H + \Gamma_L}{2}, \tag{8.30}$$

$$\Delta m = M_H - M_L, \quad \Delta\Gamma = \Gamma_L - \Gamma_H. \tag{8.31}$$

For negligible width difference $\Delta\Gamma \ll \Gamma$ – this case is realised for B_d^0 mesons – the expressions for the coefficients $g_\pm$ simplify to:

$$g_+(t) = e^{-i(m-i\frac{\Gamma}{2})t}\cos\frac{\Delta m t}{2}$$

$$g_-(t) = e^{-i(m-i\frac{\Gamma}{2})t}i\sin\frac{\Delta m t}{2} \tag{8.32}$$

$$|g_\pm(t)|^2 = \frac{e^{-\Gamma t}}{2}(1 \pm \cos\Delta m t).$$

The normalised probability $\mathcal{P}(B^0 \to B^0, t)$ to observe an initially produced B^0 after a time t in its original flavour state is thus given by[23]:

$$\mathcal{P}(B^0 \to B^0, t) = |\langle B^0|B^0(t)\rangle|^2 = \frac{e^{-\Gamma t}}{2}(1 + \cos\Delta m t). \tag{8.33}$$

23) Note that if $\Delta\Gamma \neq 0$, then for $t > 0$ the state is never again a pure flavour state.

Similarly the probability that it decays as the opposite flavour state is given by:

$$\mathcal{P}(B^0 \to \overline{B}^0, t) = |\langle \overline{B}^0 | B^0(t) \rangle|^2 = \frac{e^{-\Gamma t}}{2} \left|\frac{q}{p}\right|^2 (1 - \cos \Delta m t) \,. \tag{8.34}$$

In case that there is no CP violation in the mixing process itself, that is, $\mathcal{P}(B^0 \to \overline{B}^0, t) = \mathcal{P}(\overline{B}^0 \to B^0, t)$, one finds $|\frac{p}{q}| = 1$ simplifying the above equations. The time-dependent mixing asymmetry $A_\mathrm{mix}(t)$ is in this case a simple cosine function of the time, with the mass difference, Δm, as the oscillation frequency:

$$A_\mathrm{mix}(t) = \frac{\mathcal{P}(B^0 \to B^0, t) - \mathcal{P}(B^0 \to \overline{B}^0, t)}{\mathcal{P}(B^0 \to B^0, t) + \mathcal{P}(B^0 \to \overline{B}^0, t)} = \cos \Delta m t \,. \tag{8.35}$$

The mixing asymmetry is determined using *time-dependent decay rates* $\Gamma(t)$ of flavour-specific decays, $B^0 \to f$ and $\overline{B}^0 \to \overline{f}$ (f being different from $\overline{f}$),

$$A_\mathrm{mix}(t) = \frac{\Gamma(B^0 \to f, t) - \Gamma(B^0 \to \overline{f}, t)}{\Gamma(B^0 \to f, t) + \Gamma(B^0 \to \overline{f}, t)} \,. \tag{8.36}$$

In the Standard Model the short-distance contribution to $B^0\overline{B}^0$ mixing can be calculated from the transition amplitudes sketched in Figure 8.4. From (8.15) one finds for the mixing frequency Δm_q of B_d^0 and B_s^0 mesons ($q = d, s$),

$$\Delta m_q \propto \frac{G_F^2}{6\pi^2} M_W^2 \, S\!\left(\frac{m_t^2}{M_W^2}\right) \left|V_{tb} V_{tq}^*\right|^2 \,. \tag{8.37}$$

With V_{ts} being of $\mathcal{O}(\lambda^2)$ and V_{td} being of $\mathcal{O}(\lambda^3)$ one expects a mixing frequency $\Delta m_s \gg \Delta m_d$.

8.3.2.2 Measurement of $B^0\overline{B}^0$ Mixing

The experimental difficulty of the time-dependent mixing measurement is the determination of the B-meson's flavour at production, usually referred to as "flavour tagging", and the measurement of the B decay time which requires detectors to precisely determine the B production and decay vertex. The B-meson flavour at the time of decay is usually determined by analysing the decay particles of a flavour-specific decay such as $B_d^0 \to D^{-*}\ell^+ \nu$ or $B_s^0 \to D_s^- \pi^+$.

Flavour Tagging Most procedures to determine the initial B-meson flavour exploit the fact that the B meson is produced together with a b hadron of opposite flavour. In general the algorithms use a flavour-specific decay of the second b hadron (B_tag) to identify its flavour. The flavour of the signal B (B_signal) is then inferred from the flavour of the tagged B. If the second b hadron, referred to as the tagging B, is a neutral B meson, it also mixes and its decay flavour is in general no longer correlated with the production flavour of the signal B.

At the e^+e^- B factories the B mesons are either produced as an $B^0\overline{B}^0$ or as an B^+B^- pair, while at hadron machines the second b hadron can be any B meson or b baryon as long as the b quark content is opposite to the signal B under study. If the $B^0\overline{B}^0$ mesons at the B factories would evolve independently in

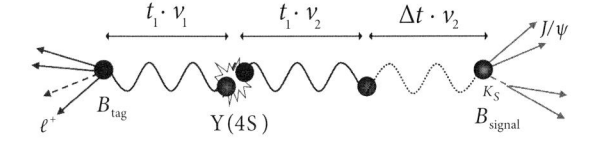

Figure 8.5 At the $\Upsilon(4S)$ the two neutral B mesons are produced in a coherent state.

time, as is the case for the two b hadrons produced in the proton collisions of the LHC, a mixing measurement at the B factories would not be possible. Instead, the $B^0\overline{B}^0$ mesons are produced in a coherent $L = 1$ state $|B_1(t)B_2(t)\rangle = 1/\sqrt{2}(|B^0(t)\rangle|\overline{B}^0(t)\rangle - |\overline{B}^0(t)\rangle|B^0(t)\rangle)$. In this state, the decay of B_1 at the time t_1 defines the flavour of B_2 at the same time, as sketched in Figure 8.5. At e^+e^- B factories the flavour of the tagging B is therefore not determined at the production time but at the time t_1 when one of the B mesons decays. The mixing is measured as a function of the decay time difference $\Delta t = t_2 - t_1$. Note that t_2 can also be smaller than t_1, which is equivalent to the signal B decaying before the tagging B.

There are different methods, with in general very different efficiencies and purities, to determine the flavour of the tagging B. The classical tagging method uses semileptonic decays $\overline{B}^0 \to X_c^+ \ell^- \overline{\nu}$, where the charge of the negative lepton is a tag for a $b \to c$ transition and identifies the tagging B as $\overline{B}^0$. Semileptonic decay channels are also used to reconstruct the flavour of the signal B. Wrongly tagged B mesons result in a reduction of the observable amplitude of the time-dependent mixing asymmetry.

Decay-Time Determination The decay rates in (8.36) have to be measured as a function of the B lifetime, if the tagging and signal B evolve independently in time, or as a function of the lifetime differences, Δt, if the two B mesons are in a coherent state. The decay time is determined from the flight distance, L, between the primary vertex and the signal decay vertex (LHC), or between the two B-decay vertices (B factories). Using the momentum, $\boldsymbol{p}_B$, of the signal B, the decay time t is given by

$$t = \frac{L}{\beta_B}\gamma = L\frac{m_B}{|\boldsymbol{p}_B|}. \qquad (8.38)$$

Note that the determination of the B momentum requires the reconstruction of all decay particles. In case of semileptonic B decays the unmeasured neutrino is missing in the reconstruction of the B kinematics. While it is possible to correct for the missing momentum on average the $\boldsymbol{p}_B$ resolution of a single event is significantly degraded. For B_s^0 mesons, for which the oscillation frequency is large, this effect prevents the usage of semileptonic decays. Instead the flavour-specific decay $B_s^0 \to D_s^- \pi^+$ is used to determine the flavour of the signal B.

Experimental Status Figure 8.6 shows the time-dependent (Δt) mixing asymmetry for B_d^0 and $\overline{B}_d^0$ decays as measured by the BABAR collaboration. The line is the fit of the expected oscillation curve which takes resolution and the wrong-tag probability into account. The world average for the mixing frequency is $\Delta m_d = (0.507 \pm 0.005)\hbar \text{ ps}^{-1}$ [6]. The B_d^0 oscillates approximately 0.12 times during its lifetime (1.5 ps).

The Tevatron experiments CDF and DØ have observed the much faster mixing of B_s^0 mesons. For the mixing frequency, the CDF experiment found the value $\Delta m_s = (17.77 \pm 0.12)\hbar \text{ ps}^{-1}$ [11] which corresponds to four full oscillation cycles per B_s^0 lifetime. The B_s^0 lifetime is close to the one of the B_d^0 mesons, about 1.5 ps. The measured ratio between the mixing frequencies of B_s^0 and B_d^0 mesons, $\Delta m_s/\Delta m_d \approx 35$, agrees well with the theoretical prediction.

8.3.3
Measurement of the CP Violating Mixing Phases

Until 2001 the neutral kaon system was the only one for which CP violation had been observed. The Standard Model predicts large CP violation through the interference between mixing and direct decay in B_d^0 decays to CP eigenstates. The experimental observation is however challenging. The different decay behaviour

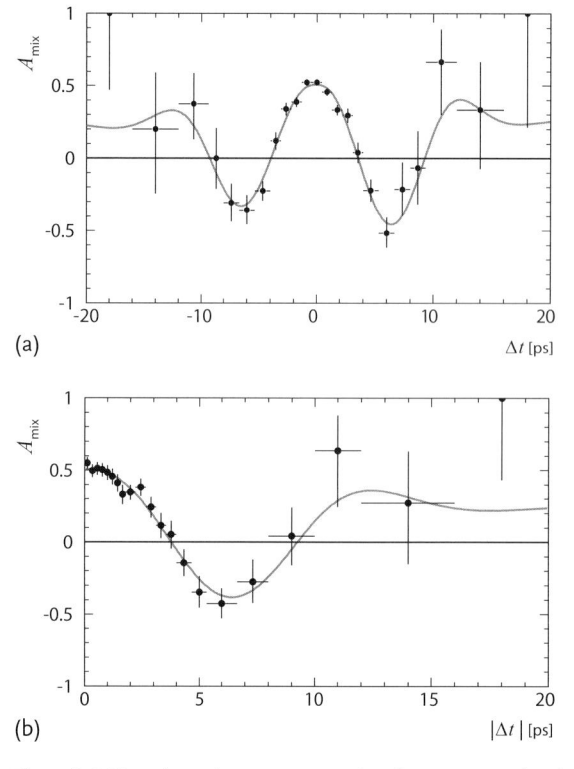

Figure 8.6 Time-dependent asymmetry A_mix between mixed and unmixed events: (a) as a function of Δt; (b) folded as a function of $|\Delta t|$ (adapted from [13]).

of B_d^0 and $\overline{B}_d^0$ mesons must be studied as a function of decay time. Furthermore, the branching fractions of the decays in question are small. It was the successful start-up of the asymmetric B factories which finally allowed the observation of CP violation in B^0 decays.

8.3.3.1 Phenomenology of CP Violation in *B*-Meson Decays

In the Standard Model the only source of CP violation is the CP violating phase of the CKM matrix. The observation of phases and thus also the observation of CP violation in meson decays requires the interference of at least two amplitudes.

For neutral B mesons CP violation can be observed as a result of three different mechanisms:

- *CP violation in mixing.* The time-dependent mixing probability of a B^0 might differ from the one of a $\overline{B}^0$, $\mathcal{P}(B^0 \to \overline{B}^0, t) \neq \mathcal{P}(\overline{B}^0 \to B^0, t)$. In this case, the ratio $\left|\frac{q}{p}\right|$ is different from one and CP violation could be observed in any flavour-specific final state of the B^0. An example is semileptonic decays, $B^0 \to \ell^+ \nu X$, for which the time-dependent decay asymmetry $A_{\text{sl}}(t)$ to "wrong-sign" leptons is given by:

$$A_{\text{sl}}(t) = \frac{\Gamma(\overline{B}^0 \to \ell^+ \nu X, t) - \Gamma(B^0 \to \ell^- \nu X, t)}{\Gamma(\overline{B}^0 \to \ell^+ \nu X, t) + \Gamma(B^0 \to \ell^- \nu X, t)}$$
$$= \frac{|p/q|^2 - |q/p|^2}{|p/q|^2 + |q/p|^2} = \frac{1 - |q/p|^4}{1 + |q/p|^4}. \quad (8.39)$$

The Standard Model prediction for A_{sl} for B mesons is small, $\mathcal{O}(10^{-4})$, and thus not easy to measure, especially as detector asymmetries, different material interaction of particles and antiparticles, and, in the case of the LHC, also the possibility of a production asymmetry between B^0 and the corresponding $\overline{B}^0$ meson have to be accounted for.

- *CP violation in decay.* For a B-meson decay to a final-state f, the ratio $|\overline{A}_{\overline{f}}/A_f|$, where A_f and $\overline{A}_{\overline{f}}$ are the amplitudes for the CP conjugated processes $B \to f$ and $\overline{B} \to \overline{f}$ respectively, is a CP observable. Clearly any deviation from unity would signal CP violation in the decay, also called *direct CP violation*. To observe CP violation in decay $|A_f| \neq |\overline{A}_{\overline{f}}|$ the $B \to f$ decay amplitude must be the sum of at least two different amplitudes, A_k, with different weak phases, ϕ_k, which, under CP conjugation, change sign[24], and with different strong phases δ_k which are CP invariant. One can thus write the total amplitude A_f and its CP conjugated counterpart $\overline{A}_{\overline{f}}$ as:

$$A_f = \sum_k A_k e^{i(\delta_k + \phi_k)}, \quad \overline{A}_{\overline{f}} = \sum_k A_k e^{i(\delta_k - \phi_k)}. \quad (8.40)$$

A prominent example is the decay $B^0 \to K^+ \pi^-$ to which a Cabibbo suppressed tree amplitude together with higher-order loop diagrams (so-called penguin di-

---

24) Complex CKM elements entering the decay amplitude A_f, enter the CP conjugate amplitude $\overline{A}_{\overline{f}}$ in the complex conjugate form.

agrams) contribute. A large CP asymmetry of about 10% is observed for this channel [6].

- *CP violation through the interference between decay and mixing.* For neutral B mesons which decay to a CP eigenstate f_{CP} such as $B^0 \to J/\psi K_S^0$, the amplitudes where the B^0 decays directly to the final state, $B^0 \to f_{CP}$, and where it first changes into a $\overline{B}^0$ before it decays, $B^0 \to \overline{B}^0 \to f_{CP}$, interfere, as shown in Figure 8.7. The interference can result in a large observable time-dependent CP asymmetry. It is useful to introduce the parameter

$$\lambda_{f_{CP}} = \frac{q}{p} \frac{\overline{A}_{f_{CP}}}{A_{f_{CP}}} = \eta_{CP} \frac{q}{p} \frac{\overline{A}_{\overline{f}_{CP}}}{A_{f_{CP}}} . \tag{8.41}$$

Here, $A_{f_{CP}}$ and $\overline{A}_{f_{CP}}$ denote the amplitudes for $B^0 \to f_{CP}$ and $\overline{B}^0 \to f_{CP}$, respectively. The second step uses the property $\overline{A}_{f_{CP}} = \eta_{CP} \overline{A}_{\overline{f}_{CP}}$, where η_{CP} is the CP eigenvalue of f_{CP} ($CP|f_{CP}\rangle = |\overline{f}_{CP}\rangle = \eta_{CP}|f_{CP}\rangle$). The amplitudes $A_{f_{CP}}$ and $\overline{A}_{\overline{f}_{CP}}$ are related by CP conjugation and differ only in the signs of their weak phase, while the CP eigenvalue, η_{CP}, of the final state is ± 1. A relative phase can result in $\lambda_{f_{CP}} \neq 1$ even in the case that both $|q/p|$ and $|A_{f_{CP}}/\overline{A}_{f_{CP}}|$ are equal to 1. A deviation of $\lambda_{f_{CP}}$ from 1 will be observable as a time-dependent CP asymmetry:

$$\begin{aligned} A_{CP}(t) &= \frac{\Gamma(\overline{B}^0 \to f_{CP}, t) - \Gamma(B^0 \to f_{CP}, t)}{\Gamma(\overline{B}^0 \to f_{CP}, t) + \Gamma(B^0 \to f_{CP}, t)} \\ &= \frac{(|\lambda_{f_{CP}}|^2 - 1) \cos \Delta m_d t + 2 \operatorname{Im} \lambda_{f_{CP}} \sin \Delta m_d t}{1 + |\lambda_{f_{CP}}|^2} . \end{aligned} \tag{8.42}$$

If $|q/p| = 1$ (no CP violation in mixing) and $|A_{f_{CP}}/\overline{A}_{f_{CP}}| = 1$ (no CP violation in decay) one obtains $|\lambda_{f_{CP}}| = 1$ and the expression for A_{CP} simplifies to

$$A_{CP}(t) = \operatorname{Im} \lambda_{f_{CP}} \sin \Delta m_d t . \tag{8.43}$$

The observed CP asymmetry in this case is a measure of the phase difference between the two interfering amplitudes. If the amplitudes $A_{f_{CP}}$ and $\overline{A}_{\overline{f}_{CP}}$ do not acquire a weak phase, the time-dependent CP asymmetry is a measure of the mixing-phase, ϕ_M, defined as

$$M_{12} = |M_{12}| e^{i\phi_M} , \tag{8.44}$$

where M_{12} is the off-diagonal element of the dispersive part of the Hamiltonian in (8.20). In the Standard Model, the mixing is a result of the $\Delta b = 2$ transition amplitudes shown in Figure 8.4, where, due to the inactive GIM suppression, only the diagrams involving a top quark contribute significantly. The mixing phases ϕ_d and ϕ_s of the $B_d^0 \overline{B}_d^0$ and $B_s^0 \overline{B}_s^0$ mixing respectively are given in the Standard Model by:

$$\phi_d^{SM} = \arg\left(\frac{V_{tb} V_{td}^*}{V_{tb}^* V_{td}}\right) \approx 2\beta \tag{8.45}$$

8.3 B-Meson Key Measurements

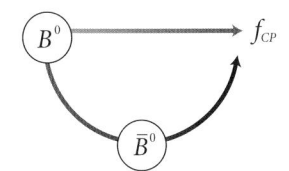

Figure 8.7 For a B^0 decay into a CP eigenstate the amplitude of the direct decay and the amplitude where the B^0 first mixes into a $\bar{B}^0$ interfere.

$$\phi_s^{SM} = \arg\left(\frac{V_{tb}V_{ts}^*}{V_{tb}^*V_{ts}}\right) \approx -2\beta_s, \quad (8.46)$$

where β is the unitarity triangle angle defined in (8.12). The angle β_s is defined in (8.19) and can be approximated by $\bar{\eta}\lambda^2$, which is very small, about 1°. The precise measurement of the time-dependent CP asymmetries for the decays of neutral B mesons to CP eigenstates offers the possibility to explore the phases of the $\Delta b = 2$ transitions, which in the Standard Model are given by the phases of the CKM matrix elements V_{td} and V_{ts}. Deviations from the expected values would signal additional contributions to the mixing amplitude beyond the description of the Standard Model and thus offer the possibility to search for new effects.

8.3.3.2 Measurement of the B_d^0 Mixing Phase Using $B_d^0 \to J/\psi K_S^0$ Decays

The decay mode $B_d^0 \to J/\psi K_S^0$, for which the Feynman diagram of the decay is sketched in Figure 8.8, is *the* prototype of the neutral B-meson decays into a CP eigenstate discussed above. For this decay no direct CP violation is expected.

The ratio of the two amplitudes (direct decay and decay after mixing) as defined in (8.41) is given by,

$$\lambda_{f_{CP}}(B_d^0 \to J/\psi K_S^0) = -\left(\frac{V_{tb}^*V_{td}}{V_{tb}V_{td}^*}\right)\left(\frac{V_{cs}^*V_{cb}}{V_{cs}V_{cb}^*}\right)\left(\frac{V_{cd}^*V_{cs}}{V_{cd}V_{cs}^*}\right). \quad (8.47)$$

The minus sign is the CP eigenvalue, $\eta_{CP} = -1$, of the final state, the first term is the result of the $B^0\bar{B}^0$ mixing, and the following term describes the ratio of the decay amplitude and its CP conjugate. The last term accounts for the fact that the K^0 and $\bar{K}^0$ mix to the observed K_S^0. Analysis of the CKM elements reveals that to $\mathcal{O}(\lambda^3)$ of the Wolfenstein parameter, the only term carrying a non-trivial phase is the mixing term with a phase $\phi_d^{SM} = 2\beta$ (see (8.45)). For the time-dependent CP asymmetry one therefore expects

$$A_{CP}(t) = \sin 2\beta \sin \Delta m_d t. \quad (8.48)$$

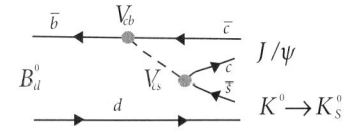

Figure 8.8 Flavour flow of the decay $B_d^0 \to J/\psi K_S^0$.

Besides its negligible theoretical uncertainties the decay mode $B_d^0 \to J/\psi K_S^0$ is also experimentally favourable. It has a relatively large branching fraction (4×10^{-4}) with readily accessible final states and small backgrounds. The measurement proceeds by first reconstructing the B meson through its decay products, tagging its flavour by using the flavour information of the partner b hadron, measuring the decay time (LHC) or the decay time difference to its partner b hadron (B factories) and finally determining the time-dependent asymmetry by comparing the decay time distribution of the B^0 and $\overline{B}^0$ mesons.

Figure 8.9a shows the BABAR measurement of the time-dependent decay rates for an initial B^0 or $\overline{B}^0$ meson. The distributions clearly differ; the resulting "raw" CP asymmetry is shown in Figure 8.9b exhibiting a clear sinusoidal behaviour. The term "raw" means that the measurement is not corrected for imperfect flavour tagging and finite time-resolution, which both result in a dilution of the observed amplitude of the time-dependent CP asymmetry. The fit to extract $\sin(2\beta)$ (the result is shown as a line) accounts for both effects, as well as for additional background contributions.

The determination of the unitarity triangle angle β from the time-dependent CP asymmetry in $B_d^0 \to J/\psi K_S^0$ decays is currently the most precise determination of any CKM phase. As discussed already, the angle agrees well with what one expects from the determination of the lengths of the triangle sides from CP-conserving observables.

8.3.3.3 Measurement of the B_s^0 Mixing Phase Using $B_s^0 \to J/\psi \phi$ Decays

While the mixing of B_d^0 mesons is very well explored and both mixing frequency as well as the mixing phase are well measured, the experimental situation for the B_s^0 system is less conclusive. After the observation of B_s^0 mixing in 2006, the Tevatron experiments CDF and DØ have also made first measurements of the B_s^0 mixing phase [15, 16]. The very fast B_s^0 mixing together with the relatively small expected

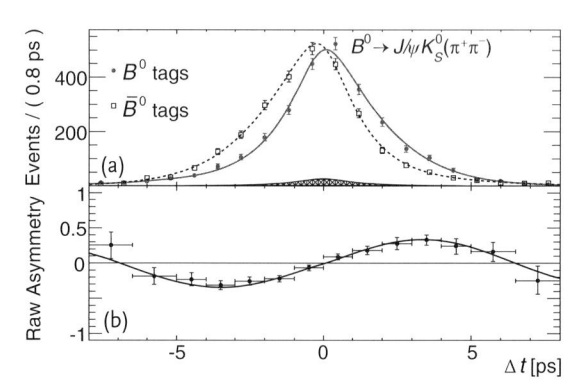

Figure 8.9 CP asymmetry for the channel $B_d^0 \to J/\psi K_S^0$ as measured by the BABAR collaboration. The closed circles show B^0 tags, while the open squares show $\overline{B}^0$ tags. (a) Decay rate; (b) "raw" asymmetry (adapted from [14]).

value of the phase $\phi_s^{SM} = -2°$ makes the observation of a time-dependent CP asymmetry in B_s^0 decays an experimental challenge.

The counterpart of the decay $B_d^0 \to J/\psi K_S^0$, used to measure the time-dependent CP asymmetry in the B_d^0 system, is the decay mode $B_s^0 \to J/\psi \phi$. The respective diagram is shown in Figure 8.10. The main theoretical treatment of the expected CP asymmetry for the $B_s^0 \to J/\psi \phi$ mode follows the above description for $B_d^0 \to J/\psi K_S^0$. No CP violation in the decay is expected and the observed time-dependent CP asymmetry is only a result of a CP violating phase in $B_s^0 \overline{B}_s^0$ mixing.

However, there are important differences from the channel $B_d^0 \to J/\psi K_S^0$: contrary to $J/\psi K_S^0$, the final-state $J/\psi \phi$ is a mixture of different CP eigenstates; the two vector particles (J/ψ and ϕ both have spin, $J = 1$) produced in the decay of the pseudoscalar B_s^0 meson can have relative orbital momentum. The CP eigenvalue of the ($J/\psi, \phi$) system depends on the relative orbital momentum, l, and is given by,

$$\eta_{CP}^l = (-1)^l \eta_{CP}(J/\psi) \eta_{CP}(\phi) . \tag{8.49}$$

The S and D-wave components with orbital momenta $l = 0, 2$ are CP even ($\eta_{CP} = +1$) while the P-wave component $l = 1$ state is CP odd ($\eta_{CP} = -1$). A measurement of the CP asymmetry, of course, demands the separation of CP-even and CP-odd states. Experimentally, only a statistical separation is possible and requires an analysis of the angular distributions of the decay particles. The total amplitude can be decomposed into three partial amplitudes which take into account the polarisation of the two vector mesons: both vector mesons are longitudinally polarised (CP even), both mesons are transversely polarised with meson spins orthogonal to each other (CP odd), and both mesons are transversely polarised with meson spins parallel to each other (CP even). The amplitudes show a large dependence on the decay product angles which can be exploited to separate the different CP states. For details see [17].

Another important difference from $B_d^0 \to J/\psi K_S^0$ is the large decay width difference, $\Delta \Gamma_s / \Gamma_s \approx 10\%$, for the B_s^0 system. As the CP violation in mixing is very small, the two mass eigenstates, $B_{H,s}$ and $B_{L,s}$ (with different lifetimes), are, to a good approximation, pure CP eigenstates. The lifetime distribution of $B_s^0 \to J/\psi \phi$ decays thus also allows a determination of the lifetime difference $\Delta \Gamma$.

As the predicted CP violation for the channels $B_s^0 \to J/\psi \phi$ is very small the channel exhibits an excellent potential to search for physics beyond the Standard Model. The observation of large CP violation would clearly signal new physics contributions to the $b \to s$ transitions in the B_s^0 mixing amplitude (see Section 8.2.2).

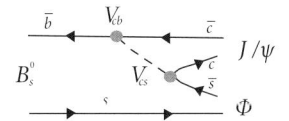

Figure 8.10 Flavour flow of the decay $B_s^0 \to J/\psi \phi$.

8.3.4
Search for Very Rare Decays

Prime examples of very rare decays are the $B^0_{d,s} \to \mu^+\mu^-$ modes. As FCNC decays, they proceed in the Standard Model only through loop diagrams. The dominant contributions come from W box and Z penguin diagrams. A Z penguin diagram for the decay $B^0_s \to \mu^+\mu^-$ is shown in Figure 8.11. The branching fraction for the decays $B_q \to \mu^+\mu^-$ is given by,

$$\mathcal{B}(B_q \to \mu^+\mu^-) = \frac{G_F^2 \alpha_{em}^2 m_{B_q}^5 \tau_{B_q} f_{B_q}^2}{64\pi^3} \left|V_{tb}V_{tq}^*\right|^2 \sqrt{1 - \frac{4m_\mu^2}{m_{B_q}^2}}$$

$$\times \left\{ \left(1 - \frac{4m_\mu^2}{m_{B_q}^2}\right) \left|\frac{C_S}{m_b}\right|^2 + \left|\frac{C_P}{m_b} + \frac{2m_\mu}{m_{B_q}^2} C_{10}\right|^2 \right\}, \quad (8.50)$$

where f_{B_q} is the B_q meson decay constant and $\tau_{B_q} = 1/\Gamma_q$ is the B_q meson lifetime. The so-called Wilson coefficients C_S, C_P and C_{10} are loop functions similar to (8.15) for the $B^0\bar{B}^0$ mixing. In the Standard Model, C_S^{SM}, $C_P^{SM} \approx 0$ while $C_{10}^{SM} \approx -4$. The strong helicity suppression $\mathcal{B}^{SM} \propto m_\mu^2/m_{B_q}^2$ leads to very small Standard Model branching fractions. For the channel $B^0_s \to \mu^+\mu^-$ the Standard Model predicts a value of 3.3×10^{-9} while the branching fraction for $B^0_d \to \mu^+\mu^-$ is further suppressed by the factor $|V_{td}|^2/|V_{ts}|^2 \approx \lambda^2 \approx 1/20$.

In many extensions of the Standard Model, loop graphs with new particles (such as charged Higgs or supersymmetric partners) contribute at the same order as the SM contribution and could significantly enhance the observable branching fraction. An example for a new physics amplitude in the case of MSSM is shown in Figure 8.11. The amplitude is enhanced for large values of $\tan\beta$, which is the ratio of the vacuum expectation values for the two Higgs doublets of the MSSM (see Chapter 7).

The Tevatron experiments can probe branching fractions which are about a factor ten higher than the Standard Model expectations for the two channels $B^0_{d,s} \to \mu^+\mu^-$. If new physics does not enhance the branching fractions significantly, the

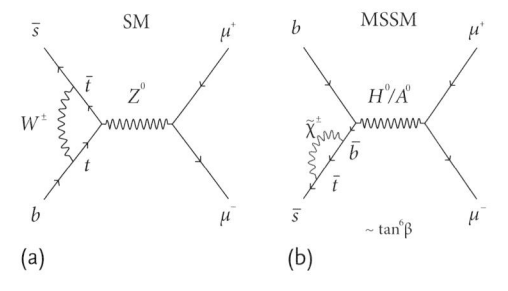

Figure 8.11 Standard Model Feynman diagram (a) and MSSM contribution (b) to the decay $B^0_s \to \mu^+\mu^-$.

observation of the rare $B^0_{d,s} \to \mu^+\mu^-$ decays is thus reserved for the LHC experiments.

8.4
Flavour at the Terascale – Outlook

In this chapter we discussed flavour physics in the quark sector. In the Standard Model, the couplings and parameters responsible for the quarks generation structure, the masses and CKM mixing originating from the Yukawa matrices, are known and to date describe consistently every observed quark flavour and CP violating phenomenon[25].

At the same time the non-observation of new physics in FCNCs implies that the flavour structure at the TeV scale must be similar to that of the Standard Model. It could well be that the physics at the Terascale reveals no further sources of flavour violation beyond those of the Standard Model. This class of models is termed *minimally flavour violating*. Well-known examples are MSSM variants with a flavour-blind supersymmetry breaking mechanism, such as gauge or anomaly mediation.

We put strong emphasis on *b* physics. It has been crucial for the identification of CKM as the, at least, dominant source of hadronic CP violation. Ongoing and future dedicated *b* initiatives will contribute in various ways to flavour physics [18–20]: besides improving the knowledge of the CKM elements, one would like to probe regions which have not been explored so far – the $B^0_s \overline{B}^0_s$ mixing phase is one very important example where striking signals of new physics can show up. Precision studies to probe the CP, chirality, or Dirac structure of the FCNCs, such as in the rare leptonic $B_{(s)} \to \mu^+\mu^-$ and semileptonic $B \to K^{(*)}\ell^+\ell^-$ decays are of great interest as well.

There are important aspects of flavours other than the *b* quark which were not discussed:

Top quark, the other third generation quark: are the top FCNCs as invisible as predicted by the Standard Model?
Charm quark, a detailed study of $D^0\overline{D}^0$ mixing, in particular CP violation: are the charm FCNCs suppressed as predicted by the Standard Model?
Kaons, measurements of the ultra-rare decays $K \to \pi\nu\nu$.

To conclude, irrespective of the details of the electroweak physics that may be revealed at the LHC, what we will be able to find out is to which extent TeV physics is minimally flavour violating or not. The observation of flavour effects that cannot be traced back to the Standard Model Yukawa matrices could point towards the origin of generational mixing and hierarchies, that is, the origin of flavour.

---

25) Modulo hints: a hint is also called an anomaly or tension. All of them refer to an intriguing, but at the moment not significant, discrepancy between data and the Standard Model.

References

1 Cabibbo, N. (1963) Unitary symmetry and leptonic decays, *Phys. Rev. Lett.*, **10**, 531, doi:10.1103/PhysRevLett.10.531.
2 Kobayashi, M. and Maskawa, T. (1973) CP violation in the renormalizable theory of weak interaction, *Prog. Theor. Phys.*, **49**, 652, doi:10.1143/PTP.49.652.
3 Wolfenstein, L. (1983) Parametrization of the Kobayashi–Maskawa Matrix, *Phys. Rev. Lett.*, **51**, 1945, doi:10.1103/PhysRevLett.51.1945.
4 Charles, J. et al. (2005) CP violation and the CKM matrix: assessing the impact of the asymmetric B factories, *Eur. Phys. J. C*, **41**, 1, doi:10.1140/epjc/s2005-02169-1, (update September 2009).
5 Buras, A.J. and Lindner, M. (eds.) (1998) *Heavy Flavours II*, World Scientific, Singapore.
6 Heavy Flavor Averaging Group (HFAG) (2006) Averages of Summer 2010, http://www.slac.stanford.edu/xorg/hfag.
7 Particle Data Group, Amsler, C. et al. (2008) Review of Particle Physics, *Phys. Lett. B*, **667**, 1
8 Inami, T. and Lim, C.S. (1981) Effects of superheavy quarks and leptons in low-energy weak processes $K_L \to \mu\bar{\mu}$, $K^+ \to \pi^+ \nu\bar{\nu}$ and $K^0 \leftrightarrow \bar{K}^0$, *Prog. Theor. Phys.*, **65**, 297, doi:10.1143/PTP.65.297.
9 Bona, M. et al. (2008) Model-independent constraints on $\Delta F = 2$ operators and the scale of new physics, *JHEP*, **03**, 049, doi:10.1088/1126-6708/2008/03/049.
10 ARGUS Collab., Albrecht, H. et al. (1987) Observation of B^0–$\bar{B}^0$ mixing, *Phys. Lett. B*, **192**, 245, doi:10.1016/0370-2693(87)91177-4.
11 CDF Collab., Abulencia, A. et al. (2006) Observation of B_s^0–$\bar{B}_s^0$ oscillations, *Phys. Rev. Lett.*, **97**, 242003, doi:10.1103/PhysRevLett.97.242003.
12 DØ Collab., Abazov, V.M. et al. (2006) First direct two-sided bound on the B_s^0 oscillation frequency, *Phys. Rev. Lett.*, **97**, 021802, doi:10.1103/PhysRevLett.97.021802.
13 BABAR Collab., Aubert, B. et al. (2002) A study of time-dependent CP-violating asymmetries and flavor oscillations in neutral B decays at the $\Upsilon(4S)$, *Phys. Rev. D*, **66**, 032003, doi:10.1103/PhysRevD.66.032003.
14 BABAR Collab., Aubert, B. et al. (2009) Measurement of time-dependent CP asymmetry in $B^0 \to c\bar{c}K^{(*)0}$ decays, *Phys. Rev. D*, **79**, 072009, doi:10.1103/PhysRevD.79.072009.
15 CDF Collab., Aaltonen, T. et al. (2008) First flavor-tagged determination of bounds on mixing-induced CP violation in $B_s^0 \to J/\psi\phi$ decays, *Phys. Rev. Lett.*, **100**, 161802, doi:10.1103/PhysRevLett.100.161802.
16 DØ Collab., Abazov, V.M. et al. (2008) Measurement of B_s^0 mixing parameters from the flavor-tagged decay $B_s^0 \to J/\psi\phi$, *Phys. Rev. Lett.*, **101**, 241801, doi:10.1103/PhysRevLett.101.241801.
17 LHCb Collaboration (2009) Roadmap for selected key measurements of LHCb, http://arxiv.org/abs/0912.4179.
18 Anikeev, K. et al. (2001) B physics at the Tevatron: Run II and beyond, http://arxiv.org/abs/hep-ph/0201071.
19 Hewett, J.L. et al. (2004) The Discovery Potential of a Super B Factory, http://arxiv.org/abs/hep-ph/0503261.
20 Artuso, M. et al. (2008) B, D and K decays, *Eur. Phys. J. C*, **57**, 309, doi:10.1140/epjc/s10052-008-0716-1.

9
Top Quarks: the Peak of the Mass Hierarchy?

Peter Uwer and Wolfgang Wagner

9.1
Introduction

In this chapter we discuss the heaviest elementary particle known so far, the top quark. With a mass of 173.1 ± 1.3 GeV [1] the top quark is almost as heavy as a gold atom (actually the mass is about the same as the mass of rhenium – but who knows rhenium?). While a gold atom has a radius of about 10^{-10} m our current understanding is that the top quark is point-like. This is experimentally probed to the incredibly small distance of 10^{-18} m. As it turns out, the top quark has many more interesting properties, making it a highly fascinating and unique research field. Since many of these properties are direct consequences of the Standard Model (SM), let us first recall the important aspects introduced in Chapter 2.

In the SM, the left-handed fields appear in doublets which transform according to the fundamental representation of the $SU(2)$ gauge group connected with the weak isospin. When the bottom quark was discovered it was thus clear that the up-type partner should also exist. Its existence and also its couplings to the electroweak gauge bosons could be predicted from the gauge structure. These couplings are a direct consequence of the top-quark quantum numbers, $I_3 = 1/2$, $Y = 1/6$ ($Q = I_3 + Y = 2/3$). For details we refer again to Chapter 2. Although the discovery of the bottom quark already suggested the existence of the top quark, it took more than 20 years until it was finally observed. A first hint was obtained at LEP. Since the energy of the collider was not high enough for direct production, the evidence was only indirect.

The SM is theoretically inconsistent without the top quark. If one removes the top quark from the theory by considering the limit $m_t \to \infty$, the results are divergent. More precisely, the correction shown in Figure 9.1 is proportional to the top-quark mass squared. Even for collider energies below the production threshold one can thus indirectly see the effect of new heavy particles through their appearance in quantum corrections, provided the effect gives a sizable contribution and the measurements are precise enough. The fact that the existence of the top quark

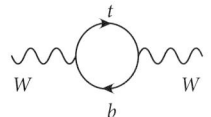

Figure 9.1 Top-quark contribution to W boson self-energy corrections.

could indeed be established through quantum effects is a great success of theoretical particle physics. It also demonstrates the need for precision measurements.

The discovery of the top quark was finally achieved at the Fermilab Tevatron in 1995 by the collider experiments CDF [2] and DØ [3]. The top-quark mass as determined from electroweak precision measurements is shown as a function of time in Figure 9.2; the figure also includes the direct measurements made at the Tevatron from 1995 to 2010.

Assuming the validity of the SM the top-quark properties are highly constrained through the gauge structure of the SM. The only free parameters that are not predicted in the SM are the mass and the relevant entries in the Cabibbo–Kobayashi–Maskawa (CKM) matrix describing the mixing between different quark flavours. Assuming the existence of only three families and the unitarity of the CKM matrix, one can determine the matrix elements V_{td}, V_{ts}, V_{tb} from indirect measurements. It turns out that $|V_{td}|$, and $|V_{ts}|$ are small while $|V_{tb}|$ is close to one. As a consequence, the dominant decay of the top quark in the SM is the decay into a W boson and a b quark. From the decay width of 1.35 GeV at $m_t = 173.1$ GeV, a mean lifetime τ of about 5×10^{-25} s is obtained. The top quark is thus extremely short-lived. Comparing the lifetime with the typical timescale for forming hadronic bound states, estimated from Λ_{QCD} or equivalently from the typical size of a hadron, the top quark decays before it has time to form hadrons. To some extent,

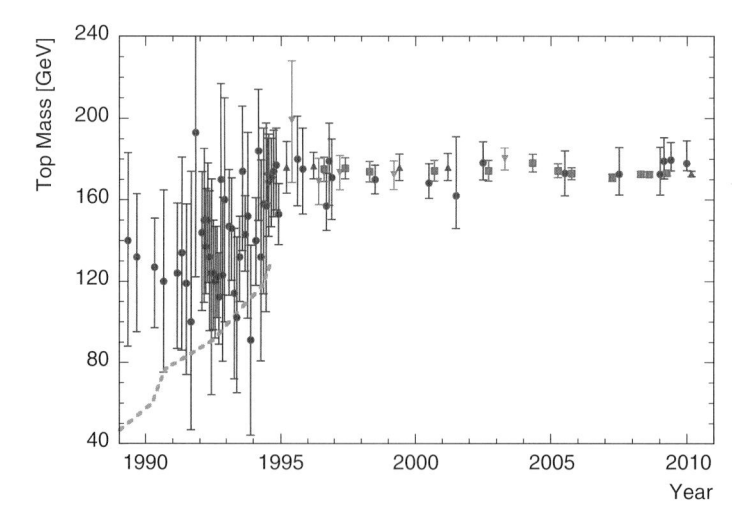

Figure 9.2 Indirect (circles) and direct (squares, triangles) determinations of the top-quark mass, compiled by Chris Quigg [4].

the top quark thus behaves like a quasi-free quark. This is a unique property of the top quark. In contrast to the lighter quarks we do not expect to observe $t\bar{t}$ bound states in nature.

The large decay width effectively cuts off non-perturbative Quantum Chromodynamics (QCD) effects, hence reliable predictions including corrections beyond the Born approximation are feasible within QCD. It is thus believed that top-quark properties can be precisely predicted in perturbation theory. This makes the top quark an ideal laboratory for precise tests of the SM and the search for new physics. A practical consequence of the fast decay is that the spin information of the top quark – a footprint of the production mechanism – is not diluted by hadronisation but transferred to the decay products. Given that the dominant decay is maximally parity violating due to the V–A structure of the Wtb vertex, one can use the angular distribution of the decay products to analyse the top-quark polarisation.

A further consequence of the V–A structure of the Wtb vertex is that the W bosons from top-quark decays are polarised. About 70% of the W bosons from top-quark decays are longitudinally polarised. The fraction of left-handed W bosons is about 30% (see also Section 9.4). The measurement of the W polarisation provides a sensitive test of the V–A structure of the corresponding vertex and gives a direct test of the gauge structure of the SM. It is also worth noting that top-quark decays provide a unique source of longitudinally polarised W bosons which are the would-be Goldstone bosons of electroweak symmetry breaking.

Since the mass of the top quark is so large, top-quark physics is an ideal laboratory to probe nature at high energy scales. This is particularly interesting because the top-quark mass is of the same order as the scale where electroweak symmetry breaking takes place. It can be said that electroweak symmetry breaking is largest in the top-quark sector since the mass is much larger than the mass of all the other elementary particles. A central question to be answered at the LHC is whether the top-quark mass, being about 40 times heavier than the next heaviest quark, is generated by the Higgs mechanism, or whether an alternative mechanism of electroweak symmetry breaking acts in the top-quark sector. A natural way to address this question is to measure the production of a top-quark pair in association with an additional Higgs boson. If the top-quark mass is generated through the Higgs mechanism, the Yukawa coupling to the Higgs boson has to be proportional to the top-quark mass. Whether this measurement will be experimentally feasible at the LHC with the required precision is at the moment unclear.

The large mass of the top quark has motivated many extensions of the SM in which the top quark plays a special role (see also Section 9.6). The generation of the large top-quark mass imposes severe constraints on any model aiming to replace the Higgs mechanism. It is thus important to measure as precisely as possible the properties of the top quark and to compare these measurements with the SM predictions, in order to see whether the top quark behaves as predicted by the SM or whether new physics effects are visible. Since the LHC is a top-quark factory many interesting measurements are possible.

9.2
Top-Quark Pair Production in Hadronic Collisions

9.2.1
Cross-Section Calculations

The dominant production mechanism of top quarks in hadronic collisions is the QCD-induced top-quark pair production. In the Born approximation the relevant parton channels are *gluon fusion* and *quark–antiquark annihilation*. The corresponding Feynman diagrams are shown in Figure 9.3. A straightforward evaluation leads to the partonic cross-sections $\hat{\sigma}_{q\bar{q}}^{(0)}$ and $\hat{\sigma}_{gg}^{(0)}$,

$$\hat{\sigma}_{q\bar{q}}^{(0)} = \frac{4}{27}\pi\alpha_s^2 \frac{1}{\hat{s}}\beta(3-\beta^2), \tag{9.1}$$

$$\hat{\sigma}_{gg}^{(0)} = \frac{1}{48}\pi\alpha_s^2 \frac{1}{\hat{s}}\left[(33-18\beta^2+\beta^4)\ln\left(\frac{1+\beta}{1-\beta}\right) - 59\beta + 31\beta^3\right], \tag{9.2}$$

where α_s denotes the QCD coupling, $\hat{s}$ is the partonic centre-of-mass energy squared and $\beta = \sqrt{1-4m_t^2/\hat{s}}$ is the velocity of the top quark in the partonic centre-of-mass system. The hadronic cross section is obtained from a convolution of the partonic cross sections with the parton distribution functions (PDFs) $F_{i/H}(x,\mu_f)$:

$$\sigma_{\text{had}} = \sum_{ij}\int dx_1 \int dx_2\, F_{i/H_1}(x_1,\mu_f) F_{j/H_1}(x_2,\mu_f) \hat{\sigma}_{ij}(\hat{s},\mu_f), \tag{9.3}$$

with $\hat{s} = x_1 x_2 s_{\text{had}}$. The sum over i, j runs over all contributing parton channels. For a comparison of different hadron colliders it is convenient to introduce the parton luminosity function $\mathcal{L}_{ij}$

$$\mathcal{L}_{ij}(\hat{s}, s_{\text{had}}, \mu_f) = \frac{1}{s_{\text{had}}} \int_{\hat{s}}^{s_{\text{had}}} \frac{ds}{s} F_{i/H_1}\left(\mu_f, \frac{s}{s_{\text{had}}}\right) F_{j/H_2}\left(\mu_f, \frac{\hat{s}}{s}\right), \tag{9.4}$$

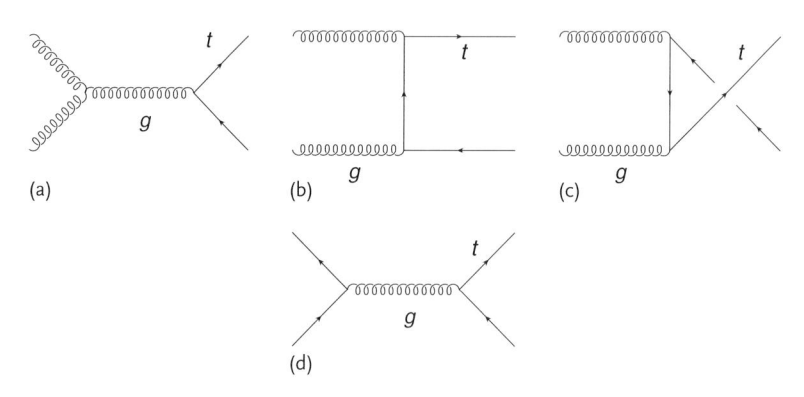

Figure 9.3 Top-quark pair production in the Born approximation. The Feynman diagrams contributing to gluon fusion are shown in (a)–(c). (d) Shows the Born diagram for top-quark pair production via quark-antiquark annihilation.

9.2 Top-Quark Pair Production in Hadronic Collisions

in terms of which the hadronic cross section reads

$$\sigma_{\text{had}}(s_{\text{had}}, m_t^2) = \sum_{i,j} \int_{4m_t^2}^{s_{\text{had}}} d\hat{s}\, \mathcal{L}_{ij}(\hat{s}, s_{\text{had}}, \mu_f)\, \hat{\sigma}_{ij \to t\bar{t}}(\hat{s}, m_t^2, \mu_f). \quad (9.5)$$

The QCD next-to-leading order (NLO) corrections to top-quark pair production have been known for more than 20 years. The results for $\mathcal{L}_{ij}$ as well as the partonic cross sections as a function of the partonic centre-of-mass energy, $\sqrt{\hat{s}}$, are shown in Figure 9.4. Figure 9.4c shows the hadronic cross section when the integration in (9.5) is performed up to a given centre-of-mass energy, $\sqrt{\hat{s}}$. Note that this is for illustrative purposes only, since the partonic centre-of-mass energy is not an observable quantity in QCD due to initial-state radiation. As a consequence of the very different collider energies, different regions of the PDFs are probed. As a result, the cross section at the LHC is dominated by gluon fusion while at the Teva-

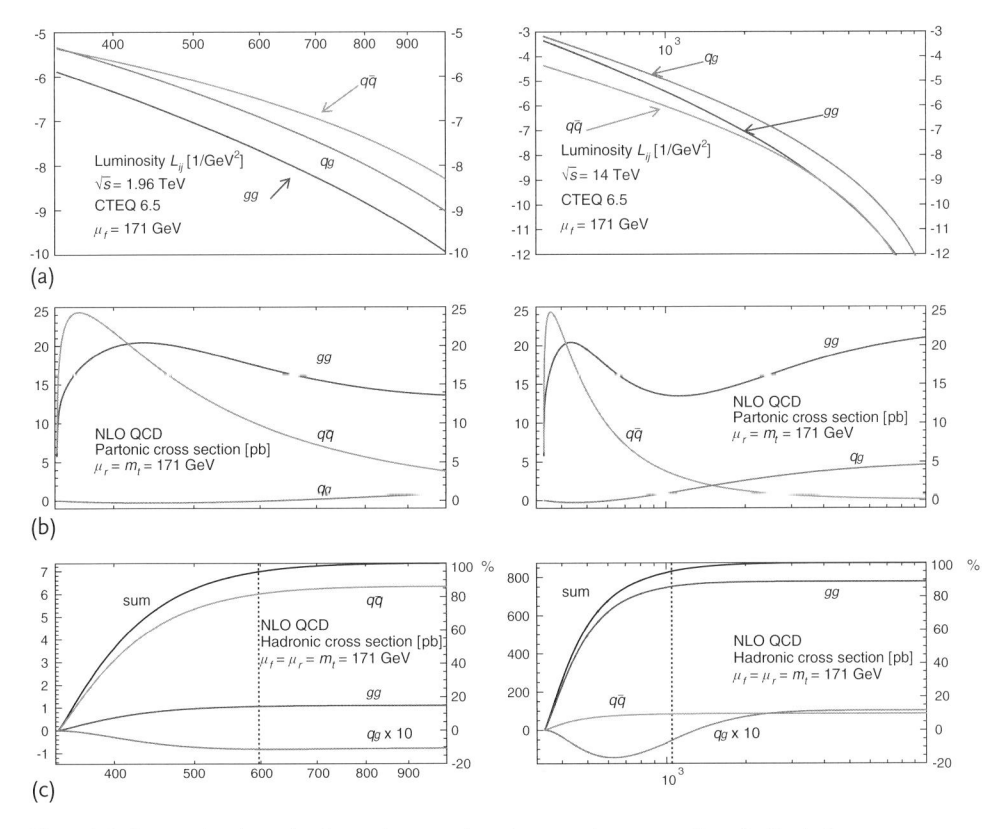

Figure 9.4 Top-quark pair production at the Tevatron and the LHC. (a) The plots show the logarithm ($\log_{10}(\mathcal{L}_{ij})$) of the parton luminosities as defined in (9.4) as a function of the partonic centre-of-mass energy. (b) The plots show the partonic cross sections for the individual parton channels again as a function of $\sqrt{\hat{s}}$. (c) The plots show the total cross section when integrating (9.5) up to a specific value for $\hat{s}$.

tron the quark channel dominates. At the LHC about 90% of the top-quark pairs are produced via gluon fusion and only 10% in quark–antiquark annihilation. At the Tevatron the situation is roughly reversed: about 85% are produced in quark–antiquark annihilation and only 15% in gluon fusion. Since a parton is not a well-defined concept beyond leading-order perturbation theory, one should read these quantities as a qualitative statement and not as a quantitative prediction. A change in the factorisation scale, for example, would move contributions from one channel to the other. The theoretical prediction for the total cross section for top-quark pair production has been continuously improved by taking various higher-order effects into account. As a reference we quote the values from [5]:

$$\sigma_{\text{LHC}} = \left(887^{+9}_{-33}(\text{scale})^{+15}_{-15}(\text{PDF})\right) \text{ pb} \quad (14 \text{ TeV}), \quad (9.6)$$

$$\sigma_{\text{Tev}} = \left(7.04^{+0.24}_{-0.36}(\text{scale})^{+0.14}_{-0.14}(\text{PDF})\right) \text{ pb} \quad (1.96 \text{ TeV}). \quad (9.7)$$

These values are obtained for $m_t = 173$ GeV. For LHC running at 10 TeV the cross section is reduced by a factor of ~ 2. At 7 TeV the cross section is reduced by a factor of ~ 5 compared to the 14 TeV cross section. The fact that the cross section at the LHC is more than a factor of one hundred larger than at the Tevatron can be easily understood as a consequence of the parton luminosities shown in Figure 9.4.

9.2.2
Top–Antitop Cross-Section Measurements

According to the SM, top quarks decay with a branching fraction of nearly 100% to a bottom quark and a W boson. Hence, the $t\bar{t}$ final states can be classified according to the decay modes of the W bosons. The most important (or so-called "golden") channel is the *lepton + jets* final state where one W boson decays leptonically into a charged lepton (electron or muon) plus a neutrino, while the second W boson decays into jets. The lepton + jets channel features a large branching fraction of about 29%, manageable backgrounds, and allows the full reconstruction of the event kinematics. Other accessible channels are the *dilepton* channel, in which both W bosons decay to leptons, and the *all-hadronic* channel, where both W bosons decay hadronically. The dilepton channel has the advantage of having a low background, but suffers on the other hand from a lower branching fraction (5%). The all-hadronic channel, on the contrary, has the largest branching ratio of all $t\bar{t}$ event categories (46%), but has the drawback of a huge QCD multi-jet background that has to be controlled experimentally. The different categories of $t\bar{t}$ final states and their branching fractions are summarised in Table 9.1.

The $t\bar{t}$ cross section has been measured in all three channels at the Tevatron. The most precise results are obtained in the lepton+jets channel in which the background is controlled either by the identification of the b-quark jets or by exploiting kinematic or topological information, see for example [6]. In the all-hadronic channel, neural networks have been used to combine several kinematic variables into one discriminant that separates the overwhelming QCD multi-jet background from $t\bar{t}$ events, see for example in [7].

Table 9.1 Categories of $t\bar{t}$ events and their branching fractions. The sum of all fractions is above 100% because of rounding effects.

W decays	$e\nu + \mu\nu$	$\tau\nu$	$q\bar{q}$
$e\nu + \mu\nu$	5%	5%	29%
$\tau\nu$	–	1%	15%
$q\bar{q}$	–	–	46%

By design the measurements in the different $t\bar{t}$ channels are statistically independent and can therefore easily be combined. A summary on measurements of the $t\bar{t}$ cross section in $p\bar{p}$ collisions at $\sqrt{s} = 1.8$ TeV in Run 1 can be found in [8, 9]. In Run 2 of the Tevatron, various measurements of the $t\bar{t}$ cross section have been performed using different amounts of recorded data. A summary of these cross-section measurements is given in [10, 11]. The combination of the Run 2 measurements at CDF yields 7.50 ± 0.31(stat.) ± 0.34(sys.) ± 0.15 (Z-theory) pb (for $m_t = 172.5$ GeV) [12], at DØ the combined cross section is found to be $8.18^{+0.98}_{-0.87}$ pb (at $m_t = 170$ GeV) [13]. All measurements are in excellent agreement with the theory prediction based on the SM, see (9.7). Note that the measured value is quoted at a specific top-quark mass, since the event detection efficiency depends on m_t and is evaluated for a preset value of the top-quark mass.

The production mechanism of $t\bar{t}$ pairs via either $q\bar{q}$ annihilation or gluon fusion as discussed in Section 9.2.1 has also been probed experimentally. One way to measure the fraction of $t\bar{t}$ pairs originating from a gg initial state exploits the fact that gg initial states produce more initial-state radiation than $q\bar{q}$ initial states. The analysis makes use of the proportionality between the mean number of low-p_T tracks in an event and the gluon content [14] and finds $\sigma(gg \to t\bar{t})/\sigma(q\bar{q} \to t\bar{t}) = 0.07 \pm 0.14$(stat.) ± 0.07(sys.). An alternative method exploits the spin information in the top-decay products employing neural networks [15].

9.2.3
Spin Correlations in Top-Quark Pair Production

Having discussed in the previous section the inclusive production of top-quark pairs we turn now to more exclusive quantities. As discussed already in the introduction, the polarisation of top quarks can be analysed through the parity-violating decay. Since the dominant production mechanism of top-quark pairs is through the strong interaction, which is parity conserving, the top quarks produced in quark–antiquark annihilation and gluon fusion are, to a very good approximation, unpolarised[26]. There is only a tiny polarisation induced through higher-order QCD and weak corrections. The fact that top quarks are unpolarised does not exclude the possibility of a correlation of the top-quark spin with the spin of the antitop quark.

---

26) Parity conservation does not forbid a polarisation transverse to the production plane. However, such a polarisation requires in addition absorptive parts and thus appears only in higher orders.

Indeed, focusing on the threshold region, that is, ignoring angular momentum, we obtain the following simple picture: in quark–antiquark annihilation the top-quark pair is in a $^3S_1$ state; in gluon fusion the top quarks are in a $^1S_0$ state, where the spectroscopic notation $^{(2S+1)}L_J$ has been used. Thus, in quark–antiquark annihilation the spins of the top and antitop quark tend to be correlated, while in gluon fusion they are anti-correlated. If we go away from the threshold, we obtain minor corrections to the simplified picture. However, the basic conclusions remain unchanged. This is also true when higher-order corrections are included. Since at the Tevatron quark–antiquark annihilation dominates, the spins are highly correlated, while at the LHC, where gluon fusion dominates, the spins are anti-correlated.

At the level of the top quarks a sensible observable to study the correlation would be

$$A = \frac{\sigma(\uparrow\uparrow) + \sigma(\downarrow\downarrow) - \sigma(\uparrow\downarrow) - \sigma(\downarrow\uparrow)}{\sigma(\uparrow\uparrow) + \sigma(\downarrow\downarrow) + \sigma(\uparrow\downarrow) + \sigma(\downarrow\uparrow)}, \tag{9.8}$$

where the arrows denote spin up or down with respect to a quantisation axis which needs to be specified. At the level of the top-quark decay products, a sensible observable is the double-differential distribution,

$$\frac{1}{\sigma} \frac{d\sigma}{\cos\theta_f \cos\theta_{\bar{f}}} = \frac{1}{4}\left(1 + B_1 \cos\theta_f + B_2 \cos\theta_{\bar{f}} - C \cos\theta_f \cos\theta_{\bar{f}}\right), \tag{9.9}$$

where $\theta_{\bar{f}}(\theta_f)$ is the angle between the charged lepton from (anti)top-quark decay and the quantisation axis. The coefficients B_1, B_2 describe the polarisation of the top quarks. As mentioned above they are tiny in the SM. The coefficient C is directly proportional to the asymmetry A defined in (9.8). An alternative observable is connected to the expectation value $\langle \mathbf{s}_t \cdot \mathbf{s}_{\bar{t}} \rangle$, which can be related to the single-differential opening-angle distribution of the leptons from top-quark and antitop-quark decay,

$$\frac{1}{\sigma} \frac{d\sigma}{d\cos\theta} = \frac{1}{2}(1 - D\cos\phi), \tag{9.10}$$

where ϕ denotes the angle between the direction of flight of the decay product of the top quark and the decay product of the antitop quark. Note that these directions are usually taken in the rest frame of the decaying particle. This leads to an easy interpretation as quantisation axis and prevents the correlation from being washed out by the additional boost into the centre-of-mass system. Using lepton+jets events, the CDF collaboration has measured the fraction f_o of $t\bar{t}$ pairs with opposite helicity ($J = 1$). The spin correlation can be expressed by the correlation coefficient κ which is related to f_o by $f_o = \frac{1}{2}(1 + \kappa)$ and is found to be $\kappa = 0.60 \pm 0.50$(stat.) ± 0.16(sys.), in good agreement with theoretical calculations [16]. More precise measurements of the $t\bar{t}$ spin correlation will be possible at the LHC.

9.2.4
Forward–Backward Charge Asymmetry

Radiative corrections to top-quark pair production can lead to interference terms in the cross section which are odd under charge conjugation of the top quarks. This effect, well known from QED, has been studied for top-quark pair production in [17]. At the Tevatron, the charge asymmetry can lead to a forward-backward charge asymmetry of the top quarks, A_{FB}. The effect is predicted to be $A_{FB} = 0.05 \pm 0.015$ [18]. The most recent measurement from the CDF collaboration, based on an integrated luminosity of $3.2\,\text{fb}^{-1}$, finds an intriguing difference: $A_{FB}^{\text{exp.}} = 0.193 \pm 0.065\text{(stat.)} \pm 0.024\text{(sys.)}$.

The interpretation of the observed discrepancy is unclear. Despite the fact that one-loop diagrams and real corrections have been calculated [17], the prediction is only at leading-order accuracy since the asymmetry appears for the first time at this order. Higher-order corrections could shift the theoretical prediction towards the experimental value. For $t\bar{t}$ production in association with an additional jet the situation is different. Here the asymmetry already appears in the Born approximation. The NLO corrections for that process [19] allow corrections to the asymmetry to be calculated. This has been studied with different values for p_T^{cut} used to define the additional jet. The main result is that the asymmetry receives large radiative corrections and is almost washed out at NLO. A possible explanation of the observed asymmetry could also be the presence of new physics (see for example [18, 20, 21]).

9.3
Single Top-Quark Production

While $t\bar{t}$ pair production via the strong interaction is the main source of top quarks at the Tevatron and the LHC, top quarks can also be produced singly via weak interactions involving the Wtb vertex. There are three production modes which are distinguished by the virtuality, Q^2, of the W boson ($Q^2 = -q^2$, where q is the four-momentum of the W):

1. *t channel* ($q^2 = \hat{t}$): a virtual, space-like ($q^2 < 0$) W boson strikes a b quark (a sea quark) inside the proton, see Figure 9.5a. The t channel is also referred to as $tq\bar{b}$ production and constitutes the dominant source of single top quarks at the Tevatron and at the LHC;
2. *s channel* ($q^2 = \hat{s}$): a production mode that is of Drell–Yan type and proceeds via a time-like W boson with $q^2 \geq (m_t + m_b)^2$, see Figure 9.5c. The s-channel mode is also called $t\bar{b}$ production.
3. *Wt production*: the top quark is produced in association with a (almost) real W boson ($q^2 = M_W^2$), see Figure 9.5d. The cross section is negligible at the Tevatron, but of considerable size at LHC energies, where associated Wt production even supersedes the s-channel cross section.

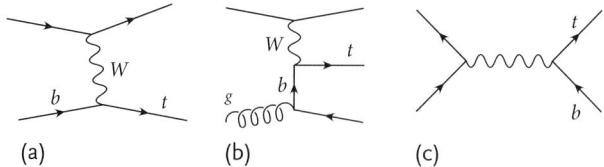

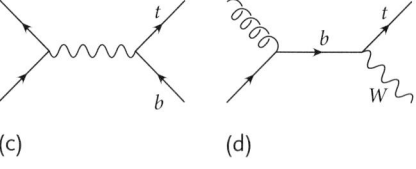

Figure 9.5 Representative Feynman diagrams for the three single top-quark production modes. The graphs show single top-quark production; the diagrams for single antitop-quark production can be obtained by interchanging quarks and antiquarks. (a) *t channel* production mode (LO graph); (b) *t channel* production mode (NLO graph); (c) *s channel* production mode; (d) *Wt production*.

Table 9.2 Predicted total cross sections for single top-quark production processes at the Tevatron and the LHC. The cross sections of the *t*-channel process are taken from [22]. The values for the *s*-channel and associated production are taken from [23]. All cross sections are evaluated at $m_t = 175$ GeV.

Process	$\sqrt{s}$	$\sigma_{tq\bar{b}}$	$\sigma_{t\bar{b}}$	σ_{Wt}
$p\bar{p} \to t/\bar{t}$	1.96 TeV	$1.86^{+0.19}_{-0.16}$ pb	1.02 ± 0.08 pb	0.25 ± 0.03 pb
$pp \to t$	14.0 TeV	149.4 ± 4.1 pb	$7.23^{+0.55}_{-0.47}$ pb	41.1 ± 4.2 pb
$pp \to \bar{t}$	14.0 TeV	88.9 ± 2.4 pb	$4.03^{+0.14}_{-0.16}$ pb	41.1 ± 4.2 pb

9.3.1
Cross-Section Calculations

Cross-section calculations exist at NLO and are summarised in Table 9.2. The cross sections given for the Tevatron are the sum of top-quark and antitop-quark production, as both contributions have the same size. At the LHC the cross section of top-quark production is about twice as big as the one for antitop-quark production, the reason being the availability of two valence up quarks leading to top-quark production and only one valence down quark leading to antitop-quark production. Traditionally, the calculations of the *t*-channel cross section at NLO use a *b*-quark PDF to control collinear singularities occurring when a gluon splits into a $b\bar{b}$ pair. In that approach Figure 9.5a describes the LO graph and Figure 9.5b an NLO correction. The most recent calculation [22], however, has in addition investigated the use of the four-flavour scheme and calculates the $g \to b\bar{b}$ splitting explicitly, leading to a reduction of the total cross section by about 5%. For $\sqrt{s} = 7$ TeV, the centre-of-mass energy of the LHC in 2010 and 2011, the cross sections are reduced by roughly a factor of four.

A very interesting feature of singly produced top quarks is their polarisation which is a result of the *V–A* structure of the weak interaction. In its rest frame the top quark is 100% polarised along the direction of the *d* quark [24]. A suitable variable to investigate the top-quark polarisation is the angular distribution of electrons and muons originating from the decay chain $t \to W^+ b$, $W^+ \to \ell^+ \nu_\ell$. If $\theta_{q\ell}$ is the angle between the charged lepton and the light-quark jet-axis in the top-quark rest frame, the angular distribution is given by

$$\frac{dN}{d\cos\theta_{q\ell}} = \frac{1}{2}(1 + \mathcal{P}\cos\theta_{q\ell}), \tag{9.11}$$

with $\mathcal{P} = 1$ corresponding to fully polarised top quarks [25].

9.3.2
First Observation of Single Top Quarks

While Run 1 (1992–1996) and early Run 2 searches for single top quarks at CDF and DØ could only set upper limits on the production cross section [26, 27], analyses using more data and advanced analysis techniques were able to observe singly produced top quarks in March 2009 with a significance of five Gaussian standard deviations [28, 29]. These analyses consider both production modes relevant at the Tevatron, t channel and s channel, as one single-top signal, assuming the ratio of t channel to s channel events to be given by the SM. This search strategy is often referred to as *combined search*. By measuring the inclusive single top-quark cross section and using theory predictions, one can deduce the absolute value of the CKM matrix element V_{tb}, without the assumption that there are only three generations of quarks. If one measured $|V_{tb}|$ to be significantly smaller than one this would be a strong indication for a fourth generation of quarks or other effects beyond the SM [30, 31].

Single top-quark events feature a $Wb\bar{b}$ (s-channel) or $Wbq\bar{b}$ (t-channel) partonic final state. The most sensitive analyses reconstruct the W boson originating from the top-quark decay in its leptonic decay modes $e\nu_e$ or $\mu\nu_\mu$, while hadronic W decays and decays to $\tau\nu_\tau$ are not explicitly considered, because of large backgrounds from QCD-induced jet production. The quarks from the hard scattering process manifest themselves as hadronic jets with large transverse momentum. Additional jets may arise from hard gluon radiation in the initial or final state. The experimental signature of SM single top quarks is therefore given by one isolated high-p_T charged lepton (electron or muon), large missing transverse energy ($\not{E}_T$), and two or three high-E_T jets, of which one or two originate from a b or $\bar{b}$ quark. CDF and DØ select events based on the described lepton + jets signature, in analogy to the selection used in $t\bar{t}$ analyses. One important difference in the jet definition in single-top analyses is the extended η range. CDF counts jets in the range $|\eta| < 2.8$, DØ in the region $|\eta| < 3.4$. This extension is essential in order to measure the forward jet in the t-channel process that originates from the light quark that couples to the virtual W boson, see Figure 9.5a. Forward jets are a distinct feature of t-channel single top-quark production. Jets produced by b quarks are identified as such by exploiting the lifetime of b hadrons and other decay properties such as the track multiplicity and the occurrence of leptons from semileptonic decays. The expected background event rates are obtained from a compound model comprising simulated events, theoretical cross sections and normalisations in background-dominated control samples. The dominating background process is $W b\bar{b}$, followed by misidentified W + light-quark jet events, $Wc\bar{c}$, and Wcj. In the $W + 3$ jets dataset $t\bar{t}$ is the most important background. The total event detection efficiency of single top-quark events is about 2% for the *t*-channel process and about 3% for the *s* channel.

Multivariate Analyses Even though the single top-quark production cross section is predicted to amount to about 40% of the $t\bar{t}$ cross section, the signal has been obscured for a long time by the very challenging background. After the cut-based event selection sketched above, the signal-to-background ratio is only about 5 to 6%, as can be seen in Figure 9.6 which shows the jet multiplicity distribution in the W + jets dataset at CDF [32]. Further kinematic cuts on the event topology proved to be prohibitive, since the number of signal events in the resulting dataset would become too small. Facing this challenge, the analysis groups in both Tevatron collaborations turned to multivariate techniques, in order to exploit as much information about the observed events as possible. The explored techniques comprise artificial neural networks, LO matrix elements, boosted decision trees, and likelihood ratios. All these techniques combine the information contained in several variables into one powerful discriminant, maximising the separation between the single top-quark signal and the backgrounds. All multivariate analyses see a signal of single top-quark production with significances ranging from 2.4 to 5.2 Gaussian standard deviations. To obtain the most precise result, all analyses of the two collaborations are combined, resulting in a single top-quark cross section of $2.76^{+0.58}_{-0.47}$ pb (at $m_t = 170\,\text{GeV}$) [33].

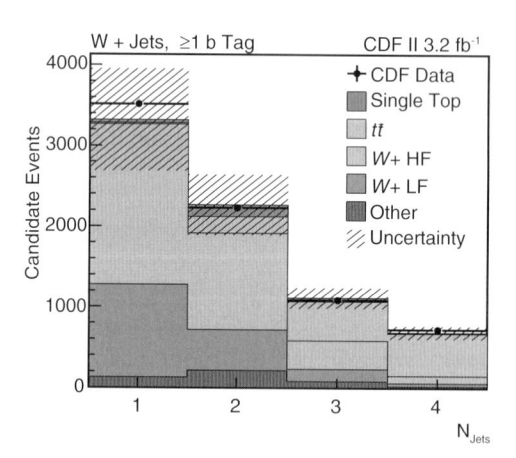

Figure 9.6 Jet multiplicity distribution in the W + jets dataset at CDF using collision data that correspond to an integrated luminosity of $3.2\,\text{fb}^{-1}$. The stacked histogram is normalised to the event rates predicted by the SM (adapted from [32]).

Determination of $|V_{tb}|$ The measured single top-quark production cross sections can be used to determine the absolute value of the CKM-matrix element V_{tb}, if one assumes $V_{tb} \gg V_{ts}$, $V_{tb} \gg V_{td}$, and a SM-like left-handed coupling at the Wtb vertex. On the basis of the combined cross-section result the Tevatron collaborations obtain $|V_{tb}| = 0.88 \pm 0.07$ [33].

Separation of t-channel and s-channel Events In addition to the combined-search analyses described in the previous section, CDF and DØ have also performed analyses to separate t-channel and s-channel single top-quark events. Both processes are treated as independent, that is, the assumption of the ratio of the two cross sections to be given by the SM is dropped. The separation of the two channels is important, because they are subject to different systematic uncertainties and they exhibit a different sensitivity to physics beyond the SM [34]. CDF measures $\sigma_{tq\bar{b}} = 0.8 \pm 0.4$ pb and $\sigma_{t\bar{b}} = 1.8^{+0.7}_{-0.5}$ pb (at $m_t = 175$ GeV), while DØ obtains $\sigma_{tq\bar{b}} = 3.14^{+0.94}_{-0.80}$ pb and $\sigma_{t\bar{b}} = 1.05 \pm 0.81$ pb (at $m_t = 170$ GeV) [35].

Single top Quarks at the LHC Collisions at the LHC will be a copious source of single top quarks. At the initial centre-of-mass energy of $\sqrt{s} = 7$ TeV the inclusive cross section for single top-quark production is expected to be a factor of 25 higher than at the Tevatron. It will therefore be possible to measure the cross sections of the different channels precisely and also to study properties of single top quarks, for example the polarisation.

9.4
Top-Quark Decay

Within the SM, top quarks decay predominantly into a b quark and a W boson, while the decays $t \to d + W^+$ and $t \to s + W^+$ are strongly CKM suppressed and can be neglected. The top-quark decay rate, including first-order QCD corrections, is given by

$$\Gamma_t = \frac{G_F m_t^3}{8\pi\sqrt{2}} |V_{tb}|^2 (1-y)^2 (1+2y) \left[1 - \frac{2}{3} \frac{\alpha_s}{\pi} \cdot f(y)\right], \quad (9.12)$$

with $y = (M_W/m_t)^2$ and $f(y) = 2\pi^2/3 - 5/2 - 3y + 9/2 y^2 - 3y^2 \ln y$ [36], yielding $\Gamma_t = 1.35$ GeV at $m_t = 173.1$ GeV. The large Γ_t implies a very short lifetime of $\tau_t = 1/\Gamma_t \approx 5 \times 10^{-25}$ s which is smaller than the characteristic formation time of hadrons $\tau_\text{form} \approx 1 fm/c \approx 3 \times 10^{-24}$ s. In other words, top quarks decay before they can couple to light quarks and form hadrons.

9.4.1
W-Boson Helicity in Top-Quark Decays

The amplitude of the decay $t \to b + W^+$ is dominated by the contribution from longitudinal W bosons, because the decay rate of the longitudinal component scales with m_t^3, while the decay rate into transverse W bosons increases only linearly with m_t. In both cases the W^+ couples solely to b quarks of left-handed chirality, which translates into left-handed helicity, since the b quark is effectively massless compared to the energy scale set by m_t. The coupling to the right-handed helicity is suppressed by m_b^2/M_W^2. If the b quark is emitted antiparallel to the top-quark spin axis, the W^+ must be longitudinally polarised, $h^W = 0$, to conserve angular momentum. If the b quark is emitted parallel to the top-quark spin axis, the W^+ boson has helicity $h^W = -1$ and is transversely polarised. W bosons with positive helicity are thus forbidden in top-quark decays due to angular momentum conservation, assuming $m_b = 0$. The fraction of longitudinally polarised W bosons f_0 and the left-handed fraction f_- are predicted to be

$$f_0 = \frac{m_t^2}{2M_W^2 + m_t^2} \simeq 0.70 \quad \text{and} \quad f_- = \frac{2M_W^2}{2M_W^2 + m_t^2} \simeq 0.30 \,. \qquad (9.13)$$

Taking the non-zero b-quark mass into account yields very small corrections, leading to a non-zero fraction of W bosons in top-quark decay with positive helicity $f_+ = 3.6 \times 10^{-4}$ at Born level [37].

The spin orientation (helicity) of the W bosons is propagated to their decay products. For leptonic W decays, the polarisation is preserved and can be measured. A useful observable for experiments is therefore the cosine of the polarisation angle, θ^*, which is defined as the angle of the charged lepton in the rest frame of the W boson measured with respect to the direction of motion of the W boson in the top-quark rest frame. The probability density ω has the following distribution:

$$\omega(\theta^*) = f_0\omega_0(\theta^*) + f_+\omega_+(\theta^*) + (1 - f_0 - f_+)\omega_-(\theta^*) \,, \qquad (9.14)$$

with $\omega_0(\theta^*) = \frac{3}{4}(1 - \cos^2\theta^*)$, $\omega_+(\theta^*) = \frac{3}{8}(1 + \cos\theta^*)^2$ and $\omega_-(\theta^*) = \frac{3}{8}(1 - \cos\theta^*)^2$. Several measurements of the helicity fractions have been performed by the CDF and DØ collaborations. One class of analyses reconstruct the $t\bar{t}$ event kinematics in lepton + jets events and measure $\cos\theta^*$ directly, for example [38], while a second class uses the square of the invariant mass of the lepton and the b-quark jet, $M_{\ell b}^2$, in the laboratory frame as the observable [39]. The variable $M_{\ell b}^2$ is closely related to $\cos\theta^*$ and has the advantage that it can also be used in $t\bar{t}$ dilepton events.

A third set of analyses [40] uses the p_T spectrum of the charged lepton, which is also sensitive to the helicity fractions because the V–A structure of the W boson decay causes a strong correlation between the helicity of the W boson and the lepton momentum. Qualitatively, this can be understood as follows: the ν_ℓ from the W^+ decay is always left-handed, the ℓ^+ is right-handed. In the case of a left-handed W^+ boson, angular momentum conservation demands therefore that the ℓ^+ is emitted in the direction of the W^+ spin, that means antiparallel to the W^+ momentum. That is why charged leptons from the decay of left-handed W bosons are softer than charged leptons from longitudinal W bosons, which are mainly emitted in the direction transverse to the W boson momentum. The spectrum of leptons from right-handed W bosons would be even harder than the one from longitudinal ones, since they would be emitted preferentially in the direction of the W momentum.

A fourth technique to measure the helicity fractions uses the LO matrix elements of $t\bar{t}$ production [41].

More recent analyses measure the fractions f_0 and f_+ simultaneously in a two-dimensional fit, while in previous analyses two fits were performed keeping one of the fractions at its SM value. CDF obtains $f_0 = 0.66 \pm 0.16$(stat.) ± 0.05(sys.) and $f_+ = -0.03 \pm 0.06$(stat.) ± 0.03(sys.) with a correlation of about -90% between f_0 and f_+ [38]. The result is in good agreement with the SM prediction.

9.4.2
Measurement of $\mathcal{R}_b$

The branching fraction $\mathcal{R}_b \equiv \mathcal{B}(t \to W b)$ is given by the ratio of the squares of the relevant CKM matrix elements, $\mathcal{R}_b = |V_{tb}|^2/(|V_{td}|^2 + |V_{ts}|^2 + |V_{tb}|^2)$. In the SM the unitarity of the CKM matrix leads to $|V_{td}|^2 + |V_{ts}|^2 + |V_{tb}|^2 = 1$ and therefore $\mathcal{R}_b = |V_{tb}|^2$.

Our present knowledge on $|V_{tb}|$ stems primarily from measurements of b-meson and c-meson decays which determine the values of the other CKM matrix elements. Using the unitarity condition of the CKM matrix one can obtain $|V_{tb}|$ in an indirect way with very high precision, $|V_{tb}| = 0.999133 \pm 0.000044$ [42]. Only recently a determination of $|V_{tb}|$ without unitarity assumption was obtained from the measurement of the single top-quark cross section, see Section 9.3.2.

In most $t\bar{t}$ cross-section analyses the assumption $\mathcal{R}_b = 1$ is made, but CDF and DØ have also performed measurements without this constraint. The W + jets dataset is split in various disjoint subsets according to the number of jets (0, 1 or ≥ 2), the charged lepton type (electron or muon), and most importantly the number of b-tagged jets. The fit yields $\mathcal{R}_b = 0.97^{+0.09}_{-0.08}$, where the statistical and systematic uncertainties have been combined [43].

9.5
Top-Quark Mass

The most important property of the top quark is its mass, m_t, as it enters the calculations of radiative corrections of many SM processes. Since m_t is about 40 times larger than the b-quark mass, these corrections are sizable and a precise knowledge of m_t is mandatory. Together with precision measurements of electroweak quantities, such as M_W, M_Z and the electroweak mixing angle θ_W, the precise determination of m_t can be used to test the consistency of the SM and to predict the mass of the elusive Higgs boson. Since the top quark does not appear as a free particle, the definition of m_t is not straightforward. From a theoretical point of view the top-quark mass has to be treated similarly to other parameters of the Lagrangian such as the coupling of the strong interaction, α_s.

Two common mass schemes are frequently used in perturbation theory. One is the *on-shell* or *pole-mass* scheme in which the mass parameter is defined as the location of the pole of the propagator calculated order by order in perturbation theory. Another scheme is the so-called *modified minimal subtraction* scheme ($\overline{\text{MS}}$). Since different renormalisation schemes should be equivalent it must also be possible to convert from one scheme to the other. This is indeed the case. The relation between the pole mass and the $\overline{\text{MS}}$ mass reads for example

$$m_{\text{pole}} = m(\mu) \left\{ 1 + \frac{\alpha_s(\mu)}{\pi} \left[\frac{4}{3} + \ln\left(\frac{\mu^2}{m(\mu)^2}\right) \right] + O(\alpha_s^2) \right\}, \qquad (9.15)$$

where m_{pole} denotes the mass in the pole-mass scheme and $m(\mu)$ is the $\overline{\text{MS}}$ mass which, like $\alpha_s(\mu)$, depends on the renormalisation scale μ. That is why this mass is often also called "running mass". In current analyses at the Tevatron, m_t is measured using the event kinematics. The measurements are calibrated based on the mass definition that is applied in the Monte Carlo event generator used to create samples of simulated events. The renormalisation scheme is thus not unambiguously fixed. It is generally believed that the measured value should be interpreted as the pole mass. There is a further subtlety which we should mention. It turns out that the pole-mass scheme has an intrinsic uncertainty of the order of Λ_{QCD} which can be attributed to so-called *infrared renormalons* – a certain class of higher-order corrections which spoil the convergence of the perturbative expansion.

From a theoretical point of view a clean approach to measure m_t would be to choose a specific observable, calculate the higher-order corrections choosing a well-defined renormalisation scheme like for example the running mass, and then to compare with the measurements. This idea has been pursued in [5]. The inclusive cross section has been used as the observable. However, due to the weak dependence of the cross section on the mass the method is not competitive with the direct measurements. Nevertheless, the method provides an independent cross check and is theoretically rather clean.

In the following we comment on the measurements of m_t performed at the Tevatron. We stress that the theoretical uncertainties described above could lead to an additional shift of the order of Λ_{QCD} which is difficult to assess theoretically.

Both Tevatron collaborations have measured m_t in three $t\bar{t}$ decay channels: the lepton + jets channel, the dilepton channel and the all-hadronic channel. A wide range of measurement techniques has been employed to obtain these results. The traditional *template method* uses an event reconstruction based on kinematic fitting and template fits to the reconstructed top-quark mass distribution, m_t^{rec}. Methods developed in recent years exploit the entire event kinematics to measure m_t either by utilising LO matrix elements, the *ideogram method* or *neural networks*. To reduce the large systematic uncertainty due to the jet energy scale, these analyses use constraints on the reconstructed invariant mass of the hadronically decaying W boson in the event. The jet energy scale and m_t are thereby simultaneously determined. Because of these advanced measurements m_t is by now the most precisely determined top-quark property. The combination of the best measurements in all channels of both Tevatron experiments yields $m_t = 173.1 \pm 0.6 \pm 1.1$ GeV [1]. The relative precision on m_t is 0.8% and m_t is thus one of the most precisely known quark masses. This constitutes one of the major achievements of Tevatron Run 2.

In the following subsection we give a brief overview of the two most important mass measurement techniques in the $t\bar{t}$ lepton+jets channel. An extensive overview on measurements of m_t in all channels can be found in [44].

9.5.1
Mass Measurement Techniques

The most precise measurements of m_t are done in the lepton + jets channel in which candidate events match the decay chain $t\bar{t} \rightarrow W^+ b\, W^- \bar{b} \rightarrow \ell \nu_\ell b q \bar{q}' \bar{b}$ and therefore feature a high-p_T electron or muon, $\not{E}_T$, and at least four high-E_T jets. An important advantage of the lepton + jets channel is its large branching fraction of about 29% providing good data statistics, while the backgrounds are moderate and can be controlled by identifying b-quark jets, cuts on topological variables and so on. An additional advantage is that the event kinematics can be fully reconstructed up to a two-fold ambiguity due to the unmeasured longitudinal momentum, p_z, of the neutrino. The four leading, that is highest-E_T, jets are assigned to the four quarks in the final state of the hard scattering. However, this jet-parton assignment has some ambiguity. If none of the jets is tagged as a b jet candidate, there are 12 possible jet-parton permutations. Adding the two-fold ambiguity of the neutrino p_z reconstruction, there are 24 combinations. With one b tag that number is reduced to 12, with two b tags there are only four possible assignments.

In the traditional template method, see for example [45], an over-constrained kinematic fit to the event kinematics is performed for each assignment by minimising a χ^2 function that implements the event hypothesis. The solution with the lowest χ^2 value is taken as the best choice, and the fit yields one value for the reconstructed top-quark mass, m_t^{rec}, per event. On the basis of samples of simulated events, template histograms of m_t^{rec} for different m_t (the true top-quark mass) are produced and a parametrisation of these templates as a function of m_t is derived. The background is included in this parametrisation as a contribution independent of m_t. Using a maximum-likelihood method the template function is then fitted to the m_t^{rec} distribution observed in collision data and m_t is determined.

The *matrix element method* was originally suggested by Kondo [46] as well as Dalitz and Goldstein [47] in the 1990s and pioneered by the DØ collaboration [48]. The method is designed to use the entire event kinematics to improve the sensitivity in m_t by exploiting the dependence of the $t\bar{t}$ cross section on m_t. The differential cross section can be used to calculate a probability density for a certain mass hypothesis. While the differential cross section is given by a phase space term and the matrix element for $t\bar{t}$ production, the matrix element is the more significant part, which motivates the name of the method. Since leading-order matrix elements are used to calculate the event weights, only events with exactly four jets are used. This jet cut minimises the effect of higher-order corrections.

The production probability $P(x, m_t)$ for a $t\bar{t}$ event with a measured set of variables x at a certain top-quark mass m_t is given as a convolution of the differential cross section, $d\sigma/dy$, with the parton distribution functions, f, and the transfer function, $W(y,x)$, that maps the measured quantities x into the quantities y at parton level:

$$P(x, m_t) = \frac{1}{\sigma_{tot}} \int dy dq_1 dq_2 \frac{d\sigma(y, m_t)}{dy} f(q_1) f(q_2) W(y, x) . \tag{9.16}$$

The parton distribution functions, $f(q_i)$, are evaluated for the incoming partons with momentum fraction q_i. The integral in (9.16) is properly normalised by dividing by the total cross section, σ_{tot}. The integration runs over 15 sharply measured variables, which are directly assigned to the parton quantities without invoking a transfer function. These variables are the eight jet angles, the three-momentum of the lepton, and four equations of energy-momentum conservation. The jet energies are not well measured and a transfer function $W_{jets}(E_{part}, E_{jet})$ is needed to map jet energies E_{jet} measured in the detector to parton-level energies E_{part}. When computing $P(x, M_{top})$ all possible 24 permutations of jet assignments and the neutrino p_z solution are considered and the average is computed.

9.6
The Top Quark as a Window to New Physics

9.6.1
Top–Antitop Resonances

The $t\bar{t}$ candidate samples offer the possibility to search for a narrow-width resonance X^0 decaying into $t\bar{t}$ pairs by investigating the $t\bar{t}$ invariant mass. In an analysis using data corresponding to 3.6 fb^{-1}, the DØ collaboration found no evidence for such a resonance and placed upper limits on $\sigma_X \cdot \mathcal{B}(X^0 \to t\bar{t})$ ranging from 1.0 pb at $M_X = 350$ GeV to 0.16 pb at $M_X = 1000$ GeV. If interpreted in a specific technicolour model that predicts a narrow leptophobic Z' a mass limit of $M(Z') > 820$ GeV at 95% C.L. is obtained [49].

9.6.2
Non-Standard Model Sources of Single Top Quarks

New physics phenomena beyond the SM may give rise to additional sources of single top quarks. One possibility is an extended Higgs sector, for example in two Higgs doublet models that include charged Higgs bosons, $H^\pm$. These can replace the virtual W boson in single top-quark production processes. The single-top candidate samples at DØ have been investigated, but no evidence for $H^\pm$ has been found in this channel [50].

Another possibility is to replace the virtual W boson in single top-quark Feynman diagrams by a heavy charged vector boson, a so-called W', that decays preferentially into a $t\bar{b}$ pair. Both CDF and DØ have searched for such processes and set limits on the W' mass between 730 and 825 GeV depending on the model assumptions [51, 52].

A further source of single top quarks could be flavour changing neutral currents (FCNC) that are absent in the SM at tree level and strongly suppressed in quantum loops via the GIM mechanism [53]. At LEP all four experiments searched for the reactions $e^+e^- \to t\bar{c}/t\bar{u}$ and presented upper limits on anomalous couplings of the Z boson and the photon, which translate into limits on the branching fractions of $\mathcal{B}(t \to u/c + \gamma) < 4.2\%$ and $\mathcal{B}(t \to u/c + Z) < 14\%$ [54]. At HERA the two experiments H1 and ZEUS searched for top quarks produced in the inclusive FCNC reaction $ep \to etX$ and set upper limits on the cross section that translate into $\mathcal{B}(t \to u + \gamma) < 0.64\%$ [55]. At the Tevatron, being a hadron collider, anomalous couplings of quarks to gluons could yield additional production channels of single top quarks. The DØ collaboration has looked for $2 \to 2$ processes, such as $q\bar{q} \to t\bar{u}$, $ug \to tg$, or $gg \to t\bar{u}$ [56], while at CDF a search in the channel $u(c) + g \to t$ has been performed [57], yielding the limits $\mathcal{B}(t \to u+g) < 3.9 \times 10^{-4}$ and $\mathcal{B}(t \to c+g) < 5.7 \times 10^{-3}$. In the future the highest sensitivity to such anomalous couplings will be reached at the LHC.

9.6.3
Search for Non-Standard Models Top-Quark Decays

Two different classes of non-SM top-quark decays have been searched for at the Tevatron: decays via FCNC and decays into a charged Higgs boson H^+.

FCNC-Induced Top-Quark Decays In the top sector FCNC are strongly suppressed with branching fractions of $\mathcal{O}(10^{-14})$. The CDF collaboration has searched for non-SM decays of the type $t\bar{t} \to ZqW^-\bar{b} \to (\ell^+\ell^-q)(q\bar{q}'\bar{b})$, but found no evidence for such events and therefore derived an upper limit on the branching fraction $\mathcal{B}(t \to Zq) < 3.7\%$ at 95% C.L. [58]. A previous Run 1 result by CDF also set a limit on FCNC branching fractions in the photon-plus-jet mode, $\mathcal{B}(t \to \gamma q) < 3.2\%$ [59].

Top-Quark Decays to a Charged Higgs Boson Supersymmetric models predict the existence of charged Higgs bosons, see Chapter 7. If the charged Higgs boson is lighter than the difference between the top-quark mass and the b-quark mass, $M_{H^\pm} < m_t - m_b$, the decay mode $t \to H^+b$ is possible and competes with the SM decay $t \to W^+b$. At the Tevatron, CDF and DØ looked for the decay modes $H^+ \to \tau^+\nu_\tau$ and $H^+ \to c\bar{s}$ but found no evidence for $t \to H^+b$ decays [13, 60]. The resulting limits on $\mathcal{B}(t \to H^+b)$ depend on $M_{H^\pm}$ and the assumptions made on the Higgs decay modes, but are typically in the range of 15 to 25%.

References

1. Tevatron Electroweak Working Group (2009) Combination of CDF and DØ Results on the Mass of the Top Quark, http://arxiv.org/abs/0903.2503.
2. CDF Collab., Abe, F. et al. (1995) Observation of top quark production in $\overline{p}p$ collisions with the collider detector at Fermilab, *Phys. Rev. Lett.*, **74**, 2626–2631, doi:10.1103/PhysRevLett.74.2626.
3. DØ Collab., Abachi, S. et al. (1995) Observation of the top quark, *Phys. Rev. Lett.*, **74**, 2632, doi:10.1103/PhysRevLett.74.2632.
4. Quigg, C. (2010) The time evolution of the top-quark mass. Extended version of "Top-ology", *Phys. Today*, **50**, 20 (1997); also circulated as arXiv: hep-ph/9704332, private communication.
5. Langenfeld, U., Moch, S., and Uwer, P. (2009) Measuring the running top-quark mass, *Phys. Rev. D*, **80**, 054009, doi:10.1103/PhysRevD.80.054009.
6. DØ Collab., Abazov, V.M. et al. (2008) Measurement of the $t\bar{t}$ production cross section in $p\overline{p}$ collisions at $\sqrt{s} = 1.96\,\text{TeV}$, *Phys. Rev. Lett.*, **100**, 192004, doi:10.1103/PhysRevLett.100.192004.
7. CDF Collab., Aaltonen, T. et al. (2010) Measurement of the Top Quark Mass and $p\overline{p} \to t\bar{t}$ Cross Section in the All-Hadronic Mode with the CDFII Detector, http://arxiv.org/abs/1002.0365.
8. Wagner, W. (2005) Top quark physics in hadron collisions, *Rept. Prog. Phys.*, **68**, 2409, doi:10.1088/0034-4885/68/10/R03.
9. Quadt, A. (2006) Top quark physics at hadron colliders, *Eur. Phys. J. C*, **48**, 835, doi:10.1140/epjc/s2006-02631-6.
10. Incandela, J.R., Quadt, A., Wagner, W., and Wicke, D. (2009) Status and prospects of top-quark physics, *Prog. Part. Nucl. Phys.*, **63**, 239, doi:10.1016/j.ppnp.2009.08.001.
11. Wagner, W. (2010) Top–antitop-quark production and decay properties at the tevatron, *Mod. Phys. Lett. A*, **25**, 1297–1314, doi:10.1142/S021773231003330X.
12. CDF Collab. (2009), Combination of CDF top quark pair production cross section measurements with up to 4.6 fb^{-1}, http://www-cdf.fnal.gov/physics/new/top/confNotes/cdf9913_ttbarxs4invfb.pdf, CDF Conf. note 9913.
13. DØ Collab., Abazov, V.M. et al. (2009) Combination of $t\bar{t}$ cross section measurements and constraints on the mass of the top quark and its decays into charged Higgs bosons, *Phys. Rev. D*, **80**, 071102, doi:10.1103/PhysRevD.80.071102.
14. CDF Collab., Aaltonen, T. et al. (2008) First measurement of the fraction of top quark pair production through gluon–gluon fusion, *Phys. Rev. D*, **78**, 111101, doi:10.1103/PhysRevD.78.111101.
15. Aaltonen, T. et al. (2009) Measurement of the fraction of $t\bar{t}$ production via gluon-gluon fusion in $p\overline{p}$ collisions at $\sqrt{s} = 1.96\,\text{TeV}$, *Phys. Rev. D*, **79**, 031101, doi:10.1103/PhysRevD.79.031101.
16. Bernreuther, W., Brandenburg, A., Si, Z.G., and Uwer, P. (2004) Top quark pair production and decay at hadron colliders, *Nucl. Phys. B*, **690**, 81, doi:10.1016/j.nuclphysb.2004.04.019.
17. Kuhn, J.H. and Rodrigo, G. (1999) Charge asymmetry of heavy quarks at hadron colliders, *Phys. Rev. D*, **59**, 054017, doi:10.1103/PhysRevD.59.054017.
18. Antunano, O., Kuhn, J.H., and Rodrigo, G. (2008) Top quarks, axigluons and charge asymmetries at hadron colliders, *Phys. Rev. D*, **77**, 014003, doi:10.1103/PhysRevD.77.014003.
19. Dittmaier, S., Uwer, P., and Weinzierl, S. (2009) Hadronic top-quark pair production in association with a hard jet at next-to-leading order QCD: Phenomenological studies for the Tevatron and the LHC, *Eur. Phys. J. C*, **59**, 625, doi:10.1140/epjc/s10052-008-0816-y.
20. Frampton, P.H., Shu, J., and Wang, K. (2010) Axigluon as possible explanation for $p\overline{p} \to t\bar{t}$ forward–backward asymmetry, *Phys. Lett. B*, **683**, 294–297, doi:10.1016/j.physletb.2009.12.043.
21. Djouadi, A., Moreau, G., Richard, F., and Singh, R.K. (2009) The forward–backward asymmetry of top quark production at the Tevatron in

warped extra dimensional models, http://arxiv.org/abs/0906.0604.

22 Campbell, J.M., Frederix, R., Maltoni, F., and Tramontano, F. (2009) NLO predictions for *t*-channel production of single top and fourth generation quarks at hadron colliders, *JHEP*, **10**, 042, doi:10.1088/1126-6708/2009/10/042.

23 Kidonakis, N. (2007) Higher-order soft gluon corrections in single top quark production at the CERN LHC, *Phys. Rev. D*, **75**, 071501, doi:10.1103/PhysRevD.75.071501.

24 Mahlon, G. and Parke, S.J. (1997) Improved spin basis for angular correlation studies in single top quark production at the Fermilab Tevatron, *Phys. Rev. D*, **55**, 7249, doi:10.1103/PhysRevD.55.7249.

25 Jezabek, M. (1994) Top quark physics, *Nucl. Phys. Proc. Suppl. B*, **37**, 197, http://arxiv.org/abs/hep-ph/9406411.

26 CDF Collab., Acosta, D.E. et al. (2005) Search for electroweak single-top-quark production in $p\bar{p}$ collisions at $\sqrt{s} = 1.96$ TeV, *Phys. Rev. D*, **71**, 012005, doi:10.1103/PhysRevD.71.012005.

27 DØ Collab., Abazov, V.M. et al. (2007) Multivariate searches for single top quark production with the DØ detector, *Phys. Rev. D*, **75**, 092007, doi:10.1103/PhysRevD.75.092007.

28 CDF Collab., Aaltonen, T. et al. (2009) Observation of electroweak single top quark production, *Phys. Rev. Lett.*, **103**, 092002, doi:10.1103/PhysRevLett.103.092002.

29 DØ Collab., Abazov, V.M. et al. (2009) Observation of single top-quark production, *Phys. Rev. Lett.*, **103**, 092001, doi:10.1103/PhysRevLett.103.092001.

30 Bobrowski, M., Lenz, A., Riedl, J., and Rohrwild, J. (2009) How much space is left for a new family of fermions? *Phys. Rev. D*, **79**, 113006, doi:10.1103/PhysRevD.79.113006.

31 Alwall, J. et al. (2007) Is $V_{tb} \simeq 1$? *Eur. Phys. J. C*, **49**, 791, doi:10.1140/epjc/s10052-006-0137-y.

32 CDF Collab., Aaltonen, T. et al. (2010) Observation of Single Top Quark Production and Measurement of $|V_{tb}|$ with CDF, http://arxiv.org/abs/1004.1181.

33 CDF and DØ Collab. (2009) Combination of CDF and DØ Measurements of the Single Top Production Cross Section, http://arxiv.org/abs/0908.2171.

34 Tait, T.M.P. and Yuan, C.P. (2001) Single top quark production as a window to physics beyond the Standard Model, *Phys. Rev. D*, **63**, 014018, doi:10.1103/PhysRevD.63.014018.

35 DØ Collab., Abazov, V.M. et al. (2010) Measurement of the *t*-channel single top quark production cross section, *Phys. Lett. B*, **682**, 363, doi:10.1016/j.physletb.2009.11.038.

36 Kuhn, J.H. (1996) Theory of top quark production and decay, http://arxiv.org/abs/hep-ph/9707321.

37 Fischer, M., Groote, S., Korner, J.G., and Mauser, M.C. (2001) Longitudinal, transverse-plus and transverse-minus W-bosons in unpolarized top quark decays at $\mathcal{O}(\alpha_s)$, *Phys. Rev. D*, **63**, 031501, doi:10.1103/PhysRevD.63.031501.

38 CDF Collab., Aaltonen, T. et al. (2009) Measurement of W-boson helicity fractions in top-quark decays using $\cos\theta^*$, *Phys. Lett. B*, **674**, 160, doi:10.1016/j.physletb.2009.02.040.

39 Abulencia, A. et al. (2007) Search for $V+A$ current in top quark decay in $p\bar{p}$ collisions at $\sqrt{s} = 1.96$ TeV, *Phys. Rev. Lett.*, **98**, 072001, doi:10.1103/PhysRevLett.98.072001.

40 CDF Collab., Abulencia, A. et al. (2006) Measurement of the helicity of W bosons in top-quark decays, *Phys. Rev. D*, **73**, 111103, doi:10.1103/PhysRevD.73.111103.

41 Abazov, V.M. et al. (2005) Helicity of the W boson in lepton + jets $t\bar{t}$ events, *Phys. Lett. B*, **617**, 1, doi:10.1016/j.physletb.2005.04.069.

42 Particle Data Group, Amsler, C. et al. (2008) Review of Particle Physics, *Phys. Lett. B*, **667**, 1, doi:10.1016/j.physletb.2005.04.069

43 DØ Collab., Abazov, V.M. et al. (2008) Simultaneous measurement of the ratio $r = B(t \to Wb)/B(t \to Wq)$ and the top-quark pair production cross section with the DØ detector at $\sqrt{s} = 1.96$ TeV, *Phys. Rev. Lett.*, **100**, 192003, doi:10.1103/PhysRevLett.100.192003.

44. Wicke, D. (2009) Properties of the Top Quark, http://arxiv.org/abs/1005.2460.
45. CDF Collab., Abulencia, A. *et al.* (2006) Top quark mass measurement using the template method in the lepton + jets channel at CDF II, *Phys. Rev. D*, **73**, 032003, doi:10.1103/PhysRevD.73.032003.
46. Kondo, K. (1991) Dynamical likelihood method for reconstruction of events with missing momentum. 2: Mass spectra for $2 \to 2$ processes, *J. Phys. Soc. Jap.*, **60**, 836, doi:10.1143/JPSJ.60.836.
47. Dalitz, R.H. and Goldstein, G.R. (1992) Analysis of top-antitop production and dilepton decay events and the top quark mass, *Phys. Lett. B*, **287**, 225, doi:10.1016/0370-2693(92)91904-N.
48. Abazov, V.M. *et al.* (2004) A precision measurement of the mass of the top quark, *Nature*, **429**, 638, doi:10.1038/nature02589.
49. DØ Collab. (2009) Search for $t\bar{t}$ resonances in the lepton+jets final state in $p\bar{p}$ collisions at $\sqrt{s} = 1.96\,\text{TeV}$. http://www-d0.fnal.gov/Run2Physics/WWW/results/prelim/TOP/T83, DØ Note 5882-CONF.
50. DØ Collab., Abazov, V.M. *et al.* (2009) Search for charged Higgs bosons decaying into top and bottom quarks in $p\bar{p}$ collisions, *Phys. Rev. Lett.*, **102**, 191802, doi:10.1103/PhysRevLett.102.191802.
51. CDF Collab., Aaltonen, T. *et al.* (2009) Search for the production of narrow tb resonances in $1.9\,\text{fb}^{-1}$ of $p\bar{p}$ Collisions at $\sqrt{s} = 1.96\,\text{TeV}$, *Phys. Rev. Lett.*, **103**, 041801, doi:10.1103/PhysRevLett.103.041801.
52. DØ Collab., Abazov, V.M. *et al.* (2008) Search for W' boson resonances decaying to a top quark and a bottom quark, *Phys. Rev. Lett.*, **100**, 211803, doi:10.1103/PhysRevLett.100.211803.
53. Glashow, S.L., Iliopoulos, J., and Maiani, L. (1970) Weak interactions with lepton–hadron symmetry, *Phys. Rev. D*, **2**, 1285, doi:10.1103/PhysRevD.2.1285.
54. Heister, A. *et al.* (2002) Search for single top production in e^+e^- collisions at $\sqrt{s}$ up to 209 GeV, *Phys. Lett. B*, **543**, 173–182, doi:10.1016/S0370-2693(02)02307-9.
55. H1 Collab., Aaron, F.D. *et al.* (2009) Search for single top quark production at HERA, *Phys. Lett. B*, **678**, 450, doi:10.1016/j.physletb.2009.06.057.
56. DØ Collab., Abazov, V.M. *et al.* (2007) Search for production of single top quarks via tcg and tug flavor-changing-neutral current couplings, *Phys. Rev. Lett.*, **99**, 191802, doi:10.1103/PhysRevLett.99.191802.
57. CDF Collab., Aaltonen, T. *et al.* (2009) Search for top-quark production via flavor-changing neutral currents in $W + 1$ jet events at CDF, *Phys. Rev. Lett.*, **102**, 151801, doi:10.1103/PhysRevLett.102.151801.
58. CDF Collab., Aaltonen, T. *et al.* (2008) Search for the flavor changing neutral current decay $t \to Zq$ in $p\bar{p}$ Collisions at $\sqrt{s} = 1.96$, *Phys. Rev. Lett.*, **101**, 192002, doi:10.1103/PhysRevLett.101.192002.
59. CDF Collab., Abe, F. *et al.* (1998) Search for flavor-changing neutral current decays of the top quark in $p\bar{p}$ collisions at $\sqrt{s} = 1.8\,\text{TeV}$, *Phys. Rev. Lett.*, **80**, 2525, doi:10.1103/PhysRevLett.80.2525.
60. CDF Collab., Aaltonen, T. *et al.* (2009) Search for charged Higgs bosons in decays of top quarks in $p\bar{p}$ collisions at $\sqrt{s} = 1.96\,\text{TeV}$, *Phys. Rev. Lett.*, **103**, 101803, doi:10.1103/PhysRevLett.103.101803.

10
Beyond SUSY and the Standard Model: Exotica

Christophe Grojean, Thomas Hebbeker and Arnd Meyer

So far the Standard Model (SM) has been desperately successful in describing any experimental data in particle physics. However, it is incomplete (quantum gravity is not part of the SM), and raises several questions about its consistency at very high energy. The mystery of the dynamics of *electroweak symmetry breaking* (EWSB), the flavour puzzles of the mixing among the three families of quarks and the absence of flavour changing neutral currents, the disappearance of antimatter to the benefit of matter only, the large abundance of the unidentified dark matter, the importance of vacuum energy in the modern cosmological history of our universe are a few clues that physicists have at their disposal to imagine what the physics beyond the SM might look like.

10.1
Alternative Higgs

Within the Standard Model, electromagnetic and weak interactions are unified. In everyday life, we are used to close encounters with photons; but we had to wait until 1983 to even detect a single Z particle. This is because the $W^\pm$ and the Z, the gauge bosons associated to weak interactions, are massive. Since these masses are not invariant under the symmetries of the original SM Lagrangian (see Chapter 2), they cannot be considered as a small perturbation, and this apparent clash calls for a spontaneous breaking of the gauge symmetry which forces physicists to amend their understanding of nature at the Terascale.

In the broken phase, a (massive) spin-1 particle describes three different polarisations: two transverse ones plus an extra longitudinal one which decouples in the massless limit. The longitudinal degrees of freedom associated with the $W^\pm$ and Z gauge bosons presumably correspond to the eaten Nambu–Goldstone bosons [1, 2] resulting from the breaking of the global chiral symmetry $SU(2)_L \times SU(2)_R$ down to its diagonal subgroup $SU(2)_V$ (the same transformation is used in the L- and R-sectors). This picture still leaves us with the question of the source

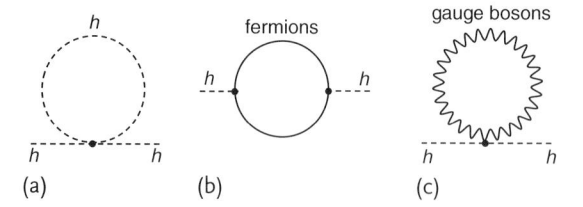

Figure 10.1 (a) one-loop corrections to the Higgs boson mass due to the Higgs boson self-interactions; (b) Yukawa interactions of the Higgs bosons with quarks and leptons; (c) the Higgs interactions with the gauge bosons. The three diagrams are quadratically divergent and make the Higgs mass highly UV-sensitive. New degrees of freedom or new symmetries are needed to cancel these divergences and to ensure the stability of the weak scale.

of the Nambu–Goldstone bosons: what is the sector responsible for the breaking[27] $SU(2)_L \times SU(2)_R \to SU(2)_V$? What are the dynamics of this sector? What are its interactions with the SM particles?

The minimum possibility consists in using four elementary scalar degrees of freedom transforming as a weak doublet. This Higgs doublet corresponds to four real scalar fields: the three eaten Nambu–Goldstone bosons plus the notorious Higgs boson [3, 4]. But the Higgs mechanism is at best a description and certainly not an explanation of EWSB since no dynamics explains the instability of the Higgs potential which is at the origin of EWSB. Moreover, it also jeopardises our current understanding of the SM at the quantum level since the Higgs potential suffers from instabilities under radiative corrections. In particular, in the presence of generic new physics, the Higgs mass becomes ultraviolet (UV) sensitive – this is the well-known hierarchy problem, see Figure 10.1 – unless a symmetry prevents it. In this case the radiative corrections are screened by the scale of the breaking of this symmetry rather than by the ultraviolet scale of new physics. Several scenarios have been developed along this line in the last 20–30 years (see [5] for a recent review and a list of references): supersymmetry, Higgs as a component of the gauge field along an extra spatial dimension, Higgs as a pseudo Nambu–Goldstone boson. In recent years, the latter possibility has been revived in many incarnations – little Higgs models and composite Higgs models formulated in five-dimensional space-time. Finally, technicolour theories avoid the hierarchy problem by breaking the electroweak symmetry via strong dynamics. Recently Higgs-less models formulated in five dimensions, which are in some sense dual to technicolour models, have been proposed.

27) Neglecting the Yukawa interactions and turning off the hypercharge gauge interaction, the SM is invariant not only under a global $SU(2)_L$ symmetry that interchanges the left-handed up-type and down-type quarks, but also under a global $SU(2)_R$ symmetry that now interchanges the right-handed up-type and down-type quarks. The $SU(2)_L \times SU(2)_R$ symmetry is broken to its diagonal subgroup $SU(2)_V$ by the Higgs vacuum value or whatever replaces it. Note that this enlarged global symmetry explains why the ρ parameter, that is, the squared ratio of the W mass over the Z mass times the cosine of the weak mixing angle, is equal to one at the classical level.

The idea behind the *little Higgs models* [6] is to identify the full Higgs doublet as a (pseudo) Nambu–Goldstone boson while keeping some sizable non-derivative interactions like the Yukawa interactions to fermions or the Higgs boson self-interactions. By analogy with Quantum Chromodynamics (QCD) where the pions $\pi^{\pm,0}$ appear as Nambu–Goldstone bosons associated to the breaking of the chiral symmetry $SU(2)_L \times SU(2)_R / SU(2)_{\text{isospin}}$, switching on some interactions that explicitly break the global symmetry will generate a mass for the would-be massless Nambu–Goldstone bosons of the order of $g \Lambda_{G/H}/(4\pi)$, where g is the coupling of the symmetry-breaking interaction and $\Lambda_{G/H} = 4\pi f_{G/H}$ is the dynamical scale of the global symmetry breaking. Therefore, obtaining a Higgs boson mass around 100 GeV would demand a dynamical scale $\Lambda_{G/H}$ of the order of 1 TeV, which is known to lead to too large disturbances to the W and Z propagators (too large oblique corrections). Raising the strong dynamical scale by at least one order of magnitude requires an additional selection rule to ensure that a Higgs boson mass is generated at the two-loop level only. The way to enforce this selection rule is through a "collective breaking" of the global symmetry via two sets of interactions, $\mathcal{L}_{1,2}$. However, each interaction $\mathcal{L}_1$ or $\mathcal{L}_2$ individually preserves a subset of the global symmetry such that the Higgs remains an exact Nambu–Goldstone boson whenever either g_1 or g_2 vanishes. A mass term for the Higgs boson can only be generated by diagrams involving both interactions simultaneously. At one-loop level, there is no such diagram that would be quadratically divergent. Explicitly, the cancellation of the SM quadratic divergences is achieved by a set of new particles around the Fermi scale (see Figure 10.2): gauge bosons, vector-like quarks and extra massive scalars, which are related, by the original global symmetry, to the SM particles with the same spin. These new particles, with definite couplings to SM particles as dictated by the global symmetries of the theory, are perfect goals for the LHC.

10.2
Technicolour, Composite Higgs and Partial Compositeness

What unitarises the WW scattering amplitude? Supersymmetric models, little Higgs models and many other models take for granted that the Higgs boson pro-

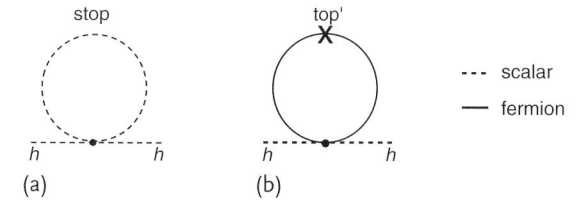

Figure 10.2 Examples of the cancellation of quadratic divergences associated to the top-quark Yukawa interactions. In supersymmetric theories, the divergence is cancelled by the scalar partners of the top quark ((a); the *stops*), while in little Higgs models it is cancelled by new heavy fermions ((b); the *top'*).

vides the answer to this pressing question. As we stressed earlier, the masses of the $W^\pm$ and Z gauge bosons are sources of trouble since they break the SM gauge symmetry. Actually, in the presence of these masses, the gauge symmetry is realised non-linearly: the longitudinal $W_L^\pm$, Z_L can be described by the Nambu–Goldstone bosons, or pions, associated to the coset $SU(2)_L \times SU(2)_R/SU(2)_{\text{isospin}}$, and the gauge-boson mass terms correspond to the pion kinetic term. Thanks to this Goldstone boson equivalence [7], the non-trivial scattering of the longitudinal gauge bosons now simply follows for the contact interactions among four pions and leads to amplitudes that grow with energy. In other words, the longitudinal polarisations of the W and Z bosons do not properly decouple at very high energy and, in the absence of any new weakly coupled elementary degrees of freedom cancelling this growth, perturbative unitarity will be lost around 1.2 TeV [8] (some people prefer to quote a scale $4\pi v \approx 3.1$ TeV) and new strong dynamics will kick in and soften the ultraviolet behaviour of the amplitude, for instance via the exchange of massive bound states similar to the ρ meson of QCD.

One particular realisation of this idea of electroweak symmetry breaking by strong dynamics is the *technicolour paradigm* [9, 10] which, in analogy to QCD, postulates the existence of new elementary degrees of freedom (often called techniquarks) which are strongly interacting and condense. Constructing an explicit and viable model along this line requires a lot of engineering: first the strong sector has be *extended* to account for the masses of the SM (light) quarks and leptons; second a special (*topcolour*) sector is needed to generate a large mass for the top quark; finally, non-trivial (*walking*, i. e. near conformal) dynamics might provide a way to avoid conflicts with electroweak precision data.

Flavour data have also been a notorious challenge for technicolour models. Actually, it is easy and instructive to understand why traditional technicolour approaches do not allow a decoupling of the flavour puzzle: the order parameter associated to the breaking of the electroweak symmetry is a bound state of two fermions and its natural scaling dimension is thus equal to 3, making the Higgs mass an irrelevant operator. However, on dimension grounds, the Yukawa interactions are expected to be as suppressed as generic four-Fermi operators. It therefore becomes difficult to generate the fermion masses without generating large flavour changing neutral currents. Walking technicolour models are an attempt to solve this issue with the use of large anomalous couplings. However, another interesting, and maybe simpler, solution emerges when the fermion masses are generated by linear mixing between an elementary fermion and a composite one [11]. This partial compositeness scenario also easily accommodates a hierarchical mass spectrum by small deviations of the anomalous dimensions of the composite operators and naturally explains the alignment between the masses and the CKM mixing angles of the three quark generations. Models with warped extra dimensions actually offer an explicit realisation of this partial compositeness idea.

In another interesting class of models in which EWSB proceeds by strong dynamics, the Higgs boson itself appears as a *composite bound state* emerging from a strongly interacting sector [12]. In order to maintain a good agreement with electroweak data, it is sufficient that a mass gap separates the Higgs resonance from the other resonances of the strong sector. Such a mass gap can naturally follow from dynamics if the strongly interacting sector possesses a global symmetry, G, spontaneously broken at a scale f to a subgroup H, such that the coset G/H contains a fourth Nambu–Goldstone boson that can be identified with the Higgs boson as in little Higgs models. Simple examples of such a coset are $SU(3)/SU(2)$ or $SO(5)/SO(4)$, the latter being favoured since it is invariant under a custodial symmetry, that is, a global symmetry that ensures that the masses of the W and Z gauge bosons are related to the weak mixing angle and therefore guarantees that the ρ parameter is equal to one at tree level as in the Standard Model. Modern incarnations of the composite Higgs idea have recently been investigated in the framework of 5D warped models [13] where the holographic composite Higgs boson now originates from a component of a gauge field along the fifth dimension with appropriate boundary conditions. The composite Higgs models offer a nice and continuous interpolation [14] between the SM and technicolour-type models. The dynamical scale f defines the compositeness scale of the Higgs boson: when $\xi = v^2/f^2 \to 0$, the Higgs boson appears essentially as a light elementary particle (and its couplings approach the ones predicted by the SM) while the other resonances of the strong sector become heavier and heavier and decouple; on the other hand, when $\xi \to 1$, the couplings of the Higgs boson to the W_L bosons go to zero and unitarity in gauge-boson scattering is ensured by the exchange of the heavy resonances [15].

10.3
Extra Dimensions, Strings and Branes

Physicists are perpetual explorers of the space-time in which they live, in which they work. The more they know, the more questions they ask. There are many good reasons to believe that space-time is four-dimensional: from the experimental evidence of Gauss' law and the $1/r^2$ strength of the electromagnetic and gravitational forces to the theoretical dogma of renormalisability of gauge theories. Nonetheless, living in more than four dimensions can have some advantages far beyond making hide-and-seek games more fun. In the quest for unification of the fundamental interactions, extra dimensions have been a key player (see [16] for a recent introduction and a list of references) since the pioneering works of Kaluza and Klein in the 1920s who discovered that Einstein's gravity in five dimensions was very much like Einstein's gravity plus electromagnetism in the usual four dimensions when the extra spatial dimension is compactified on a small circle. Later, with the advance of string theory, it was discovered that reconciling gravity and quantum mechanics requires some extra space: indeed an important classical symmetry needed to remove unphysical states from the spectrum of the string excitations survives at the quantum level only in the presence of 6, 7 or 22 extra space dimensions. While a "why not?" answer is perfectly legitimate when asking "why extra dimensions?", the proof of the undeniable necessity of some extra dimensions to quantise gravity has acted as a catalyst, triggering a lot of interest and leading to the emergence of new revolutionary ideas. For instance, in the late 1990s, it was realised that extra dimensions could also help to solve the mystery why gravity is so weak compared to the other fundamental interactions. Two plausible pictures have been proposed: (i) gravity is weak because it is diluted in large extra dimensions not directly accessible to matter or (ii) gravity is weak because it is localised away from the SM matter that only feels the tail of the graviton wave function. Stimulated by these two paradigms, the imagination of theorists and model builders has flourished. For instance it was learnt how to break (gauge or global) symmetries by suitable boundary conditions, how to explain the dark matter abundance, maybe also how to accelerate the expansion of the universe without the help of a cosmological constant.

But the range of applications of extra dimensions exploded when it was realised that these extra dimensions do not have to be real ones but could simply be a tool to deal with strongly coupled systems in standard four-dimensional space-time. This new chapter of the story started with the audacious conjecture [17] put forward in 1997 by a young string theorist, J. Maldacena: in the same way that in quantum mechanics particles and waves are two descriptions of the same reality, in some cases a new force could be equivalently described in terms of a new spatial dimension. This idea found immediate application to the breaking of the electroweak symmetry and to the construction of the *dual* five-dimensional description [15] of the dynamical technicolour models of the 1970s. Later, it was also applied to study the properties of QCD itself or to describe a strongly interacting plasma in the hydrodynamics limit. Today this duality has disseminated away from particle physics and allows the physics of strongly interacting superconductors and superfluids to be studied using methods that standard weak-coupling BCS-like techniques would not allow one to grasp.

Even if real, the putative extra dimensions of our universe are certainly not identical to the ones we experience daily. One way to *hide* them is to assume that they are compact. For instance, in the simplest case of a single extra dimension, our world would be the direct product of the usual four-dimensional Minkowski space times a circle of radius R, and any field would be periodic in the coordinate, z, along the circle,

$$\phi(x_\mu, z) = \phi(x_\mu, z + 2\pi R), \quad \mu = 0, 1, 2, 3. \tag{10.1}$$

This periodicity in z allows a Fourier decomposition,

$$\phi(x_\mu, z) = \sum_n \phi_n(x_\mu) e^{inz/R}. \tag{10.2}$$

Using the 5D equations of motion, the mass of the 4D field, ϕ_n, is found to be $|n|/R$. A 5D field is just equivalent to an infinite tower of 4D fields more and more massive: they are the *Kaluza–Klein excitations* of the massless mode, ϕ_0.

For an observer unable to resolve distances smaller than the size of the extra dimensions, the world would essentially appear as four dimensional. It is only when the energy at her/his disposal will be greater than the mass gap of the first Kaluza–

Klein excitations that she/he will start experiencing the new dimensions of space. What are the sizes of the extra dimensions? Will they ever be observable at the LHC? For merely 80 years after the pioneering work of Kaluza and Klein, it was assumed that these extra dimensions could only be accessible to a Planck observer, that is, someone able to probe distances as small as 10^{-33} cm (or energies as high as 10^{19} GeV). The dogma radically changed in the 1990s with the discovery of *branes*, some solitonic objects of string theory which correspond to hypersurfaces with fewer dimensions than the space in which they are embedded and which confine matter and forces. If all the SM particles are confined on a *D3 brane*, the extra dimensions accessible to gravity only could be as large as one twentieth of a millimetre without conflicting with any experimental result. This "large extra dimension" idea was put forward by Arkani-Hamed, Dimopoulos and Dvali and is therefore named "ADD". In this paradigm [18], gravity is strengthened and some quantum effects could even manifest themselves at the Terascale, where the LHC will hear the echoes sent by particles moving in the hidden dimensions. The production of microscopic black holes in colliders could be a spectacular manifestation. These mini black-holes would evaporate rapidly via Hawking radiation and would manifest themselves as final states with energetic jets, leptons, photons as well as missing energy carried away by neutrinos.

In the large extra dimension scenario, the brane where the SM world is confined has no impact on the rest of the universe. On the contrary, in the *warped extra dimension* setup first proposed by Randall and Sundrum [19], the brane strongly curves the extra dimensions: the gravitational effects of the brane are finely balanced by an extra dimensional vacuum energy, resulting in a universe that seems static and flat for any observer on our brane. The warping of the space acts as a powerful gravitation well that strongly distorts the spectrum of the Kaluza–Klein excitations, which become physically equivalent to the bound states emerging from a strongly interacting sector. This is the limit in which a new spatial dimension is just another description of a new force.

In the *ADD model* [18], the graviton states are too close in mass to be distinguished individually, and the coupling remains small. However, the number of accessible states is very large. It is therefore possible to produce gravitons which immediately disappear into bulk space, leading to an excess of events with a high transverse energy jet and large missing transverse energy, the monojet signature: $b\bar{b} \to gG$, $qg \to qG$ and $gg \to gG$, where G is the emitted graviton. The dominant Standard Model backgrounds are the production of Z or W bosons plus jets, with the Z boson decaying to a pair of neutrinos or the lepton from the W decay escaping detection. Similarly, one can search for an excess of events with a high transverse energy photon and large missing transverse energy. Tevatron searches have set limits [20, 21] of up to $M_D > 1.4$ TeV and $M_D > 0.94$ TeV on the effective Planck scale for a number of extra dimensions n_d ranging from 2 to 6, where the sensitivity in the monojet signature is typically somewhat better than in the monophoton channel. At the LHC the sensitivity is quickly extended, and one can hope for a signal even if $M_D \simeq 3.6(2.6)$ TeV for $n_d = 2$ (4) after accumulating 1 fb^{-1} of data at $\sqrt{s} = 14$ TeV [22].

The virtual exchange of Kaluza–Klein gravitons would modify the cross sections for SM processes like the production of fermion or boson pairs. The sensitivity is expressed in terms of the scale M_s, which is expected to be close to M_D. For example, DØ [23] has investigated the high-mass dielectron and diphoton mass spectrum, and has found no indications for large extra dimensions. The resulting limits on M_s are $M_s > 2.1(1.6, 1.3)$ TeV for $n_d = 2\,(4,\,7)$ at 95% confidence level. A similar sensitivity has been achieved in a dijet angular analysis by DØ [24]. The LHC reach in the dilepton and diphoton channels is expected to be $M_s \simeq 9(6)$ TeV for $n_d = 3\,(6)$ with 100 fb^{-1} of data at $\sqrt{s} = 14$ TeV [25].

In the simplest version of the Randall–Sundrum model, gravitons are the only particles that can propagate in the extra dimension. The gravitons appear as towers of Kaluza–Klein excitations with masses and widths determined by the parameters of the model. These parameters can be expressed in terms of the mass of the first excited mode of the graviton, M_1, and the dimensionless coupling to the Standard Model fields, $k\sqrt{8\pi}/M_{pl}$. If it is light enough, the first excited graviton mode could be resonantly produced at colliders. It is expected to decay to fermion–antifermion and to diboson pairs. The latest Tevatron search from DØ [26] analyses the e^+e^- and $\gamma\gamma$ final states. Using invariant-mass distributions as shown in Figure 10.3, upper cross-section limits are derived, which are then translated into lower mass limits for the lowest excited mode of Randall–Sundrum gravitons. For a coupling parameter $k\sqrt{8\pi}/M_{pl} = 0.1\,(0.01)$, masses $M_1 < 1050(560)$ GeV are excluded at the 95% confidence level. At the LHC it is expected that the theoretically interesting regime will be almost entirely covered already with ~ 10 fb^{-1} of data at $\sqrt{s} = 14$ TeV [25, 27].

10.4
Grand Unified Theories

Every breath of fresh air comes with one of the deepest questions agitating particle physics: the inhaled molecules of air are electrically neutral, which requires a

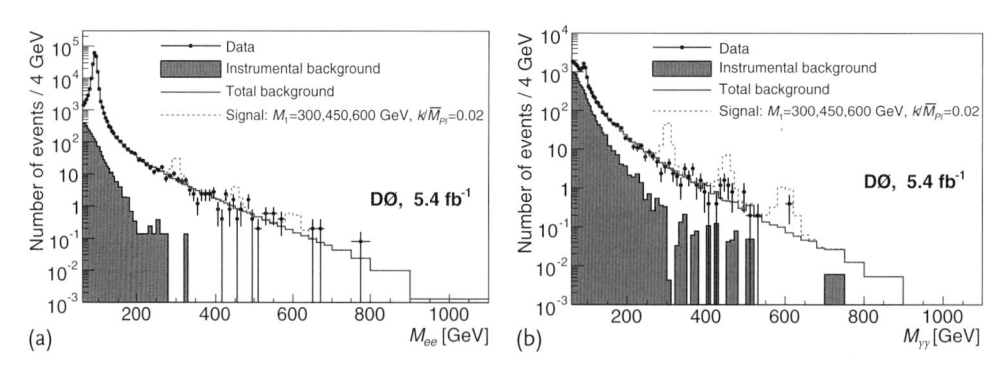

Figure 10.3 Invariant e^+e^- (a) and $\gamma\gamma$ (b) mass spectra used by the DØ collaboration in the search for Randall–Sundrum gravitons (adapted from [26]).

delicate balance between the electric charge of the electrons and the ones of the quarks inside the protons and the neutrons of the nuclei. Why is it the case that the charges of all observed particles are simple multiples of a fundamental charge? Or, in more pedantic terms, why is the electric charge quantised? Following Dirac's superb insight, such a quantisation could be accounted for by the existence of even a single magnetic monopole. Another plausible explanation comes naturally with the idea that the fundamental interactions of the SM could unify and be described by a single non-Abelian gauge group (see [28] for an exhaustive exposure to the topic). The seemingly different particles of the SM would have to organise themselves into irreducible representations of this grand unified gauge group; then electric charge quantisation, as well as the exact cancellation of the anomalies between the leptons and the quarks, which ensures the quantum consistency of the SM gauge interactions, would automatically follow.

Mathematically, the goal of grand unification theories (GUT) is to embed the SM gauge groups $U(1)_Y \times SU(2)_L \times SU(3)_c$ into a larger simple or semi-simple gauge group, which is spontaneously broken at a very high energy in such a way that the three SM interactions happen to be different, due to a dynamical accident at low energy:

$$G_{\text{GUT}} \xrightarrow{M_{\text{GUT}}} U(1)_Y \times SU(2)_L \times SU(3)_c \xrightarrow{M_{\text{weak}}} U(1)_{\text{em}} \times SU(3)_c . \tag{10.3}$$

The trace of the electric charge operator, Q, being one of the generators of G_{GUT}, must vanish. If we further assume that no new particle beyond the SM ones fills in the same representation of G_{GUT}, we are bound to conclude that each irreducible representation of G_{GUT} must contain both quarks and leptons, which means that there must exist some gauge bosons of G_{GUT} which can change a lepton into a quark and vice versa. This inevitably leads to an instability of the proton, for instance via the decay channel $p \to e^+ \pi^0$, with a lifetime of the order of $\tau_p \sim M_{\text{GUT}}^4 / m_p^5$ (m_p denotes the mass of the proton). From the current experimental bounds, $\tau_p \geq 10^{32 \div 33}$ years, we conclude that $M_{\text{GUT}} \geq 10^{15}$ GeV. Remarkably this energy scale is very close to the estimate of the unification scale obtained by running the low-energy gauge couplings: as a matter of fact, the particle content of the theory determines the logarithmic running of the couplings and the SM particles allow for the three gauge couplings to almost become equal at an energy around 10^{16} GeV. The unification of the gauge couplings is even better with the particle content of the MSSM. But adding particles is not the only way to ameliorate this coupling unification. Removing the contributions of the Higgs boson and the right-handed top quark, which could well be some composite particles above the weak scale, leads to a gauge coupling unification almost as precise as in the MSSM.

The simplest example of grand unification is the one proposed by Georgi and Glashow [29] based on $G_{\text{GUT}} = SU(5)$. The gauge bosons belong to the 24-dimensional adjoint representation, which decomposes under $SU(3)_c \times SU(2)_L \times U(1)_Y$ as

$$24 = \underbrace{(2,3)_{5/6} + (2,\bar{3})_{-5/6}}_{X,Y} + \underbrace{(1,8)_0}_{\text{gluons}} + \underbrace{(3,1)_0 + (1,1)_0}_{\gamma, Z, W^{\pm}} . \tag{10.4}$$

The 12 new gauge bosons, called X and Y, with electric charges 4/3 and 1/3, respectively, will have to acquire a mass at least of the order of 10^{15} GeV to suppress the proton decay amplitude. The $SU(5)$ model does not fully exhaust the GUT agenda: the 15 particles of a SM family fill not a single but two irreducible representations of $SU(5)$:

$$\bar{5} = \underbrace{(1,\bar{3})}_{d_R^c} + \underbrace{(2,1)}_{\nu,e_L} \quad \text{and} \quad 10 = \underbrace{(2,3)}_{Q_L} + \underbrace{(1,\bar{3})}_{u_R^c} + \underbrace{(1,1)}_{e_R^c} \,. \tag{10.5}$$

In this regard, the gauge anomaly cancellation still happens as an accidental cancellation between the contributions of the two irreducible representations. Anomaly cancellation is automatic in the $SO(10)$ GUT model where an additional state, a right-handed neutrino, is added to fill in a single 16-dimensional irreducible representation. Not a flaw but rather a virtue, a right-handed neutrino allows for an elegant generation of the neutrino spectrum through the seesaw mechanism.

Beyond its indisputable conceptual elegance, grand unification might also serve to explain cosmological puzzles like the abundance of matter over antimatter, baryon number violation being one of the necessary ingredients for a dynamical baryogenesis.

10.5
Extra Gauge Bosons

A possible way of resolving the inherent problems of the Standard Model is by extending the gauge sector of the theory. New heavy gauge bosons are predicted in many extensions of the Standard Model. The *sequential standard model*, where the couplings to quarks and leptons are as in the SM, may not be gauge invariant, but it serves as a good benchmark for comparisons of results. The Tevatron experiments have excluded a sequential Z' or W' in direct searches up to masses of about 1 TeV in decays into e^+e^-, $\mu^+\mu^-$ or $e\nu$, respectively [30–33]. Mass limits in other decay modes are typically weaker, while indirect limits from LEP are somewhat stronger.

The LHC experiments substantially extend the mass reach for a discovery, to a $\sim$2 TeV sequential Z' with 1 fb^{-1} of data at $\sqrt{s} = 14$ TeV in the decay modes to dielectrons or dimuons, and an ultimate sensitivity up to about 5 TeV with more than 100 fb^{-1} of data. For W' searches in the e/μ plus $\not{E}_T$ final state, the required integrated luminosity for the same mass reach is even lower [25, 27].

10.6
Leptoquarks

Leptoquarks are particles that can interact simultaneously with quarks and leptons. They appear naturally in theories which unify quarks and leptons into a single multiplet, and as such are expected to be gauge bosons as well (scalar leptoquarks are also expected in technicolour-like theories). While their masses may at first sight be expected to be of the order of the unification scale, in some models they can be relatively light. Experimentally, it is customary to consider one leptoquark per generation. These are assumed to be very short-lived and decay to a quark and a lepton. The branching fraction to a charged lepton and a quark is then denoted as β. At hadron colliders, leptoquarks can be pair-produced through the strong interaction, or are singly produced. In the latter case the production cross section depends on the (unknown) quark–lepton coupling, which is generally taken to be of the same order of magnitude as the fine-structure constant.

At the Tevatron searches for the pair-production of leptoquarks of all three generations have been performed, and for different regions of β, that is, different dominant decay modes. Exclusion limits are presented as a function of the leptoquark mass and β, often both for the scalar and vector case. For low values of β, the final state consists of two acoplanar jets and $\not{E}_T$, and single-generation leptoquarks with a mass below 205 GeV have been excluded by DØ [34]. For larger values of β, final states with a single charged lepton and jets, or two charged leptons and jets, become important. Searches including e, μ, τ final states as well as b-quark jets have been performed by DØ and CDF, and the excluded scalar leptoquark masses exceed 300 GeV (see e.g. [35–37]). Searches by the HERA experiments H1 and ZEUS are complementary to the Tevatron. A multitude of models has been investigated, and beyond the leptoquark mass the obtained limits (see e.g. [38]) depend here on the size of the R-parity violating coupling, λ. In particular, single leptoquarks can be produced resonantly at HERA in reactions like $ep \to eX$ and $ep \to \nu X$. For $\lambda \ll 1$ the limits are typically close to the kinematic limit of HERA of about 300 GeV. At the LHC, an integrated luminosity of about $1\,\text{fb}^{-1}$ at $\sqrt{s} = 14\,\text{TeV}$ is sufficient to discover first- and second-generation scalar leptoquarks in dilepton plus jets final states up to a mass of about 800 GeV [27].

10.7
Unexpected Physics: Hidden Valley, Quirks, Unparticles...

The quest for new physics at the LHC is an art. Because of the overwhelming SM background, identifying an interesting signal requires first an intimate knowledge of the signal sought for and second a dedicated study to extract the salient characteristics of this signal and distinguish it from the background. However, does the complexity of these analyses leave any room for a fortunate stroke of serendipity? In other words, is it possible to discover something if it is not looked for? Are we sure that because of finely-tuned cuts and triggers, we are not going to miss any *unexpected* physics – like some physics that does not have to be there but that simply happens to be there and enriches the landscape of particle physics?

The hierarchy problem, that is, the question of the stability of the electroweak scale at the quantum level, together with the evident need for a large amount of dark matter have shaped our vision of physics beyond the Standard Model. Weakly or strongly interacting with the SM states, new particles are populating the dreams of the physicists at the Terascale: not only superpartners or Kaluza–Klein excitations but also heavy Z', leptoquarks or even an entire fourth generation.

A lot of attention has recently been devoted to a class of new physics that features a *hidden sector* with new low-mass states that would escape the traditional search techniques (see [39] for a recent review). The hidden sector will have its own dynamics. Typical examples have been considered. In *hidden-valley models*, the hidden sector confines in a way analogous to QCD, that is, there are some *quarks* which are lighter than the confinement scale. In contrast, *quirk models* have only quarks heavier than the confinement scale, which would result in stable and long strings connecting two quarks and oscillating inside the detector. Finally, the hidden sector of *unparticle models* is conformal in the infrared: there is no mass gap but only a continuum of states which could be seen as the limit of a spectrum of excitations of models with warped extra dimensions.

10.8
Model-Independent Search for New Physics

It is not clear which effects of new physics will appear in the first data, and which theory beyond the Standard Model will describe them. This is the main idea behind a systematic analysis of the data with as little bias as possible. For example at CMS, a special algorithm called "MUSiC" (Model Unspecific Search in CMS) has been developed [40]. Similar strategies have already been applied successfully by other collaborations (see e.g. [41–44]). Events are classified into event classes which group events according to their final-state topology. Each class is defined by the number of physics objects in the event, for example, 1μ 3jet. Exclusive and inclusive event classes are considered: in the inclusive classes a minimal number of objects (e.g. 1μ 3jet $+ X$, at least one muon and three jets) is required. Depending on which objects are considered, this can lead to more than 1000 event classes. It is then often assumed that new physics will appear in events with high-p_T objects. Critical for the successful implementation is a solid understanding of all objects over the entire phase space of such an analysis, and a consistent Monte Carlo description of the Standard Model.

In the initial MUSiC algorithm, three distributions are investigated for each event class: the scalar sum of the transverse momentum, $\sum p_T$, of all physics objects; the invariant mass, M_{inv}, of all physics objects (transverse mass, M_T, for classes with $\slashed{E}_T$); and for classes with missing transverse energy, $\slashed{E}_T$. All distributions are scanned systematically for deviations, comparing the Monte Carlo (MC) prediction with the measured data. The deviations are quantified with the help of MC experiments and including the best knowledge of systematic uncertainties. Benchmarking is possible with "known" models of new physics, and such benchmarks indicate that the model-independent approach is able to find signals with somewhat lower sensitivity than an optimised search for a certain channel – but has the advantage of being sensitive in many channels simultaneously, also including the channels that may have been omitted in a dedicated search. Tests show that some

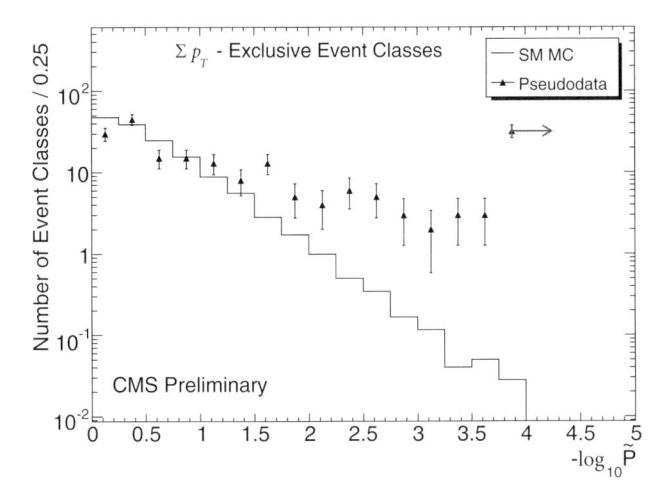

Figure 10.4 CMS model-independent search: frequency distribution of *p* values corrected for a "look-elsewhere-effect" using only exclusive event classes with a high p_T lepton and using only the $\sum p_T$ distribution. Histogram: averaged CMS experiment (SM only), points: a single pseudo-experiment with a certain low mass supersymmetry benchmark (LM4). The last point shown includes all values above itself, as indicated by the horizontal arrow (adapted from [40]).

models of new physics like supersymmetry can lead to complex multiple deviations, in tens or even hundreds of final states, as illustrated in Figure 10.4.

References

1 Nambu, Y. (1960) Axial vector current conservation in weak interactions, *Phys. Rev. Lett.*, **4**, 380, doi:10.1103/PhysRevLett.4.380.
2 Goldstone, J. (1961) Field Theories with 'Superconductor' Solutions, *Nuovo Cimento*, **19**, 154, doi:10.1007/BF02812722.
3 Englert, F. and Brout, R. (1964) Broken Symmetry and the Mass of Gauge Vector Mesons, *Phys. Rev. Lett.*, **13**, 321, doi:10.1103/PhysRevLett.13.321.
4 Higgs, P.W. (1964) Broken symmetries and the mass of gauge bosons, *Phys. Rev. Lett.*, **13**, 508, doi:10.1103/PhysRevLett.13.508.
5 Grojean, C. (2009) New theories for the Fermi scale, http://arxiv.org/abs/0910.4976.
6 Arkani-Hamed, N., Cohen, A.G., Katz, E., and Nelson, A.E. (2002) The littlest Higgs, *JHEP*, **07**, 034, doi:10.1088/1126-6708/2002/07/034.
7 Chanowitz, M.S. and Gaillard, M.K. (1985) The TeV physics of strongly interacting W's and Z's, *Nucl. Phys. B*, **261**, 379, doi:10.1016/0550-3213(85)90580-2.
8 Lee, B.W., Quigg, C., and Thacker, H.B. (1977) The strength of weak interactions at very high-energies and the Higgs boson mass, *Phys. Rev. Lett.*, **38**, 883, doi:10.1103/PhysRevLett.38.883.
9 Weinberg, S. (1976) Implications of dynamical symmetry breaking, *Phys. Rev. D*, **13**, 974, doi:10.1103/PhysRevD.13.974.
10 Susskind, L. (1979) Dynamics of spontaneous symmetry breaking in the Weinberg Salam theory, *Phys. Rev. D*, **20**, 2619, doi:10.1103/PhysRevD.20.2619.
11 Kaplan, D.B. (1991) Flavor at SSC energies: A new mechanism for dynam-

ically generated fermion masses, *Nucl. Phys. B*, **365**, 259, doi:10.1016/S0550-3213(05)80021-5.
12. Georgi, H. and Kaplan, D.B. (1984) Composite Higgs and custodial $SU(2)$, *Phys. Lett. B*, **145**, 216, doi:10.1016/0370-2693(84)90341-1.
13. Agashe, K., Contino, R., and Pomarol, A. (2005) The minimal composite Higgs model, *Nucl. Phys. B*, **719**, 165, doi:10.1016/j.nuclphysb.2005.04.035.
14. Giudice, G.F., Grojean, C., Pomarol, A., and Rattazzi, R. (2007) The strongly-interacting light Higgs, *JHEP*, **06**, 045, doi:10.1088/1126-6708/2007/06/045.
15. Csaki, C., Grojean, C., Pilo, L., and Terning, J. (2004) Towards a realistic model of higgsless electroweak symmetry breaking, *Phys. Rev. Lett.*, **92**, 101802, doi:10.1103/PhysRevLett.92.101802.
16. Shifman, M. (2010) Large extra dimensions: becoming acquainted with an alternative paradigm, *Int. J. Mod. Phys. A*, **25**, 199, doi:10.1142/S0217751X10048548.
17. Maldacena, J.M. (1998) The large N limit of superconformal field theories and supergravity, *Adv. Theor. Math. Phys.*, **2**, 231. http://arxiv.org/abs/hep-th/9711200.
18. Arkani-Hamed, N., Dimopoulos, S., and Dvali, G.R. (1998) The hierarchy problem and new dimensions at a millimeter, *Phys. Lett. B*, **429**, 263, doi:10.1016/S0370-2693(98)00466-3.
19. Randall, L. and Sundrum, R. (1999) A large mass hierarchy from a small extra dimension, *Phys. Rev. Lett.*, **83**, 3370, doi:10.1103/PhysRevLett.83.3370.
20. CDF Collab., Aaltonen, T. *et al.* (2008) Search for large extra dimensions in final states containing one photon or jet and large missing transverse energy produced in $p\overline{p}$ collisions at $\sqrt{s} = 1.96\,\text{TeV}$, *Phys. Rev. Lett.*, **101**, 181602, doi:10.1103/PhysRevLett.101.181602.
21. DØ Collab., Abazov, V.M. *et al.* (2008) Search for large extra dimensions via single photon plus missing energy final states at $\sqrt{s} = 1.96\,\text{TeV}$, *Phys. Rev. Lett.*, **101**, 011601, doi:10.1103/PhysRevLett.101.011601.
22. CMS Collab. (2008) Search for Mono-Jet Final States from ADD Extra Dimensions, http://cdsweb.cern.ch/record/1198687.
23. DØ Collab., Abazov, V.M. *et al.* (2009) Search for large extra spatial dimensions in the dielectron and diphoton channels in $p\overline{p}$ collisions at $\sqrt{s} = 1.96\,\text{TeV}$, *Phys. Rev. Lett.*, **102**, 051601, doi:10.1103/PhysRevLett.102.051601.
24. DØ Collab., Abazov, V.M. *et al.* (2009) Measurement of dijet angular distributions at $\sqrt{s} = 1.96\,\text{TeV}$ and searches for quark compositeness and extra spatial dimensions, *Phys. Rev. Lett.*, **103**, 191803, doi:10.1103/PhysRevLett.103.191803.
25. CMS Collab., Bayatian, G.L. *et al.* (2007) CMS Physics Technical Design Report, Volume II: Physics Performance, *J. Phys. G*, **34**, 995, doi:10.1088/0954-3899/34/6/S01.
26. DØ Collab., Abazov, V.M. *et al.* (2010) Search for Randall–Sundrum gravitons in the dielectron and diphoton final states with 5.4 fb^{-1} of data from $p\overline{p}$ collisions at $\sqrt{s} = 1.96\,\text{TeV}$, *Phys. Rev. Lett.*, **104**, 241802, doi:10.1103/PhysRevLett.104.241802.
27. ATLAS Collab., Aad, G. *et al.* (2009) Expected Performance of the ATLAS Experiment – Detector, Trigger and Physics, http://arxiv.org/abs/0901.0512.
28. Ross, G.G. (1984) Grand Unified Theories, Benjamin/Cummings, Menlo Park, CA, USA, 497 p., http://www.slac.stanford.edu/spires/find/hep/www?key=1451618, (Frontiers in Physics, 60).
29. Georgi, H. and Glashow, S.L. (1974) Unity of all elementary particle forces, *Phys. Rev. Lett.*, **32**, 438, doi:10.1103/PhysRevLett.32.438.
30. CDF Collab., Aaltonen, T. *et al.* (2009) Search for high-mass e^+e^- resonances in $p\overline{p}$ collisions at $\sqrt{s} = 1.96\,\text{TeV}$, *Phys. Rev. Lett.*, **102**, 031801, doi:10.1103/PhysRevLett.102.031801.
31. CDF Collab., Aaltonen, T. *et al.* (2009) Search for high-mass resonances decaying to dimuons at CDF, *Phys. Rev. Lett.*, **102**, 091805, doi:10.1103/PhysRevLett.102.091805.
32. DØ Collab., Abazov, V.M. *et al.* (2008) Search for W' bosons decaying to an

electron and a neutrino with the DØ detector, *Phys. Rev. Lett.*, **100**, 031804, doi:10.1103/PhysRevLett.100.031804.

33 DØ Collab., Abazov, V.M. *et al.* (2011) Search for a heavy neutral gauge boson in the dielectron channel with $5.4\,\text{fb}^{-1}$ of $p\bar{p}$ collisions at $\sqrt{s} = 1.96\,\text{TeV}$, *Phys. Lett. B* **695**, 88–94.

34 DØ Collab., Abazov, V.M. *et al.* (2008) Search for scalar leptoquarks and T-odd quarks in the acoplanar jet topology using $2.5\,\text{fb}^{-1}$ of $p\bar{p}$ collision data at $\sqrt{s} = 1.96\,\text{TeV}$, *Phys. Lett. B*, **668**, 357, doi:10.1016/j.physletb.2008.09.014.

35 DØ Collab., Abazov, V.M. *et al.* (2009) Search for pair production of second generation scalar leptoquarks, *Phys. Lett. B*, **671**, 224, doi:10.1016/j.physletb.2008.12.017.

36 DØ Collab., Abazov, V.M. *et al.* (2008) Search for third generation scalar leptoquarks decaying into τb, *Phys. Rev. Lett.*, **101**, 241802, doi:10.1103/PhysRevLett.101.241802.

37 DØ Collab., Abazov, V.M. *et al.* (2009) Search for pair production of first-generation leptoquarks in $p\bar{p}$ collisions at $\sqrt{s} = 1.96\,\text{TeV}$, *Phys. Lett. B*, **681**, 224, doi:10.1016/j.physletb.2009.10.016.

38 H1 Collab., Aktas, A. *et al.* (2005) Search for leptoquark bosons in e p collisions at HERA, *Phys. Lett. B*, **629**, 9, doi:10.1016/j.physletb.2005.09.048.

39 Zurek, K.M. (2010) TASI 2009 Lectures: Searching for Unexpected Physics at the LHC, http://de.arxiv.org/abs/1001.2563.

40 CMS Collab. (2008) MUSiC – An Automated Scan for Deviations between Data and Monte Carlo Simulation, http://cdsweb.cern.ch/record/1152572, PAS EXO-08-005.

41 DØ Collab., Abbott, B. *et al.* (2001) Quasi-model-independent search for new physics at large transverse momentum, *Phys. Rev. D*, **64**, 012004, doi:10.1103/PhysRevD.64.012004.

42 H1 Collab., Aaron, F.D. *et al.* (2009) A general search for new phenomena at HERA, *Phys. Lett. B*, **674**, 257, doi:10.1016/j.physletb.2009.03.034.

43 CDF Collab., Aaltonen, T. *et al.* (2009) Global search for new physics with 2.0/fb at CDF, *Phys. Rev. D*, **79**, 011101, doi:10.1103/PhysRevD.79.011101.

44 DØ Collab. (2009) Model Independent Search for New Physics at DØ in Final States Containing Leptons, http://www-d0.fnal.gov/Run2Physics/WWW/results/prelim/NP/N65/N65.pdf, DØ Note 5777-CONF.

11
Forward and Diffractive Physics: Bridging the Soft and the Hard
Jochen Bartels and Kerstin Borras

11.1
Introduction

Forward physics spans a broad range of high energy physics topics from fundamental properties of Quantum Chromodynamics (QCD) to new physics phenomena. *Diffraction* is one central physics theme addressed within forward physics and can be measured with very high precision right at the beginning of the LHC datataking. Diffraction in high energy physics refers to collisions where at least one of the two projectiles stays intact and moves into the forward direction. This implies that the final states contain regions in the detector devoid of particles or jets, so-called *rapidity gaps*. The dynamics of such scattering processes are closely related to the binding forces between the constituents of hadrons. The study of diffraction provides insight into the transition from perturbative hard physics at short distances to non-perturbative, soft long-distance physics. In addition, studies have shown that diffractive final states offer the possibility of searches for new physics.

In this chapter we summarise important aspects of forward physics, both in electron–proton and in proton–proton collisions. This includes:

- the total cross section and elastic scattering in proton–proton and electron–proton scattering;
- parton densities, small-x physics and BFKL dynamics;
- saturation;
- diffractive final states;
- multiple interactions, the underlying event and AGK rules.

11.2
Cross Sections in pp and ep Scattering

When addressing proton–proton collisions at high energies, important observables are the total cross section, $\sigma_{\text{tot}}(pp)$, and the amplitude for elastic scattering,

$T_{\text{elastic}}(s, t)$ (see Figure 11.1a). There is a close connection between these two quantities through the optical theorem:

$$\sigma_{\text{tot}}(s) = \frac{1}{s}\Im(T_{\text{elastic}}(s, t = 0)). \tag{11.1}$$

Here s is the square of the total centre-of-mass energy and $t = q^2$ is the squared four-momentum transfer. The elastic cross section,

$$\frac{d\sigma_{\text{elastic}}}{dt}(s, t) = \frac{1}{16\pi s^2}|T_{\text{elastic}}(s, t)|^2, \tag{11.2}$$

measures the probability that in a collision of two protons both protons stay intact: this reflects the binding forces which keep the constituents of the proton together, that is, the interactions between quarks and gluons. In a highly energetic collision, the direction of the incoming protons defines a longitudinal axis; the binding forces act in the direction transverse to this axis. We can visualise the scattering process in a space-time picture: in their centre-of-mass frame, both incoming protons see each other as a collection of partons, travelling along the longitudinal direction. The break-up of a proton into its partonic constituents, described by the quantum mechanical wave function, starts long before the protons hit each other: the higher the energy, the earlier the break-up starts. The transverse distance b between the colliding protons is called the *impact parameter*. In most cases in the course of the collision, the two protons fragment into many jets and particles; there is, however, a finite fraction of events (for example slightly above 20% for $p\bar{p}$ events at the Tevatron) in which the protons stay intact and continue their paths in the forward direction, that is with a small deflection from the incoming direction. These are the events which are counted by the elastic cross section. For elastic scattering, the scattering angle θ is measured through the momentum transfer, t. Since the impact parameter b is the Fourier conjugate of the transverse component of the four-momentum q, the t dependence of the elastic cross section measures the average transverse size of the colliding system, $\langle b^2 \rangle$. Experiments have shown that this transverse extension grows logarithmically with energy, $\langle b^2 \rangle \sim \alpha' \ln s$, that is, the scattering system formed by the colliding protons slowly expands in the transverse directions, and the parameter $\alpha' \approx 0.25\,\text{GeV}^2$ measures the strength of the binding forces in the transverse direction.

The other piece of information which is derived from the measurement of σ_{tot} and of the elastic cross-section $d\sigma_{\text{elastic}}/dt$ is the energy dependence. Measurements of σ_{tot} (see Figure 11.2 for an overview) reveal an increase which can be parametrised either by a small power $s^{\alpha(0)-1}$ with $\alpha(0) - 1 \approx 0.08$ [1] or, alternatively, by a factor $\ln^2 s$. This energy dependence seems to be rather universal; it holds for pp, $p\bar{p}$, and γp scattering. Measurements at the LHC will continue the curve of the total cross section towards higher energies. In the past 50 years enormous efforts have been made to develop a theory for the elastic scattering of hadrons at high energies, especially to find a description within QCD. Historically, in the 1960s, the high energy behaviour of hadron–hadron scattering has been attributed to the exchange to a particular Regge pole, named the *pomeron*. The derivation of

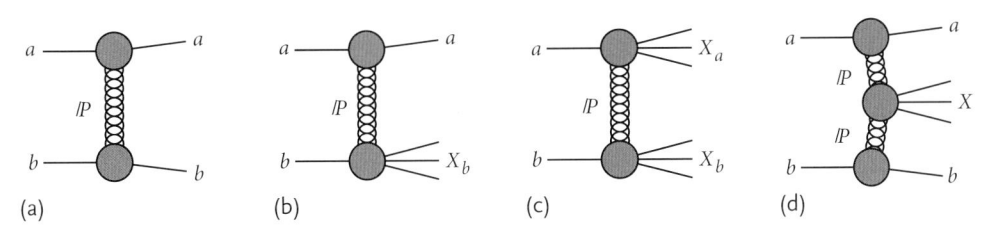

Figure 11.1 Diffractive processes in hadron–hadron scattering: (a) elastic scattering; (b) single-diffractive dissociation (sd); (c) double-diffractive dissociation (dd); (d) double-diffractive central production via double pomeron exchange (DPE).

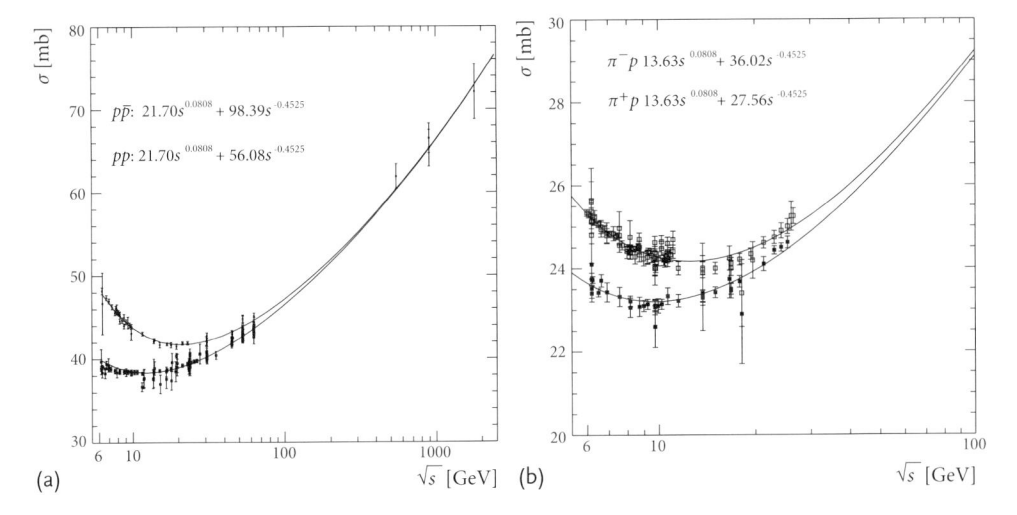

Figure 11.2 Fits to the total cross section in $p\bar{p}/pp$ (a) and $\pi^+ p/\pi^- p$ (b) scattering as a function of the centre-of-mass energy (adapted from [1]).

this pomeron from QCD remains a topic of strong interest. Further details can be found in [2].

The exchange of a pomeron determines also other processes, for example *single-diffractive* (sd) scattering, Figure 11.1b, *double-diffractive* (dd) scattering, Figure 11.1c and *double-diffractive central production* via double pomeron exchange (DPE), Figure 11.1d. The last process attracts interest in connection with the search for "new physics", in particular the Higgs boson(s) and supersymmetric particles. Experimentally all these processes have in common that the final state contains rapidity gaps.

Compared to proton–proton scattering described above, the scattering of a virtual photon, γ^*, off a proton represents an interesting variant. HERA has measured deep inelastic electron–proton (ep) scattering: $ep \rightarrow e'X$. Here the incoming electron emits a photon with virtuality Q^2 which scatters off the proton. In this way the process of $\gamma^* p$ scattering appears as a subprocess of deep inelastic ep scattering. The total cross section of this subprocess, $\sigma_{\text{tot}}(\gamma^* p)$, is parametrised in terms of the parton densities of the proton:

$$\sigma_{\text{tot}}(\gamma^* p) = \frac{4\pi^2 \alpha_{\text{em}}}{Q^2} F_2(Q^2, x),\quad (11.3)$$

where $F_2(Q^2, x) = F_T(Q^2, x) + F_L(Q^2, x)$ denotes the sum of the transverse and longitudinal structure functions, $-Q^2 = q^2$ is the virtuality of the photon, and the Bjorken scaling variable x is related to the squared energy of the photon–proton subsystem via $W^2 = Q^2(\frac{1}{x} - 1)$.

The total cross section, σ_{tot}, depends on the transverse radius of the photon, which is proportional to $\sqrt{1/Q^2}$, and can be varied by selecting the momenta of the outgoing electron. Hence, in contrast to pp scattering where two objects with rather complicated substructures collide with each other, in $\gamma^* p$ scattering a small-size photon with a simple substructure probes the composition of the proton. With this property deep inelastic electron–proton scattering is the ideal tool to explore the structure of a proton at high energies [3]. It is not surprising that the energy dependence of $\sigma_{\text{tot}}(\gamma^* p)$ [4] has been found to be quite different from $\sigma_{\text{tot}}(pp)$. The fact that the virtual photon has a small transverse size and hence probes only small regions inside the proton allows the use of perturbative QCD: as a result, the theoretical understanding of $\sigma_{\text{tot}}(\gamma^* p)$ is more advanced than that of $\sigma_{\text{tot}}(pp)$.

In analogy to pp scattering, one would like to measure also the elastic scattering of the $\gamma^* p$ system. Since the cross section for the elastic process $\gamma^* p \to \gamma p$ is very small, it is useful to study the quasi-elastic process $\gamma^* p \to V p$ where V denotes a vector particle with quantum numbers identical to the photon [5]. Examples include the neutral ρ or the J/ψ meson. The energy dependence of the integrated cross section

$$\int dt \frac{d\sigma_{\gamma^* p \to V P}}{dt} \quad (11.4)$$

has been found to differ from the corresponding cross section of elastic proton–proton scattering. Also the t dependence of the elastic pp cross section and of the quasi-elastic $\gamma^* p \to V p$ cross sections, as well as their profiles in impact parameter b, are different from each other.

11.3
Parton Densities, Small-x and BFKL Dynamics

It has already been mentioned that the total cross section for $\gamma^* p$ scattering is conveniently parametrised in terms of the structure function $F_2(Q^2, x)$, which can be written as the sum of quark and gluon densities, $q(Q^2, x)$ and $g(Q^2, x)$. They denote the probability density of finding – per rapidity interval – a quark or gluon with virtuality Q^2 and momentum fraction x. The virtuality Q^2 determines the transverse size which is of the order $\sqrt{1/Q^2}$. The accurate determination of these parton densities represents one of the highlights of the HERA measurements and was shown in Figure 1.1. Most striking is the observed strong rise of the gluon density at small x values, which was discovered at HERA. The precise knowledge of these structure functions is of fundamental importance for analysing LHC data. Vice versa the LHC will allow the kinematic regions of the parton densities to be extended (see Figure 4.4).

On the theoretical side, the parton densities are the basic building blocks of QCD at high energies. On general grounds it can be shown that the property of asymptotic freedom (see Chapter 4) allows the use of perturbation theory for quantities which are governed by large momentum scales or, equivalently, small distances. For the parton densities it is the virtuality, Q^2, which defines the large momentum scale. The dependence on Q^2 is described by a set of stochastic evolution equations, the *Dokshitzer–Gribov–Lipatov–Altarelli–Parisi* (DGLAP) equations which are of the form (for further details see Section 4.5)

$$Q^2 \frac{\partial}{\partial Q^2} f_i(Q^2, x) = \sum_j \int_x^1 \frac{dz}{z} P_{ij}(z) f_j(x/z, Q^2) , \qquad (11.5)$$

where the index i denotes the species of the parton (quark or gluon) and the kernels $P_{ij}(z)$ describe the probability of finding a parton of species i inside the parton j. They have been computed in perturbation theory up to order α_s^3 (next-to-next-to-leading order (NNLO)). Together with suitable initial conditions (which have to be modelled and adjusted to experimental data), these equations define the parton densities as functions of Q^2 and x and are essential for the computation of scattering cross sections in high energy collisions. As an example, the cross section for the production of a pair of jets with momentum scale μ^2 in pp collisions can be written as

$$d\sigma = \sum_{i,j} \int dx_1 dx_2 \, f_i(x_1, \mu) f_j(x_2, \mu) d\hat{\sigma}_{ij}(x_1, x_2, \mu) , \qquad (11.6)$$

where x_1 (x_2) denotes the momentum fraction of the parton i (parton j) inside the incoming proton 1 (incoming proton 2), f_i (f_j) the corresponding parton densities, and $\hat{\sigma}_{ij}(x_1, x_2, \mu)$ the cross section for the partonic subprocess parton i + parton j → pair of jets. We illustrate this formula in Figure 11.3a. It is important to

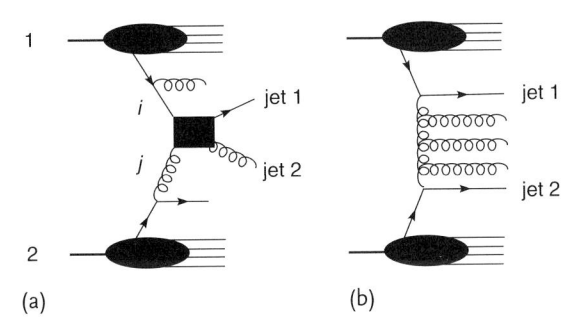

Figure 11.3 Illustration of the factorisation formula (11.6) for (a) inclusive jet production and (b) the production of Mueller–Navelet jets.

note that this simple formula only applies to inclusive cross sections: the momenta of the two partons (jets) in the centre of the final state are kept fixed, whereas all additional partons (i.e. their number, and their momenta as well as their fragments) are not specified and simply summed over (see also discussion in Chapter 4).

For this QCD-improved parton picture to be valid one needs the momentum scale μ^2 to be large ($\mu^2 > \mathcal{O}(\text{few GeV}^2)$) and the momentum fraction x not too small and not too close to 1. In deep inelastic scattering the large momentum scale is defined by the virtuality of the photon, Q^2. The region of small x ($x < 10^{-2}$) and low Q^2 ($Q^2 < \mathcal{O}(\text{few GeV}^2)$) has attracted particular interest. Small x corresponds to large energies of the γ^*–proton system, while at low Q^2 the incoming photon is no longer small in the transverse direction. Therefore in both kinematic regions, the QCD parton picture, based on the DGLAP evolution equations, requires corrections. In the small-x region, another form of evolution equations becomes applicable – the Balitsky–Fadin–Kuraev–Lipatov (BFKL) equations [6] which describe the evolution in x of the gluon structure function:

$$\frac{\partial}{\partial \ln 1/x} F(x,k) = \int d^2 k' \, K_{\text{BFKL}}(k,k') F(x,k') \,, \tag{11.7}$$

where K_{BFKL} denotes the BFKL kernel which has been computed up to the order α_s^2 (NLO) [7]. From $F(x,k)$ the gluon density $f_g(x, Q^2) = g(x, Q^2)$ is obtained by integrating over the transverse momenta k: $x g(x, Q^2) = \int^{Q^2} dk^2 F(x,k)$. This equation leads to a power-like growth of the gluon density at small x: detailed investigations indicate that this increase is compatible with the rise of F_2 observed at HERA. Within the last few years attempts have been made to find, within the DGLAP formalism, small-x corrections which incorporate essential features of the BFKL equation. These corrections improve the description in the small-x region.

Both evolution equations (DGLAP and BFKL) can be described in terms of intuitive physical pictures. As illustrated in Figure 11.3a, the partons i and j which enter the hard subprocess result from evolution processes which have started inside the upper and lower protons. This evolution process consists of a chain of parton emissions, as a result of which the "parent parton" loses a fraction of its momentum.

In the case of the DGLAP evolution, the virtualities of the partons (proportional to the square of the transverse momentum) are strongly ordered. When leaving the proton, the virtuality is small (of the order of a few GeV^2). When entering the hard subprocess, the virtuality has reached its maximal value (for our example, the production of a pair of jets, it is of the order of the jet transverse momenta). During this sequence of steps of emissions, the fractions of longitudinal momenta carried by the partons, decrease: in the beginning, that is when leaving the proton, the parton carries a finite momentum fraction of the large longitudinal proton momentum. At each subsequent step of emission the parton then loses a fraction of its longitudinal momentum until, when reaching the hard subprocess, it has slowed down to the value x_i (or x_j).

There are situations in which it is more appropriate to use the BFKL evolution equation, instead of the DGLAP equations. An example is shown in Figure 11.3b.

This final state has a two-jet configuration where the jets are widely separated in rapidity, but have comparable transverse momenta (*Mueller–Navelet jets*). This is where the BFKL evolution applies. In this case the parton chain between the two jets does not evolve in virtuality (or transverse momentum), but in rapidity (or in the logarithm of the longitudinal momenta). In particular, there is no ordering in transverse momenta. Deep inelastic electron–proton scattering at small x and low Q^2 presents a limiting case where DGLAP evolution with BFKL-based corrections has been used. Dedicated searches for BFKL applications include the scattering of two virtual photons in a high energy e^+e^- collider, in particular the measurement of the total cross section, $\sigma_{tot}(\gamma^*\gamma^*)$: it is dominated by BFKL exchange and hence provides a clean environment for BFKL searches [8].

11.4
Saturation

It has already been mentioned that this description of cross sections in terms of partons and scattering subprocesses at small momentum fractions, x, and low momentum scales begins to require corrections and modifications. In the following we mention one particular scenario which is known as *saturation*. In deep inelastic electron–proton scattering, the virtual photon emitted by the incoming electron interacts with partons inside the proton whose properties are specified by the kinematics of the photon. In particular, the transverse size of the partons is inversely proportional to the square root of the virtuality of the photon: $\langle x_T^2 \rangle \sim 1/Q^2$. The deep inelastic cross section, parametrised through parton densities (quarks and gluons) thus "counts" the number of partons per rapidity interval. For sufficiently large photon virtualities and not too small x, the QCD parton picture works well, because the partons inside the proton, on the distance scale defined by the small photon, are dilute: the density of partons is small, and they interact only weakly. This is a direct consequence of the property of asymptotic freedom which makes the strong coupling, α_s, small. This diluteness condition is not satisfied if the density of partons grows, as is the case if either the number of partons increases (large structure function) or the interaction between the partons becomes strong (large α_s). The former situation is realised at small x since the parton densities grow for small x, the latter for smaller photon virtualities, Q^2, which set the scale of the strong coupling, $\alpha_s(Q^2)$.

This simple qualitative argument shows that corrections to the standard QCD parton picture can be described in terms of quarks and gluons and their interactions as long as Q^2 is not too small ($\alpha_s(Q^2) < 1$) and the gluon density is large (small x). Combining these two conditions one arrives at the picture shown in Figure 11.4: there is a straight line in the $\ln Q^2$–$\ln 1/x$ plane below which the partons are dilute, and the standard QCD parton picture applies. In the vicinity of the line the QCD corrections become important and above the line partons are in a state of high density. Further above and to the left of the line, interactions become strong and QCD perturbation theory is no longer applicable.

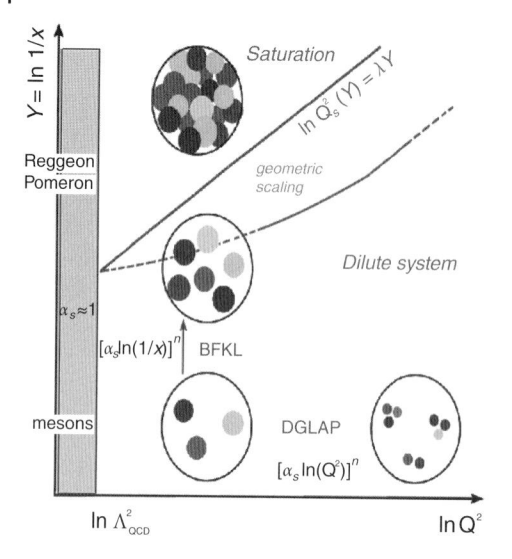

Figure 11.4 Saturation QCD "phase diagram" in the ln Q^2, ln $1/x$ plane (each dot represents a parton with transverse area $\approx 1/Q^2$ and fraction x of the hadron momentum). The different evolution regimes (DGLAP, BFKL, saturation) as well as the curves between the dense and dilute domains labelled "Saturation" and "geometric scaling" are indicated (adapted from [9]).

Within this picture one easily understands which type of corrections can be expected. Once the density of gluons increases, it becomes probable that, prior to their interaction with the photon, two gluons inside the proton recombine into one. The probability for this to happen is proportional to the probability of finding a second gluon, to the gluon density of the proton and to the cross section of this partonic subprocess:

$$P \sim xg(x, Q^2) \frac{\alpha_s(Q^2)}{Q^2} . \tag{11.8}$$

Approximating the growth of the gluon density at small x by a power law, $xg(x, Q^2) \sim (1/x)^\lambda$, the requirement that the probability of recombination is of order unity leads to

$$Q_{\text{sat}}^2 \sim Q_0^2 (1/x)^\lambda . \tag{11.9}$$

The exponent $\lambda \approx 0.3$ can be derived from QCD, whereas the scale Q_0^2 has to be measured by experiment.

The discussion of saturation has strongly been stimulated by the HERA measurements; therefore two pieces of experimental evidence for saturation are briefly mentioned [10]. Most striking is the scaling property of the deep inelastic structure function F_2: saturation predicts that $F_2(x, Q^2)$, in fact, depends only upon the single variable $Q^2/Q_{\text{sat}}^2(x)$ where $Q_{\text{sat}}^2(x) = Q_0^2(1/x)^\lambda$. This type of "geometric" scaling has clearly been seen in data [11]. Another striking feature has been observed in the diffractive final states: at fixed photon virtuality, Q^2, and fixed squared mass, M^2, of the diffractively produced system, the ratio of the diffractive deep inelastic cross section over the total cross section, $\sigma_{\text{diff}}/\sigma_{\text{tot}}$, does not vary with the total energy, W^2. Saturation provides a natural explanation for this behaviour [12, 13]. While the observed geometric scaling can also be explained within the DGLAP framework, the behaviour of the diffractive cross section appears to have no other explanation.

The LHC probes kinematic regions where parton densities have not been measured before, including low-x regions. It will be an important task to look for any deviations from DGLAP predictions, including signals for saturation. Most promising are Drell–Yan lepton pairs near the forward direction, where one of the x variables can be as low as 10^{-6}.

Saturation plays an important role in the understanding of heavy ion collisions [14]: it is believed that the initial state of incoming ions contains gluonic components of high density, which belong to the saturation regime described above. This state defines the initial conditions for the further evolution of the strong colour field which determines the formation of the quark–gluon plasma.

11.5
Diffractive Final States

Final states in hadron–hadron scattering at high energies contain, in general, large numbers of jets and particles, and their distribution in angle and momentum show quite characteristic patterns. In contrast to the inclusive jet cross section described before, the theoretical description of the full event structure is very complicated, and at present it can be achieved only by means of models and sophisticated computer simulations. There exist, however, specific classes of final states which show very distinct features and attract strong attention, namely the diffractive events with rapidity gaps. As mentioned above, the final state of these events exhibit empty regions of significant size. Astonishingly, the number of events does not decrease when the total energy increases.

Let us return first to deep inelastic electron–proton scattering which, as we have said before, can be interpreted as the scattering of a virtual photon off a proton. The observed partonic substructure of the proton might suggest that, in a typical event, the highly virtual and energetic photon would completely destroy the structure of the proton and create a large number of particles and jets in the final state. Nevertheless, in more than 10% of the events it was observed that the proton stays intact and that the photon creates an excited final state with quantum numbers identical to those of the photon. Examples of these excitations include vector mesons and pairs of jets. Between the fragments of these excitations and the proton there is the region devoid of particles or jets. Since these final states continue to exist also at high energies, their appearance has been attributed to the same mechanism which is also responsible for the elastic scattering of two protons, the exchange of a pomeron. It was, however, found that the energy dependence in this case slightly differs from that seen in proton–proton scattering, suggesting that this *hard pomeron* is different from the usual *soft pomeron* in hadron–hadron scattering.

A convenient way of describing these rapidity gap events is the notion of a *diffractive structure function*. In analogy with the structure function F_2 which expresses the total $\gamma^* p$ cross section, $\gamma^* + p \to X$, one introduces a structure function F_2^D for the restricted class of final states: $\gamma^* + p \to X'$, where the symbol X' indicates that only those final states are included which contain a rapidity gap of a given size. These structure functions have been measured and diffractive parton densities, f_i^D, have been extracted using the DGLAP equations [15–17]. Their shapes differ from the "normal" structure functions, and F_2^D is gluon dominated.

Final states with rapidity gaps are observed also in proton–proton collisions. Let us begin with *soft diffraction*. The most prominent example is elastic scattering of the two protons (Figure 11.1a). As a first generalisation, it may happen that one proton turns into a state with slightly higher mass and with quantum numbers identical to those of the proton (*diffractive excitation*, see Figure 11.1b). If both protons are diffractively excited the events are categorised as double-diffractive scattering (Figure 11.1c). For $p\bar{p}$ collisions at the Tevatron all these events constitute about 30% of the total cross section. In a second step of generalisation, one proton (or both) may be excited into a state of mass much larger than the proton mass, for example $pp \to pX$: here the state X consists of unobserved soft final-state particles. These *soft high-mass diffractive* states contribute a small percentage to the total cross section. As has been said before, the mechanism responsible for these soft diffractive states, summarised under the term pomeron physics, cannot (yet) be described by tools of QCD and has attracted theoretical interest for many years.

A second class of rapidity gap events exhibit final states with hard jets, heavy quarkonia, a W boson or a Z boson: they have been named *hard diffraction*, where the hard scale is given by the high p_T of the jets or by the heavy mass. In analogy with HERA physics, one again defines diffractive parton densities: starting from (11.6), we define an inclusive jet cross section. But in contrast to (11.6), we impose the additional requirement that a rapidity gap of a defined size is present in the parton chain j in Figure 11.3a. Equation 11.6 is then replaced by

$$d\sigma = \sum_{i,j} \int dx_1 dx_2\, f_i(x_1, \mu^2)\, f_j^D(x_2, \mu^2) d\hat{\sigma}_{ij}(\mu^2)\,, \qquad (11.10)$$

where the superscript in f_j^D denotes the imposed restriction that the final states inside the lower structure function must contain the prescribed rapidity gap. Measurements at the Tevatron show that these diffractive parton densities obtained from proton–proton scattering are not the same as those measured in deep inelastic scattering: they are smaller than the HERA densities by a suppression factor between 10 and 20 [18]. This shows that in proton–proton scattering rapidity gaps in hard final states are much more strongly suppressed than in $\gamma^* p$ collisions. The reason is that in pp scattering multiple interactions (see below) are much stronger than in deep inelastic scattering. Gluon radiation from additional parton chains tend to fill the rapidity gaps. This suppression is commonly referred to as a *small rapidity gap survival probability*.

Diffractive final states are of interest also in the context of searches for new physics, in particular for the Higgs boson. For the most interesting process, the exclusive Higgs production $pp \to pHp$, the cross section has been estimated, for a Higgs mass of 120 GeV, to be of the order of 3 fb [19]. This production process has the advantage of the spin selection rule $J_Z = 0$ which, by suppressing the QCD production of $b\bar{b}$ pairs, allows the $b\bar{b}$ decay mode of the Higgs particle to be observed. By measuring the outgoing protons in the forward detectors it is possible to reconstruct the Higgs mass with an accuracy of a few percent.

11.6
Multiple Scattering, Underlying Event and AGK

The previous paragraphs make it clear that the description of a single event, from a theoretical point of view, is much more complicated than the calculation of an inclusive cross section where most of the final state is summed over. Within the parton picture one can approach the structure of single events by generalising the single-chain picture shown in Figure 11.3 to multi-chain configurations and initial-state and final-state interactions.

Here it is assumed that in the centre of each parton chain there is a hardest parton with sufficiently large transverse momentum which allows the use of QCD perturbation theory. The radiation between these hardest partons and the protons above and below is described by DGLAP splitting functions. The soft interaction in the initial state (denoted by the grey blob) has to be modelled and cannot be calculated within QCD.

It is a remarkable feature of deep inelastic scattering, where a small-size virtual photon scatters off the proton, that such multi-chain configurations are suppressed: it is only when the virtuality of the photon gets small and the photon begins to resemble a hadron (more precisely: a vector meson) that the photon emits more than one parton.

Taking into account the presence in hadron–hadron scattering of such multi-chain configurations one can understand, at least qualitatively, why, in hadron–hadron collisions, rapidity gaps in the final state are seen less frequently than in ep collisions. Let us assume that a single chain has produced a partonic final state with a rapidity gap. Then, with a certain probability (which for hadron–hadron scattering is larger than for the scattering of a virtual photon off a proton) there is a second chain which again fills the gap. As a result, the observed final state does not contain the empty region characteristic for a rapidity gap.

Theoretical investigations (known as factorisation theorems, see Chapter 4) state that in a single-inclusive jet cross section, which is obtained by summing over all final states except for the jet, extensive cancellation between the multiple chains and the initial-state interactions (Figure 11.5) take place such that, at the end, the simple factorisation formula (11.6) holds. As a generalisation, for a double-inclu-

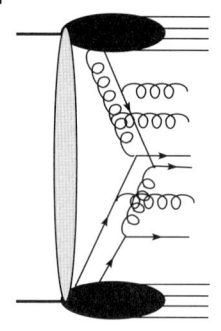

Figure 11.5 Example of a two-chain process with initial-state interaction.

sive jet cross section, after summation over all partons other than the two observed jets, one finds two contributions. In the first term, both jets are produced from the same parton chain (see (11.6)); in the second contribution, the two jets come from different chains, leading to a formula of the form

$$d\sigma = \sum_{i,j} \int dx_1 dy_1 dx_2 dy_2 \, f_{ii'}(x_1, y_1, \mu_1^2, \mu_2^2) \, f_{jj'}(x_2, y_2, \mu_1^2, \mu_2^2) \, d\hat{\sigma}_{ii',jj'},$$

(11.11)

where $f_{ii'}(x_1, y_1)$ denotes a double parton density, that is the probability of finding two partons of species i and i' with momentum fractions x_1 and y_1.

The fact that when going from single events to inclusive cross sections, there are such strong cancellations and simplifications has been formulated in terms of the *Abramovsky–Gribov–Kancheli* (AGK) cutting rules [20]. As a simple example, let us consider, in pp scattering, the inclusive production of a single pair of jets, and in addition to the parton chain which emits the observed jets we include a second chain which contributes to the unobserved particles and jets.

Figure 11.6a describes the emission of partons from two chains. The observed pair of jets (denoted by the black ring) is produced by the central chain. Figure 11.6b,c describe absorptive corrections to the usual single-chain jet production. One can show that in the sum all three contributions cancel. As a result, in this inclusive jet cross section the corrections due to a second chain cancel, and we are back to our single-chain formula (11.6).

A second example for the application of the AGK cutting rules is given in $\gamma^* p$ scattering (Figure 11.6d–g). The contribution shown in Figure 11.6d denotes the diffractive process with a rapidity gap between the diffractive excitation of the photon and the proton; the contributions shown in Figure 11.6e,f contain absorptive corrections, and the contribution shown in Figure 11.6g describes the emission of partons from two chains. The AGK rules state that all these contributions are equal, up to overall factors:

$$(d) : ((e) + (f)) : (g) = 1 : (-4) : 2,$$

(11.12)

and the total sum is equal to

$$(d) + (e) + (f) + (g) = -(d).$$

(11.13)

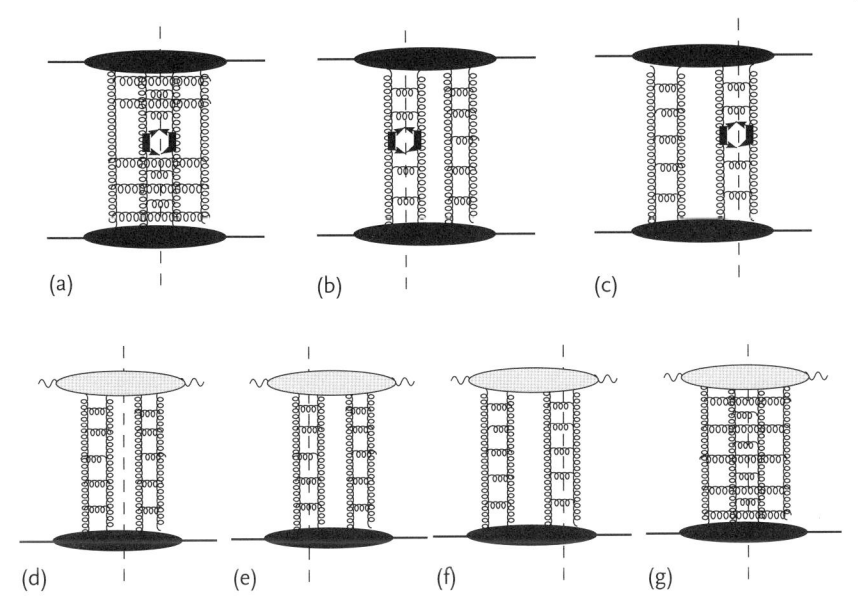

Figure 11.6 Illustration of the AGK cutting rules. First row, (a–c): cancellation of two chain contributions to the inclusive production of a single pair of jets (dark ring); second row, (d–g): two-chain contributions to the $\gamma^* p$ scattering).

This remarkable counting rule, when applied to the measurement of diffractive final states in ep collisions, allows the size of all two-chain corrections to $\gamma^* p$ scattering to be predicted. This discussion applies also to proton–proton scattering. The AGK cutting rules establish a close connection between the diffractive (rapidity gap) events, the usual inclusive events and multiple scattering.

11.7
Necessary Instrumentation at the LHC

For the discussion in this section it is convenient to use the pseudorapidity $\eta = -\log \tan(\theta/2)$, where θ is the scattering angle with respect to the beam axis. The quantity η is the analogue of the rapidity, y, used in Chapter 4 for massless particles. A scattering angle of $90°$ corresponds to $\eta = 0$ and denotes the direction transverse to the beam direction. Particles with high p_T are predominantly produced in the central region of the detector ($|\eta| < 2.5$). Forward physics concentrates on regions close to the beam directions with small scattering angles of $\theta \approx 0$ or π, which corresponds to large values of $|\eta|$.

The Interaction Point 5 of the LHC is the host of the experimental setup for two collaborations, as displayed in Figure 11.7a. The CMS detector (see Chapter 13) covers the interaction point with tracking devices up to roughly $|\eta| < 2.5$. This is complemented by calorimetry up to $|\eta| < 3$ and is further extended by the Hadron-

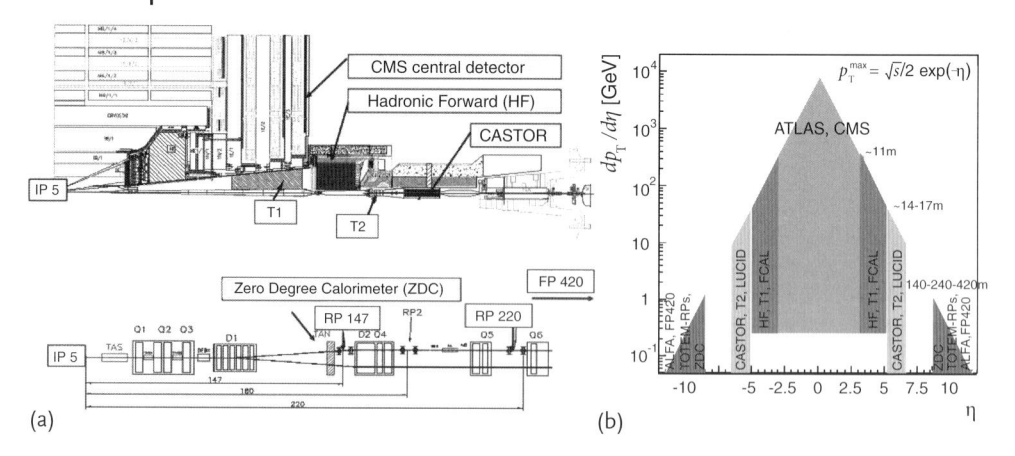

Figure 11.7 (a) forward detectors in the CMS interaction region; (b) covered phase space in the ATLAS and CMS experiments in units of η and p_T (adapted from [21]).

ic Forward calorimeter (HF) in the region of $3 < |\eta| < 5$, by the CASTOR calorimeter in $(5.3 < |\eta| < 6.6)$ and for neutral particles by the Zero Degree Calorimeter (ZDC) starting at about $|\eta| > 8$.

The detectors of the TOTEM experiment (see Chapter 17) complement this calorimetric coverage with tracking stations (T1 and T2) and with the installation of *roman pot* (RP) stations at 147 m and 220 m away from the interaction point.

In addition, within the ATLAS and CMS collaborations discussions for the installation of forward proton tagging devices at 420 m are ongoing and efforts are underway to realise these detectors.

Concentrating first on luminosity measurements (see Chapter 17) the ATLAS collaboration installed a Cerenkov counter (LUCID) at a position at $\eta \approx 6$, similar to the CASTOR calorimeter in CMS, and roman pots 240 m away from the interaction point. The goal is to obtain an absolute luminosity by an elastic scattering measurement with the roman pots down to the Coulomb region. This measurement is then used for the cross-calibration of LUCID, which will serve for a relative luminosity measurement at higher beam luminosities. For later measurements of hard diffraction, plans exist in the ATLAS collaboration to exchange the scintillating fibre detectors in the roman pots with more radiation-hard silicon strip detectors. The ATLAS forward detectors will be complemented by a Zero Degree Calorimeter in a design similar to that used by PHENIX at RHIC. These studies will provide valuable input for the tuning of forward particle production, used, for example, to cross-calibrate results of highly energetic cosmic ray shower experiments.

In total, the LHC experiments are well equipped with detectors in the forward region. In Figure 11.7 this is demonstrated with the coverage of the p_T–η plane. It is an unprecedented coverage of the forward region at hadron colliders and opens up the unique possibility to study the physics topics which were briefly outlined in this chapter. It is hoped that this extraordinary access to new kinematic regions will allow us to unravel and solve several mysteries.

References

1. Donnachie, A. and Landshoff, P.V. (1992) Total cross sections, *Phys. Lett. B*, **296**, 227, doi:10.1016/0370-2693(92)90832-O.

2. Barone, V. and Predazzi, E. (2002) *High Energy Particle Diffraction*, Springer Verlag, Berlin Heidelberg.

3. Bartels, J. and Kowalski, H. (2001) Diffraction at HERA and the confinement problem, *Eur. Phys. J. C*, **19**, 693, doi:10.1007/s100520100613.

4. H1 Collab., Adloff, C. et al. (2001) On the rise of the proton structure function F2 towards low x, *Phys. Lett. B*, **520**, 183, doi:10.1016/S0370-2693(01)01074-7.

5. Levy, A. (2007) Exclusive vector meson electroproduction at HERA, http://www.slac.stanford.edu/spires/find/hep/www?eprint=arXiv:0711.0737.

6. Forshaw, J.R. and Ross, D.A. (1997) Quantum chromodynamics and the pomeron, *Cambridge Lect. Notes Phys.*, **9**, 1, http://www.slac.stanford.edu/spires/find/hep/www?j=00385,9,1.

7. Fadin, V.S. and Lipatov, L.N. (1998) BFKL pomeron in the next-to-leading approximation, *Phys. Lett. B*, **429**, 127, doi:10.1016/S0370-2693(98)00473-0.

8. Kwiecinski, J. and Motyka, L. (2000) Theoretical description of the total $\gamma^*\gamma^*$ cross-section and its confrontation with the LEP data on doubly tagged e^+e^- events, *Eur. Phys. J. C*, **18**, 343, doi:10.1007/s100520000535.

9. d'Enterria, D.G. (2007) Low-x QCD physics from RHIC and HERA to the LHC, *Eur. Phys. J. A*, **31**, 816, doi:10.1140/epja/i2006-10206-6.

10. Bartels, J. (2005) Parton evolution and saturation in deep inelastic scattering, *Eur. Phys. J. C*, **43**, 3, doi:10.1140/epjc/s2005-02216-y.

11. Stasto, A.M., Golec-Biernat, K.J., and Kwiecinski, J. (2001) Geometric scaling for the total γ^*p cross-section in the low x region, *Phys. Rev. Lett.*, **86**, 596, doi:10.1103/PhysRevLett.86.596.

12. Golec-Biernat, K.J. and Wusthoff, M. (1998) Saturation effects in deep inelastic scattering at low Q^2 and its implications on diffraction, *Phys. Rev. D*, **59**, 014017, doi:10.1103/PhysRevD.59.014017.

13. Golec-Biernat, K.J. and Wusthoff, M. (1999) Saturation in diffractive deep inelastic scattering, *Phys. Rev. D*, **60**, 114023, doi:10.1103/PhysRevD.60.114023.

14. Gelis, F., Iancu, E., Jalilian-Marian, J., and Venugopalan, R. (2010) The Color Glass Condensate, http://www.slac.stanford.edu/spires/find/hep/www?eprint=arXiv:1002.0333.

15. H1 Collab., Aktas, A. et al. (2006) Measurement and QCD analysis of the diffractive deep-inelastic scattering cross section at HERA, *Eur. Phys. J. C*, **48**, 715, doi:10.1140/epjc/s10052-006-0035-3.

16. H1 Collab., Aktas, A. et al. (2007) Dijet cross sections and parton densities in diffractive DIS at HERA, *JHEP*, **10**, 042, doi:10.1088/1126-6708/2007/10/042.

17. ZEUS Collab., Chekanov, S. et al. (2010) A QCD analysis of ZEUS diffractive data, *Nucl. Phys. B*, **831**, 1, doi:10.1016/j.nuclphysb.2010.01.014.

18. CDF Collab., Affolder, A.A. et al. (2000) Diffractive dijets with a leading antiproton in $\bar{p}p$ collisions at $\sqrt{s} = 1800$ GeV, *Phys. Rev. Lett.*, **84**, 5043, doi:10.1103/PhysRevLett.84.5043.

19. Kaidalov, A.B., Khoze, V.A., Martin, A.D., and Ryskin, M.G. (2003) Central exclusive diffractive production as a spin parity analyser: From hadrons to Higgs, *Eur. Phys. J. C*, **31**, 387, doi:10.1140/epjc/s2003-01371-5.

20. Abramovsky, V.A., Gribov, V.N., and Kancheli, O.V. (1973) Character of inclusive spectra and fluctuations produced in inelastic processes by multi-Pomeron exchange, *Yad. Fiz.*, **18**, 595. http://www.slac.stanford.edu/spires/find/hep/www?j=YAFIA,18,595.

21. Bartels, J. et al. (2009) Proceedings of the 38th International Symposium on Multiparticle Dynamics (ISMD08), http://www.slac.stanford.edu/spires/find/hep/www?eprint=arXiv:0902.0377.

Part Two   The Technology

12
Accelerators: the Particle Smashers

Helmut Burkhardt, Jean-Pierre Delahaye and Günther Geschonke

12.1
Introduction

Experimental elementary particle physics at the high energy frontier is done largely with powerful particle accelerators: charged particles are accelerated to very high energies and brought into collision either with a *fixed target* or with particles of a second beam, travelling in the opposite direction. The latter method is used for the highest centre-of-mass collision energies and these machines are generally referred to as *colliders*. In these collisions, either all or a fraction of the energy of the particles is available to create new particles. In LEP, for example, the colliding particles had an energy more than 200 000 times their rest energy! The major machines today either accelerate leptons (electrons/positrons) or hadrons (protons/antiprotons and ions up to lead). Often these machines are built as *storage rings*, where the two counter-rotating high energy beams are kept circulating and the same beams can be brought into collision at each revolution. Since only a very small fraction of the particles actually collide in each crossing and the loss of particles due to other mechanisms, like collisions with the residual gas in the vacuum chambers, can be made small, the beams can be stored for many hours.

In the following, we present a few examples of accelerators at the present state of the art with their technology and performance. Finally, we sketch the work in progress for even higher energies.

Within the scope of this book it is impossible to give a comprehensive review of all existing machines, which would easily fill a book on its own. We rather highlight some features of the state of the art of major high energy machines. As examples we use LEP, the Large Electron–Positron collider (CERN, 1989–2000), the electron–proton collider HERA (DESY, 1992–2007), the proton–antiproton collider Tevatron (Fermilab, in operation since 1987), and the LHC, Large Hadron Collider (CERN, in operation since 2008). Finally we also discuss the possible future linear colliders ILC and CLIC.

12.1.1
Why Are Existing Accelerators So Big?

Electron storage rings need a large bending radius because of *synchrotron radiation emission*. Charged particles lose energy in the form of synchrotron radiation when their trajectory is deflected. Depending on the particle energy and bending angle this radiation ranges from visible light to gamma rays. In a circular storage ring, a particle with charge e, Lorentz factor $\gamma = E/mc^2$ (E is the beam energy and m the mass of the particle) loses an energy of

$$U_0 = \frac{e^2}{3\epsilon_0} \frac{\gamma^4}{\rho} \propto \frac{1}{\rho} \frac{E^4}{m^4} \tag{12.1}$$

per turn, where $e^2/(3\epsilon_0) = 6.032 \times 10^{-9}$ eV m.

The energy loss of the circulating particles increases with the fourth power of the beam energy and decreases only linearly with the bending radius! This is why the bending radius of LEP was chosen to be about 3 km, leading to a circumference of the tunnel of about 26 km. At a beam energy of about 100 GeV, the circulating electrons and positrons nevertheless lost about 3% of their energy on a single turn! This energy had to be replaced by a very powerful acceleration system, otherwise the beams would have spiralled into the vacuum chamber and got lost.

For protons, which are about 2000 times heavier than electrons, synchrotron radiation, which scales inversely with the fourth power of the particle mass, is not a problem. Here, the limit in energy comes from the magnetic bending field. For proton machines at high energy, superconducting magnet technology is favoured in order to provide high magnetic bending at affordable cost and electricity consumption.

12.1.2
Energy and Luminosity

Over the years particle physics demanded higher and higher beam energies both for lepton and proton colliders. Technological developments such as superconductivity for both high-current magnets and radio-frequency (RF) accelerating systems, as well as a steady advance of the understanding of beam dynamics, computing techniques, beam diagnostics and beam control have made this possible.

Figure 12.1 shows the increase in beam energy over the years. The graph shows both lepton and hadron machines. A rough rule of thumb suggests that in order to reach the same discovery potential, hadron accelerators need a factor six higher energy than lepton machines. This is because leptons are elementary particles, whereas protons consist of three quarks and several gluons which are all involved in the collisions. The dashed lines show an exponential increase in energy by a factor of 4 every 10 years over four decades. Proton and lepton accelerators complement each other: proton machines are generally considered as "discovery" machines, whereas lepton colliders tend to target precision measurements.

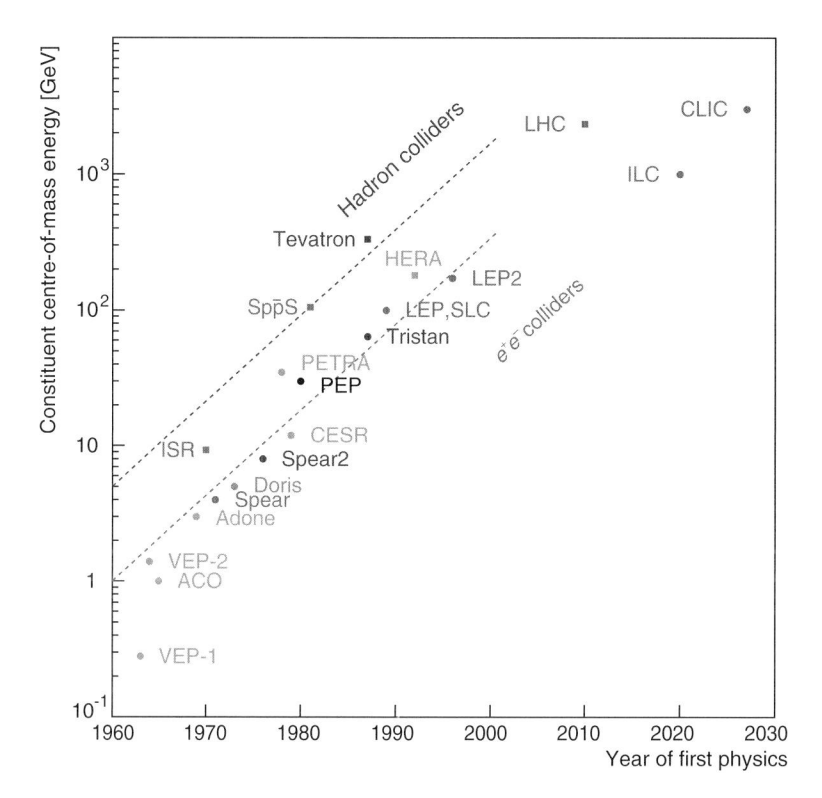

Figure 12.1 Growth in collider energies with time, often referred to as the Livingston plot.

Beam energy is not the only figure of merit of an accelerator for elementary particle physics research. Equally important is *luminosity*, $\mathcal{L}$, a parameter which defines the number of reactions of a certain type in the collisions. Each reaction has a certain probability, given by its cross section, σ. The number of events for the reaction is then $L\sigma$, where $L = \int \mathcal{L} dt$. Typical luminosities of LEP were of the order of 10^{31} cm^{-2} s^{-1} and the design luminosity of LHC is 10^{34} cm^{-2} s^{-1}. For higher energies, the point-like reaction cross sections decrease as $1/E^2$. Maintaining the same event rates requires that the luminosity scales with the square of the beam energy. The demands on accelerator designers become more difficult with increasing energy. A linear collider of the future (see Section 12.6) in the TeV energy range should have a luminosity of 10^{34} cm^{-2} s^{-1}.

In modern colliders, the particles are grouped in *bunches*. In LEP, for example, the beams consisted of four bunches of electrons and four bunches of positrons only, which were brought into collision in four points. The luminosity $\mathcal{L}$ of a collider is given by

$$\mathcal{L} = \frac{N_b^2 k_b f}{A} H_D, \tag{12.2}$$

where N_b is the number of particles per bunch, k_b is the number of bunches (per train in case of a linear collider), f is the revolution frequency in case of a ring and the bunch-train crossing frequency in case of a linear collider. A is the effective beam overlap cross section at the interaction point. For Gaussian-shaped beams with horizontal and vertical r.m.s. beams sizes of σ_x, σ_y colliding head on $A = 4\pi\sigma_x\sigma_y$. The beam sizes are determined by the *beam emittance*, ϵ, and the *β function* which describes the local focusing properties of the accelerator, $\sigma_{x,y} = \sqrt{\beta_{x,y}\epsilon_{x,y}}$. Strong quadrupole magnets are used around the interaction regions to focus the beams down to small values of the β functions at the interaction point (called β^*) to get small beam sizes and high luminosity. H_D corresponds to the modification of the effective beam sizes by the electromagnetic fields of the colliding bunches, the so-called *beam–beam effect*. In a storage ring, where the same bunches collide many times, the beam–beam effect must be kept small ($H_D \approx 1$).

The collision of bunches of oppositely charged particles results in a reduction of the beam sizes because the colliding bunches see each other's electromagnetic fields. In linear e^+e^- colliders, this so-called *pinch effect* helps to increase the luminosity. The effect must however be limited, as in common with any other bending of high energy electrons, synchrotron radiation photons will be emitted. The synchrotron radiation in the pinch effect is known as *beamstrahlung* and results in unwanted effects like a broadening of the energy spectrum and backgrounds in the experimental detectors.

12.2
LEP

LEP was the highest energy e^+e^- collider ever built. It produced its first collisions in August 1989, less than six years after ground was broken in September 1983. The near-circular tunnel with a circumference of 26.7 km extends from the foothills of the Jura mountains to Geneva airport and straddles the border between France and Switzerland, see Figure 12.2.

The maximum beam energy and total beam intensity in LEP were strongly limited by synchrotron radiation: at the highest energy of 209 GeV in the centre-of-mass system, the beam lost about 3% of its energy per turn. In order to compensate this energy loss and also to accelerate the particles from their injection energy (22 GeV) to top energy, LEP had a powerful RF acceleration system. In its final configuration it provided 3660 MV to each beam per turn.

The typical operation of LEP over a day is shown in Figure 12.3. Beams were injected at 22 GeV. Once the desired beam current was accumulated, the beam energy was ramped to about 100 GeV, where the beams were stored for typically about 3 h. The decay of the circulating beam current together with associated reduction in luminosity is seen. Then the beams are dumped and the cycle starts again.

LEP went through a number of upgrade phases between its initial operation in 1989 and its shut-down at the end of 2000. Initially it was operated at around 2×45 GeV centre-of-mass energy to produce Z bosons and study their proper-

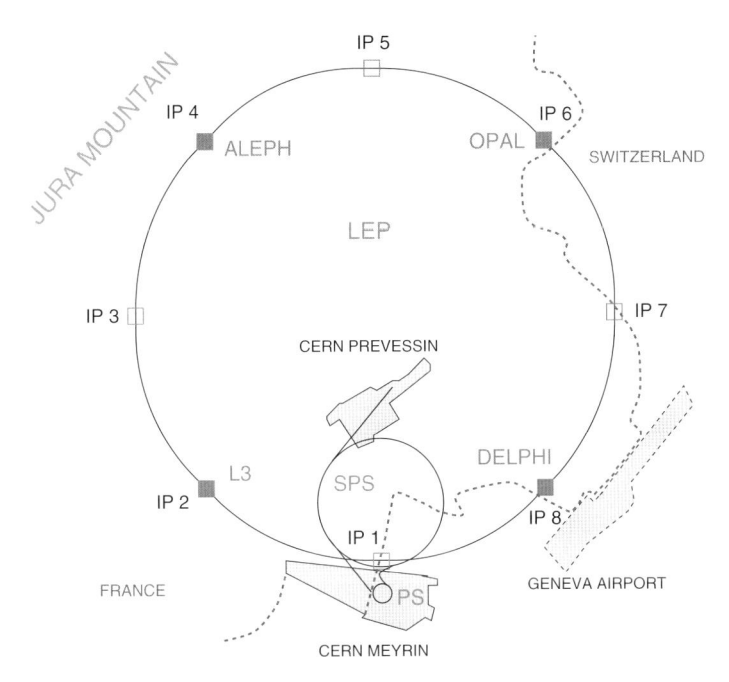

Figure 12.2 Layout of the LEP ring at CERN on the border between France and Switzerland. The eight access points are denoted IP1 through IP8. The four LEP experiments L3, ALEPH, OPAL and DELPHI were installed at the even-numbered access points. Positrons travelled clockwise, electrons anti-clockwise. The locations of two LEP injectors, the SPS (Super Proton Synchrotron) and the PS (Proton Synchrotron) are also indicated (Source: CERN).

ties. Already at this energy, a total accelerating voltage of 360 MV per turn was installed. It was provided by a system of 128 copper accelerating cavities, driven by 16 klystrons with a continuous RF power of 1 MW each. Since an energy upgrade of the copper RF system would have been too costly in power consumption, it was clear from the beginning that superconducting RF acceleration technology had to be used for higher energies. From 1995 onwards, more superconducting cavities were installed every year. Eventually, the acceleration system provided a circumferential voltage of 3660 MV, more than a factor 10 above the initially installed system. This was the biggest super-conducting acceleration system built up to then.

As luminosity scales with the bunch intensity squared (see (12.2)), LEP was operated with a relatively small number of bunches per beam ($k_b = 4$–12) and instead rather high single-bunch intensities of up to $N_b = 4.2 \times 10^{11}$ particles, close to the limit of single-bunch transverse mode coupling instability at injection [1]. In LEP, both beams circulated in the same beam-pipe in opposite directions. Local orbit bumps produced by electrostatic separators were used to avoid unwanted collisions, for example during injection and ramping.

LEP was operated with colliding beams at beam energies around 45 GeV (LEP 1), 65 GeV and 80 to 104 GeV (LEP 2), thus covering over a factor of two in beam

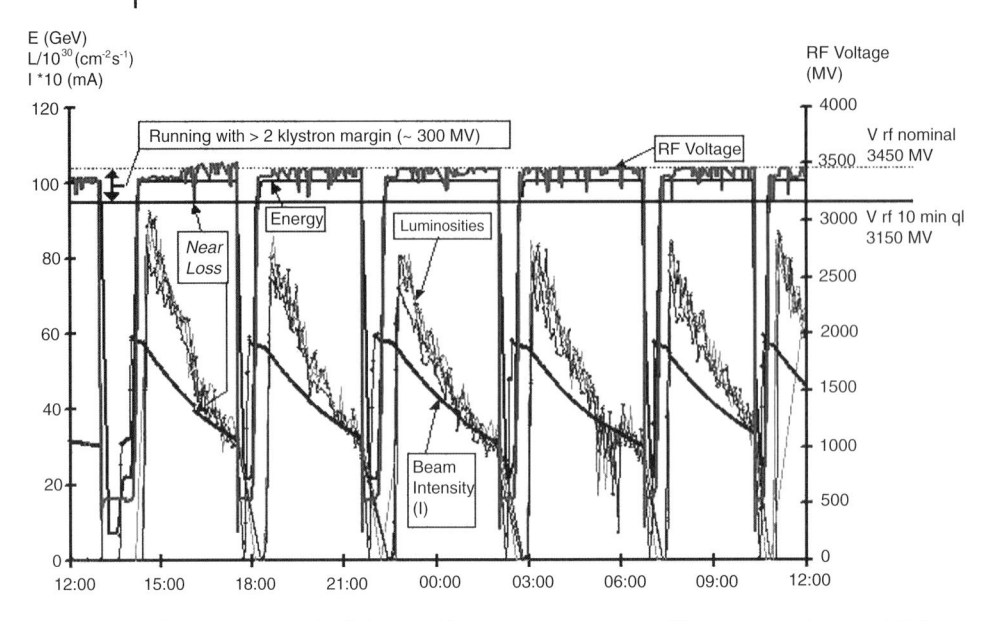

Figure 12.3 Example of the actual beam intensity, energy and luminosity evolution in LEP for a good day of operation in the year 2000.

energy. The increased energy loss at higher energy also has one beneficial effect: it results in a short *damping time*, which is basically the time it takes a particle to radiate its own energy in synchrotron radiation. Hence, the 3% energy loss at LEP 2 corresponds to a damping time of 30 turns. This strong damping reduces any blow-up due to noise and resonances.

A main limitation in colliding storage rings is the beam–beam interaction. This is the effect of the electromagnetic fields of one bunch on the other as they cross each other in the collision point. The strength of the beam–beam interaction can be quantified by the *linear beam–beam tune shift parameter*, ξ. At LEP the increased synchrotron radiation damping at higher energies resulted in a smaller increase of beam sizes due to the beam–beam effect and correspondingly allowed operation at higher beam–beam tune shift parameters, resulting in higher luminosities. The horizontal tune shift, ξ_x, was generally significantly lower than the vertical one, ξ_y. Table 12.1 lists the LEP beam parameters at the times of maximum vertical beam–beam parameters. The table also gives the main optics parameters, the betatron tunes, Q, the synchrotron tune, Q_s, and the values of the β function at the interaction points β^*.

Figure 12.4 presents an overview of the LEP performance during its 11 years of operation. The beam–beam effect was a strong limitation at LEP 1 and resulted in significant blow-up of the beam sizes. LEP 1 was initially operated with four bunches in each beam. Adding extra bunches required an upgrading of the *separation scheme* to avoid collisions outside the four interaction points. Two different separation schemes were successfully used in LEP: the *Pretzel scheme*, which us-

Table 12.1 LEP beam parameters corresponding to the best performances at three different energies. The luminosities and beam–beam tune shifts are averaged over a time interval of 15 min. For each beam energy, the first line corresponds to the horizontal, the second line to the vertical plane.

E_b [GeV]	N_b [×10¹¹]	k_b	$\mathcal{L}$ [cm⁻² s⁻¹]	Q_s	Q	β^* [m]	ϵ [nm]	σ [μm]	ξ
45.6	1.18	8	1.51×10^{31}	0.065	90.31	2.0	19.3	197	0.030
					76.17	0.05	0.23	3.4	0.044
65	2.20	4	2.11×10^{31}	0.076	90.26	2.5	24.3	247	0.029
					76.17	0.05	0.16	2.8	0.051
97.8	4.01	4	9.73×10^{31}	0.116	98.34	1.5	21.1	178	0.043
					96.18	0.05	0.22	3.3	0.079

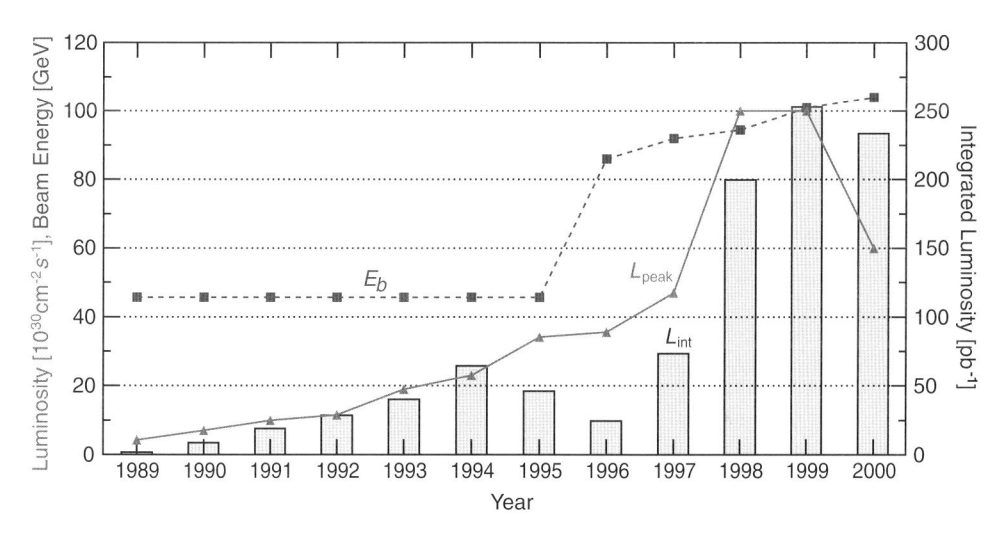

Figure 12.4 Peak luminosity, beam energy and integrated luminosity per experiment for each year of LEP operation.

es opposite horizontal separation in the arcs; and *bunch trains*[28] using additional separation around the interaction regions.

After installation of extra superconducting cavities, the beam energy in LEP could be raised to above the W-pair production threshold, which is referred to as LEP 2. The increased damping decreased the beam–beam blow-up and allowed higher bunch intensities to collide. The energy loss increased to 3 GeV per particle and turn. The total beam currents stored at LEP 2 had to be limited to 6 mA to keep the total loss in synchrotron radiation below 20 MW (3 GV acceleration

---

28) In the bunch-train scheme each of the four equidistant single bunches of electrons and positrons was replaced by a sequence of up to four bunches spaced by 75 m.

voltage ×6 mA = 18 MW). The maximum single-bunch intensity that could be accumulated at injection energy and safely be collided at LEP 2 energies is 0.75 mA. This corresponded to 4.2×10^{11} particles, just enough for four bunches in each of the two beams (42 × 4 × 0.75 mA = 6 mA).

The electrons for LEP were produced by a heated cathode and accelerated to 600 MeV in a standard linear accelerator (linac). Positrons were generated via pair production from gamma radiation coming from electrons at 200 MeV hitting a tungsten block. The positrons were then accelerated to 600 MeV and both electrons and positrons were accumulated in a dedicated (126 m circumference) accumulator ring. Further acceleration then made use of already existing accelerators on the CERN site, first the Proton Synchrotron (PS) up to 3.5 GeV, then the Super Proton Synchrotron (SPS) up to 22 GeV. Both these accelerators had been modified such that the electron and positron bunches could be accelerated and extracted between the normal proton cycles for fixed-target physics.

LEP stopped operations in 2000 to make space for its successor, the LHC. More details on LEP and further references can be found in review articles [2–5]. Accelerator physics aspects are summarised in [1] and details on the RF system can be found in [6].

12.3
Tevatron

Up to the beginning of LHC operation, the Tevatron was the hadron collider with the highest energy in the world. Heavy ion colliders like RHIC[29], which reach even higher total collision energies, are not considered here.

The circumference of the Tevatron of 6.28 km is just below the size of the SPS, which has a circumference of 6.9 km. It was the first particle accelerator which made large-scale use of superconducting magnets, paving the way for other applications of this technology, such as RHIC, HERA and LHC. The use of superconducting dipole magnets (772 in total) with a bending field of 4.4 T allows a beam energy of up to 980 GeV, compared to 450 GeV for the normal-conducting SPS. The focusing magnets, quadrupoles and sextupoles, are also superconducting. Construction was completed in 1983 and the Tevatron is scheduled to run at least until the end of fiscal year (end September) 2011, depending on physics results and LHC progress.

The Tevatron was first used for fixed-target physics at an energy up to about 800 GeV. Now it runs exclusively in collider mode. In this mode protons and antiprotons circulate in opposite directions in the same vacuum chamber and use the same magnetic lattice. This makes the sum of the beam energies of both beams available in the centre-of-mass system. The two beams collide in the centre of two detectors, CDF and DØ, with a centre-of-mass energy of 1.96 TeV.

The Tevatron is part of a complex chain of accelerators, shown in Figure 12.5. Particles are injected into the Tevatron at 150 GeV, accelerated to 980 GeV and stored

29) http://www.bnl.gov/rhic

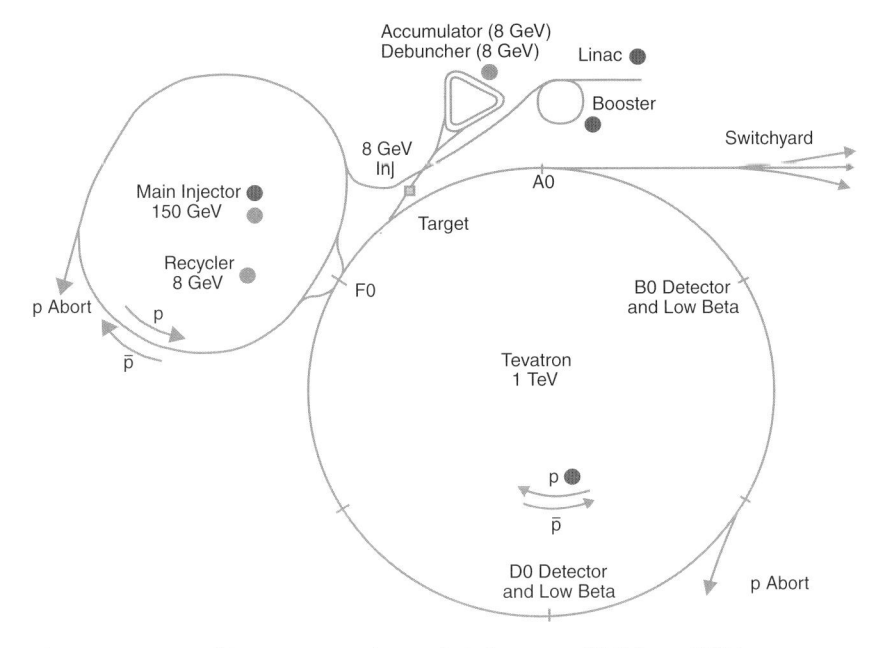

Figure 12.5 Layout of the Tevatron accelerator chain (courtesy of V. Shiltsev, FNAL).

at this energy for typically 16 h. Both beams consist of 36 bunches, which are each about 50 cm (r.m.s.) long. In the two experiments, the transverse beam dimensions are focused down to about 30 μm using low β insertions. At the beginning of a fill, between 25×10^{10} and 30×10^{10} protons per bunch are stored, and up to 9×10^{10} antiprotons per bunch. This corresponds to an average stored beam current of 77 mA for protons and 23 mA for antiprotons. The dipole magnet coils are made of NbTi and are operated at a temperature of 4.2 K at a current of up to about 4350 A, cooled by liquid He. The cryogenic cooling system was the biggest ever built at the time; it has a capacity of 23.5 kW at 4.2 K and has in addition the capacity to produce 1000 l of liquid He per hour.

12.3.1
Performance

Figure 12.6 shows the Tevatron integrated luminosity as a function of time starting from the year 2000 [7]. The whole Tevatron system has gone through a major process of performance improvements [7, 8], which have resulted in the parameters shown in Table 12.2[30].

---

30) For performance records see also http://tomato.fnal.gov/ops/records.php.

12 Accelerators: the Particle Smashers

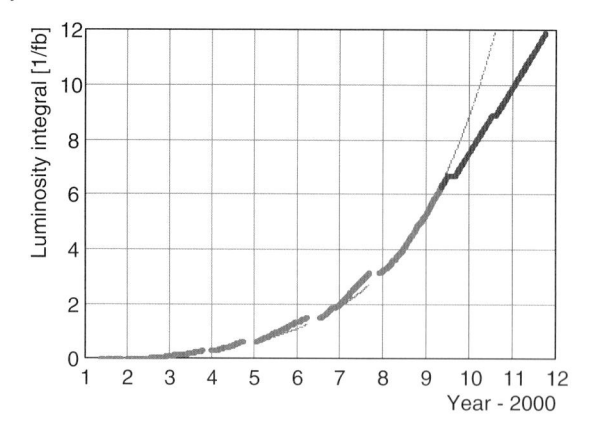

Figure 12.6 Tevatron Run 2 integrated luminosity. Actual in light grey, projected in dark grey (adapted from Lebedev, PAC 2009).

Table 12.2 Tevatron parameters [7].

	Value	Units
Average antiproton production rate (record)	25	10^{10} per hour
Number of protons per bunch	25–30	10^{10}
Number of antiprotons per bunch	8.3	10^{10}
Proton emittance, normalised	18	mm mrad (95%)
Antiproton emittance, normalised	8	mm mrad (95%)
Proton bunch length	50	cm
Antiproton bunch length	45	cm
Luminosity at beginning of fill (record)	355	10^{30} cm^{-2} s^{-1}
Store duration	16	h
Shot setup time	1.5	h
Number of stores/week	5–8	
Weekly luminosity (record)	73	pb^{-1}

12.3.2
Where Do the Protons and Antiprotons Come from?

Protons   In a gas discharge negatively charged H$^-$ ions are produced from hydrogen gas in a magnetron source and accelerated to 750 keV by a Cockcroft–Walton electrostatic accelerator. The negative ions are then accelerated to 400 MeV in a linear accelerator, before being accelerated to 8 GeV in a synchrotron, called the *Booster*. Before injection into the booster synchrotron, the negative hydrogen ions pass through a stripper foil, where they lose their electrons and become positively charged. This technique is called *charge exchange injection*. Since during the stripping the particles change their charge, Liouville's theorem on the conservation of the beam density in phase space does not apply, and high intensity bunches can be obtained.

After acceleration to 8 GeV, the protons are transferred to the *Main Injector*, where they are accelerated to 150 GeV. Then the protons are transferred to the Tevatron.

Antiprotons Antiprotons are generated by protons at 120 GeV coming from the Main Injector, hitting a target made of INCONEL (a special NiCr-based superalloy). The antiproton beam generated here has a large emittance and needs to be cooled and accumulated before it can be injected into the Tevatron. This is accomplished in a series of rings.

The *Debuncher Ring* captures the antiprotons from the target; in the subsequent *Accumulator* they are stored and accumulated. Before injection into the Main Injector for acceleration to 150 GeV, they are stored again in the *Recycler*. The Recycler is a fixed energy (8.9 GeV) storage ring, which has permanent dipole magnets. The first step of "cooling" the emittance of antiprotons is stochastic cooling, a technique developed at CERN when using the SPS as a proton–antiproton collider. This work was honoured by a Nobel prize to Simon van der Meer in 1984, which he shared with Carlo Rubbia. A position signal is generated in a beam pick-up and a correction signal is applied to the same particles at the appropriate phase and amplitude further downstream in the ring. This is done in two stages: first in the Debuncher, then in the Accumulator, which are both housed in the same delta-shaped tunnel. In the Accumulator the antiprotons are stacked in order to increase their beam current.

In the Recycler both stochastic and electron cooling are applied. In the latter technique, a beam of electrons passes parallel to the antiproton beam and some of the transverse momentum and of the energy spread of the antiprotons is transferred to the electrons, thus reducing the emittance of the antiproton beam.

Finally, a normalised emittance as small as 4 to 7 π mm mrad can be achieved. This is the "95% emittance" ($\pm 3\sigma$). The 1σ r.m.s. value is six times smaller.

The antiprotons then follow the same path as the protons: acceleration to 150 GeV in the Main Injector, then injection into the Tevatron.

12.4 HERA

HERA started operation in 1992 as the first and so far only lepton–hadron collider in the world. Two species of charged particles, protons and electrons, were stored in two different accelerator rings constructed on top of each other in the same tunnel, and brought into collision at two interaction points (IPs). The main design parameters are summarised in Table 12.3.

During the first years of operation, the luminosity increased significantly. This was mainly the result of an increase in the number of colliding bunches (9 in 1992, 84 in 1993 and 174 in 1995) and to some extent also from a decrease of the beam

Table 12.3 HERA design parameters [9].

	p	e	Units
Beam energy, E_b	820	30	GeV
Circumference	6335.83		m
Bending field at design energy	4.53	0.185	T
Revolution frequency, f_{rev}	47 317		Hz
Number of bunches, k_b	210		
Beam currents	163	58	mA
Bunch population	10^{11}	3.6×10^{10}	
Transverse emittance $\epsilon_{x/y}$	8/3.4	41/5.1	nm
β's at IP, $\beta^*_{x/y}$	10/1	2/0.7	m
Beam sizes at IP, $\sigma_{x/y}$	280/58	286/60	µm
Beam–beam tune shift, $\xi_{x/y}$	0.0010 / 0.0005	0.016 / 0.028	
Bunch length, σ_z	19	0.85	cm
Luminosity, $\mathcal{L}$	1.6×10^{31}		cm^{-2} s^{-1}

sizes at the IPs and an increase in single-bunch intensities. From 1995 on, through the upgrade until the end of the operation of HERA in 2007, the number of bunches remained rather constant at typically 170 colliding and some extra non-colliding bunches. In 1997 the peak luminosity reached a value of $\mathcal{L} = 1.4 \times 10^{31}$ cm^{-2} s^{-1}, close to the design goal and it appeared difficult to gain much more by optimisation of the existing machine. An upgrade programme was launched aiming at a fourfold increase in luminosity. It implied a redesign of the interaction region, using an early separation dipole within the detectors and a first element to focus the protons starting at 11 m from the IP compared to 27 m before the upgrade. It allowed the β functions at the IPs to be decreased by nearly a factor of three [10]. A further gain came from a decrease of the transverse electron emittance by an increase of the phase advance per standard cell in the optics of the electron ring and a moderate increase in intensities.

The HERA beam parameters [11] before and after the upgrade are shown in Tables 12.4 and 12.5. The beam energy (920 GeV for the protons and 27.5 GeV for the electrons) remained the same before and after the upgrade.

The parameters in Table 12.4 used before the upgrade are in fact rather close to the design values quoted in Table 12.3. In addition, the β^* values had already been decreased a bit before the upgrade. This allowed the design luminosity to be reached at slightly reduced intensities.

As planned, the HERA upgrade resulted in smaller beam sizes at the IP necessary for the increase in luminosity. The upgrade, however, also resulted in a significant increase of beam-induced backgrounds in the detectors. The levels were such that the tracking detectors could initially not be turned on at full beam intensity [12]. Through extra measures and natural out-gassing, on the timescale of

Table 12.4 HERA parameters in 2000, before the luminosity upgrade.

	p	e	Units
Beam currents	100	45	mA
Bunch population	7.3×10^{10}	3.5×10^{10}	
Transverse emittance $\epsilon_{x/y}$	5.1/5.1	36/4.3	nm
β's at IP, $\beta^*_{x/y}$	7.0/0.5	1.0/0.6	m
Beam sizes at IP, $\sigma_{x/y}$	190/50	190/50	μm
Luminosity, $\mathcal{L}$	1.7×10^{31}		$\mathrm{cm}^{-2}\,\mathrm{s}^{-1}$

Table 12.5 HERA parameters after the luminosity upgrade.

	p	e	Units
Beam currents	140	48	mA
Bunch population	1.03×10^{11}	4.0×10^{10}	
Transverse emittance $\epsilon_{x/y}$	5.1/5.1	20/3.6	nm
β's at IP, $\beta^*_{x/y}$	2.45/0.18	0.63/0.26	m
Beam sizes at IP, $\sigma_{x/y}$	112/30	112/30	μm
Luminosity, $\mathcal{L}$	7×10^{31}		$\mathrm{cm}^{-2}\,\mathrm{s}^{-1}$

over a year, the backgrounds were reduced to a level that allowed the experiments to fully profit from the luminosity increase of the upgrade, see Figure 12.7.

12.5 LHC

The Large Hadron Collider is installed in the 26.7-km long tunnel previously used by LEP. The LHC is built with two beam-pipes which cross at four interaction regions, see Figure 12.8. This allows particles of the same charge – proton–proton or heavy ions – to be accelerated and collided. The mass of these particles is much higher than that of electrons. Despite this, synchrotron radiation from protons at LHC energies becomes noticeable, but is not yet a limitation. The maximum beam energy in the LHC, or more precisely the beam momentum p, is given by the maximum bending field strength, B, according to $p = B\rho$, where the bending radius $\rho = 2804$ m is essentially given by the tunnel geometry. Numerically, $p[\mathrm{GeV}] = B[\mathrm{T}]\rho[m]/3.336$. The LHC is built with superconducting NbTi magnets operated at a superfluid helium temperature of 1.9 K, allowing for fields up to $B = 8.33$ T and $p = 7$ TeV.

Table 12.6 shows a comparison of beam parameters of the LHC with LEP 2.

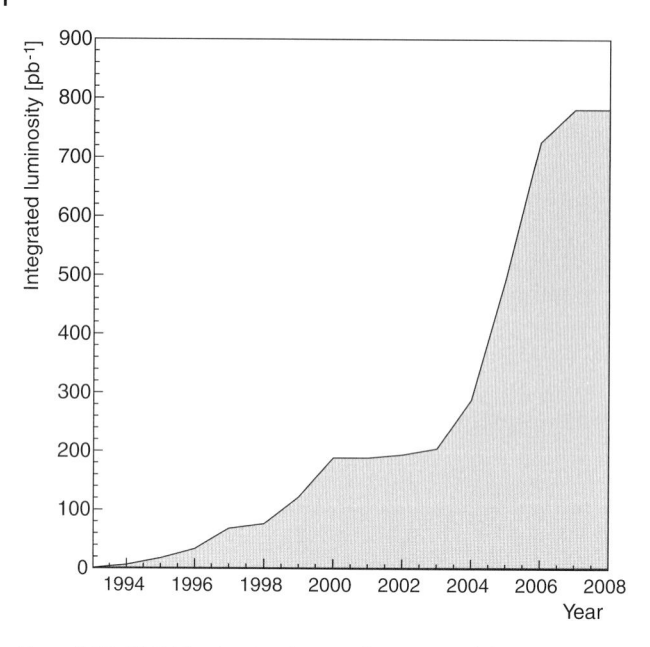

Figure 12.7 HERA luminosity showing the average of the delivered luminosities to H1 and ZEUS.

The LHC magnet system stores a large amount of energy. A current of 12 kA is used to reach the design field of 8.33 T in the LHC dipoles. The inductance of a LHC dipole is typically $L = 100$ mH. The energy stored in a single dipole at full current is $I^2 L/2 = 7.2$ MJ, which adds up to 1.1 GJ for the 154 dipoles in a sector

Table 12.6 Comparison of LHC and LEP 2 beam parameters.

	LHC	LEP 2	Units
Momentum at collisions	7	0.1	TeV
Design peak luminosity	10^{34}	10^{32}	cm^{-2} s^{-1}
Dipole field at top energy	8.33	0.11	T
Number of bunches, each beam	2808	4	
Total beam current	1.16	0.006	A
Particles/bunch	1.15×10^{11}	4.2×10^{11}	
Typical beam size in the ring	200–300	1800/140 (H/V)	μm
Beam size at IP	16	200 / 3 (H/V)	μm
Fraction of energy lost in sync. rad. per turn	10^{-9}	3%	
Total power radiated in synchrotron radiation	0.0078	18	MW
Total energy stored in each beam	360	0.03	MJ

12.5 LHC

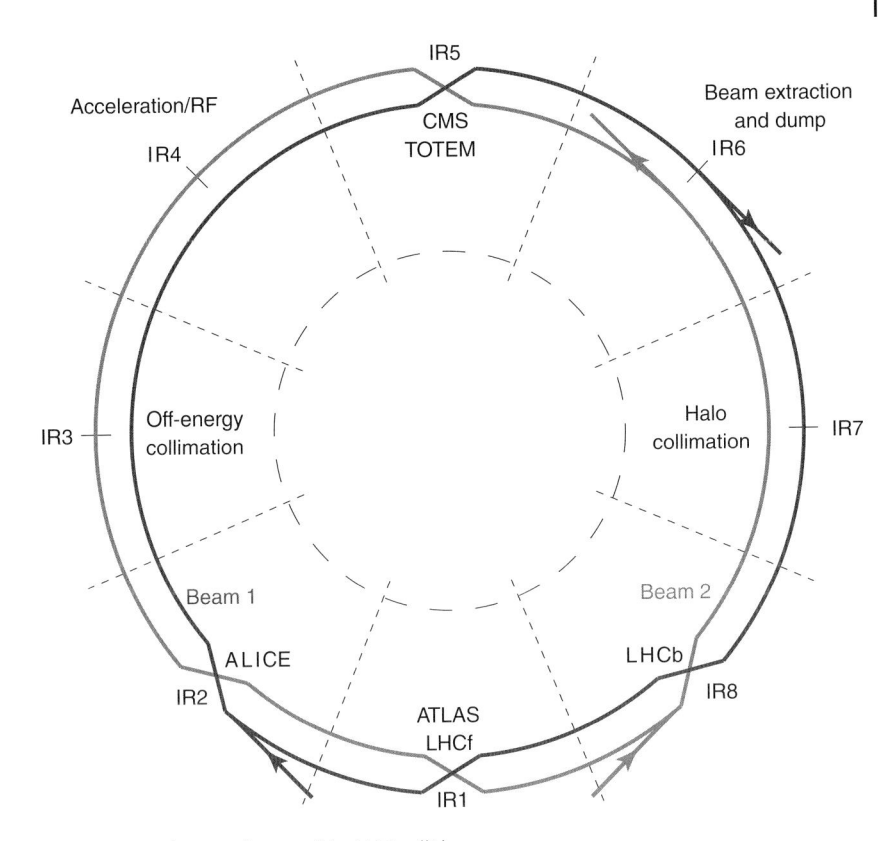

Figure 12.8 Schematic layout of the LHC collider.

or nearly 10 GJ for the entire LHC. This is comparable to the kinetic energy of a large passenger plane at full speed.

The LHC is being commissioned in steps. For the early commissioning at the end of 2009, the dipole currents were limited to 2 kA corresponding to 1.18 TeV beam energy, which reduced the power stored in the magnets by a factor of 36 compared to nominal. In 2010, the magnets are operated at half of the design current, allowing a beam energy of 3.5 TeV.

The main path to very high luminosities in the LHC is to use many bunches. The frequency of the LHC RF system is 400 MHz which corresponds to 2.5-ns long "buckets" of 75 cm wavelength. A bucket is the longitudinal space available for storing bunches as they travel along the circumference. The bucket boundaries are the limits of stable longitudinal particle oscillations. The LHC circumference corresponds to 35 640 RF wavelengths or buckets. Filling all buckets would produce collisions every 37.5 cm. The minimum realistic bunch spacing is 10 RF buckets or 25 ns. With a crossing angle of 0.3 mrad, this allows a 9σ beam separation for the parasitic beam encounters at multiples of 3.75 m from the interaction points. Filling every tenth RF bucket would allow 3564 bunches spaced by 25 ns. In order

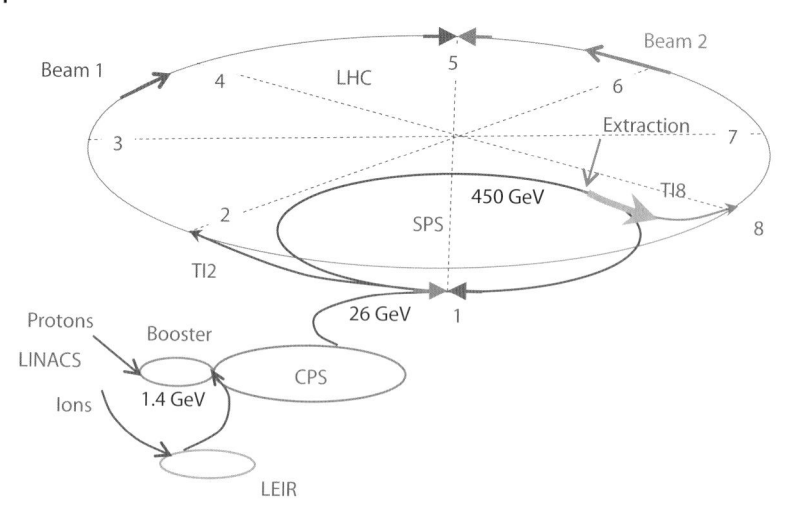

Figure 12.9 Schematic view of the LHC with its injectors.

to fill the LHC, gaps between batches of bunches of 225 ns, corresponding to the injection kicker rise time, are required. A single 3 μs gap is required to allow the particles in the LHC to be cleanly dumped. This limits the maximum number of bunches in the LHC per ring to 2808.

Filling the LHC with protons or ions requires a chain of pre-accelerators. The first stage is a linear accelerator, followed by the Booster, PS and SPS rings, see Figure 12.9.

The total energy in the beam reaches 360 MJ. This is several orders of magnitude higher than in other machines and well above the damage level for uncontrolled beam loss. The LHC is equipped with a fast beam protection system. Beam losses are monitored by several thousand monitors all around the ring. In case of magnet trips or abnormal beam losses, the LHC beams will be dumped within a few turns.

The design peak luminosity of the LHC of 10^{34} cm^{-2} s^{-1} is very challenging. Current plans are to limit the total beam intensity to 10% in the first year of operation, then 30%, reaching the nominal intensity after 4 years of operation.

Further information about the LHC can be found in [13, 14].

12.6
Linear Collider

As was pointed out earlier, proton and electron colliders complement each other. There is a consensus in the international physics community that the LHC results need to be refined by experiments at a lepton collider. As shown in (12.1), synchrotron radiation precludes circular electron–positron colliders at much higher energies than LEP. Two alternatives exist to overcome this limitation and build a Terascale lepton collider:

a) Since the energy loss is proportional to $1/m^4$, one can avoid excessive synchrotron radiation by using muons instead of electrons. The principles of *muon colliders* are being extensively studied. The feasibility of *muon cooling* remains to be proven and many years of R&D effort are required to get to the technical design stage of a muon collider. A very comprehensive summary of ongoing work can be found in [15].
b) Avoid bending the particle trajectories by using linear colliders, a novel technique successfully developed in the SLC at SLAC (http://www-sldnt.slac.stanford.edu/alr/slc.htm): two opposing linear accelerators accelerate the particles to their final energy in one pass, with the collision point at their centre.

Two approaches towards a new generation of linear colliders are presently being pursued by world-wide collaborations, the *International Linear Collider* (ILC) and *Compact Linear Collider* (CLIC). The ILC aims for a beam collision energy of 0.5 TeV, upgradeable to 1 TeV in the centre of mass, whereas CLIC extends the linear collider energy reach into the multi-TeV range, nominally 3 TeV. This leads to different technologies. ILC is based on superconducting RF acceleration technology with high RF-to-beam efficiency; CLIC takes advantage of a novel scheme of *two-beam acceleration* with normal-conducting copper cavities at high frequency and high accelerating field. Collaboration between the two design groups on technical subjects with large synergies is ongoing.

More information about the ILC can be found on the ILC web page[31] and in the ILC Reference Design Report[32]. Information about CLIC can be found on its web pages[33]. Here we highlight the major features of these two approaches.

12.6.1
Acceleration

Since the total beam energy in a linear collider has to be reached in a single pass, the accelerating gradient in the linac has to be as high as possible. In order to keep the power consumption small, the RF-to-beam power efficiency has to be maximised. In an accelerating cavity the voltage, U, is given by

$$U = \sqrt{\frac{R}{Q} \omega W} \ . \qquad (12.3)$$

R/Q is a structure parameter which depends only on the cavity geometry, ω is the RF frequency and W is the stored RF energy in the cavity. W is given by

$$W = \frac{Q}{\omega} P \ , \qquad (12.4)$$

where Q is the quality factor of the cavity and P is the RF input power required to provide the stored energy and to replace the power dissipated in the cavity walls due to RF currents.

31) www.linearcollider.org
32) http://www.linearcollider.org/cms/?pid=1000437
33) http://clic-study.web.cern.ch/CLIC-Study

12.6.2
Superconducting Accelerating System

Superconducting cavities can have Q-values of the order of 10^{10}, copper cavities typically of the order of 10^4. Hence, superconducting cavities are favoured for efficient acceleration. Losses due to wall currents do, however, appear at cryogenic temperature where niobium is superconducting. In addition the accelerating gradient achievable in superconducting cavities is intrinsically limited to about 50 MV/m (material dependent).

ILC is based on accelerating cavities made of solid Nb, operating at 2 K at a gradient of 31.5 MV/m at 500 GeV and 35 MV/m at 1 TeV. The low power dissipation in the cavities allows the RF to be kept on for a relatively long time, which in turn allows the acceleration of long bunch trains. The duty cycle is finally a compromise between heat dissipation in the cryogenic system, luminosity and detector technology. In ILC this has led to a design with 5680 bunches in trains of 1 ms with a 5-Hz repetition frequency.

12.6.3
Normal-Conducting Acceleration System

The high CLIC beam energy requires a higher accelerating gradient than achievable with superconducting cavities, while keeping the total accelerator length within reasonable limits. CLIC is based on copper travelling-wave accelerating structures, developed from "conventional" linac structures, which consist of a chain of coupled pill-box cells at a specially high RF frequency of 12 GHz. The accelerating gradient is pushed to the limit of sustainable fields with acceptable breakdown rates.

In order to generate an electric field of 100 MV/m the accelerating structures have to be supplied with RF power of 275 MW/m per metre of structure length! This is achievable only for short pulses of 240 ns at a repetition frequency of 50 Hz [16]. It is hard to imagine that individual RF power sources could be used in this case. This led to the development of the two-beam concept described below.

The basic design features of ILC and CLIC are shown in Table 12.7.

It is interesting to note that even though the peak RF power in the two concepts is quite different, the average RF power per unit length is very similar.

12.6.4
Achieving the Luminosity

The beams have to be focused to very small dimensions at the interaction point, see (12.2). The generation of beams with low emittance and the preservation of emittance all the way up to the collision point is an important requirement for linear colliders. Both machines require powerful damping rings in order to achieve the low emittance required and a sophisticated final focus system. Emittance

Table 12.7 Basic parameters for ILC and CLIC.

	ILC	CLIC
Centre-of-mass energy	500 GeV (upgradeable to 1 TeV)	3 TeV
Luminosity	2×10^{34} cm^{-2} s^{-1}	2×10^{34} cm^{-2} s^{-1}
Accelerating gradient	31.5 MV/m	100 MV/m
RF frequency	1.3 GHz	12 GHz
RF peak power per metre	0.37 MW/m, 1.6 ms, 5 Hz	275 MW/m, 240 ns, 50 Hz
RF average power	2.9 kW/m	3.7 kW/m
Total length	31 km	48.4 km
Total AC site power	230 MW	415 MW
Particles per bunch	20×10^9	3.7×10^9
Number of bunches per pulse	2625/pulse of 0.96 ms	312/pulse of 156 ns
Bunch spacing	396 ns	0.5 ns
Beam size horizontal/vertical:	640 nm/5.7 nm	40 nm/1 nm
Normalised emittance $\gamma \epsilon_x / \gamma \epsilon_y$	10 000 nm rad/40 nm rad	660 nm rad/20 nm rad
Beam power	20 MW	28 MW

preservation implies tight control of wakefields all along the linac as well as control of vibrations. In the case of CLIC, at 3 TeV, these tolerances are particularly severe. The final focus quadrupoles have to be stabilised to about 0.3 nm for frequencies above 4 Hz in the vertical and 2 nm in the horizontal plane. All quadrupoles in the linac have to be stabilised to 1 nm and 5 nm above 1 Hz in the vertical and the horizontal planes, respectively. For ILC with its bigger beam dimensions and the use of intra-train feedbacks, the tolerances of the final focus quadrupole stability are relaxed to about 50 nm.

12.6.4.1 ILC Base-Line Design

The base-line layout of ILC is shown in Figure 12.10. It is based on a two-tunnel layout with one tunnel housing the main accelerator and the second one running parallel containing the klystrons and other equipment. At the time of writing this text, other alternatives are being evaluated, possibly using one single tunnel housing the main linac, supplied with RF power provided by clusters of 35 klystrons on the surface. The positron source uses a beam from the first part of the electron linac which is sent through a helical ondulator to produce photons, which in turn are converted to positrons in a rotating target.

12.6.4.2 CLIC Base-Line Design

The CLIC scheme is based on normal-conducting travelling-wave accelerating structures, which requires high peak RF power to generate the accelerating gradient of 100 MV/m. A two-beam scheme is being developed, where the power

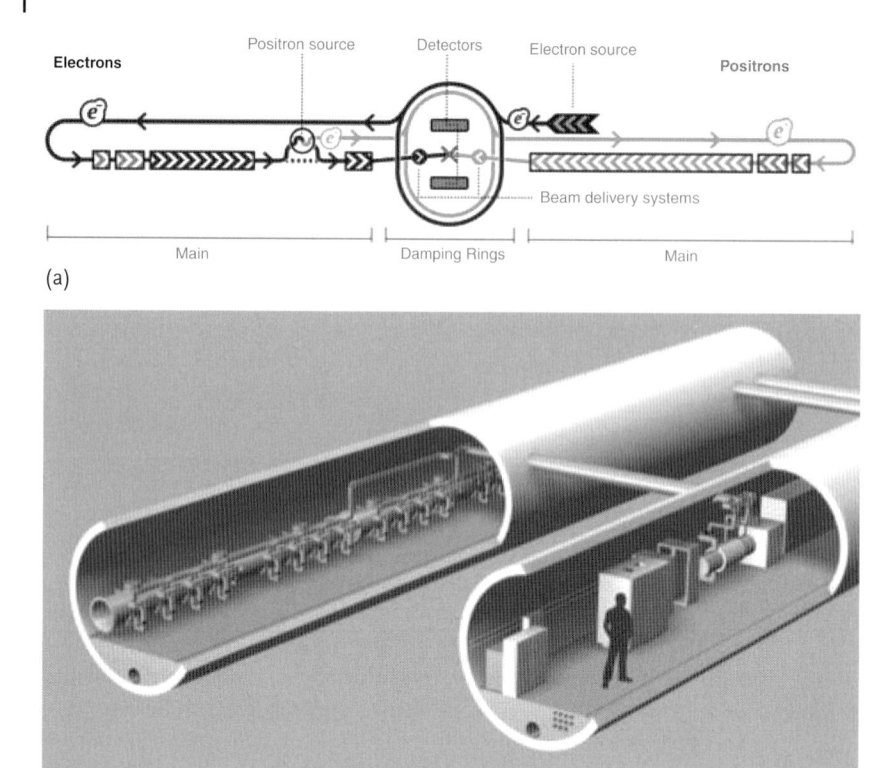

Figure 12.10 (a) ILC base-line design; (b) the two-tunnel layout with the main accelerator tunnel on the left, the klystron and service tunnel on the right (Source: ILC, Fermilab).

necessary for acceleration is generated by a secondary electron beam, the *drive beam*, which runs parallel to the main beam. Its beam power is converted to 12 GHz RF power in special RF structures, *Power Extraction and Transfer Structures* (PETS) and is transported to the accelerating structures by short waveguides. This is shown in Figure 12.11a. The generation of the drive beam with its very high beam current of 100 A over the pulse and a bunch repetition frequency of 12 GHz is done using a novel scheme of pulse compression and frequency multiplication with RF deflectors for bunch interleaving in several rings. The full scheme of CLIC is shown in Figure 12.11b.

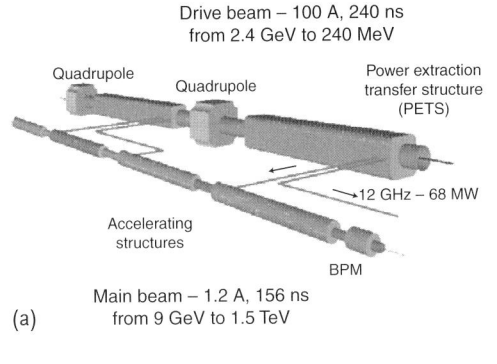

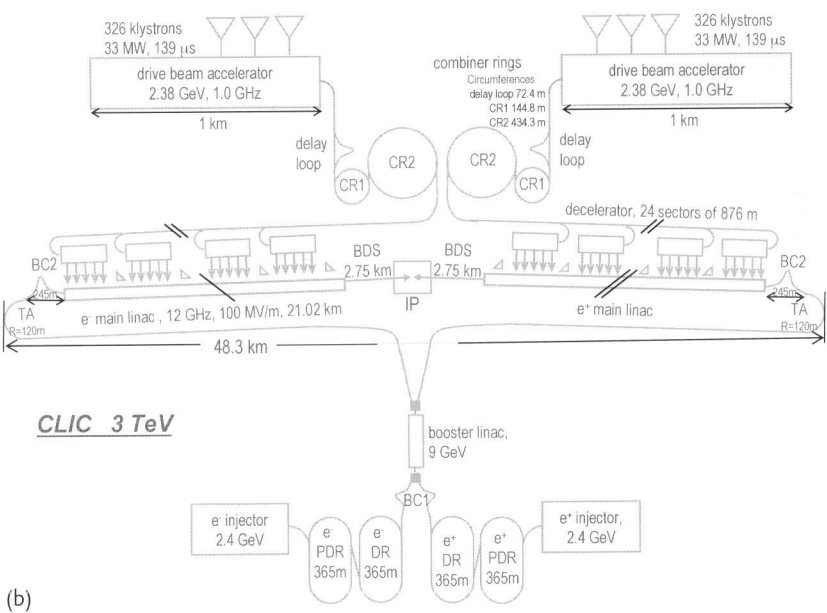

Figure 12.11 (a) CLIC two-beam scheme. (b) shows the complete complex with the drive beam generation complex on top and the main beam generation at the bottom (adapted from CERN Courier, Volume 48, Number 7, Sep. 2008).

References

1. Brandt, D., Burkhardt, H., Lamont, M., Myers, S., and Wenninger, J. (2000) Accelerator physics at LEP, *Rept. Prog. Phys.*, **63**, 939, doi:10.1088/0034-4885/63/6/203.
2. Assmann, R., Lamont, M., and Myers, S. (2002) A brief history of the LEP collider, *Nucl. Phys. Proc. Suppl. B*, **109**, 17.
3. Bailey, R. *et al.* (2002) The LEP collider, *Comptes Rendus Physique*, **3**, 1107, doi:10.1016/S1631-0705(02)01402-0.
4. Hubner, K. (2004) Designing and building LEP, *Phys. Rept.*, **403–404**, 177, doi:10.1016/j.physrep.2004.09.004.
5. Burkhardt, H. and Jowett, J.M. (2009) A retrospective on LEP, *ICFA Beam Dyn. Newslett.*, **48**, 143, http://www-spires.

fnal.gov/spires/find/hep/www?j=00376, 48,143.

6 Butterworth, A. *et al.* (2008) The LEP2 superconducting RF system, *Nucl. Instrum. Meth. A*, **587**, 151, doi:10.1016/j.nima.2007.12.038.

7 Lebedev, V. (2009) Status of Tevatron Run II, Particle Accelerator Conference (PAC 09), Vancouver, BC, Canada, 4–8 May 2009. http://accelconf.web.cern.ch/AccelConf/PAC2009/papers/mo1bci02.pdf.

8 Shiltsev, V., private communication.

9 Wiik, B.H. (1985) HERA: The accelerator and the physics, *Acta Phys. Polon. B*, **16**, 127.

10 Schneekloth, U. (ed.) (1998) The HERA luminosity upgrade, DESY-HERA-98-05.

11 Holzer, B. (2009) HERA: Lessons Learned from the HERA Upgrade, http://cdsweb.cern.ch/record/1204549, Proc. CARE HHH-2008.

12 Niebuhr, C. (2009) Background at HERA: Perspective of the Experiments, http://cdsweb.cern.ch/record/1184439/files/p14.pdf, CERN-2009-003 p.14.

13 Evans, L. and Bryant, P. (eds.) (2008) LHC Machine, *JINST*, **3**, S08001, doi:10.1088/1748-0221/3/08/S08001.

14 Evans, L. (2009) *The Large Hadron Collider: a Marvel of Technology*, EPFL Press, Lausanne, Switzerland.

15 Low emittance muon collider workshop. http://www.muonsinc.com/lemc2009, June 8–12, 2009, Fermilab.

16 Delahaye, J.P., Guignard, G., Raubenheimer, T., and Wilson, I. (1999) Scaling laws for e^+/e^- linear colliders, *Nucl. Instrum. Meth. A*, **421**, 369, doi:10.1016/S0168-9002(98)01132-2.

13
Detector Concepts: from Technologies to Physics Results

Ian C. Brock, Karsten Büßer and Thomas Schörner-Sadenius

13.1
Introduction

Since the 1970s, where the story of this book begins, the world has seen several generations of accelerators at the energy frontier together with their associated detectors. From generation to generation the detectors grew in size and complexity, in order to be able to study, in as much detail as technically possible, the particles produced by the accelerators. At a very early stage of this history most of the detectors became *omni-purpose detectors* combining a large variety of detectors in a single apparatus, thus allowing as many physical quantities as possible to be measured for each collision. The most important quantities are the momentum and energy of the created particles, the position of the interaction point, the coordinates of possible particle decays and the identity of the particles, for example electrons, muons, photons and charged hadrons. It is only the combination of all of the information provided by the different detectors inside the same apparatus that makes it possible to explore fully the physics of the collisions in the accelerator. In the literature (and even in this book) the entire apparatus is sometimes called a detector with its components being subdetectors, or it is called an experiment consisting of several detectors. In this chapter we will stick to the latter definition.

13.2
Technical Concepts

For the description of the technical concepts of some experiments we will restrict ourselves to omni-purpose experiments at the modern accelerators LEP, HERA, Tevatron and LHC, which were discussed in the previous chapter. At the end a section is dedicated to the design of a possible detector at the future accelerator ILC. At all of the accelerators discussed there is more than one omni-purpose experiment. This is because these apparatus are very complicated and the accelerators are unique, and having a second experiment is the only way to have independent

Physics at the Terascale, First Edition. Edited by Ian C. Brock and Thomas Schörner-Sadenius
© 2011 WILEY-VCH Verlag GmbH & Co. KGaA, Weinheim.
Published 2011 by WILEY-VCH Verlag GmbH & Co. KGaA.

measurements of the same quantity. There are many factors that influence the choice of the technology for the different detectors within the experiment. At first there are the physics questions which the contributing institutes rate of highest importance to address. Very important is also the know-how and tradition concerning detector development of different members of the collaboration. Therefore, the detectors that different collaborations have built for the same accelerator are always sufficiently different that the measurements have different systematic uncertainties and can be regarded as independent. But all experiments have some common general concepts and we will start with a description of these aspects.

For all experiments the interface between the accelerator and the experiment is the beam-pipe. The beam-pipe separates the ultra-high vacuum of the accelerator from the innermost layer of the first detector. Beam-pipes must fulfil several partly conflicting requirements: they must be rigid and able to withstand strong forces from the vacuum; they should be able to span large distances with as little support as possible; they must be vacuum-tight and they should be as transparent as possible for particles created at the collision point. For the last requirement, the material used close to the interaction point is usually beryllium or a beryllium–aluminium alloy, aluminium or a carbon–aluminium compound. Further away from the interaction point stainless steel is frequently used. In the detector region the radius of the beam-pipe has to be chosen such that any background will be safely caught by collimators and not hit the pipe, in order to avoid background spraying into the detector. Furthermore, if the radius of the beam-pipe changes inside the detector region one has to take care not to create traps for the electromagnetic waves used to accelerate the beam. These traps, called *higher-order mode* or RF (radio frequency) losses, can lead to an uncontrolled heating of the beam-pipe. Because the beam-pipe has to run through the detector all collider experiments have cylindrical symmetry.

To precisely measure the position of the interaction point and the direction of particles coming from it, it is important to place the first detector layer as close as possible to the beam-pipe. Modern detectors use for this component, often called *vertex detector*, several layers of silicon detectors with pixels or strips. With these detectors the path of charged particles can be measured with a precision of a few microns. Silicon detectors became popular as vertex detectors from the second half of the 1980s onwards, when the technology became reliable and affordable. Before that *wire chambers* were used as vertex detectors.

Vertex detectors only have a small volume and a few sensitive layers to track the particles. Further out in radius charged particles are usually measured by wire chambers. The details of both are discussed in Chapter 14. Wire chambers can be built with many layers and for large volumes, allowing a particle to be tracked along its path through the detector. The CMS experiment at the LHC is the first and up to now only experiment that has an all-silicon tracker, consisting of a pixel detector as the innermost part and a strip detector for the outer parts. It uses more than 200 m^2 of silicon as the active detector material. For future high-intensity experiments and upgrades of existing ones this tracking technology is the favoured one.

All trackers are built as lightweight as possible so that particles lose as little energy as possible and are not scattered on their path through the detectors. This is important because the next detectors the particles will hit are calorimeters, which measure their energy. The smaller the correction one has to make for energy losses in front of the calorimeter the more precise the energy measurement will be.

Except for the initial Run 1 version of the DØ experiment at the Tevatron, all high energy physics experiments discussed here have a solenoid magnet surrounding the tracker. The magnetic field parallel to the beam axis does not disturb the beam of the accelerator, but it bends the tracks of outgoing particles in the plane perpendicular to the beam. This allows the momentum of the particle to be measured, as the bending radius is inversely proportional to the particle's momentum.

For the placement of the magnet two options exist: either one can place it directly outside the tracker, or one can also put the calorimeters inside the magnet. Both options have been realised and they have their specific advantages and problems.

The first option requires a smaller magnet, which is easier and cheaper to build. The calorimeters behind the magnet have neither space restrictions nor restrictions on the use of materials and electronics, which otherwise have to be able to function in a strong magnetic field. However, the particles must traverse the material of the magnet before reaching the calorimeter. Energy losses must be corrected for and multiple scattering can confuse the association of a track with the position of the energy deposit in the calorimeter. To minimise these effects, the coils are built as thin and lightweight as technically possible. The energy correction can be improved by installing *pre-shower* or *pre-sampler detectors* which are position sensitive. These can help to decide whether or not an individual particle has already lost energy because it has started showering.

For the second option, there is no additional material between the tracker and the calorimeter, but the calorimeters have to be extremely compact, their absorber material must be non-magnetic and all on-board electronics must work in strong magnetic fields.

The calorimeters of high energy physics experiments measure the energy of neutral and charged particles by absorbing them. The physics processes and the technologies used for this purpose are discussed in Chapter 15. Here we will only distinguish between the different types of calorimeters. The energy of photons and electrons is measured in so-called *electromagnetic calorimeters*, by inducing showers of bremsstrahlung and electron–positron pairs. The energy of hadrons is measured by producing *hadronic showers* via nuclear interactions in heavy material.

There are two classes of electromagnetic calorimeters. The first class – *homogeneous calorimeter* – is made of crystals in which scintillation light is produced, the intensity of which is proportional to the energy of the originating particle. The second class – *sampling calorimeter* – consists of sandwiches of a heavy absorber and an active detection layer. Typical absorbers are iron, copper, brass, lead, tungsten and depleted uranium; the active layers can be any detector able to count charged particles, mostly used are scintillators or liquid argon.

Crystal calorimeters have excellent energy resolution, but it is difficult and expensive to build them with very high granularity, as this requires a large number of crystals of exactly the same quality. Every crystal has to be read out separately. A high granularity is much easier to realise in sampling calorimeters and, depending on the judgement of the importance of energy resolution versus granularity, different solutions can be chosen.

Up to now all hadron calorimeters are sampling calorimeters with sometimes different materials from those used for electromagnetic sampling calorimeters. The heavier the absorber the more compact the calorimeter can be built.

Because the cross section for e^+e^- pair production and bremsstrahlung is much higher than that for nuclear interactions, electromagnetic calorimeters are more compact than hadron ones and must be placed in front of them.

Having a strong solenoid either around the tracking detectors or around the calorimeters requires an iron yoke around the entire ensemble of detectors to close the magnetic field lines in a controlled way.

The only known particles that penetrate all detectors and the yoke without being absorbed are neutrinos and muons. Neutrinos manifest themselves in the experiment as *missing energy* and *missing momentum*. Because their interaction cross section with matter is so small, there is no chance to measure them directly in any high energy physics detector. For muons the situation is different: because of their charge, they produce a track in the inner detector but only lose very little energy in the calorimeters. With another layer of position-sensitive charged-particle detectors on the outside of the yoke one can identify muons unambiguously.

For this reason, all high energy physics experiments are enclosed in muon detectors. The detection techniques differ and will be discussed in more detail in Chapter 16. Some experiments have separate muon spectrometers with magnets, which allow the momentum to be measured once more with high precision, others instrument the iron of the return yoke with wire chambers. If the iron is sufficiently magnetised the muon track will be bent when passing through the iron, also allowing a second momentum measurement to be made.

In short, the general layout of the central part of an omni-purpose detector for particle colliders looks a bit like a "cake roll" with a hole in the centre for the beam-pipe. The layers are from inside out: a vertex detector, a tracker, an electromagnetic calorimeter, a hadron calorimeter and a muon detector. Either around the tracker or around the hadron calorimeter is a solenoid. The length of this "cake-roll" cylinder defines the smallest angle under which a particle from the interaction point can reach all layers of detectors. Making the cylinder very long compromises the precision of the measurements as the particles have to traverse more and more material as the angle with respect to the beam axis becomes smaller. Therefore the cylinders are closed by layers of endcap disks, equipped with detectors in the same sequence as the barrel part: vertex detector, tracker, electromagnetic calorimeter, hadron calorimeter and muon detectors. With that the experiment is hermetically closed, except for the region of the beam-pipe hole. Some experiments have added separate detectors downstream close to the beam-pipe to measure particles under extremely small angles.

13.3
Infrastructure

All experiments have in common that they require infrastructure to be able to work. Since the experiments have become more complex, the demands on infrastructure are today much larger than for example in the 1980s. A typical high energy experiment at this time had some 10^5 electronics channels, whereas the LHC experiments have almost 10^8 channels. Readout through optical fibres, which require much less space than traditional cables, made this increase possible without requiring much more space. Copper cables are usually used in current-day experiments only to power detectors or to supply high voltage. Data are usually read out through optical fibres.

Modern LHC experiments have big extra caverns to house the electronics for their detectors. There are several thousand kilometres of cables and fibres laid between the detectors and their electronics racks. Several hundred kilometres of these cables run inside and along the experiment itself. All these cables have to be laid along planned routes, they have to be inventoried in order to allow repairs in case of later problems, and many of them have strict limitations on possible overlength.

Large electrical installations always require cooling. The total length of all cooling lines is approximately one order of magnitude smaller than those of the cables, still adding up to more than 100 km. The cooling systems also became more and more sophisticated; some are running with normal water, others with demineralised water or with specific cooling fluids, which allow temperatures far below $0°C$.

Many systems have highly elaborate steering and control systems to meet the very specific cooling requirements of modern detectors in terms of flux, pressure difference and temperature difference. The cooling power of the LHC experiments alone amounts to several megawatts.

In the mid-1980s high-power magnets using superconducting technology became affordable which, however, require large cryogenic installations. In order to avoid quenches of the magnet, that is, loss of superconductivity and subsequent failure of the magnet, the cryogenics must be extremely reliable. Any failure will shut down the experiment – in the best case for several hours, normally for days to allow all components to be cooled down again. Therefore the cryogenic installations have an enormous amount of redundancy built in and use very sophisticated monitoring systems to guarantee stable operation over several months.

ATLAS uses liquid argon as the active medium of their electromagnetic calorimeter. There is a total of 83 m^3 of liquid argon in three cryostats. A large and complicated cryogenic system guarantees that the argon always stays liquid and is kept at around 87 K. Argon is heavier than air; a large leak can become very dangerous as it can fill the lower parts of the experimental cavern. Therefore the handling of large amounts of liquid argon in underground areas requires a special safety system with many surveillance detectors.

Several detectors in modern experiments must be filled with special gases for operation. Most of these detectors have very large volumes and are sensitive to pressure, flux and temperature variations. Usually the gases are mixtures of several components, adjusted for the detector and its operation mode. Small changes in the mixture can change the calibration or make the operation unreliable. Therefore modern gas systems have to be stable and reliable. For safety reasons one tries to use non-flammable gases wherever possible. However, many non-flammable gas mixtures contain flammable or explosive components. Therefore the mixing of the gases is done in special gas rooms on the surface, far away from the detectors, and then the non-flammable mixture is brought through pipes to the detector. The gas systems have built-in safety features to ensure that the mixture reaching the detector can under no circumstances become flammable or explosive.

Another extremely important part of any experiment infrastructure is safety. Of course highest priority is given to personnel safety. Electricity, fire, smoke, radioactivity, cryogenic liquids and gases are the most obvious risks. Large systems of monitors, emergency switches, alarms and evacuation routes are installed in all experimental caverns to minimise these risks. These systems are installed, operated and maintained according to the safety rules of the host laboratory of the experiment. Furthermore, all detectors have their own controls and safety systems to be able to properly react to any unsafe operation condition such as failures of cooling, electricity or ventilation.

The planning, assembly and operation of this infrastructure is a major enterprise requiring a large number of well-trained specialists. This includes a 24/7 on-call service for all installations mandatory for running the experiment.

13.4
Organisation

Already for many decades, high energy physics experiments have been too complex and too expensive to be built and operated by a single institute or university. Therefore the field has a long tradition of working in collaborations of several (nowadays many) universities and institutes to construct, assemble, commission, operate and maintain large experiments. Most of these collaborations were from the beginning international. With the complexity of the experiments also the size of the collaborations grew. For example, the collaboration that operated the TASSO experiment at the PETRA storage ring at DESY in Hamburg between 1979 and 1986 consisted in 1984 of 110 physicists from nine institutes in four countries. The CMS collaboration that works today at the LHC collider at CERN consists of 3170 physicists and engineers including about 800 students, coming from 169 institutes in 39 countries from all over the world. To work in collaborations of this size requires a sophisticated organisation to guarantee a high level of quality control and quality assurance. Detector parts are delivered from all over the world and installed by teams from the producing institutes. As the central coordination team has no formal authority to issue directives to members of foreign institutes, the collaboration, from the planning up to the day-to-day work at the detectors, has to be organised in mutual agreement and with extremely flat hierarchies. The efficiency and the success of this working model are absolutely remarkable – the largest experiments operating

Table 13.1 Comparison of the main solenoids of the omni-purpose experiments discussed in this chapter.

Accelerator	Experiment	Diameter	Length	Field	Type
LEP	ALEPH	5.3	6.6	1.5	superconducting
LEP	DELPHI	5.2	7.4	1.2	superconducting
LEP	L3	11.4	11.9	0.5	normal-conducting
LEP	OPAL	4.4	6.3	0.44	normal-conducting
HERA	H1	6.0	5.8	1.2	superconducting
HERA	ZEUS	1.9	2.5	1.8	superconducting
Tevatron	CDF	3.0	5.0	1.4	superconducting
Tevatron	DØ	1.1	2.73	2.0	superconducting
LHC	ATLAS	2.5	5.8	2.0	superconducting
LHC	CMS	6	13	3.8	superconducting
ILC	ILD	8	8.5	3.5	superconducting

at the edge of technology have been built and operated in this way successfully. These aspects are discussed in more detail in Chapters 20 and 22 in Part Three of the book.

In the subsequent sections we give a short description of the omni-purpose experiments at LEP, HERA, Tevatron and LHC, and we give an overview of their common features and specific characteristics. The solenoid magnets of the experiments are compared in Table 13.1.

13.5
ALEPH, DELPHI, L3 and OPAL at LEP

The LEP accelerator was operated from 1989 to 1994 with beam energies of slightly above 45 GeV to produce real Z bosons. From 1995 to 2000 it was operated with beam energies around 100 GeV to be able to produce pairs of W bosons. In the course of searching for the Higgs boson and other new physics the energy of LEP was raised in steps to 104.5 GeV per beam towards the end of its operation time.

LEP had four interaction regions all equipped with large omni-purpose experiments. At the time when these experiments were designed, in the early 1980s, superconducting coils and silicon detectors were cutting-edge technology. L3 and OPAL chose a more conservative approach and used normal-conducting coils, while ALEPH and DELPHI had superconducting coils. In their first versions the experiments did not have silicon vertex detectors; however, starting in 1991 all four were upgraded with silicon microstrip detectors. In 2001 all four experiments were dismantled to allow the construction of the LHC.

Precision measurements of the absolute cross section were one of the main physics goals at LEP. For normalisation purposes the luminosity was measured through low-angle Bhabha scattering with the highest possible precision. Therefore all four experiments were equipped with special luminosity monitors measuring the position and energy of low-angle electrons. These detectors are discussed in Chapter 17.

13.5.1
ALEPH

The name of this experiment [1, 2] stands for "A detector for LEP PHysics". The design was optimised for as complete as possible coverage of the solid angle, a good momentum resolution for charged particles made possible by using a strong magnetic field and a good electron and photon identification with a high-granularity calorimeter. The tracking detectors and the electromagnetic calorimeter were placed inside the coil. Figure 13.1a shows a cut through the experiment with its different components.

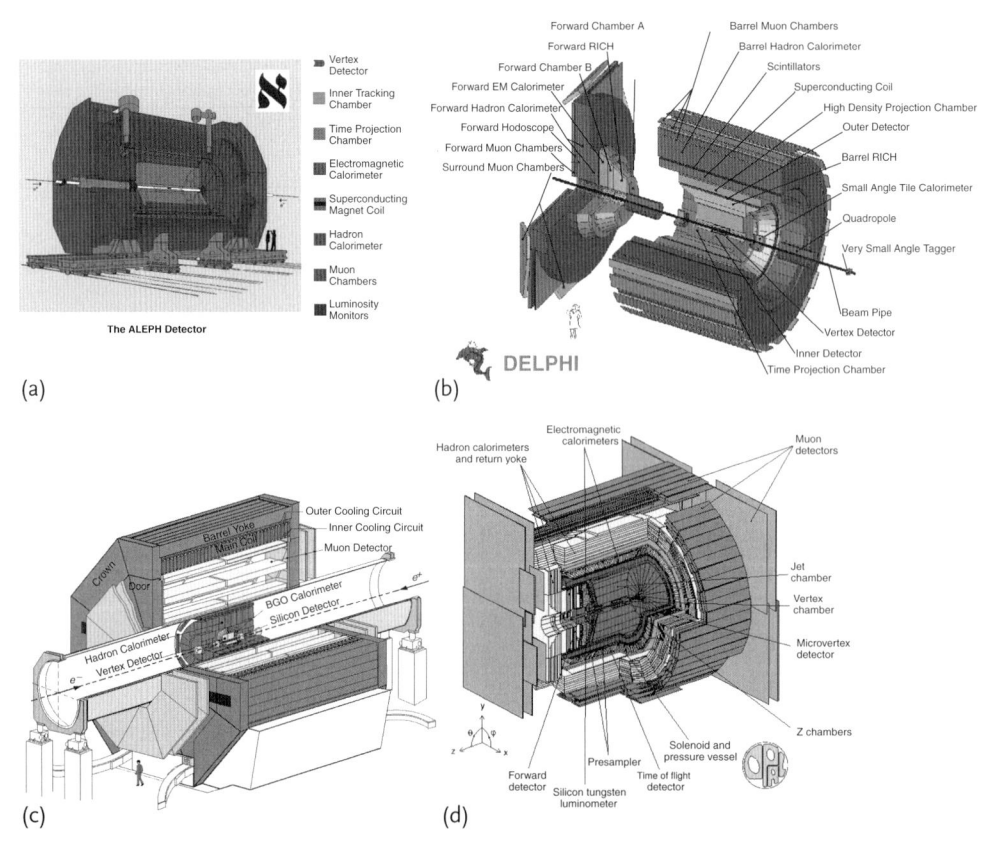

Figure 13.1 Three-dimensional views of the four LEP experiments: (a) ALEPH; (b) DELPHI; (c) L3; (d) OPAL (Source: CERN).

The innermost detector was a silicon microstrip detector. It consisted of two layers of double-sided silicon microstrip sensors. This detector was upgraded in 1996 by a longer one with improved electronics [3].

Outside the silicon detector followed the central tracking chambers. An inner multi-wire drift chamber covered the radii from 16 cm to 29 cm over a length of 2 m. It provided up to eight r–φ coordinates per track with a precision of about 100 μm and served for tracking and triggering. It was followed by a very large three-dimensional imaging time projection chamber (TPC) with a diameter of 3.6 m and a length of 4.4 m. Besides precise tracking of charged particles it allowed particles to be identified by measuring the energy loss, dE/dx, from up to 340 samples per track.

The electromagnetic calorimeter consisted of alternating layers of lead sheets and proportional tubes, read out in 73 728 projective towers, each subdivided into three longitudinal sections.

The iron yoke outside the magnet served as absorber for the hadron calorimeter and for muon identification. Iron plates were interleaved with limited streamer tubes, which were read out in 4609 projective towers. The tubes also gave digital signals for muons which penetrated the iron. The thickness of the iron amounted to 1.2 m. Muons were again measured outside the iron in two layers of limited streamer tubes, separated by 50 cm.

13.5.2
DELPHI

DELPHI was the second experiment at LEP with a superconducting coil. The name is an acronym for "DEtector with Lepton, Photon and Hadron Identification". DELPHI was a general-purpose experiment with special emphasis on particle identification [4, 5], for which it had *ring-imaging Cerenkov counters* (RICH). The DELPHI experiment provided three-dimensional information with high granularity in most components and allowed precise vertex determination with a silicon strip detector. The components of this experiment are shown in Figure 13.1b.

The superconducting solenoid housed the tracker, the barrel RICH and the electromagnetic calorimeter. Outside the coil, a *time-of-flight system* for particle identification, the hadron calorimeter and the muon detectors were installed.

The electromagnetic calorimeter used lead sheets as absorber material, interleaved by gas volumes as active detector elements. The electrons were collected at the ends using the time-projection principle. This gave the device its name: high density projection chamber (HPC). The hadron calorimeter was a sampling calorimeter with 20 layers of limited streamer tubes inserted into slots of the return yoke.

The muon chambers had two layers: one inside the return yoke after about 90 cm of iron, and one outside. As can be seen from Figure 13.1b, DELPHI had tracking, particle identification, calorimeters and muon chambers both for the barrel part and the endcaps. With its large number of different components and its specialised detectors for particle identification, DELPHI was the most complex LEP experiment.

13.5.3
L3

The third experiment at LEP was called L3. Its main construction goals were high-resolution measurements of the energy of electrons, photons and muons [6]. All detectors were installed inside the large-volume normal-conducting magnet. A layout of the L3 experiment is shown in Figure 13.1c.

The muon chambers were placed inside the coil to be able to very precisely measure the muon momentum independently from the central tracker. There were three layers of multi-wire drift chambers mounted on two large Ferris wheels. They allowed a relative momentum measurement with a precision of 2% for muons with momenta of 50 GeV.

The barrel part of the tracking detectors and the calorimeters were mounted inside a support tube of 32 m length and 4.45 m diameter that rested on adjustable jacks. The tube was concentric to the beam and facilitated a precise alignment of the detector with respect to the beam.

The L3 tracker was a *time expansion chamber* (TEC), a special type of drift chamber measuring the r and φ components of the tracks. The z component was measured with a set of drift chambers covering the outer part of the TEC. In 1991 the beampipes in the LEP interaction regions were reduced in diameter from 16 to 11 cm, giving space for a silicon microvertex detector, which in the case of L3 was added in 1993 [7].

The electromagnetic calorimeter consisted of about 11 000 crystals of bismuth germanate (BGO) pointing to the interaction point. At the time of the detector design these crystals were very new detector material. It provided an excellent energy resolution of 5% at 100 MeV and less than 1% for energies above 2 GeV.

The hadron calorimeter was a sampling calorimeter using depleted uranium as absorber and proportional wire chambers as active detectors. The high density of the uranium made this calorimeter very compact. On the outside of the hadron calorimeter a muon filter made of several layers of brass interleaved with proportional chambers completed the detector.

For triggering purposes a layer of scintillators was installed between the electromagnetic and the hadron calorimeters.

The L3 magnet has been re-used for the ALICE experiment at the LHC [8].

13.5.4
OPAL

The experiment name stands for Omni Purpose Apparatus at LEP [9]. At OPAL a normal-conducting solenoid surrounded the tracking detectors. Outside of the coil, the electromagnetic calorimeter made of lead-glass crystals was installed, followed by an iron yoke instrumented as a hadron calorimeter and four layers of muon chambers. To cover as much as possible of the solid angle the experiment had a barrel and an endcap part. The general layout of OPAL is shown in Figure 13.1d.

The tracking part of OPAL originally consisted of a drift chamber as the precision vertex detector, a large-volume jet chamber and drift chambers on the outside to measure the z component of the tracks. All tracking chambers were mounted inside a pressure vessel and operated at 4 bar overpressure. The 1991 reduction of the beam-pipe radius gave space for a two-layer silicon strip detector [10, 11]. In its first version it measured only the r–φ coordinates. In 1993 it was replaced by a detector with two silicon diodes glued back-to-back with perpendicular strip orientation in each layer and improved electronics. This detector provided r–φ and z coordinates of the tracks. The large-volume jet chamber had 159 layers of wires allowing particle identification by precise dE/dx measurements.

The calorimeters were installed outside of the pressure vessel and the solenoid. In front of them was a set of scintillation counters used to measure the time-of-flight, again allowing particle identification. Behind the time-of-flight and directly in front of the calorimeters, thin high-gain gas chambers were installed. These so-called pre-samplers were used to correct for energy losses by electromagnetic showers that started in the material in front of the calorimeter.

The electromagnetic calorimeter consisted of about 11 700 lead-glass blocks covering the barrel and the endcap regions. Every block was read out by its own photomultiplier. Lead glass has an excellent intrinsic energy resolution varying between 5% at 1 GeV and 0.5% at 100 GeV.

The energy of hadrons emerging from the lead-glass calorimeter was measured in the hadron calorimeter that was integrated into the iron yoke. The latter was segmented into layers interleaved with limited streamer tubes.

The outside of the yoke was covered by large-area drift chambers for muon identification. About 93% of the solid angle was covered by at least one layer of muon chambers. Over almost the entire solid angle particles had to traverse at least the equivalent of 1.3 m iron before reaching the muon chambers. This reduced the probability for punch-through from pions to less than 0.001.

Of the four LEP experiments, OPAL was regarded the most conservative and conventional one, as most of its detectors were further developments and improvements from predecessors at earlier accelerators.

13.6
H1 and ZEUS at HERA

HERA was the first and up to now only electron–proton collider in the world. Its details are described in the previous chapter. Most of the time, HERA was operated with 920 GeV protons, colliding with electrons or positrons of about 27.5 GeV. The accelerator was in operation from 1992 to 2007 with a major upgrade in 2000 to increase the luminosity and to install spin rotators that provided longitudinally polarised lepton beams.

HERA was the first accelerator at which the time between two bunch crossings was much too short to make a trigger decision. Therefore the analogue information from the detectors had to be buffered until the trigger decision had been made while the next collisions were taking place and the detectors continued recording them. This required a completely new architecture of the trigger and data acquisition systems, which can be regarded as the predecessor of today's state-of-the-art systems used at the LHC (see Chapter 18).

At HERA there were two omni-purpose experiments: H1 and ZEUS. They differed from all other experiments described here by their geometry. The collision of high energy protons with lower energy electrons leads to a strong boost of the reaction products into the proton flight direction, which at HERA is called the "forward" direction. Therefore this forward region of both experiments had much deeper calorimeters than the backward region. Both experiments also had muon spectrometers in the forward region, separated from the main detector. In addition, both experiments were equipped with several small detectors close to the beam-pipe inside and outside of the yoke to register particles at small angles, in particular for background and luminosity measurements, but also to detect very low-angle particles from the electron–proton interaction.

It should be mentioned that there were two more interaction regions at HERA, used by the fixed-target experiments HERMES and HERA-B which only made use of the electron or the proton beam, respectively. These two experiments are described elsewhere [12, 13].

13.6.1
H1

One of the most important design criteria of H1 was the precise and unambiguous electron identification over a large range of energies and angles [14]. Therefore the tracker and the electromagnetic and hadron calorimeters were placed inside

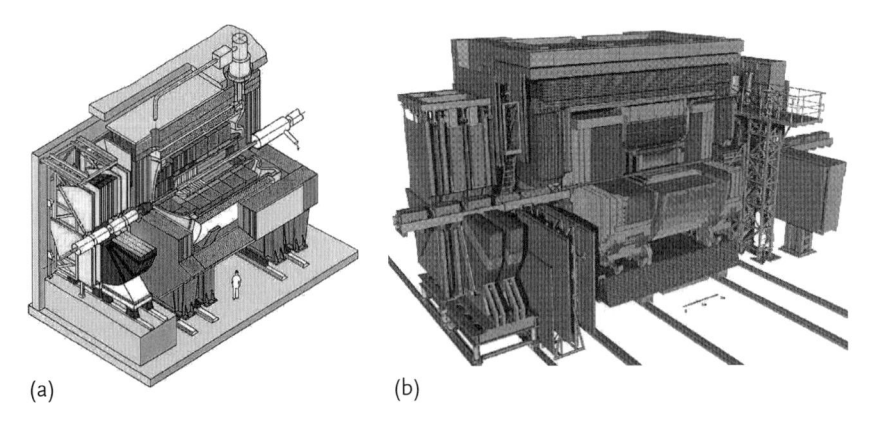

(a)   (b)

Figure 13.2 Three-dimensional views of the two HERA experiments: (a) H1; (b) ZEUS (Source: DESY).

a large superconducting coil [14]. The general layout of the detector is shown in Figure 13.2a. A two-layer double-sided silicon strip detector was installed in 1997 as the innermost detector [15, 16]. It was followed by central and forward tracking systems consisting of several layers of drift chambers and trigger proportional chambers. The absorber material of the electromagnetic calorimeter was lead, that of the hadron calorimeter was steel. Absorber plates were interleaved with readout gaps, filled with liquid argon as the active detector material. The entire calorimeter was installed inside a cryostat which surrounded the tracker.

The particular strength of the H1 calorimeter was its high resolution and fine granularity in the electromagnetic part, allowing electron recognition from very low energies. The return yoke of the magnet was laminated and filled with limited streamer tubes. It served to detect hadronic showers leaking out of the hadron calorimeter and to track muons. Inside and outside the iron yoke there were additional wire chambers for muon identification. High energy muons in the forward direction were measured separately in the field of a toroidal magnet with sets of drift chambers on both ends. To close remaining holes in the geometric acceptance, warm calorimeters were added in the forward and the backward regions. In 1997 several layers of silicon strip detectors were added to improve the tracking in the backward direction.

13.6.2
ZEUS

The ZEUS experiment (Figure 13.2b) had its tracking detectors inside a thin superconducting coil [17]. The tracking system was originally made of a vertex chamber and a large central drift chamber. During the luminosity upgrade of HERA in 2000, the vertex chamber was replaced by a three-layer silicon strip detector, having two single-sided detectors per layer with orthogonal strip orientation [18]. Four layers of wedge-shaped detectors were added to measure forward-going tracks.

The heart of the ZEUS experiment was its calorimeter, a sandwich of depleted uranium plates as absorber and scintillator tiles as active detector material. The calorimeter had an electromagnetic section with small tower size in front, followed by two hadronic sections, and was segmented into a barrel, a forward and a rear part. It was especially designed to give exactly the same light response to electromagnetically and hadronically interacting particles. Such a calorimeter is called *compensating* and is particularly suited to precisely measure jet energies and energy flows without the need to distinguish the different particles hitting the calorimeter. With an energy resolution of $35\%/\sqrt{E}$ for hadrons, the ZEUS calorimeter was one of the best ever built.

In order to improve the identification of electrons inside jets, a layer of silicon pad detectors was inserted into the forward and the rear calorimeter at the depth of the maximum of an electromagnetic shower. The pad size of this detector was an order of magnitude smaller than the granularity of the calorimeter.

The iron of the return yoke was instrumented with aluminium tubes used in proportional mode to measure muons and late-showering particles. In the barrel and rear regions muons were identified by limited streamer tubes mounted in front of and behind the iron yoke. In the forward region muons were identified and measured in a separate forward muon spectrometer, using drift chambers and limited streamer tubes, mounted on the magnetised iron yoke and between magnetised iron toroids.

13.7
CDF and DØ at the Tevatron

The Tevatron $p\bar{p}$ collider has been in operation since 1983 and hosts the oldest running high energy physics experiments: CDF and DØ. Since its beginning, the machine has been upgraded several times, the most important addition to the accelerator complex being the *Main Injector* which was built from 1994 to 1999.

After the installation of DØ in the beginning of the 1990s, the Tevatron had two running periods: Run 1 (1992 to 1996) and Run 2 (2002 up to now). During these periods, both the delivered luminosity and the energy of the $p\bar{p}$ collisions were substantially increased – the latter to now 1.96 TeV in the centre-of-mass. With this high energy, the Tevatron was the strongest hadron collider in the world, until the beginning of LHC operation, and provided (and still keeps providing) the best limits for many "new physics" observables. A limiting factor for luminosity production is the use of antiprotons which are difficult to accumulate in sufficient quantities.

Besides providing excellent limits on new physics, the Tevatron experiments contributed substantial results to electroweak physics (e.g. excellent W-boson mass measurements) and to QCD (e.g. studies of jet production at the highest energy scales). In addition, several discoveries were made at the Tevatron – most importantly the discovery of the top quark in the mid-1990s. Furthermore, in recent years the first measurement of mixing in the B_s^0 system was achieved by the two experiments.

The Tevatron is currently foreseen to run at least until the end of 2011 – a time when in many physics channels the LHC will have become competitive.

13.7.1
CDF

The CDF experiment ("Collider Detector at Fermilab") [19, 20] took first data in 1987 and since then has been upgraded several times. The Run 2 detector has been in operation since 2001, and an overview is shown in Figure 13.3a.

In Run 2, CDF provides a tracking system contained inside the superconducting solenoid consisting of a silicon vertex detector of 96 cm length and an open-cell wire drift chamber. Together these two detectors provide r–φ and z measurements in the pseudorapidity regions $|\eta| < 2$ and $|\eta| < 1.7$, respectively. The CDF calorimeters lie outside the solenoid and are physically divided into a central region ($|\eta| < 1.1$) and an upgraded plug region ($1.1 < |\eta| < 3.6$). The electromagnetic part is realised as a lead–scintillator sandwich calorimeter, the hadronic part is made from iron and

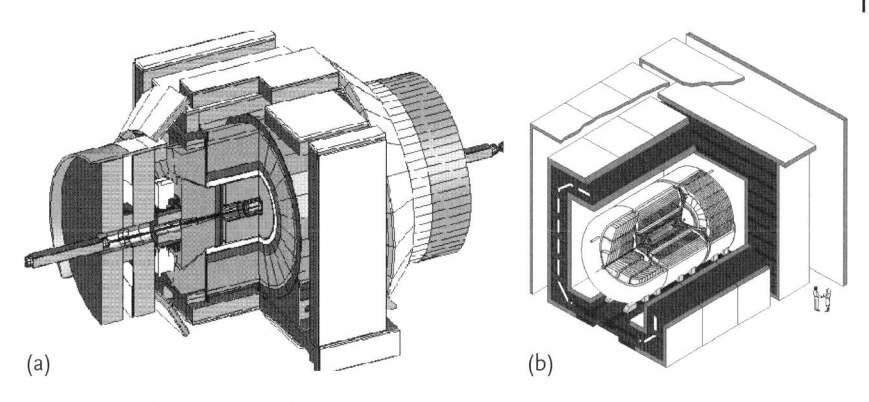

Figure 13.3 Three-dimensional views of the two Tevatron experiments: (a) CDF; (b) DØ (Source: Fermilab).

scintillator. Finer position information for electrons and photons is obtained from proportional chambers located approximately at the position of the electromagnetic shower maximum inside the calorimeter.

The CDF muon system consists of drift chambers surrounding the calorimeter in the range $|\eta| < 1$: a central part outside the central hadron calorimeters covers $|\eta| < 0.6$, and an additional system made from four layers of single-wire drift tubes is shielded by 60 cm of steel. In addition, the region $0.6 < |\eta| < 1.0$ is equipped with drift tubes.

13.7.2
DØ

In Run 1 the DØ detector [21] (see Figure 13.3b) had no magnetic field within the tracking volume (similar to UA2). The tracking consisted of a vertex drift chamber, a transition-radiation detector, a central and two forward drift chambers. The tracking volume was surrounded by a liquid-argon calorimeter, which was housed in a barrel and two endcap cryostats and had a coverage up to $|\eta| \approx 4$. The absorber was depleted uranium, with the exception of the coarse hadron calorimeter (stainless steel or copper). Because of the need for a missing momentum measurement in the central detector, the calorimeter had to exhibit a good energy resolution for electrons, photons and jets. The achieved resolution for the endcap calorimeter for electrons was $\sigma/E = 15.7\%/\sqrt{E}$ and for pions $\sigma/E = 41\%/\sqrt{E}$.

The muon spectrometer surrounding the calorimeter consists of five separate iron toroid magnets (central, two endcap and two small-angle magnets), instrumented with proportional drift tube chambers. The muon system covers $|\eta| \leq 3.6$.

For Run 2 the detector was substantially upgraded [22]. A superconducting solenoid of 2 T was installed within the bore of the central calorimeter cryostat. The central detector was completely replaced by a silicon vertex detector (microstrips) with four layers in the central region from 2.7–10.5 cm, eight discs in the forward region (up to $|\eta|$ of nearly 4) and a surrounding scintillating fibre tracker

($|\eta| < 1.7$). Because of the small diameter of the magnet the momentum resolution is only moderate. In 2005, an additional microstrip silicon detector (Layer 0) with an inner radius of $R = 1.61$ cm was added directly onto the beam-pipe and helped to improve the b-tagging capability of DØ [23].

Between the solenoid and the calorimeter and in front of the endcap cryostats a pre-shower detector has been installed. The instrumentation of the muon system was replaced by mini drift tubes and scintillator tiles, which are utilised for an effective muon trigger. In addition, a new forward proton detector (roman pots) allows diffractive physics to be studied.

The read-out electronics of the detector were replaced in order to cope with the reduced bunch-crossing time at Run 2 and the higher occupancy due to the higher luminosity.

13.8
ATLAS and CMS at the LHC

The LHC proton–proton collider is discussed in detail in Chapter 12. It is installed in the former LEP tunnel, has a circumference of about 27 km and is designed for proton energies of up to 7 TeV per beam. Since November 2009 it has been providing collisions, currently with 3.5 TeV energy per beam. The beam energy is expected to be ramped up to 7 TeV in 2013. The LHC is the highest energy particle accelerator in the world. Beside protons the LHC can also accelerate heavy ion beams.

There are four interaction regions, two of them equipped with omni-purpose experiments, specialised for proton–proton interactions. One interaction region hosts an experiment specialised for heavy ion physics called ALICE [8], and the fourth houses the LHCb experiment [24] that studies CP violation in B-hadron decays with a specialised forward spectrometer. Beside these large experiments there are two small ones – TOTEM [25] and LHCf [26] – that are dedicated to very specific measurements and are not discussed here.

LHC is designed to provide the highest energy collisions at a much higher luminosity than any of its predecessors. As a consequence, if operated at nominal performance, there will be more than 20 interactions on average per bunch crossing. This leads to more than 1000 high energy particles which have to be measured and distinguished per bunch crossing. Therefore the experiments have to have an unprecedented size and granularity. This can also be seen from the number of readout channels, which is more than one order of magnitude larger than those of previous experiments and amounts to close to 10^8. The large majority of the readout channels (7×10^7) for both experiments actually are from the relatively small silicon pixel detectors mounted close to the beam-pipe.

Because of the high energy and high luminosity, the radiation environment at the LHC will be extremely hostile for all detectors close to the interaction point. This is one of the main reasons why already now major work in planning and R&D is going on in both experiments to upgrade and replace some of the detectors which will be most exposed to radiation when the LHC reaches its ultimate performance.

13.8.1
ATLAS

The name of the experiment is an acronym for "A Toroidal LHC ApparatuS". With a height of 25 m and a length of 44 m it is the largest high energy physics experiment ever built; it has a mass of about 7000 t [27]. The general layout is shown in Figure 13.4a.

The design of ATLAS is determined by its magnet configuration which consists of a superconducting solenoid and three large superconducting toroids, one for the barrel and two for the endcaps. The tracking detectors are mounted inside the central thin solenoid. The electromagnetic and hadron calorimeters surround the solenoid and are followed by a large muon spectrometer inside the strong toroidal field.

The inner detector consists of a silicon pixel detector, a silicon strip tracker and a large-volume transition radiation tracker (TRT) made of so-called straw-tube detectors. All three tracking detectors consist of a barrel and two endcaps, with the pixel detector having three barrel layers and two times three endcap disks while the strip tracker has four barrel layers and nine endcap disks on each end.

The electromagnetic calorimeter is a high-resolution and high-granularity liquid-argon sampling calorimeter. The absorber material is lead. It consists of two identical half-barrels separated by a small gap at the interaction point. Each endcap is segmented into two coaxial wheels covering different polar angles.

There is a system of hadron calorimeters covering as much as possible of the solid angle. The barrel part is 11 m long and divided into three barrel modules with an inner radius of 2.28 m and an outer radius of 4.25 m. It is a sampling calorimeter with steel as absorber and scintillating tiles as active detectors. The ends are closed with two wheels of liquid-argon calorimeters, placed outside the electromagnetic endcap calorimeters, sharing the same cryostat. The absorber material

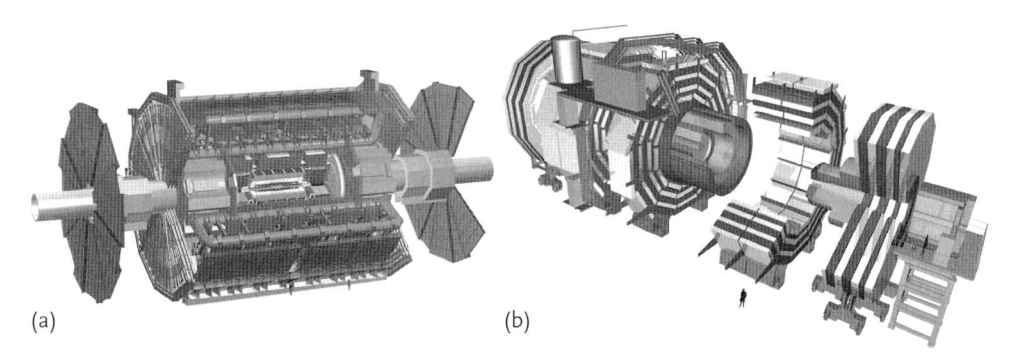

(a)　　　　　　　　　　　　　(b)

Figure 13.4 Three-dimensional views of the two omni-purpose experiments at the LHC: (a) ATLAS; (b) CMS (Source: CERN).

of the hadron endcap calorimeter is copper. The region around the beam-pipe is closed with a liquid-argon forward calorimeter, again sharing the cryostat with the endcap calorimeters. The first few absorber plates are copper, for measuring electromagnetic showers; the deeper ones are made of tungsten in order to rapidly stop hadrons.

The calorimeters are surrounded by the muon spectrometer, which defines the enormous size of this experiment. The idea is to deflect the muon tracks in a large-volume magnetic field, provided by strong superconducting air-core toroid magnets, instrumented with high-precision tracking chambers and separate trigger chambers.

In the barrel region the magnetic field is provided by eight large coils. In the endcap region the muon tracks are bent by two smaller toroids inserted into both ends of the barrel toroid. The field configuration is chosen to be usually orthogonal to the muon trajectory. In the barrel region the detector chambers are mounted in three cylindrical layers, in the endcap regions they are arranged perpendicular to the beam axis in three planes of disks.

13.8.2
CMS

The name of this experiment means "Compact Muon Solenoid" [28]. Compared with ATLAS it is indeed compact, 15 m in diameter and 21 m long, but still significantly larger than any previous omni-purpose experiment. With a weight of about 14 000 t CMS is the heaviest LHC experiment. Conceptually CMS differs a lot from ATLAS. A general layout of the experiment is shown in Figure 13.4b. The superconducting coil contains the inner detector and both the electromagnetic and hadron calorimeters. The coil is surrounded by a massive iron yoke, instrumented with muon chambers. A hadron forward calorimeter outside the return yoke completes the experiment.

CMS was the first large experiment that was constructed on the surface and then brought in pieces to its final destination in an underground cavern. This was possible because the yoke is arranged in 11 big slices which can be separated. An interesting feature of this construction is the possibility to open up the detector and be able to reach any detector component quickly for possible repair.

The tracking detector of CMS consists of a pixel detector with three barrel layers and two endcap wheels on each end. The pixel detector has about 66 million channels and covers an area of about 1 m^2. The radial range from 20 cm to 116 cm is occupied by a silicon strip detector, consisting of several separate barrel and endcap parts. It uses about 200 m^2 active silicon area, has 9.3 million channels and provides ten measurements per track. CMS has the largest silicon tracker ever built and is the only experiment that exclusively uses silicon tracking detectors.

The tracker is surrounded by an electromagnetic calorimeter consisting of about 68 500 crystals made of lead tungstate ($PbWO_4$), providing an energy resolution of about 0.4% for electrons above 75 GeV. In the endcap region a pre-shower detector in front of the electromagnetic calorimeter improves the ability to separate photons from neutral pions. The electromagnetic calorimeter is surrounded by a hadron calorimeter. It is a sampling calorimeter, using brass as absorber and scintillating tiles as active material.

The superconducting coil outside the hadron calorimeter provides a field of 3.8 T parallel to the beam. This is the largest solenoid of this strength that has ever been built. The stored magnetic energy amounts to about 2.65 GJ (for comparison, the energy stored in the ALEPH solenoid amounted to 130 MJ). The magnetic field saturates about 1.5 m of iron, enough to install four layers of muon stations for precise muon tracking and independent momentum measurement. Each muon station consists of several layers of drift tubes in the barrel region and cathode-strip chambers and resistive-plate chambers in the endcaps. The thick iron can only be traversed by muons, making the tracking relatively simple and clean.

13.9
ILD – a Detector Concept for the International Linear Collider

As discussed in the previous chapter, there is a common understanding in the high energy physics community that the LHC data need to be complemented with high-precision data from a lepton collider. The most advanced concept of an electron–positron linear collider in the TeV range is the International Linear Collider (ILC) which would allow e^+e^- collisions at 500–1000 GeV with high luminosities of the order of 2×10^{34} cm^{-2} s^{-1}. The ILC acceleration technology relies on the use of superconducting cavities, which allow for high acceleration efficiencies and excellent beam conditions. The CLIC concept of a normal-conducting linear collider powered with a two-beam acceleration scheme would open perspectives to the multi-TeV energy range. However, the technologies are not as far advanced and require further research and development. While the ILC could be built in close timely relation to the running of the LHC, CLIC would be a project of future decades.

The physics menu of the International Linear Collider promises a plethora of measurements and possible discoveries in the TeV energy range. The LHC will certainly discover the Higgs particle – if it exists – and will give hints of physics beyond the Standard Model. The ILC will then offer the environment for precision studies of the Higgs and of other hypothetical particles for example as predicted in supersymmetry. In addition Standard Model physics like precision top-quark measurements are possible at a linear collider. At the heart of the experimental success of a linear collider research programme is therefore not only the development of a machine design which fulfils all requests on reliable initial-state conditions of the collisions, but also the development of detectors which reliably measure the final states of the colliding particles with adequate accuracies.

13.9.1
Requirements and Rationales

Experiments at the ILC need to match the required precision of the physics programme with their detector technologies. The detectors need to be able to reconstruct multi-jet final states with high accuracy, for example the jet energies need to be measured precisely enough that the hadronic decays of W and Z bosons can be separated. This is usually translated into a jet energy-resolution requirement of $30\%/\sqrt{E}$. New measurement concepts need to be exploited to reach this challenging goal. At ILC energies and taking into account the expected event topologies, the paradigm of particle flow is the most promising concept to reach these challenging goals, see also Section 15.5.

The rationale behind the *particle-flow* idea is to measure the energy of each single particle in a jet, while in more classical approaches the whole jet energy is measured in the calorimeter system. The idea is to always use the most performant detector component for the energy measurement. For instance the tracking system is often much more precise in the measurement of the energies of charged particles than the calorimeters. Therefore its information should be used wherever possible.

In a typical 100-GeV particle jet, about 60% of the energy is carried by charged particles (mostly hadrons), about 30% by photons and roughly 10% by long-lived neutral hadrons. In classical approaches, about 70% of the jet energy would therefore be measured with combined electromagnetic and hadron calorimeters. As the hadronic energy resolution usually is relatively poor, around $60\%/\sqrt{E}$, the envisaged goal for the ILC detectors could not be approached by this. In a particle-flow detector, the energies of all charged particles would be measured with the excellent resolution of the tracking system. The calorimeters would then only be needed to measure the photons (ECAL) and the neutral hadrons (HCAL). The relatively poor energy resolution of a HCAL would therefore be mitigated. However, it is mandatory to measure the four-vectors of as many visible single particles in a jet as possible. It is therefore essential to match the hits in the calorimeters to the tracks in the tracking system. The energy depositions in the calorimeters which can be assigned to tracks would then be subtracted from the calorimeter responses, while the remaining hits would be associated to neutral particles. Of course the big risk is that mis-matching of calorimeter hits to tracks would deteriorate the achievable resolutions. Mandatory ingredients are therefore highly granulated calorimeters (ECAL and HCAL). The concept of particle flow is described in detail for example in [29].

13.9.2
The ILD Detector Concept

Two complementary detector concepts are under study for the International Linear Collider: the "International Large Detector", ILD, and the "Silicon Detector", SiD. Both concepts have submitted Letters of Intent to the ILC Research Director in

13.9 ILD a Detector Concept for the International Linear Collider

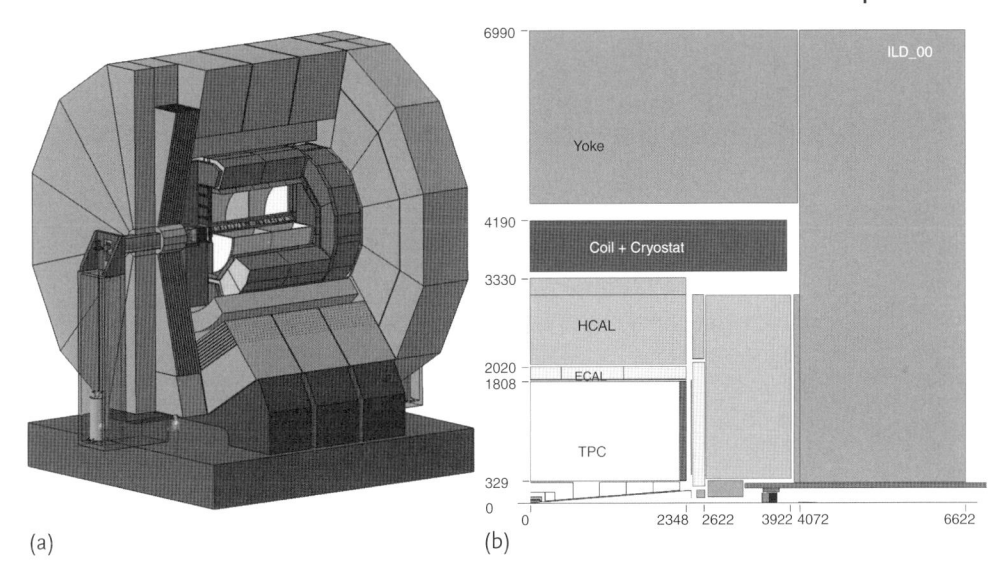

Figure 13.5 (a) Engineering model of the ILD detector; (b) Quadrant view of the ILD subdetector systems. The radii are given in millimetre (adapted from [30]).

2009 which were evaluated and finally validated by an international panel of detector experts, IDAG (International Detector Advisory Group). Both detector concepts are evolving to a detailed technical design which will accompany the Technical Design Report of the ILC, a construction proposal to be published in 2012.

While the nature of the envisaged physics programme suggests similar solutions in both detector concepts, SiD and ILD are complementary in their key design features. ILD has chosen a relatively large detector design (hence the name), whereas SiD is more compact and has a higher magnetic field for compensation. While SiD is aiming for a tracking system of all-silicon detectors, ILD has chosen a large time projection chamber (TPC) as the main tracking device. As both detector concepts are following the particle-flow idea, they rely on highly granular sampling calorimeters, following either an analogue or digital paradigm. In an analogue calorimeter the deposited energies are measured in sampling layers; in a digital calorimeter only the number of hits on the layers are counted. If the readout system is sufficiently granular, the pure number of hits is a good measure for the energy loss of the particle shower in the sampling layer.

For illustration the ILD detector concept is discussed in a bit more detail in this section. ILD is designed as a multi-purpose detector with high precision track and energy measurements over a large solid angle. A 3D model and a quadrant view of the ILD detector are shown in Figure 13.5.

The main subcomponents of the ILD concept are:

- a high-precision pixel vertex detector (VTX) comprising three double layers close to the beam-pipe;

- an intermediate silicon-based tracking system of two cylindrical silicon (SI) strip layers (SiT) and seven SI pixel and strip forward disks (FTD) in each direction for forward-angle coverage;
- a large volume time projection chamber with up to 224 3D points per track and minimum material in the end-plates;
- an outer tracking system (SET) of silicon strip detectors surrounding the TPC which provides a high-precision entry point to the ECAL and improves the overall momentum resolution;
- a highly segmented ECAL (either silicon-tungsten or scintillator-tungsten) with up to 30 samples in depth and small cell sizes;
- a highly segmented HCAL, either analogue with $3 \times 3\,\text{cm}^2$ scintillator tiles or (semi-)digital with a gas-based readout system with $1 \times 1\,\text{cm}^2$ cell geometries;
- forward calorimeters for the determination of the luminosity via Bhabha scattering (LumiCal), for the determination of the beam parameters from pair-radiation background (Beam Cal) and a low-angle hadron calorimeter (LHCAL). The forward calorimeters provide also hermiticity at very low polar angles (below 5 mrad);
- a superconducting solenoid magnet providing a magnetic field of 3.5 T;
- an iron return yoke instrumented to serve as a muon detector and tail catcher; the yoke needs to confine the magnetic fringe fields outside of the detector to minimum levels as the second ILC detector would reside in the same experimental hall (cf. Section 13.9.3).

13.9.3
The Challenge of Push–Pull Operations

In contrast to experiments at storage rings like the LHC, in linear colliders the total integrated luminosity does not scale with the number of interaction regions. As each accelerated bunch can only be used once and is disposed of in the beam dump afterwards, a second interaction region does not *per se* increase the physics reach of the machine. Nevertheless, it is a common understanding that the beneficial effects from having two experiments that are run by two collaborations are large. Obvious effects come from the use of complementary detectors which might be optimised following different design paradigms. Also the overall reduction of systematic effects in the combination of results from mostly independent experiments is a benefit. In addition "soft factors" like the competition between two collaborations running their own detectors are known to be relevant.

The obvious solution for the realisation of two interaction regions at a linear collider is the installation of two independent interaction beamlines that originate from a common beam switchyard at the exit of the linear accelerators. Unfortunately the interaction beamlines at TeV energy linear colliders are complicated systems as they need to accommodate the full beam delivery system with the focusing optics, beam diagnostics and collimation systems. The two beam delivery systems at the ILC have a length of about 2.2 km each and are complicated and costly devices. Therefore the ILC design tries to combine cost awareness with the benefits of having two independent detectors. In this design two detectors share one interaction beamline in a push–pull configuration[34].

The success of push–pull operations relies on a fast, but safe and reliable, movement system. The turn-over times for the exchange of the two detectors need to be small in comparison to the total operation time of the machine, because otherwise too much integrated luminosity would be lost. Assuming data-taking periods of the order of one month per detector, the luminosity-to-luminosity transition needs to be done in the order of a few days. The biggest challenge is the safe movement and subsequent alignment of the $\approx$ 15 000 t massive detectors. Some experience could be gained from the movement of existing detectors. If properly designed, the movement itself should be feasible in less than one day. This would include the disconnection of the vacuum beam-pipes with automated valves, the de-energising of the detector solenoid magnet, and the mechanical movement of the detector together with its nearby services for example on a movable concrete platform. It is planned to keep the detector connected to as many services (e. g. cryogenics supply, high-voltage and data cables) as possible via the use of flexible cable chains. Additional time will, however, be needed for the detector that moves back into the beam position for alignment and eventual re-calibration before the data-taking can start. Interferometric laser positioning systems will be part of the integral detector design for the monitoring and re-alignment of the detector components.

Also for the machine itself, a push–pull system is a challenge of its own. Especially in the region of the machine–detector interface new concepts need to be found on how to reliably establish fast luminosity-to-luminosity transitions. Most important for a fast establishment of the luminosity will be the reproducible alignment of the final focus quadrupoles. As their focal length is typically in the range of only a few metres, they are integral parts of the detectors themselves. So both detectors will carry their own quadrupole magnets which need to be mechanically aligned to orders of some tens of μm with respect to the static beam magnets. The beam-based alignment procedures and the feedback systems of the beam delivery system then need to control the alignment of the quadrupoles to better than ± 200 nm during the luminosity runs.

As the technical design studies for a push–pull system are ongoing and have resulted in no show-stoppers for a realistic design so far, the real challenge of a push–pull system lies in sociological issues. A reliable common understanding of the two collaborations needs to be found about how to share the beam time. If one assumes that the linear collider would run in "discovery mode" at the highest energies, no experiment should be allowed to gain enough luminosity in one run to make a discovery of new physics alone. Under these conditions one would probably decide to choose a high push–pull frequency. However, if the physics programme foresees a phase of precision measurements with a low "risk" of new discoveries, one could allow each experiment to be longer in the beam, without giving it too much of an unfair advantage over the other detector.

---

34) This notion is also followed in the current developments of the CLIC baseline.

References

1. ALEPH Collab., Decamp, D. *et al.* (1990) ALEPH: A detector for electron–positron annihilations at LEP, *Nucl. Instrum. Meth. A*, **294**, 121, doi:10.1016/0168-9002(90)91831-U.
2. ALEPH Collab., Buskulic, D. *et al.* (1995) Performance of the ALEPH detector at LEP, *Nucl. Instrum. Meth. A*, **360**, 481, doi:10.1016/0168-9002(95)00138-7.
3. Creanza, D. *et al.* (1998) The new ALEPH silicon vertex detector, *Nucl. Instrum. Meth. A*, **409**, 157, doi:10.1016/S0168-9002(97)91255-9.
4. DELPHI Collab., Aarnio, P.A. *et al.* (1991) The DELPHI detector at LEP, *Nucl. Instrum. Meth. A*, **303**, 233, doi:10.1016/0168-9002(91)90793-P.
5. DELPHI Collab., Abreu, P. *et al.* (1996) Performance of the DELPHI detector, *Nucl. Instrum. Meth. A*, **378**, 57, doi:10.1016/0168-9002(96)00463-9.
6. L3 Collab., Adeva, B. *et al.* (1990) The construction of the L3 experiment, *Nucl. Instrum. Meth. A*, **289**, 35, doi:10.1016/0168-9002(90)90250-A.
7. L3 SMD Collab., Acciarri, M. *et al.* (1994) The L3 silicon microvertex detector, *Nucl. Instrum. Meth. A*, **351**, 300, doi:10.1016/0168-9002(94)91357-9.
8. ALICE Collab., Aamodt, K. *et al.* (2008) The ALICE experiment at the CERN LHC, *JINST*, **3**, S08002, doi:10.1088/1748-0221/3/08/S08002.
9. OPAL Collab., Ahmet, K. *et al.* (1991) The OPAL detector at LEP, *Nucl. Instrum. Meth. A*, **305**, 275, doi:10.1016/0168-9002(91)90547-4.
10. OPAL Collab., Allport, P.P. *et al.* (1993) The OPAL silicon microvertex detector, *Nucl. Instrum. Meth. A*, **324**, 34, doi:10.1016/0168-9002(93)90964-J.
11. OPAL Collab., Allport, P.P. *et al.* (1994) The OPAL silicon strip microvertex detector with two coordinate readout, *Nucl. Instrum. Meth. A*, **346**, 476, doi:10.1016/0168-9002(94)90583-5.
12. HERMES Collab., Ackerstaff, K. *et al.* (1998) The HERMES Spectrometer, *Nucl. Instrum. Meth. A*, **417**, 230, doi:10.1016/S0168-9002(98)00769-4.
13. HERA-B Collab., Hartouni, E. *et al.*, HERA-B Design Report, DESY-PRC-95-01.
14. H1 Collab., Abt, I. *et al.* (1997) The H1 detector at HERA, *Nucl. Instrum. Meth. A*, **386**, 310, doi:10.1016/S0168-9002(96)00893-5.
15. Pitzl, D. *et al.* (2000) The H1 silicon vertex detector, *Nucl. Instrum. Meth. A*, **454**, 334, doi:10.1016/S0168-9002(00)00488-5.
16. List, B. (2001) The H1 central silicon tracker, *Nucl. Instrum. Meth. A*, **501**, 49, doi:10.1016/S0168-9002(02)02009-0.
17. ZEUS Collab., Holm, U. *et al.*, ZEUS Status Report 1993, http://www-zeus.desy.de/bluebook/bluebook.html.
18. ZEUS Collab., Polini, A. *et al.* (2007) The design and performance of the ZEUS Micro Vertex detector, *Nucl. Instrum. Meth. A*, **581**, 656, doi:10.1016/j.nima.2007.08.167.
19. CDF Collab., Abe, F. *et al.* (1988) The CDF detector: an overview, *Nucl. Instr. Meth. A*, **271**, 387, doi:10.1016/0168-9002(88)90298-7.
20. CDF Collab., Abulencia, A. *et al.* (2007) Measurements of inclusive W and Z cross sections in $p\overline{p}$ collisions at $\sqrt{s} = 1.96$ TeV, *J. Phys. G*, **34**, 2457, doi:10.1088/0954-3899/34/12/001.
21. DØ Collab., Abachi, S. *et al.* (1994) The DØ detector, *Nucl. Instrum. Meth. A*, **338**, 185, doi:10.1016/0168-9002(94)91312-9.
22. DØ Collab., Abazov, V.M. *et al.* (2006) The upgraded DØ detector, *Nucl. Instrum. Meth. A*, **565**, 463, doi:10.1016/j.nima.2006.05.248.
23. Angstadt, R. *et al.* (2009) The Layer 0 Inner Silicon Detector of the DØ Experiment, http://arxiv.org/abs/0911.2522.
24. LHCb Collab., Alves, A.A. *et al.* (2008) The LHCb detector at the LHC, *JINST*, **3**, S08005, doi:10.1088/1748-0221/3/08/S08005.
25. TOTEM Collab., Anelli, G. *et al.* (2008) The TOTEM experiment at the CERN Large Hadron Collider, *JINST*, **3**, S08007, doi:10.1088/1748-0221/3/08/S08007.

26 LHCf Collab., Adriani, O. *et al.* (2008) The LHCf detector at the CERN Large Hadron Collider, *JINST*, **3**, S08006, doi:10.1088/1748-0221/3/08/S08006.

27 ATLAS Collab., Aad, G. *et al.* (2008) The ATLAS Experiment at the CERN Large Hadron Collider, *JINST*, **3**, S08003, doi:10.1088/1748-0221/3/08/S08003.

28 CMS Collab., Adolphi, R. *et al.* (2008) The CMS experiment at the CERN LHC, *JINST*, **3**, S08004, doi:10.1088/1748-0221/3/08/S08004.

29 Thomson, M.A. (2009) Particle flow calorimetry and the pandoraPFA algorithm, *Nucl. Instrum. Meth. A*, **611**, 25, doi:10.1016/j.nima.2009.09.009.

30 The ILD Concept Group (2009) The International Large Detector – Letter of Intent, DESY-2009-87, KEK 2009-6. http://arxiv.org/abs/1006.3396.

14
Tracking Detectors: Following the Charges

Jörn Große-Knetter, Rainer Mankel and Christoph Rembser

14.1
Introduction

What makes up the beauty of an event display, the visualised collision of two high energy particles? What is it that gives us a glimpse of what has happened shortly after the Big Bang, of the production and the decay of heavy (and hopefully up to now undiscovered) particles? It is the tracks of particles painting lines on our screens, showing particle jets, beautiful patterns of particles spraying away from the interaction point, revealing details like the decay of lambda particles or kaons, showing the creation of an electron–positron pair by a photon or hadronic interactions of particles with the detector material.

For more than a century gaseous detectors have been used by physicists to detect and to study ionising radiation, starting with H. Geiger in 1908 with his cylindrical single-wire counter. Nowadays gaseous detectors with several hundred thousand individual channels record the paths of particles and allow us to follow the tracks over several tens of meters with amazing precision of 50 µm. The detectors make use of ionisation signals left by charged particles when traversing matter. Position-sensitive detection of such signals allows individual space points to be determined along the particle track. In recent history, semiconductor detectors complemented gaseous detectors in their complex task. They are often found in the innermost layers of a tracking detector, which are most crucial for vertex reconstruction. In outer layers, both gaseous and semiconductor detectors are found. A typical layout of a tracking detector is shown in Figure 14.1. The measured points along a track are indicated, illustrating the way track reconstruction works, which can often even be seen by eye.

In this chapter these two different categories are discussed, gaseous detectors in the first section and semiconductor detectors in the second section. The remaining sections address software reconstruction methods in three different aspects: the track reconstruction deals with all aspects of determining the trajectories of the particles and estimating their kinematic properties. The quality of this reconstruction depends crucially on an accurate determination of the geometrical alignment

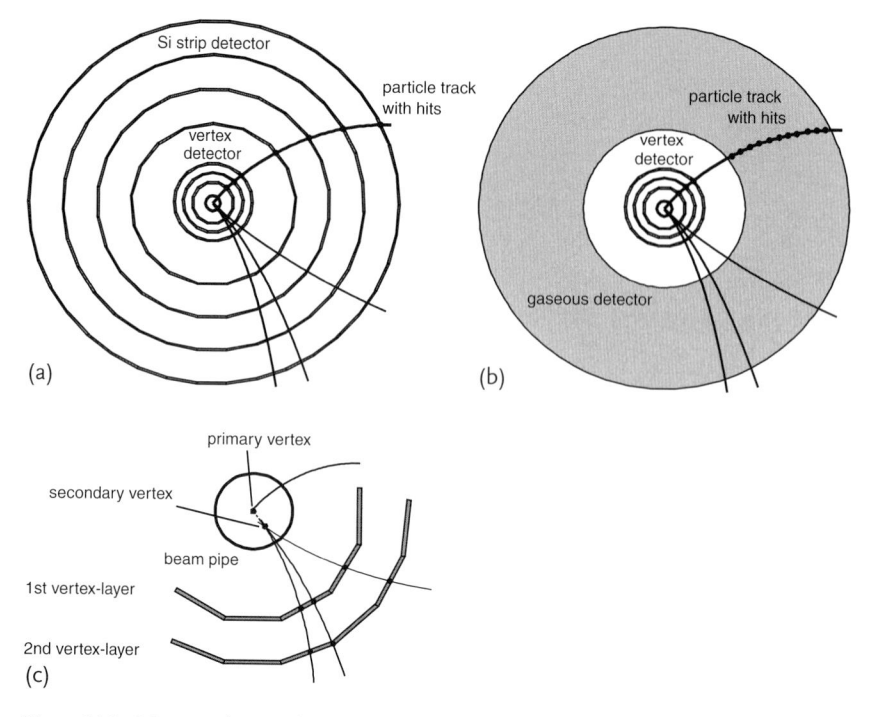

Figure 14.1 Schematic layout of a modern tracking detector with a track and its hits in the detector material being indicated. (a) Sketch of an all-silicon detector, where the vertex detector is composed of pixelated silicon tiles and the outer tracking layers consist of strip silicon detectors. (b) Sketch of a silicon pixel vertex detector combined with a large-volume gaseous tracking detector in the outer part. (c) Zoom-in of the vertex region, illustrating the importance of good resolution close to the interaction point (primary vertex) in order to extrapolate tracks back to a possible decay vertex (secondary vertex).

of the detector elements. Finally, the tagging of b flavour makes use of lifetime signatures that are detectable with high-resolution vertex detectors.

14.2
Gaseous Detectors

14.2.1
Working Principle

The working principle of all gaseous and solid-state detectors is similar. Particles traversing matter leave excited atoms, electron–ion pairs (in gaseous detectors) or electron–hole pairs (in solid-state detectors) behind. Excited atoms emit photons which in transparent materials can be detected with photon detectors such as photomultipliers or semiconductor photodetectors. Ionisation electrons and ions are guided by an electrostatic field in the detector volume to metal electrodes (pads or wires) where they induce electric signals. These signals are then read out by dedicated electronics.

The noise of electronic amplifiers determines whether the signal can be registered. Very good recent amplifiers need a charge signal of about 50 electrons at the input to produce an output signal which is equal to the noise: the *Equivalent Noise Charge* (ENC) is 50 e^-. Typical amplifiers used in current experiments at the Terascale have an ENC of 1000 e^-. In order for the signal to be registered the input charge has to be much larger than the ENC. About 100 electron–ion pairs per cm path length are typically produced in gaseous detectors. This is small compared to the signal of a solid-state detector with a density about 100 times higher and an ionisation energy which is 5 to 10 times lower. Thus, gaseous detectors need an internal amplification in order to be sensitive to a single particle track.

An internal electron multiplication is reached with a simple trick: at a sufficiently high electric field of roughly 100 kV/cm an electron gains energy in addition to the ionisation energy which leads to further ionisation of the gas atoms. The secondary ionisation electrons again gain energy, ionise other atoms – an avalanche has started and sufficiently many electrons reach the amplifier.

The gain of ionisation electrons dN with respect to the number of primary electrons, N, per unit path length, dx, by the avalanche effect is described by the Townsend coefficient, A, by $dN = N_0 A dx$ or $N(x) = N_0 \exp(Ax)$. The Townsend coefficient is a function of the gas composition, the pressure of the gas inside the detector and other experimental conditions. It describes the amplification, the gas gain of a gaseous detector.

14.2.2
Operation Modes of Gaseous Detectors

The gas gain certainly also depends heavily on the strength of the electrostatic field. Usually detectors are operated at such a field that the internal amplification occurs close to the anode, for example in typical wire chambers close to the wire, which has a diameter of 20–30 μm. In less than a nanosecond the electrons move to the wire surface inducing the characteristic fast rise of the signal, a spike. The ions slowly drift towards the detector cathode inducing a long tail of typically several 100 μs. While the ions drift, they shield the anode, changing the electrostatic field. This is called the *space–charge effect* and has an important effect on the design and performance of the detector depending on the choice and performance of the electrostatic field. Examples of choices for configurations of the electrostatic field for gaseous detectors are shown in Figure 14.2. Figure 14.2a,b shows configurations for wire chambers, while Figure 14.2c–f shows configurations for micro-pattern gas detectors.

Modern trigger chambers are operated at very high electrostatic fields, that is, at very high internal gain. An amplification factor of 10^9 is commonly used to ensure that the induced signal exceeds the ENC threshold of the amplifiers. This high-gain operation mode, because it was also the operation mode of the first gaseous detectors in physics experiments, is called the *Geiger–Müller mode*. If the space charge

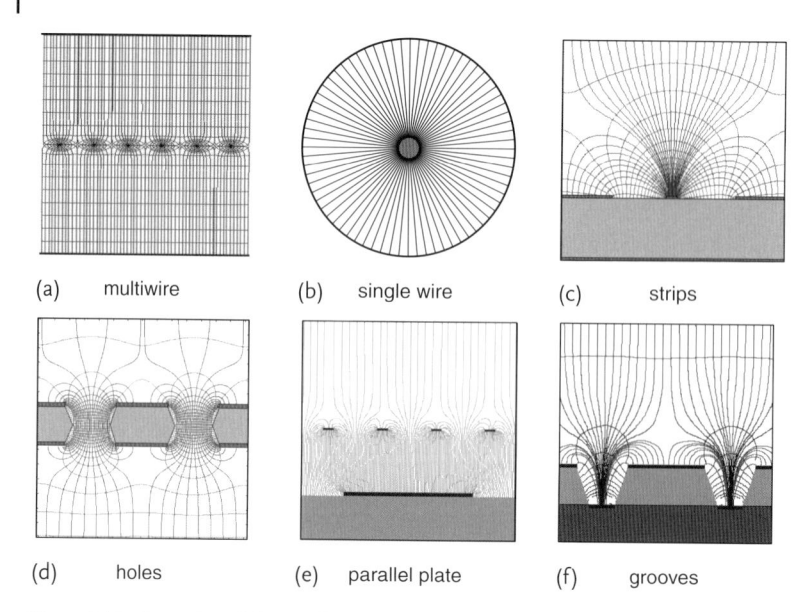

(a) multiwire (b) single wire (c) strips
(d) holes (e) parallel plate (f) grooves

Figure 14.2 Examples of choices for configurations of the electrostatic field for gaseous detectors. (a,b) Configurations for wire chambers. In modern gaseous detectors the wires are replaced by microstructures for which different configurations are shown (c–f) (adapted from [1]).

inside the avalanche is strong enough to shield the external field, a recombination of ions and electrons occurs resulting in photon emission from the atoms. The emitted photons again ionise gas atoms and new avalanches are created. This process can propagate, and once an ion column links anode and cathode a spark discharge occurs. Signals by traversing particles are thus fast and large; this operation mode is thus very useful for trigger detectors. However, the space charge has to disappear before more particle tracks can be recorded accurately under stable conditions – the dead-time will be high if no additional measures are taken. An example for such devices are *resistive-plate chambers* (RPCs) – two parallel electrode plates spaced by a thin gap, with highly resistive plates in front of the electrodes. The resistive plates are made from Bakelite or window glass with a resistance of 10^{10} to 10^{13} Ω cm. A discharge is prevented from propagating through the whole gas because, due to the high resistivity of the electrode, the electric field is suddenly switched off in a limited area around the point where the discharge occurred. In combination with a component of the gas that absorbs ultra-violet photons, further discharges are stopped and more tracks can be recorded. RPCs are used in all current LHC experiments as trigger counters in the outer muon system.

With improving readout electronics, gaseous detectors in big high energy physics experiments in the 1970s were operated with a lower gas gain of about 10^6 to avoid discharges between anode and cathode. However, at these levels of amplification the induced charge is still independent of the number of primary ionisations created by a traversing particle.

It is actually the ability to measure the number of primary ionisations which allows gaseous detectors to contribute more information than just the exact path of a particle and its time with respect to a certain event: measuring the number of primary ionisations or the amount to which the particle ionises the matter tells us about the particle type – whether it is an electron, a pion, a kaon or a heavy and new supersymmetric particle. The amount of ionisation is described by the *Bethe–Bloch formula*, and for momenta of the particles between a few GeV and a few 10 GeV the specific ionisation energy loss per unit of track length differs for various particles. This feature can only be exploited when operating the gaseous detector in the so-called *proportional mode*, which is reached at gas amplifications of 10^3 to 10^4 (in a restricted way also at 10^4 to 10^5, the semi-proportional region where space-charge effects already have an influence on the measurement).

14.2.3
Gas Mixtures in Gaseous Detectors

In principle all gases and mixtures could be used for generating avalanches if the electric field near the sensors is strong enough. However, depending on the operation mode and the intended use of the detector (e.g. fast response and high gain for trigger purposes, good signal proportionality for particle identification) the choice of the appropriate mixture is limited.

In noble gases multiplication occurs at lower fields than in gases with complex molecules. Thus, noble gases are usually the main component of detector gas. To avoid that the detector enters into the regime of discharges, which is caused by recombination of ions with electrons and the resulting emission of photons which themselves ionise the gas and create new avalanches, *quench gases* are added to absorb the photons in the relevant energy range. These quenchers are usually polyatomic gases, organic compounds from the hydrocarbon or alcohol families. A classical gas mixture for proportional counters is, for example, 90% Ar + 10% CH_4.

A very important property of the gas mixture used is the velocity with which electrons and ions drift towards the electrodes. Good drift properties allow gaseous detectors to significantly improve the position resolution of the particle tracks, meaning the determination of the exact position where the particle crossed the chamber. By measuring the difference when the signal arrives at the electronics with respect to an event trigger, for example the time of a particle collision in a collider experiment, the position of the ionisation can be determined if the drift velocity is known. Using this feature, modern gaseous detectors reach a position resolution of about 100 μm even though the segmentation of the electrodes is of the order of a few tens of mm or even larger. The choice of the appropriate mixture is delicate. The drift velocity should be high to avoid a large dead-time, important for experiments like those at the LHC with high counting rates, but at the same time it should be slow enough to minimise the effects of timing errors on the position resolution. The gas mixture should also guarantee small diffusion of the ionisation electrons along their drift to the electrode, also in order to minimise errors on the resolution.

Different requirements apply to chambers with long drift time; they include particularly good drift properties: gas purity is important, and special attention must be paid to the drift velocity. If the chamber is to operate at high counting rates, the drift velocity should be high to avoid losses due to dead-time. Additives to the gas mixture which ensure a good position resolution are gases like dimethylether (DME) or CO_2, a typical mixture (used in the ALICE time projection chamber) is 90% Ne + 10% CO_2. The drift velocity depends on the electric field as well as on the composition of the gas mixture. Typical values chosen are 5 to 10 cm/µs.

A type of gaseous detector which brings the use of the drift properties in gases to perfection is the *time projection chamber* (TPC). In a large drift distance of up to 250 cm electrons drift to the sensor electrode arranged on one endface. There the timing and spatial information of the signals induced in pads, strips or wires can be used to obtain three-dimensional information on the particle tracks. In collider experiments that operate their tracking detectors within magnetic fields to determine the momentum and charge of the traversing particles, the drift direction is often along the magnetic field lines, thus reducing the diffusion of the drifting electrons. Long drift time and the difficulty of shaping the field are drawbacks as space charge builds up, and inhomogeneities in the field can cause serious degradation of the precision. To compensate for this, ion-stopping grids (so-called *gates*) and a careful tuning of the drift field, sometimes by an additional potential wire plane, are introduced.

Many more details about gaseous detectors, details about their working principle and examples of the various designs can be found for example in [2] or in [3]. An exhaustive description of the ionisation process and the transport of electrons and ions in gases under various experimental conditions is given in [4]. In recent years a lot of progress in the development of new types of gaseous detectors has been made. Articles like [5] summarise what has been achieved (and presented at conferences), describe state-of-the-art detectors and give more references.

14.2.4
Gaseous Detectors at Modern Experiments

The compelling scientific goals of physics experiments at the Terascale demand huge research and development efforts for innovative concepts of gaseous detectors in order to deal with the challenges of high collision rates and a harsh environment. The current four big LHC experiments, ALICE, ATLAS, CMS and LHCb all use gaseous detectors for trigger and bunch-crossing tagging (resistive-plate chambers, thin-gap chambers) and high-rate tracking and particle identification (straw-tube detectors, time projection chambers, drift tubes, cathode-strip chambers).

The choice for gaseous detectors of the different types is based on the following advantages and features:

- gas is light, and together with support structures out of modern materials such as carbon fibres the amount of material influencing the measurement of particle tracks and properties is minimal;

- gases are comparatively cheap, they can easily be exchanged and do not suffer radiation damage;
- gaseous detectors have little sensitivity to gamma particles and neutrons and thus do not suffer from background radiation;
- through the application of strong electric fields, noiseless internal gas amplification is possible. Thus, power to the signal amplifiers can be reduced, thus minimising the required cooling infrastructure;
- gaseous detectors do not need to be operated in a cooled environment, which again reduces the complexity of the detector infrastructure.

A consequence of these advantages is that large-area detectors can be built, a mandatory requirement for the muon systems, for example, which include trigger chambers like RPCs and standalone precision trackers like drift tubes. At the LHC, the RPC systems of all four detectors cover an area of more than $15\,000\,m^2$, a challenge which other detector technologies for various reasons would not master. For future experiments with higher event rates, even harsher radiation environment and the demand of even better track resolution gaseous detectors face a number of difficulties:

- operation at high gas gains demands high electric fields. Discharges can destroy the detector;
- under high radiation the detector and the materials suffer ageing, leading to degraded detector performance or even to the destruction of a detector.

With the intensive research and development programme for the LHC detectors, which began even before 1990, world-wide coherent and longer-term approaches started to address the above difficulties. Often the efforts were organised in own collaborations like the so-called RD (for Research & Development) experiments. An example is the RD32 collaboration that developed the ALICE TPC from 1992 until 2004. We now know much more about detector ageing where measurement campaigns in test-beams and irradiation facilities tested and studied the impact of radiation on materials in high electrical fields. All materials for the LHC detectors were tested. Also many campaigns to study detector gases and gas mixtures were carried out, leading to a much better qualitative and quantitative knowledge about the transport of electrons and ions in gases. As a consequence the simulation of gaseous detectors has significantly improved and modern simulation programs allow the performance of detectors with new geometries and gas mixtures to be studied.

Next to these developments and efforts another kind of revolution is much more important and will shape the design of all gaseous detectors used at future experiments at the Terascale: modern photolithography. The advance of this technology has enabled a series of inventions of novel detectors: the family of *micro-pattern gas detectors* (MPGD). Although first small-area prototypes were already built and tested about 15 to 20 years ago, the technology is now mature enough to produce large-area detectors at relatively low costs; micro-structures simply replace the wires.

14.2.5
Micro-Pattern Gas Detectors

The first such structure was the *microstrip gas chamber*. In principle the detector resembles a wire chamber, with fine printed strips instead of thin wires. Anode strips are typically of the order of 10 µm thick, spaced by about 200 µm and interleaved with wider (about 80 µm) cathode strips. When an electrostatic field is applied from another cathode plane located a few hundred µm away from the plane of the fine anode and cathode strips, most of the ions created in an avalanche close to the anode drift towards the thicker cathode strips and the short drift path overcomes the space-charge effect. The technology of the microstrip gas chamber pushes the limit of the rate capability to about 10^6 particles per mm^2/s – 2 to 3 orders of magnitude higher than that of multi-wire proportional chambers – and shows that the MPGD technology is very suitable for future high-rate experiments. However, the principle of the microstrip gas chamber also shows new limitations most of which are common to all MPGD devices. Discharges can damage the structures of strips and lead to long-term damages and failures of the detectors.

Other principles of MPGDs are the *gas electron multipliers* (GEM) and the *Micromegas*. Both technologies have in common that they only provide an amplification structure, which is independent of the choice of the readout structure.

A GEM is a copper-clad polyimide foil with a regular pattern of densely spaced holes. Applying a voltage between top and bottom electrodes a dipole field is formed which focuses inside the holes where it is then strong enough for sufficient gas amplification. Typically the polyimide foil has a thickness of 50 µm, diameters of the holes are typically of the order of 70 µm. The distance between the centre of two holes is twice the hole diameter.

Micromegas have a parallel-plate geometry with an amplification gap between a micromesh and the readout board. Special for the Micromegas is the very narrow amplification gap of 50 to 100 µm which ensures fast signals and high rate capability. The micromesh is supported by regularly spaced pillars which maintain the accurate spacing.

To overcome the problem of discharges destroying the detector structure, a simple trick is applied. Two or more layers of the amplification structure are used, spaced by a small amount, for example for multi-layer GEMs by a few mm. Each layer contributes to the gas amplification, allowing the field for each layer to be reduced which avoids any discharges.

Besides their high position resolution and high rate tolerance, MPGDs provide a further advantage: with their very small segmented structures they can use finely segmented chips, identical to the chips of solid-state pixel detectors, as readout structures. These chips allow an ultra-high precision and can provide pictures of particle tracks like an electronic bubble chamber, even resolving the individual ionisations along the particle track. This will allow gaseous detectors to open a window for many new applications, especially for collider experiments at the Terascale, as trigger chambers, detection systems of large volume trackers like a time projection chamber, for the detection of low-energy particles, photons and particle identification.

14.3
Semiconductor Detectors

Tracking and in particular the measurement of primary and secondary vertices requires detector elements close to the interaction point. This results in exceptionally high hit rates and high radiation doses which the detector has to cope with. At the same time, precise momentum and vertex measurement requires a good spatial resolution of the sensing elements and thus a segmentation of the readout electrodes to values below 100 µm. Semiconductor materials are well suited to this due to their fast charge carrier transport, relatively easy processing with photolithographic techniques and their comparatively high radiation tolerance.

Semiconductor tracking detectors are typically assembled of many thin tiles (*sensors*), in almost all cases made of silicon, which themselves must be segmented at the electrodes and implants in order to obtain position-sensitive information. The *pitch* between the segments is closely related to the desired position resolution, which is typically (several) 10 µm. There is a variety of strategies for this aspect with implications for the design of the device.

The most widely used segmentation is that into strips, that is, rectangular-shaped electrodes with the long side being basically as long as the silicon sensor itself (of the order of cm), and the short side being of the order of several 10 µm. This segmentation provides one-dimensional position information of high resolution and is relatively easy to read out. Two-dimensional information about the position at which a particle traverses the detector is obtained by using two layers of strips with the strips being rotated with respect to each other. The advantage of strip detectors is the moderate effort needed for production of individual parts, resulting in lower costs compared to more complex semiconductor detector concepts. They are thus nowadays often used in large-scale tracking detectors such as the ATLAS SCT [6–8] or the CMS strip detector [9, 10]. The disadvantage of strip detectors is their limitation at high rates due to high occupancy in individual channels and ambiguities in hit reconstruction as discussed below.

Segmentation into pixels, that is, electrodes with edges of similar length, allows a two-dimensional position to be measured without any ambiguities, with spatial resolutions similar to that of detectors with crossed strips. By far the most important aspect of the pixel geometry is the reduced hit rate. It goes essentially with the area of the electrode and is thus reduced by several orders of magnitude for pixels with respect to strips. Pixels allow the electronics more time for the readout per channel and thus offer the only possible geometry that is still operational in a high-rate environment like that provided by the LHC close to the beam line. Depending on the actual requirements, two different techniques are currently available. *Hybrids* of sensors and readout electronics offer the fastest possible data processing for the price of more material per detector layer. This design is used in the innermost layers of the ATLAS [8, 11] and CMS [10, 12] tracking devices. In contrast, *active pixel detectors* combine the first part of the readout with the sensor and thus allow a design with less material than hybrid pixel detectors. The readout is done row-by-row and is consequently slower than that of hybrid pixels. Different concepts of active pixel detectors have been proposed for use as vertex detectors at ILC (for a recent summary see for example [13]). Both hybrid and active pixel detectors are discussed in more detail below.

14.3.1
Silicon as Detector Material

The primary process used for detection of charged particles in silicon is ionisation similar to that described for gaseous detectors in Section 14.2.1, in which one electron–hole pair is released on average per $W_{Si} = 3.6\,\text{eV}$ of deposited energy. For example, a *mip*[35] in silicon of $d = 300\,\mu\text{m}$ thickness deposits on average an energy of $\langle \Delta E \rangle \approx 0.1\,\text{MeV}$, corresponding to $\langle \Delta E \rangle / W_{Si} \approx 29 \times 10^3$ electron–hole pairs. For electronics purposes, this charge is sufficient to be processed. However, in order to detect such a charge, the semiconductor must initially be depleted of intrinsic free charge carriers. Also, an external field is needed in order to separate the two charge carrier types, which then induce a signal on the electrodes while drifting towards them.

To achieve this, most silicon detectors consist of a *pn-junction* operated under reverse bias: at the junction, the difference in charge carrier concentration results in a diffusion of the majority charge carriers into the oppositely doped side to recombine with the other charge carriers. This diffusion process is compensated by an electric field caused by the remaining space charge, leaving the immediate vicinity of the pn-junction depleted of any free charge carriers (*depletion zone*) [14, 15].

The largest fraction of the depletion zone extends into the weakly doped material. Therefore, a typical design would use a lightly doped substrate with a highly doped implant in a relatively thin layer on the surface. With N_{eff} being the effective doping concentration[36] of the substrate, an external voltage U_{ext} applied in addition to the built-in potential[37] U_{bi} will increase the depth of the depletion zone to

$$d \approx \sqrt{\frac{2\varepsilon_{Si}\varepsilon_0(U_{ext} + U_{bi})}{e|N_{eff}|}}, \quad (14.1)$$

where ε and ε_0 denote the dielectric constant of the semiconductor and the permittivity of free space, respectively. For typical values in silicon like $N_{eff} = 10^{12}\,\text{cm}^{-3}$ for the bulk and a detector thickness of 300 μm, a bias voltage of $U_{dep} \approx 70\,\text{V}$ is

35) A *mip* refers to a minimum ionising particle, that is, a particle that deposits the minimum possible amount of energy in matter.
36) Because of impurities, actual semiconductors always contain both acceptors and donors, so that commonly the donor concentration N_D and the acceptor concentration N_A are replaced by the effective doping concentration $N_{eff} = N_D - N_A$, being positive for n-type material and negative for p-type material.
37) The potential U_{bi} corresponds to the electric field created by diffusion at the pn-junction.

needed for full depletion, that is, the full deprivation of charge carriers of the entire bulk.

Transport of charge carriers in silicon can proceed through drift or diffusion [14]. In the presence of an electric field, the charges will drift to either electrode of the silicon device, the process usually desired for measuring the created charge (*charge collection*). As soon as the charge carriers begin to drift in the electric field, they induce a current signal on the electrodes following Ramo's theorem [16]. The total measured charge is the full total charge of the drifting carriers, if it is measured throughout the entire drift. The drift of charge carriers can be stopped for example by trapping, an effect which is expected to be caused or enhanced by radiation damage (see Section 14.3.5).

14.3.2
Strip Detectors

In order to obtain position sensitivity, the signal obtained from drifting charge carriers must be measured on segmented electrodes. The easiest way to achieve this is a segmentation of the pn-junction into strips. Effectively, this is realised by implanting strips of highly doped regions into a lightly and oppositely doped bulk. The electrodes on such implants are connected to a charge-sensitive amplifier followed by further processing (e. g. analogue-to-digital conversion), one-by-one, so that the amount of charge seen on each strip is read out. This requires a high density of connections and readout electronics, made possible by wire bonds – few 10-μm thin wires connected to electrodes by means of ultrasonic welding – and modern micro-electronics (*application-specific integrated circuits* or ASICs).

This concept of a strip sensor with ASIC-based readout delivers one-dimensional hit positions within the depth of the sensor. For a track measurement, at least two such layers have to be crossed in order to obtain a two-dimensional position. The fact that two silicon strip sensors are needed can be avoided by double-sided sensor segmentation [15]. Therefore, the material of the detector is reduced, which is beneficial in terms of multiple scattering. For this purpose also the doping of the side opposite to the pn-junction side can be segmented. This side has already a highly doped implant to obtain a good ohmic contact, so segmenting this into strips is relatively little extra effort. However, since this doping and that of the bulk is of the same type and just differs in concentration, the resulting implants are at first not isolated from each other and position-sensitive information will not be obtained. This is avoided by adding small implants of opposite doping type between the strips (e. g. p-implants between high-concentration n-type strips in an n-type bulk) resulting in small additional pn-junctions between each strip pair. The depletion zone around those provides the necessary interstrip isolation.

As straightforward as this concept of two sets of crossed strips is, it becomes problematic at high rates. When more than one particle traverses the detector at a time, a hit is detected in several strips of each orientation, and each pair of strips must be considered a possible two-dimensional hit candidate. From the readout, this can be combined into several possible two-dimensional hit positions. Generally, for n simultaneous hits, n^2 reconstructed hits are obtained out of which $n^2 - n$ are not real ("ghost hits"). This problem is avoided by using crossing angles smaller than 90°, effectively reducing the area in which ghost hits can be created. This comes at the price of reduced resolution in one of the directions. Alternatively, more layers are added under varying crossing angles increasing the amount of material and thus the amount of multiple scattering.

14.3.3
Hybrid Pixel Detectors

The *hybrid pixel detector* design is closely related to the strategy commonly used for strip detectors [16]: each pixel is realised by an individually pixelated implant on a lowly doped substrate. Each pixel is individually connected to a charge-sensitive amplifier followed by a zero suppression such that only signals above a certain threshold are read out. This reduces the amount of data to be handled, making it possible to deal with a large number of pixels even for high collision rates. The threshold of the zero suppression can typically be adjusted individually for each pixel, thus allowing a homogeneous response across the detector to be achieved. The signals which pass discrimination are buffered and sent to an off-detector readout upon a trigger request. This concept requires separate devices for detection and readout which are then connected in a high-density technique called *bump-bonding*. The idea is illustrated in Figure 14.3. Small metal spheres or cylinders are placed on contacts to the pixel cells of the silicon sensor and on the other side on contacts to the pixel cells of the readout ASIC(s). The bump-bonds are placed on the ASIC and/or the sensor using photolithographic methods and (electro-)chemical deposition techniques. ASIC and sensor are then put together with high demands on alignment and, depending on the bump-bond material (indium, silver-tin or lead-

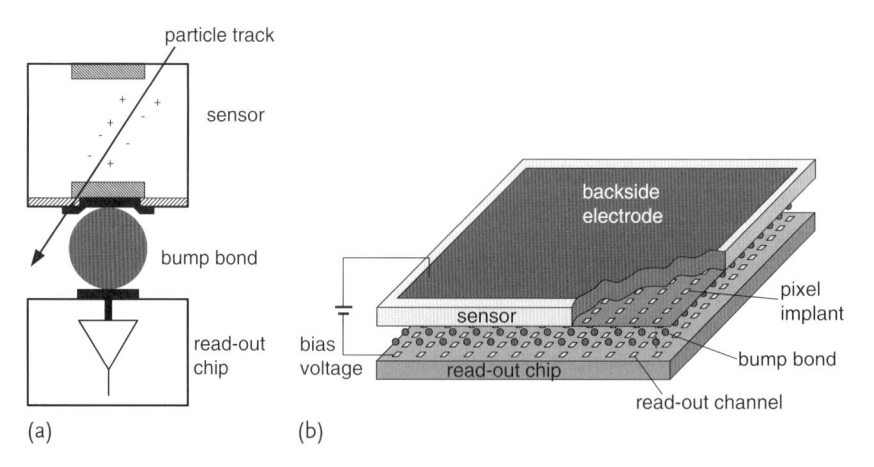

Figure 14.3 Principle of bump-bond connection in hybrid pixel detectors. (a) Cross-sectional view of the vicinity of a bump-bond connection; (b) schematics of a sensor-readout hybrid indicating the large number of bump connections needed.

tin are currently the most common types), soldered together with explicit heating or thermo-compression. Self-adhesion helps in the final alignment of the sensor-ASIC sandwich.

14.3.4
Active Pixel Detectors

In contrast to hybrid pixels with separate detection and readout layers, *active pixels* aim at an integration of sensing and amplification structures in one silicon substrate and have become a serious competitor to CCDs. This kind of integration allows for less material per detecting element and is therefore of interest to detectors which have to minimise the effect of multiple scattering. Typically, also the heat production of such detectors is reduced by a lower power consumption compared to hybrid detectors, therefore also allowing less complicated cooling techniques to be realised, reducing the amount of material in the detector volume further. ILC vertex detectors require a minimum amount of material; apart from CCDs, concepts using *monolithic active pixel sensors* (MAPS) and *depleted field effect transistor* structures (DEPFET) have been proposed [17–19].

MAPS rely entirely on highly doped substrate material as used for CMOS chips [20]. Their design is similar to that of a photodiode, readily available in CMOS technology. The photodiode includes a lightly doped epitaxial layer (epi) of silicon in which the charge carriers released from traversing particles are collected. The epi layer is grown on the substrate and covered by a layer of highly doped silicon, both of which act as a reflective barrier due to the differences in doping concentration. Diffusion in combination with this barrier structure guides the generated charge carriers towards an n-implant in the top p-layer, see Figure 14.4a. The smaller charge signal due to the relatively low thickness of the epi layer (between 12 and 16 µm) compared to hybrid detectors is counter-balanced by lower noise due to the integration of amplification structures. The latter is easily realised since the detecting elements are already based on a CMOS process. Therefore, an acceptable signal-to-noise ratio and thus detection efficiency is achieved.

The DEPFET detector integrates a MOS field effect transistor onto a fully depleted substrate [21] as shown in Figure 14.4b. Additional n-implants near the transistor act as a trap for charge carriers created in the substrate (*internal gate*), so that they are collected beneath the transistor gate. The transistor current is therefore controlled by the amount of collected charge. A clear contact allows the removal of the collected charge. Compared to MAPS, the DEPFET structure uses the (thicker) substrate for charge collection, so that the signal to be detected is larger. The integrated amplification reduces the noise compared to hybrid detectors, so that a thinner substrate with a similar signal-to-noise ratio is possible.

Both MAPS and DEPFET detectors are read out in a fashion similar to CCDs: individual pixels or at most an entire column or row have to be addressed and processed off the detector devices, while all other pixels are not (yet) read out at that moment. A zero suppression is only possible after this readout step, thus making the active pixel strategy slower in readout speed compared to hybrid pixel detectors.

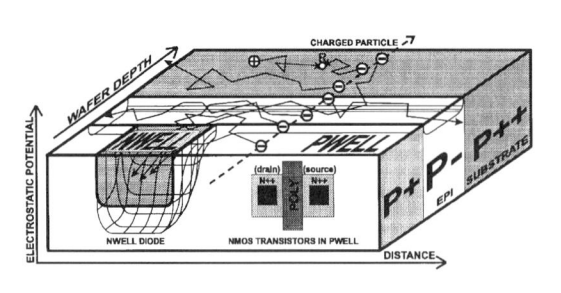

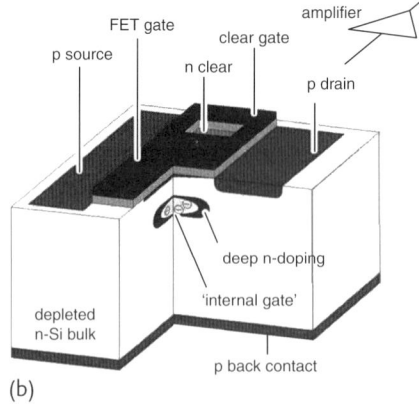

Figure 14.4 (a) Sketch of the internal structure of a pixel for charged particle tracking with MAPS. The circuit's three transistors are integrated in the p-well (p^+) while the charge-collecting element is an n-well diode (n^+) on the p-epitaxial layer. (b) The DEPFET detector and amplification structure is based on a p-channel MOSFET structure on a completely depleted substrate. A deep n-implant forms a potential minimum for electrons which are collected there. The accumulated charge in this internal gate modulates the transistor current. The charge can be removed by the clear contact (adapted from [17, 20], with kind permission of Elsevier).

14.3.5
Radiation Tolerance

Naturally, silicon detector devices suffer from radiation damage due to the particles produced in the collisions, an issue which is particularly important at LHC experiments due to the unprecedented doses in the tracking systems. Radiation damage can occur inside the bulk (*lattice damage*), which is of more importance to the sensor performance, or on the surface, with more importance for the readout electronics.

Surface damage is generally caused by ionising radiation. It has most noticeable effects inside the oxide layer of MOS transistor structures: holes created by the radiation are trapped at the silicon–oxide boundary and attract electrons inside the silicon to the boundary. This leads to parasitically conducting channels, influencing the transistor properties if inside it, or causing shorts if located between transistors. Tunnelling of electrons into the oxide cures the positive oxide charge in a thin layer close to the silicon. Chip technologies with a thin oxide layer are thus intrinsically radiation harder than those with thick oxide layers.

If the incident particle has sufficient energy it can transfer enough energy to a silicon atom to remove it from the lattice. Such transfers are mediated mainly by the Coulomb interaction in case of charged particles or ions and by nuclear forces in case of neutrons. Silicon atoms that were removed and have a sufficient amount of kinetic energy can remove other atoms from the lattice, resulting in a cascade of interactions.

These defects are of importance for the detector properties because such modifications to the ideal semiconductor lattice create additional energy levels in the gap between the valence and conduction bands. Levels in the middle of the gap act mainly as generation and recombination centres, given a similar distance to either band and thus similar access probability for electrons and holes. In contrast, levels close to the valence or conduction band act as trapping centres, that is, electrons or holes are captured from the relevant band and released with a delay. The trapping modifies the effective doping concentration, N_{eff}, since free charge carriers are removed from the material. The combination of donor or acceptor atoms with vacancies or interstitials into stable defects acts in a similar direction. Both result in a change of the effective doping concentration. This has a direct consequence on the depletion depth described by (14.1) and thus on the voltage needed for full depletion. For large radiation doses, the depletion voltage may therefore exceed values permitted for safety reasons (cabling, sparks and so on), typically around several 100 to 1000 V.

The charge trapping also has a consequence for the signal that is measured: the trapped charge carriers are released with a delay of the order of μs, which is much longer than the duration of the pulse created by the un-trapped charge (typically a few ns to a few 10 ns). The trapped charge is therefore not measured as part of the signal, the size of which is consequently smaller than for devices that are not irradiated, reducing the detector performance.

Finally, the generation–recombination centres effectively lead to free electron–hole pairs in addition to those present intrinsically, resulting in an increased volume current. The exponential temperature dependence of the volume current will lead to an avalanche effect: a rise in temperature increases the current, which in turn increases the temperature again, and so on (*thermal runaway*). Therefore sufficient cooling must be provided, with low temperatures also reducing the extent of radiation damage.

Bulk radiation damage diminishes in time after irradiation stops (*annealing*) for example by filling of a vacancy with a silicon atom or by transformation of one stable defect into another, changing the properties of the defect. The progress of annealing depends strongly on the temperature of the device. Given that the second effect can result in larger defects over the long term, the initially beneficial effect (e.g. reduced volume current) can turn into its opposite (reverse annealing). It is therefore beneficial to keep the temperature of the device low in order to slow down the progress of reverse annealing. However, short warm periods may still be desirable to make use of the beneficial annealing.

Various strategies are pursued in order to improve sensor performance after radiation. New materials are being considered which are intrinsically more radiation hard (e.g. diamond) or different types of silicon (different crystal growing, different substrate doping). The problem of charge trapping is specifically addressed by "3D" designs, using columns as electrodes and thereby reducing the drift distance between them. Alternatively, higher fields may yield charge multiplication and thus at least partially recover some of the trapped signal. For a summary of current activities, see [22, 23].

14.4
Track Reconstruction

Tracking detectors measure the kinematic parameters of charged particles by reconstructing the trajectory from the hit information in the sensitive elements traversed. The accuracy of this reconstruction has a large influence on the parameter resolution achieved by the tracking device and thus on its performance for physics analysis. The reconstruction of a charged particle track addresses three main aspects. Firstly, the signals delivered by the detector component are processed to form a *local coordinate*. In solid-state detectors, this *local reconstruction* usually involves identifying clusters of strips or pixels originating from the passage of a particle and estimating the accurate position by the evaluation of the registered pulse heights. In drift chambers the measured drift time of primary charges between the point of ionisation and the anode is converted into a spatial drift distance. The next stage is the *track pattern recognition* in which the available local coordinate measurements caused by the same particle are grouped together. Finally, the *track fit* determines the parameters of the particle's trajectory from the associated local coordinate measurements. In practice, pattern recognition and track fit are often not clearly separated, as for example pattern recognition algorithms may use track-fitting methods internally at some affordable level of accuracy. Track reconstruction is then followed by the *vertex reconstruction*, which associates tracks with their point of origin. Reviews on track reconstruction in particle physics experiments can be found for example in [24–26].

14.4.1
Track Pattern Recognition

Track pattern recognition is a very challenging task under high track densities, and a main consumer of CPU resources in the overall event reconstruction. Figure 14.5 shows a simulated $t\bar{t}$ event in the ATLAS inner tracking system [27, 28]. Various concepts have been developed and implemented in the last decades, and it is beyond the scope of this book to cover the range of technologies. In the last generation of experiments, a widely used approach is based on *arbitrated track following with concurrent evolution of track candidates* [29], which is commonly referred to as the *combinatorial Kalman filter* (CKF).

The first step in a track following approach is the generation of track seeds, made by combining a small number of usually nearby hits sufficient to create a starting set of track parameters. Seeding usually starts at one logical border of the tracking system, frequently in the innermost detector layers that have the highest granularity, and is often complemented by a second seeding pass from the other end. In the second step a seed enters an iterative procedure. In each iteration the track candidate is extrapolated to the next active detector volume, generating a prediction window for the next hit on the trajectory. A compatible hit is added to the trajectory, and the process continues as long as possible until the end of the tracking volume is reached. An ideal underlying mathematical machinery supporting a very effi-

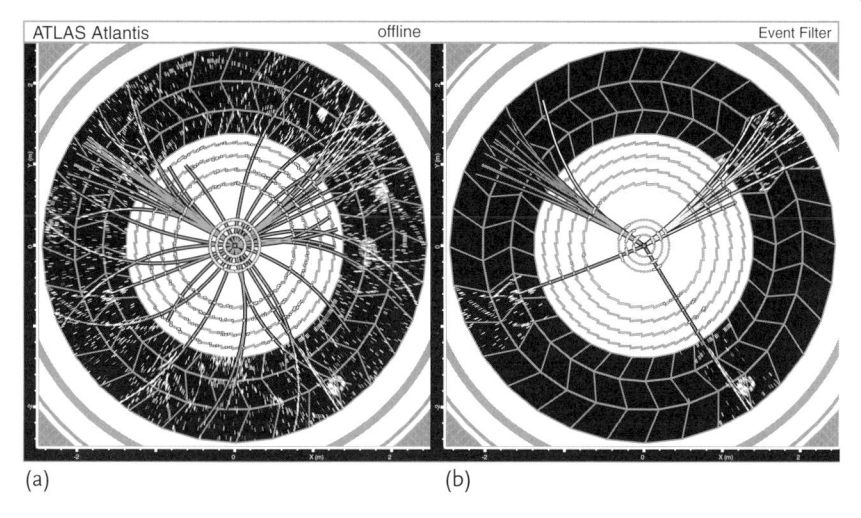

Figure 14.5 (a) Display of reconstructed tracks in a simulated ATLAS event. (b) Same event with display restricted to tracks already found at the level of the online filter (adapted from [27]).

cient implementation is the Kalman filter, which will be discussed in more detail in Section 14.4.2.

The CKF is a sophisticated implementation of track following which addresses ambiguities – the main complication of track pattern recognition in complex events. Ambiguities arise from the passage of several particles through the same detector areas or the presence of noise hits, which can both result in several hits being compatible with the prediction. Also the possibility that the expected hit is missing due to limited efficiency must be considered. Picking up a wrong hit would result in skewing the track parameters with the danger of missing hits further down the trajectory, or removing hits from the pool that would be needed to build other tracks. Forcing an immediate decision would dilute the track purity and reduce the track finding efficiency as the particle density increases. The solution is to defer the decision in case of ambiguities and to build up a tree of track candidates whose branches are followed up concurrently [29]. Each branch is assigned a quality derived typically from the number of good hits, the number of missing hits and the goodness of fit. Combinatorics are controlled by trimming the tree, keeping only a maximum number of branches for further propagation at each step. A schematic view of such a method [29] is shown in Figure 14.6.

14.4.2
Track Fitting

Although pattern recognition often already provides some estimate of track parameters, there is usually a subsequent track fitting step in which the ultimate reconstruction of track parameters is performed. This part can include much more time-consuming procedures than those affordable in the nested loops of the pattern

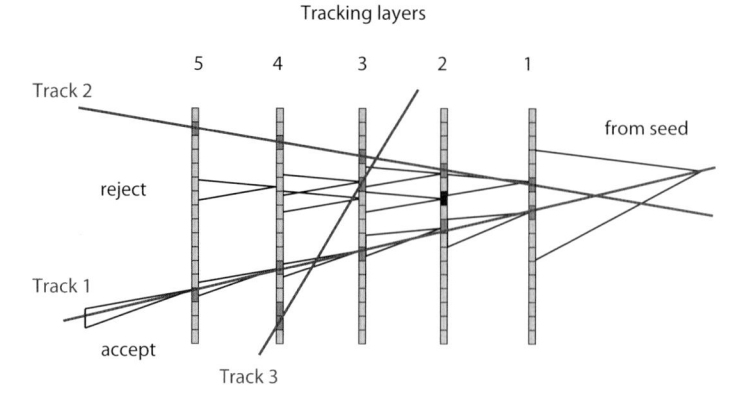

Figure 14.6 Schematic view of pattern recognition with an arbitrated track following method, illustrating the principle of the combinatorial Kalman filter. A seed originating from track 1 is propagated through five symbolic detector layers. Products from two other tracks crossing the same detector area create ambiguities that intermittently give rise to additional fake track candidates, which the algorithm subsequently discards in favour of the correct track (adapted from [29]).

recognition step. The trajectory of a particle consists in general of a deterministic and a stochastic component. The deterministic aspect of the trajectory depends strongly on the magnetic field: it is helical if the field is homogeneous and can be much more complex in inhomogeneous fields. While this part of the trajectory is entirely determined by the equations of motion, material effects like multiple scattering or energy loss give rise to stochastic components. Since many detectors operating at the Terascale feature considerable amounts of material in the tracking area, it is essential to account properly for stochastic effects in order to obtain an unbiased estimation of the track parameters and, very importantly, their covariance matrix.

A widely used track-fitting algorithm in experiments at the Terascale is the Kalman filter [30], which is (in a nutshell) a progressive method of performing a least-squares fit. Starting from an initial seed, the method proceeds iteratively by first predicting the trajectory state at the detector element that contains the next local coordinate measurement, and subsequently updating the trajectory state by adding the information of this measurement. This step is called *filtering*. In this way, the track parameters become increasingly more accurate as more and more hits are added. The Kalman filter also provides a very elegant way of treating stochastic perturbations, like multiple scattering or energy loss, as *process noise* without introducing explicitly additional free parameters. Finally, a *smoothing* step can be executed to propagate the full information of the track fit to all intermediate points on the trajectory.

Various extensions of the Kalman filter are in use to address particular challenges relevant for Terascale detectors. The *deterministic annealing filter* (DAF) addresses the problem of outlier hits by a down-weighting technique, improving the robustness beyond the simple least-squares estimator. The *Gaussian sum filter* (GSF) [31] is able to deal with profoundly non-Gaussian random variables, which makes it an attractive solution to address radiative energy loss of electrons, at a significant expense of computing power.

14.5 Alignment

The full physics performance of tracking detectors depends on the achieved spatial resolution, in particular near the vertex region where it is common to target accuracies in the order of 10 µm and less. The essential element in achieving this level of accuracy is the precise alignment of the detector modules. Beyond the initial mounting precision, surveys *in situ* provide the first correction of module positions and orientations. Many modern detectors are equipped with hardware-based alignment systems, largely based on lasers and digital sensors, which have the potential advantage of measuring movements in real-time. Especially in the vertex region, however, the ultimate precision must be achieved using the reconstructed tracks themselves. The track model itself, for example the helix in a homogeneous magnetic field, is used as a constraint. The distance of a measured hit position from the fitted trajectory is called *hit residual*. Systematic deviations of residuals of a certain detector module from zero can be indicative of misalignment. Since the track used as a probe is reconstructed from hits in other modules, which in turn have their own misalignment, a formidable coupled mathematical problem needs to be solved.

Among many varieties, there are two structurally different methodologies commonly used to address this coupled problem. In the *local method*, the set of residuals from a sample of tracks passing a certain detector module enters a χ^2 estimator which is a function of the alignment parameters of this module. Solving for minimal χ^2 gives an estimate of the alignment parameters. While this can be done for all modules within the detector, the tracks need to be refitted due to the change in geometry, which in turn may result in a new set of residuals. The procedure is repeated until convergence is achieved, which can require a very sizable number of iterations depending on the complexity of the problem. The *global method*, on the other hand, aims to solve the whole problem simultaneously with at most a small number of iterations. For a complex detector, this can involve the determination of the order of 100 000 alignment parameters together with millions of track parameters in the same fit. The Millepede program[38] is specifically designed for solving this kind of problem.

In the commissioning phase of the LHC experiments, cosmic rays have been used to align the detectors. While cosmic muons provide very useful track signatures connecting opposite sides of the detectors, the geometry in underground caverns tends to be less favourable for vertically mounted detector planes. Figure 14.7 shows results from aligning the CMS pixel tracker with cosmic ray events [32]. The

---

38) https://www.wiki.terascale.de/index.php/Millepede_II

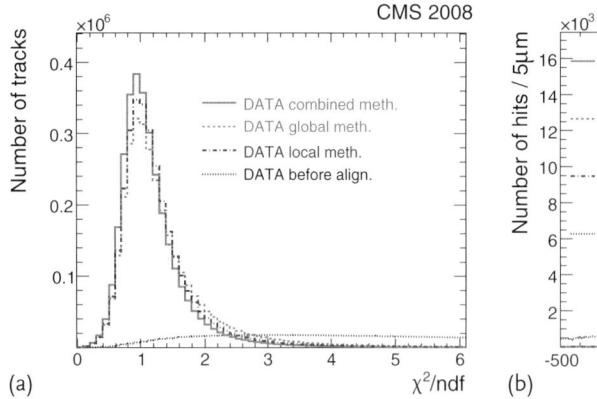

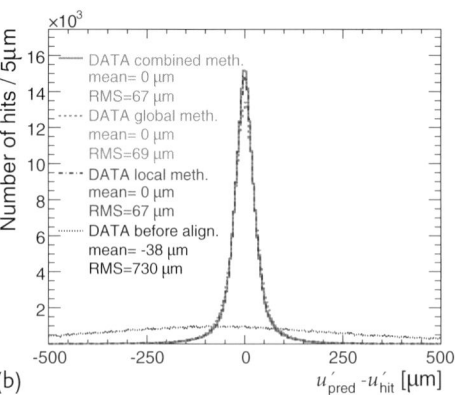

Figure 14.7 Alignment of the CMS tracker modules with cosmic rays using a combination of the local and the global method. (a) Goodness of fit before alignment and after alignment with the global and local method and a combination of both; (b) Hit residuals in the tangential coordinate of the pixel barrel tracker before and after alignment (adapted from [32]).

histogram in Figure 14.7a shows the improvement of the goodness of fit due to the alignment with different methods. Both the local and the global methods give very good results, and the best performance is achieved by a combination of both methods. The histogram in Figure 14.7b shows the hit residuals in the tangential measurement coordinate of the pixel barrel detector. The originally very wide residual distribution in the unaligned case is narrowed down by more than a factor of ten by the alignment procedure and comes very close to the ideally expected residual resolution, which is dominated by multiple scattering at the typical momenta of cosmic muons in the CMS underground cavern.

14.6
Tagging of Heavy Flavours

A key application of high-resolution tracking detectors is the tagging of heavy-flavour decays, in particular of b quarks. Within a mean lifetime of about 1.5 ps, b quarks confined in energetic hadrons can travel distances that modern vertex detectors are capable of resolving if their sensors are close to the interaction point and their spatial resolution is sufficiently high.

The major signatures directly related to the b lifetime are secondary vertices and impact parameters. These can be combined with other features of heavy-flavour decays related to the large b-quark mass, which reveals itself for example in an excess of the mean particle momentum transverse to the jet axis and increased track multiplicities. Since the spatial resolution is crucial, the achieved level of detector alignment (see Section 14.5) is the decisive factor for the final performance. The significance of each lifetime-related observable depends directly on the resolution

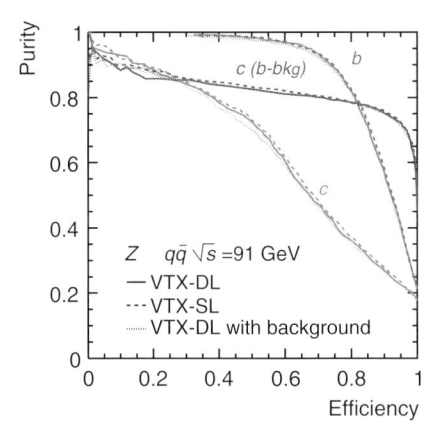

Figure 14.8 Flavour-tagging performance of the ILD detector in vertex detector scenarios with three double-sided ladders (VTX-DL) and five single-sided ladders (VTX-SL) for $Z \to q\bar{q}$ events at 91 GeV (adapted from [33]).

in the respective event topology. In order to compute the resolution reliably, not only the intrinsic detector resolution but also the remaining alignment uncertainty must be known accurately, which represents a particular challenge.

Heavy-flavour tagging performance is also a major aspect of designing new detectors, for example for future linear colliders. Besides the spatial resolution also the material in the innermost detector area is essential, since multiple scattering will dilute the resolution of lifetime signatures. Figure 14.8 shows results from a performance study for the ILD detector design for the International Linear Collider (see Section 13.9), where an artificial neural network has been trained to combine the various input observables in an optimal way. It shows that the studied detector designs are already highly optimised to achieve a high purity of b-tagged events at relatively moderate expense in terms of efficiency.

References

1 Duarte Pinto, S. (2009) Contribution to the Proceedings of the XLVII International Winter Meeting on Nuclear Physics, Bormio, 2009.
2 Blum, W., Riegler, W., and Rolandi, L. (2008) *Particle Detection with Drift Chambers*, Springer Verlag, Berlin, Heidelberg.
3 Grupen, C. and Schwartz, B. (2008) *Particle Detectors*, 2nd edn., Cambridge Univ. Press, Cambridge.
4 Sitár, B. et al. (1997) *Ionization measurements in high energy physics*, Springer Verlag, Berlin, Heidelberg.
5 Titov, M. (2007) New developments and future perspectives of gaseous detectors, Nucl. Instrum. Meth. A, **581**, 25, doi:10.1016/j.nima.2007.07.022.
6 Abdesselam, A. et al. (2006) The barrel modules of the ATLAS semiconductor tracker, Nucl. Instrum. Meth. A, **568**, 642, doi:10.1016/j.nima.2006.08.036.
7 Abdesselam, A. et al. (2007) The ATLAS semiconductor tracker end-cap module, Nucl. Instrum. Meth. A, **575**, 353, doi:10.1016/j.nima.2007.02.019.
8 ATLAS Collaboration (2008) The ATLAS Experiment at the CERN Large

Hadron Collider, *JINST*, **3**, S08003, doi:10.1088/1748-0221/3/08/S08003.

9 CMS Collaboration (1998), The CMS tracker system project: technical design report, http://cmsdoc.cern.ch/ftp/TDR/TRACKER/tracker_tdr_files.html.

10 CMS Collaboration (2008) The CMS experiment at the CERN LHC, *JINST*, **3**, S08004, doi:10.1088/1748-0221/3/08/S08004.

11 Aad, G. *et al.* (2008) ATLAS pixel detector electronics and sensors, *JINST*, **3**, P07007, doi:10.1088/1748-0221/3/07/P07007.

12 Kästli, H. *et al.* (2007) CMS barrel pixel detector overview, *Nucl. Instrum. Meth. A*, **582**, 724–727, doi:10.1016/j.nima.2007.07.058.

13 Damerell, C.J.S. (2008) Pixel-based Vertex and Tracking Detectors for ILC, *PoS(VERTEX 2008)*, 029, http://cdsweb.cern.ch/record/1198252.

14 Sze, S. (1985) *Semiconductor Devices*, John Wiley & Sons, Inc., New York.

15 Lutz, G. (1999) *Semiconductor Radiation Detectors*, Springer Verlag, Berlin, Heidelberg.

16 Rossi, L., Fischer, P., Rohe, T., and Wermes, N. (2006) *Pixel Detectors*, Springer Verlag, Berlin, Heidelberg.

17 Lutz, G. *et al.* (2007) DEPFET-detectors: New developments, *Nucl. Instrum. Meth. A*, **572**, 311–315, doi:10.1016/j.nima.2006.10.339.

18 Moser, H.G. *et al.* (2007) DEPFET Active Pixel Sensors, http://cdsweb.cern.ch/record/1187333.

19 Değerli, Y. *et al.* (2005) A fast monolithic active pixel sensor with pixel-level reset noise suppression and binary outputs for charged particle detection, *IEEE Trans. Nucl. Science*, **52/6**, 3186, doi:10.1109/TNS.2005.862931.

20 Turchetta, R. *et al.* (2001) A monolithic active pixel sensor for charged particle tracking and imaging using standard VLSI CMOS technology, *Nucl. Instrum. Meth. A*, **458**, 677, doi:10.1016/S0168-9002(00)00893-7.

21 Kemmer, J. and Lutz, G. (1987) New detector concepts, *Nucl. Instrum. Meth. A*, **253**, 365, doi:10.1016/0168-9002(87)90518-3.

22 RD50 Collaboration (2008) RD50 Status Report 2007 – Radiation hard semiconductor devices for very high luminosity colliders, http://cdsweb.cern.ch/record/1082083.

23 Moll, M. (2006) Radiation tolerant semiconductor sensors for tracking detectors, *Nucl. Instrum. Meth. A*, **565**, 202–211, doi:10.1016/j.nima.2006.05.001.

24 Mankel, R. (2004) Pattern recognition and event reconstruction in particle physics experiments, *Rept. Prog. Phys.*, **67**, 553, doi:10.1088/0034-4885/67/4/R03.

25 Strandlie, A. (2004) Track reconstruction – from bubble chambers to the LHC, *Nucl. Instrum. Meth. A*, **535**, 57, doi:10.1016/j.nima.2004.07.112.

26 Bock, R. *et al.* (2000) *Data Analysis Techniques for High-Energy Physics Experiments*, Cambridge University Press.

27 ATLAS Collab. (2008) The expected performance of the inner detector, CERN-OPEN-2008-020.

28 Cornelissen, T. *et al.* (2008) The new ATLAS Track reconstruction (NEWT), *J. Phys. Conf. Ser.*, **119**, 032014, doi:10.1088/1742-6596/119/3/032014.

29 Mankel, R. (1997) A concurrent track evolution algorithm for pattern recognition in the HERA-B main tracking system, *Nucl. Instrum. Meth. A*, **395**, 169, doi:10.1016/S0168-9002(97)00705-5.

30 Fruhwirth, R. (1987) Application of Kalman filtering to track and vertex fitting, *Nucl. Instrum. Meth. A*, **262**, 444, http://adsabs.harvard.edu/abs/1987NIMPA.262..444F.

31 Fruhwirth, R. (1997) Track fitting with non-Gaussian noise, *Comput. Phys. Commun.*, **100**, 1, doi:10.1016/S0010-4655(96)00155-5.

32 CMS Collab., Chatrchyan, S. *et al.* (2010) Alignment of the CMS silicon tracker during commissioning with cosmic rays, *JINST*, **5**, T03009, doi:10.1088/1748-0221/5/03/T03009.

33 ILD Concept Group (2010) The International Large Detector. Letter of Intent, http://www.ilcild.org/documents/ild-letter-of-intent/LOI_final.pdf/view.

15
Calorimetry: Precise Energy Measurements
Felix Sefkow and Christian Zeitnitz

15.1
Introduction

The measurement of the energy of particles is one of the central ingredients of modern high energy physics experiments. In order to determine the energy, a particle has to be absorbed completely in a block of material. Calorimeters are devices that ideally measure the total energy lost in this absorption process. This is especially important for neutral particles, which are not detectable by other means. Jets are usually composed of all possible particles, including neutrals. Therefore, a calorimeter is the only device to determine the actual jet energy.

Most real calorimeters measure only a fraction of the actually absorbed energy. Hence the resolution of the energy determination of the initial particle is inherently limited. The absorption process goes hand in hand with the production of secondary particles and is very different for different particle types (charged, neutral, leptons, hadrons, etc.). The understanding of the different interactions of particles with matter is therefore a prerequisite in order to understand the function of calorimeters.

The acceptance of the calorimeters in typical high energy physics experiments has to cover as much of the solid angle as possible in order to not only measure the total energy released in an event, but to also measure the missing energy. Missing energy could be due to undetectable particles like neutrinos or yet unknown and non-interacting or weakly interacting particles (e. g. SUSY LSPs).

15.2
Basic Principles of Particle Detection

The energy loss of particles in a block of material depends on the particle type and the material parameters. In order to measure the deposited energy a medium is needed that allows a signal to be extracted that is proportional to the energy loss. This so-called *active medium* measures directly only the energy loss of charged par-

Physics at the Terascale, First Edition. Edited by Ian C. Brock and Thomas Schörner-Sadenius
© 2011 WILEY-VCH Verlag GmbH & Co. KGaA, Weinheim.
Published 2011 by WILEY-VCH Verlag GmbH & Co. KGaA.

ticles by means of charge collection (ionisation chamber) or by light measurements (scintillation or Cerenkov light). Neutral particles generate signals only indirectly in this medium.

In the following, first the particles that lose their energy only via the electromagnetic interaction are discussed. These are electrons, positrons, muons and photons. Afterwards the absorption of hadrons, which undergo inelastic nuclear interactions, is described.

15.2.1
Energy Loss of $e^{\pm}$ and Photons

A highly energetic electron or positron, $e^{\pm}$, which passes through a medium will lose its energy mainly due to *bremsstrahlung* and *ionisation*. Bremsstrahlung, which originates from the deceleration of the particle in the electric field of the atoms, is dominant for energies above the critical energy, E_c, at which both processes yield the same energy loss. The value of E_c depends on the charge number, Z, of the material and can be parametrised for solids and liquids by $E_c = 800 \text{ MeV}/(Z + 1.2)$ [1].

The energy loss due to bremsstrahlung is proportional to the energy of the $e^{\pm}$,

$$-\frac{dE}{dx} = \frac{E}{X_0}, \qquad (15.1)$$

where X_0 denotes the so-called *radiation length*, which denotes the mean path length of an $e^{\pm}$ until its energy is reduced to E/e by bremsstrahlung processes. X_0 depends on $1/Z^2$ of the material and the square of particle mass, hence in most cases, only energy loss via bremsstrahlung needs to be considered for $e^{\pm}$. Bremsstrahlung produces photons with a $1/E$ energy spectrum.

Since all material dependence is parametrised in terms of the radiation length, X_0, this quantity is a good scale to describe the energy loss of high energetic $e^{\pm}$ and photons as a function of the depth of the calorimeter. A rule-of-thumb formula for the radiation length (2.5% precision for $Z > 2$) is [1]

$$X_0 = \frac{716.4 A}{Z(Z+1) \ln(287/\sqrt{Z})} \left[\frac{\text{g}}{\text{cm}^2}\right]. \qquad (15.2)$$

The ionisation of atoms in the absorber material is due to the interaction of the incoming charged particle with the electrons in the atomic shell. If the energy transfer exceeds the binding energy of the electron the latter will be liberated. The energy loss depends on the square of the speed of the particle, β^2, on the square of the particle's charge and on the square of the atomic number of the material. The mean energy loss is given by the *Bethe–Bloch formula* [1], which is different for light ($e^{\pm}$) and for heavy charged particles. The energy loss as a function of β has a minimum at approximately $\beta\gamma = 4$ and rises again for higher speeds due to the relativistic distortion of the electric field of the particle within the medium. A particle losing the minimal possible energy due to ionisation processes is called a *mip* (minimum *ionising particle*). A *mip* is often used as a reference to describe the properties of a calorimeter. The value of the minimal energy loss can be parametrised as (10% precision for $Z > 1$)

$$-\frac{dE}{dx}\bigg|_{mip} = 6\frac{Z}{A} - 1.25 \left[\frac{\text{MeV}}{(\text{g/cm}^1)}\right]. \qquad (15.3)$$

The typical energy loss in a gas is 1–3 keV/cm and for silicon 3.9 MeV/cm. The mean energy, W, to create an electron–ion pair is higher than the binding energy of the outermost electron in the atomic shell because all energy levels in the shell are involved. For gases the value of W is between 20 eV and 40 eV. In silicon only $W = 3.6$ eV are required to produce an electron–hole pair. From the total energy loss and the value of W one can estimate the signal (charge) in the active material.

For highly energetic photons the *conversion into* e^+e^- *pairs* is the dominant process. The path length until such a process occurs can be related to the radiation length X_0: the mean free path length until a *pair-production* process occurs is 9/7 of a radiation length. For energies below the threshold of 1 MeV, the photon can only be absorbed by or scattered off electrons in the atomic shell (*photoelectric effect* and *Compton scattering*).

Up to now $\mu^\pm$ have not been mentioned. Because of their high mass, which is approx. 200 times larger than the mass of the electron, and the $1/m^2$ dependence of the cross section, bremsstrahlung is heavily suppressed. Hence $\mu^\pm$ loses energy mainly due to ionisation. This leads to the effect that the muons travel long distances even in dense media (see also the discussion in Chapter 16).

Overall, only a few processes determine the main features of the energy loss of electromagnetically interacting particles in matter.

15.2.2
Interaction of Hadrons with Matter

Charged hadrons also continuously lose energy due to ionisation processes, as described in the previous section. Because of the high particle masses, bremsstrahlung is of no importance for the energy loss. The main energy loss for highly energetic charged and neutral hadrons originates from inelastic interactions with the nucleus of the atoms. In so-called *spallation* processes – collisions of the incoming hadrons with one of the nucleons in the nucleus – the transfer of energy leads to subsequent collisions of the struck nucleon with other nucleons within the nucleus (*intra-nuclear cascade*). In these interactions pions (π^0, $\pi^\pm$) and other hadrons can be produced which can escape the nucleus if their energy is high enough. In addition nucleons can be emitted. The nucleus will very likely stay in an excited state after the spallation process has occurred. The residual nucleus (or nuclei in case of a fission process) will undergo a de-excitation process which involves an *evaporation step*, that is, the emission of nucleons, photons and heavy particles (e.g. α, deuterons).

The mean free path length of protons, before undergoing an inelastic interaction, is the so-called *nuclear interaction length*, λ_I, which can be parametrised by ($\pm 1\%$

for $A > 7$)

$$\lambda_I = 20 A^{0.4} + 32 \left[\frac{\text{g}}{\text{cm}^2} \right]. \tag{15.4}$$

Be aware that the interaction length for pions is larger. For high-A materials the hadronic interaction length is much larger than the radiation length. For example for Pb the radiation length is $X_0 = 5.6\,\text{mm}$, but $\lambda_I = 18\,\text{cm}$. For low-$A$ materials these two characteristic values are very similar. For the efficient absorption of particles in a compact detector, materials like Fe and Pb are utilised. This leads to the effect that, compared to electromagnetic interactions, the hadronic interactions happen on a substantially larger scale; hence a greater material depth is required for hadron than for electromagnetic calorimeters.

The π^0 mesons produced in a spallation process will immediately decay into a pair of photons and contribute to the electromagnetic fraction of the deposited energy in the medium. Since the nucleus is, at least partially, disintegrated binding energy is lost in the process.

Some of the produced particles have a very limited lifetime and can decay before undergoing a subsequent inelastic interaction. Part of the resulting decay products (often neutrons, muons and neutrinos) can escape the volume of the calorimeter.

A special case is the neutron which can travel long distances within the calorimeter. It can undergo elastic scattering or inelastic interactions. Once slowed down sufficiently, the cross section for *capture processes* increases for some materials (e.g. hydrogen, cadmium). Most capture processes will lead to the emission of photons. In the scattering process on hydrogen the proton will recoil and subsequently lose its energy by ionisation. For large-A materials (e.g. uranium) *induced fission* is a possible process. The residual nuclei emit neutrons in order to make the transition to more stable isotopes. As a consequence the yield of secondary neutrons and low-energy photons is high even for relatively low neutron energies below 1 MeV. All other hadrons will no longer produce secondaries in this energy regime.

15.3
Particle Showers

Highly energetic particles entering a block of material will undergo multiple interactions with the material and create a multitude of secondary particles. In the following this so-called particle cascade or *shower* is described for electromagnetically and hadronically interacting particles.

15.3.1
Electromagnetic Cascades

When electrons or photons enter a dense material, the combination of bremsstrahlung and pair production gives rise to a chain reaction in which a cascade of secondary particles is created: the radiated photons convert, and the conversion

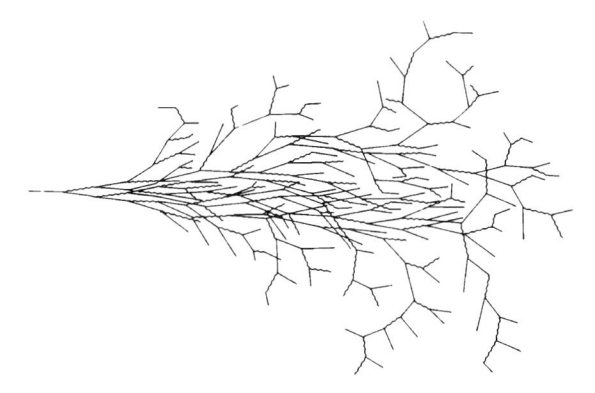

Figure 15.1 Schematic of an electromagnetic cascade developing in a block of material (courtesy of C. Grupen).

pairs radiate. A schematic view of such a cascade is shown in Figure 15.1. In a simple model the particle multiplicity doubles and the average energy per particle is halved within one radiation length. Multiplication stops at the critical energy, when ionisation takes over. Thus $N \approx E_0/E_c$ particles are produced, most of them being electrons and photons, while positrons are produced only above a threshold of 1 MeV; they account for about 25% of the energy deposition.

Showers initiated by electrons and positrons start immediately on their entrance into the material, while for photons the onset follows an exponential absorption law and occurs, on average, after $9/7\,X_0$.

If expressed in terms of radiation length, the shapes of electromagnetic showers are roughly independent of the material, and in the longitudinal direction they scale logarithmically with energy. The longitudinal profile can be parametrised as

$$f(t) = \frac{dE}{dt} = a t^\omega \cdot \exp(-bt) \tag{15.5}$$

(t in units of X_0), and the maximum of the multiplicity and energy deposition occurs at $t_{max} = \ln(E_0/E_c) - 0.5$. Since E_c decreases with Z, showers in heavy absorbers reach their maximum later and decay more slowly. Another effect of the lower E_c is that the number of soft photons becomes very large, which can introduce a dependence of the energy response on the step in the cascade.

For an accurate energy measurement it is critical that the calorimeter is deep enough to (almost) fully contain the shower. In copper, for example, for 99% containment one needs $16\,X_0$ at 1 GeV and $27\,X_0$ at 1 TeV. The numbers are smaller for lighter absorbers and higher for dense materials.

The transverse extension of electromagnetic showers is characterised by the *Molière radius*, ρ_M, which corresponds to a cylinder containing 90% of the energy. It scales with X_0,

$$\rho_M = 21.2 \text{ MeV} \cdot X_0/E_c \,, \tag{15.6}$$

but is independent of energy. Therefore, showers become increasingly elongated with higher incident energy.

It is important to note that the shower shape and composition fluctuate from event to event, mostly due to statistical variations in the early shower stage. Therefore it is not sufficient to probe the energy deposition at one position, for example in the maximum. Instead, one has to integrate the whole energy or sample it at sufficiently many representative positions. Fluctuations in longitudinal leakage, for example, are considerably larger than the mean value of the leakage itself.

15.3.2
Hadronic Cascades

Compared with electromagnetic showers, hadron showers are much more complex and diverse, due to the large variety of the underlying physics processes. The multiplicity in a single interaction is larger, but the number of "generations" in the cascade and the total number of particles are much smaller, such that the inevitable fluctuations have a stronger impact. Typically, depending on the material, there is around one charged pion per GeV and ten times as many (softer) nucleons. Figure 15.2 shows a schematic view of a hadronic shower. At each nuclear interaction, charged and neutral hadrons are produced, together with π^0 and η mesons which immediately decay into photons and give rise to electromagnetic showers near the production point. Soft short-range hadrons also deposit energy locally, while the harder fragments travel further until they initiate another nuclear interaction. The shower topology is thus described by two scales: the overall evolution is governed by the nuclear interaction length, λ_I, while the substructure is characterised by the radiation length, X_0.

The transfer of energy into electromagnetic sub-showers is a "one-way" street, since π^0 mesons are likely to be produced at each new nuclear interaction, whereas no hadrons emerge from π^0 decays. Therefore, the fraction of electromagnetic energy, f_{em}, increases with the number of steps in the cascade and therefore with

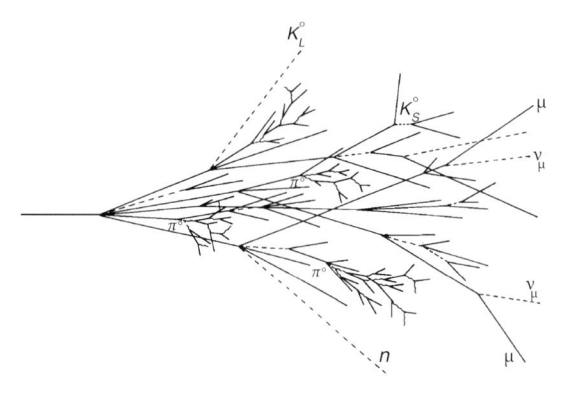

Figure 15.2 Schematic of a hadronic cascade developing in a block of material (courtesy of C. Grupen).

the energy of the incident particle. Ultra-high energy cosmic hadrons asymptotically produce purely electromagnetic showers. f_{em} also depends on the incident particle type; protons, for example, produce less π^0 mesons due to baryon conservation.

Around two thirds of the energy of the non-electromagnetic part is deposited via ionisation by charged hadrons, and 5–10% is released as kinetic energy of evaporation neutrons. The remainder is "invisible": nuclear binding energy and target recoil do not contribute to the detector signal.

The composition of the shower depends on the development stage (shower depth) and on the radial distance from the line of impact. There is a core enriched with electromagnetic sub-showers, surrounded by a more hadron-rich halo. Thermal neutrons are found in all directions at distances of several metres ("neutron gas").

The longitudinal shower profile reaches its maximum about $1\lambda_I$ after the position of the first nuclear interaction, which itself is exponentially distributed. Variations in the onset are the biggest single source of fluctuations in longitudinal shape and leakage. In iron, about $8\lambda_I$ are required to contain the energy of a shower initiated by a pion of 100 GeV. Transversely, roughly 90% of the energy is contained in a cylinder with radius λ_I; this fraction is, however, energy-dependent.

15.4
Calorimeters: Response and Resolution

The determination of the energy deposited by an electromagnetic or hadronic shower within the calorimeter volume is done by measuring the generated charged particles in the active medium. This medium is either identical to the absorber (*homogeneous calorimeter*), or interleaved with the absorber material (*sampling calorimeter*). Any medium which allows the energy loss of charged particles to be measured can be utilised (e. g. scintillators, liquid argon, crystals, silicon). For a detailed discussion of active media see [2].

The number of charged particles in a shower is, to first order, proportional to the total deposited energy. The signal measured in the active material is therefore proportional to the energy, E_0, of the primary particle. In the previously mentioned simple model of an electromagnetic shower the sum of the track lengths of all particles is given by

$$T_d[X_0] = F \cdot E_0 / E_c . \tag{15.7}$$

Because of the proportionality between T_d and E_0 the former quantity is a good estimator for the energy of the particle. The factor F accounts for a lower cut-off energy for the measurability of a track. The fluctuation of the total track length, T_d, limits the energy resolution. An additional contribution to the measured resolution is the actual statistical fluctuation of the ionisation, scintillation or Cerenkov light process (charge production and photon statistics).

The number of generated charged particles, N, varies statistically from shower to shower. The intrinsic resolution, σ, is therefore proportional to $\sqrt{N}$. Since N is proportional to E_0, the resolution can be parametrised by $\sigma = A\sqrt{E_0}$. The relative resolution σ/E therefore improves with $A/\sqrt{E_0}$. The parameter A denotes the so-called *stochastic* constant of the resolution, which is determined by the statistical fluctuation of the measured signal.

The readout system of the active medium will contribute noise to the resolution, $\sigma_N = B$, which does not depend on energy but on the number of involved electronic channels. Imperfections of the calorimeter (mechanical and density variations, inefficiencies of the readout, incorrect calibration of channels) will further worsen the resolution. The corresponding contribution $\sigma_I = C \cdot E$ scales with the total deposited energy. Adding up all contributions in quadrature yields the standard parametrisation of the energy resolution of a calorimeter:

$$\frac{\sigma}{E} = \frac{A}{\sqrt{E}} \oplus \frac{B}{E} \oplus C . \tag{15.8}$$

In the following only the stochastic term – which is determined by the calorimeter design in terms of material and geometry – is considered in order to describe the properties of a calorimeter. In practice the energy resolution of a calorimeter at high energies is limited by the constant term C.

15.4.1
Response and Resolution of a Sampling Calorimeter for Electromagnetic Particles

A sampling calorimeter consists of layers of absorber material interleaved with active detector material. This leads to the effect that only a fraction of the deposited energy is actually measured, while the rest is lost in the absorber plates. The so-called *sampling fraction* can be calculated from the fraction of the energy loss of a *mip* in the active layers. A good measure of a *mip* is a muon passing through the calorimeter.

The response of the calorimeter to electrons is usually given with respect to the *mip* signal and called *e/mip*. This ratio is usually less than 1 if the absorber has a substantially higher Z than the active medium. The reason is that an electron will lose substantially more energy in the high-Z absorber than a *mip* does. This is due to the fact that the cross section of the bremsstrahlung process is proportional to Z^2/m^2, leading to the production of a high number of low energy photons with $E_\gamma < 1\,\mathrm{MeV}$ in the absorber material. Since the absorption of these photons is dominated by the photoelectric effect, which exhibits a Z^5 dependence, most photons will already be stopped in the absorber plates. The value of *e/mip* depends on the difference in Z between the absorber and the active layers. For absorber thicknesses in excess of $1 X_0$ the value is nearly constant. Typical values are 0.6 for a liquid-argon/lead calorimeter and 0.8 for liquid-argon/iron [3].

In the following we consider, as an example, the ATLAS electromagnetic barrel calorimeter (which covers pseudorapidities $\eta < 0.8$) [4, 5] with 1.53-mm Pb absorbers (covered by 0.2-mm stainless steel plates) and a 2.1-mm liquid-argon gap

Figure 15.3 An ATLAS electromagnetic barrel calorimeter module during assembly of a prototype (courtesy of the ATLAS collaboration).

as the active medium. Figure 15.3 shows one of the modules of this calorimeter during construction. The sampling fraction for this calorimeter is $\approx 17\%$. The particles enter the sampling layers under an angle of $\approx 45°$ (in reality this angle varies between 46° and 34°) due to the accordion structure of the calorimeter. The effective thickness of the absorber and active material is therefore increased by a factor $\sqrt{2}$. The accordion structure allows the calorimeter to be built as a barrel without any transition cracks between the modules, hence avoiding inactive areas in the ϕ direction.

The energy resolution of this calorimeter for a particle of energy E_0 can be estimated from the total track length, T_d, of all particles in the electromagnetic shower. For the total track length defined above in (15.7) and assuming, that all tracks are measurable ($F = 1$) one obtains $T_d = E_0/E_c$. To first order, the active layers of the calorimeter measure only the number of crossing particles. The number of crossings, N_s, can be calculated as $N_s = T_d/d$, where d denotes the thickness of a single sampling layer (active and passive) in X_0. Since N_s is subject to statistical fluctuations (assuming Poisson statistics), one obtains for the relative energy resolution,

$$\frac{\sigma}{E} = \sqrt{\frac{E_c \cdot d}{E_0}}. \tag{15.9}$$

The stochastic term alone is then given by $A = \sqrt{E_c \cdot d \cdot 10^{-3}}$, where E_c and d are given in units of MeV and X_0, respectively. Plugging in the numbers of the ATLAS barrel calorimeter ($d = 0.31 \cdot \sqrt{2} X_0$, $E_c = 12.6\,\text{MeV}$), where E_c has been determined as the thickness-weighted average of the involved materials, one obtains $A = 7.4\%\sqrt{\text{GeV}}$. The measured resolution of the calorimeter is $\approx 10\%\sqrt{\text{GeV}}$ [4, 5], which reflects the fact that the assumed model is too simple since it ignores effects like $e/mip < 1$ and the non-zero crossing angle of particles. Additional contributions from instrumental effects (charge-collection efficiency, energy cut-off for particle detection) reduce the signal amplitude and worsen the resolution. However, the obtained result provides a lower bound for the achievable energy resolution.

15.4.2
Homogeneous Calorimeters

In order to achieve a very good energy resolution for electrons and photons, different homogeneous calorimeters have been realised. Since the whole energy of the initial particle is deposited in the active medium, only fluctuations of the measurable signal will limit the resolution. This can be photon statistics in the case of light detection (scintillation or Cerenkov light), or fluctuations of the liberated charge in ionisation chambers (e. g. in liquid noble gas – see also the detailed discussion in [2]).

The NA48 experiment [6] at CERN utilised liquid krypton ($X_0 \approx 4.7$ cm) in order to absorb the energy of photons and $e^{\pm}$. The liquid noble gas allows the liberated charge from the ionisation processes to be directly measured. The charge is collected by embedding electrodes into the liquid and applying a high electric field. The active area covers laterally 128 cm in radius and 120 cm in depth ($26 X_0$). The geometrical arrangement of 2×2 cm^2 electrodes allows the location of the shower to be measured with a precision of 1 mm by the centre-of-gravity method. An energy resolution of $3.2\%/\sqrt{E}$ has been achieved. Homogeneous liquid noble gas calorimeters are only suitable for fixed-target experiments, where the space requirement is no issue. The described liquid-krypton calorimeter will be re-used for the NA62 experiment at CERN.

A different approach is the utilisation of dense crystals, which facilitates extremely compact homogeneous calorimeters. Crystal materials like $PbWO_4$ have been available for a number of years and have a very small radiation length ($X_0 = 8.9$ mm), which allows their use in a collider experiment. The disadvantage with respect to crystals like CsI is the very low light yield of only 80 γ/MeV, which requires a very low-noise readout and makes these crystals only usable for highly energetic particles. The readout can be done either with photomultipliers or with avalanche photodiodes. The latter is a suitable choice if the calorimeter sits within the magnetic field of the detector.

An example where $PbWO_4$ crystals have been utilised as the electromagnetic part of the calorimeter is the CMS detector [7] at CERN. This calorimeter consists in the barrel part of 61 200 $PbWO_4$ crystals with a size of $22 \times 23 \times 230$ mm^3 and in the endcap of 14 648 crystals with a size of $30 \times 30 \times 220$ mm^3. The length of the crystals corresponds to $\approx 26 X_0$. The energy resolution (see Figure 15.4) has been determined in test-beam measurements to be $2.8\%/\sqrt{E}$ and the spatial resolution is found to be 1 mm at an energy of 100 GeV.

15.4.3
Response and Resolution of Hadron Calorimeters

The response of a calorimeter to hadronically interacting particles is substantially more complex than for $e^{\pm}$ and photons. The complexity of the hadronic cascade with a mixture of charged hadrons, photons, neutrons and invisible energy leads to a response which depends on material, geometry and energy. The fluctuation of the

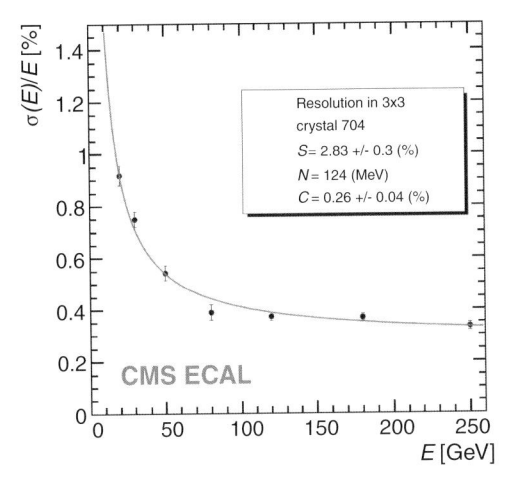

Figure 15.4 The electromagnetic energy resolution of the CMS PbWO$_4$ calorimeter (adapted from [7]).

Resolution in 3x3 crystal 704
S = 2.83 +/- 0.3 (%)
N = 124 (MeV)
C = 0.26 +/- 0.04 (%)

large fraction of the invisible energy, f_{inv}, (up to 40%) ultimately limits the energy resolution of the calorimeter. Because of f_{inv} the signal obtained from primary hadrons is usually smaller than from an electron with the same energy. Hence the ratio of the signals of electrons and hadrons, the so-called e/h ratio, is in most calorimeters larger than 1.

The different contributions to the signal of a hadron in a sampling calorimeter consist of

- the intrinsic signal from all charged hadrons in the shower (fraction f_{ion}) due to ionisation;
- the electromagnetic component from π^0 and η decays, which depends on energy and material. It increases logarithmically with the energy and, for a calorimeter with lead as an absorber, amounts to $\approx$ 30% at 10 GeV ($\approx$ 55% at 100 GeV). A more detailed description can be found in [2] and [8].
- the fraction, f_γ, of photons mainly originating from nuclear de-excitation;
- the indirect signal from neutrons (fraction f_n) liberated in nuclear processes.

The signal, without the π^0 contribution, for a hadron entering the calorimeter is therefore composed of $f_{had} = f_{ion} + f_\gamma + f_n$. The dominant contribution comes from f_{ion}. The actual composition depends on the materials and geometry of the absorber and active layers.

The ionisation signal is mainly due to the protons and charged pions produced in the spallation processes. The amplitude of the signal is equal to a *mip* (ion/*mip* = 1) for high energies ($E > 1$ GeV) but varies substantially for lower energies [3]. Since a big contribution originates from low energy protons, one has to take saturation effects in the active medium into account. The contribution from low energy photons from nuclear de-excitation, f_γ, again depends on the difference in Z of the absorber and the active medium. Therefore, for most real calorimeters one finds $\gamma/mip < e/mip$.

The fraction of neutrons from the inelastic interactions is proportional to the number of interactions and proportional to the energy lost in these interactions. Especially in heavy media (e. g. lead or uranium) the number of produced neutrons is substantial. f_n therefore depends critically on the material choice and the readout of the calorimeter (integration time). This is due to the fact that the neutrons are usually fast ($E > 1$ MeV) when emitted. The recoiling protons from scattering processes on hydrogen (if present) contribute to f_{ion}. The scattering leads to the moderation of the neutrons. The neutron capture cross section increases substantially for slow neutrons ($E < 1$ eV). The subsequent emission of photons is again detectable and contributes to f_γ.

In summary, the hadronic response had/mip, without the π^0 contribution, is, in most cases, smaller than e/mip, but can be tuned by the choice of the calorimeter's material and geometry.

The response of a calorimeter to an electron with respect to a hadron with the same energy is given by

$$\frac{e}{h} = \frac{\frac{e}{mip}}{f_{em}\frac{e}{mip} + f_{had}\frac{had}{mip}} . \qquad (15.10)$$

It depends on the energy of the initial hadron and in most cases is substantially larger than 1 (typically 1.3). This leads to two effects:

- the response of the calorimeter is not linear in energy (increase of f_{em} with E_0);
- the energy resolution deteriorates, because the fluctuation of f_{em} results in a fluctuation of the signal amplitude.

Both effects would not be present for $e/h = 1$. To equalise the signal of electrons and hadrons (so-called *compensation*), two options exist:

- lowering e/mip by increasing the thickness of the absorber;
- increasing had/mip by enhancing the visible signal from neutrons.

Lowering e/mip requires a certain choice of materials in order to reach $e/h = 1$ and in addition worsens the resolution for electrons (more details can be found in [3]). The compensation of the energy losses in hadronic showers requires the moderation of the neutrons as described above. This is best achieved by materials containing a substantial amount of hydrogen, like scintillator. Choosing an absorber which produces abundant neutrons, like uranium, allows the e/h to be tuned by adjusting the signal integration time of the readout. This is needed because the moderation, capture and photon emission occurs on a time scale of a few 100 ns.

The ZEUS calorimeter utilised 3.3-mm depleted uranium absorber plates (cladded with stainless steel plates) with a scintillator readout (thickness 2.6 mm).

For 200-ns signal integration time an e/h ratio close to unity and an energy resolution for single pions of $\sigma/E = 34\%/\sqrt{E}$ ($E > 10\,\text{GeV}$) could be achieved in test-beam measurements [9, 10]. To date this is the best resolution ever achieved for a hadron calorimeter.

An alternative method to the above-described *hardware* compensation is the weighting of electromagnetic subshowers caused by the π^0 decays. The photons create a very localised energy deposition, which can be identified in a calorimeter with high lateral and longitudinal granularity. With an energy-dependent weighting function of these signals, the e/h ratio can be equalised in software.

This method has been pioneered by the CDHS [11] collaboration and further been improved by the H1 [12] and ATLAS [13] experiments for their liquid-argon calorimeters.

15.4.4
Spatial Resolution

For a good spatial resolution of a calorimeter a granularity with a cell size below the characteristic shower size is required ($1X_0$ for electromagnetic and $1\lambda_I$ for hadronic showers). This allows the energy sharing between neighbouring cells to be exploited in a centre-of-gravity calculation. The resolution scales roughly with square-root of the cell size. The achievable resolution is of the order of 1 mm for electrons, positrons and photons. Because of the strong fluctuations of hadronic showers their resolution is substantially worse.

15.5
New Concepts

Future experiments at the Terascale pose new demands on calorimetry, in particular at a future e^+e^- linear collider. Heavy bosons like W and Z or possibly H must be identified on the basis of the invariant mass reconstructed in dijet decays. To achieve a 3σ separation between W-boson and Z-boson signals, a relative jet energy resolution of about 3.5% is required, which exceeds that of existing collider detectors by roughly a factor 2[39].

Hadron calorimetry is therefore the subject of intense R&D; two main directions are being followed: one aims at an event-by-event decomposition of the shower energy by using multiple signatures from the detecting media (*dual readout*), the other optimises the combination of the calorimeter information with tracking information by reconstructing each particle individually (*particle flow*), using high granularity. In principle these two methods complement each other and could ultimately be combined.

---

39) This is true even for the ZEUS detector, because the jet energy and the invariant-mass resolution are in general worse than the single-hadron energy resolution.

15.5.1
Dual-Readout Calorimetry

In non-compensating hadron calorimeters fluctuations in the electromagnetic energy fraction, f_{em}, represent the biggest single contribution to the resolution. The *dual-readout method* (DREAM) [14] corrects for this by measuring f_{em} on an event-by-event basis in parallel to the total deposited energy. In practice this is done by either using two different active media, scintillator and quartz, to register scintillation and Cerenkov light, respectively, or by using heavy crystals and disentangling the optical signals from the two processes. The method is based on the fact that only the electromagnetic part of the shower contains relativistic particles, and only these produce Cerenkov light, while the signal of the hadronic part is mostly due to non-relativistic protons.

The method was first tested with the DREAM module, a 1-t copper matrix with embedded quartz and scintillating fibres. The signals of the two fibre types, on electromagnetic scale, are given by

$$Q = [f_{em} + (h/e)_Q(1 - f_{em})]E , \qquad (15.11)$$

$$S = [f_{em} + (h/e)_S(1 - f_{em})]E , \qquad (15.12)$$

which can be solved for f_{em} and E:

$$E = (S - \chi Q)/(1 - \chi) , \qquad (15.13)$$

where $\chi = [1-(h/e)_S]/[1-(h/e)_Q]$ was about 0.3 in the DREAM module. A resolution of about 5% was obtained for 200 GeV "jets" produced on a target. Because of the small size of the module, this presumably includes contributions from transverse leakage and thus underestimates the potential of the method. Tests with larger prototypes are being prepared.

Another limitation of the DREAM module resolution is related to photoelectron statistics, because the light yield in the Cerenkov fibres corresponds to only eight photoelectrons per GeV. This can be improved using dense crystals like $PbWO_4$ or BGO. Here one registers scintillation light (as a measure of the total energy) and Cerenkov light (the electromagnetic part) from the same medium. Several methods have been successfully used in beam tests to disentangle the two sources. The directionality of the Cerenkov light was exploited by measuring the asymmetry of optical signals read at opposite ends of a crystal bar. Alternatively UV filters have been used to enhance the Cerenkov signal over the longer-wavelength scintillation light. Finally, the time structure of the signal has been used with dedicated electronics: Cerenkov light is prompt, while scintillation is typically delayed by several tens of nanoseconds due to the molecular de-excitation process.

Using the time structure gives the potential of decomposing the deposited energy even further. The neutron signal is delayed by nuclear de-excitation which shows up as a tail in the scintillation light spectrum. Since the visible neutron energy is correlated with the invisible energy used for nuclear excitation and break-up, this delayed part of the signal can be used to correct on an event-by-event basis for fluctuations in the invisible energy. This is complementary information to the electromagnetic fraction extracted from Cerenkov light and was shown to improve the resolution further. The experimental studies with crystals have been done using the blocks stand-alone or as an "electromagnetic section" in front of the DREAM module. Test-beam studies with a more realistic prototype are a target of future developments.

A relatively new idea to use the concept of dual readout and to overcome the limitations of a module with fibres is the so-called *total absorption calorimeter*, a homogeneous electromagnetic and hadron calorimeter, segmented into crystals finely enough to even apply particle-flow methods (see next section). This is a promising concept, but has many open issues for R&D such as the development of dense, bright and affordable crystals, efficient UV-sensitive and compact photosensors and an overall compact system design.

15.5.2
Particle-Flow Calorimetry

The particle-flow approach [15–17] starts from the observation that in a typical jet 62% of the energy is carried by charged hadrons, 27% by photons, 10% by neutral hadrons and 1.5% by neutrinos. In the classical calorimetric approach to jet energy reconstruction, where all calorimeter signals are summed up, 72% of the measurement suffers from the poor hadronic energy resolution of the combined electromagnetic (ECAL) and hadron (HCAL) calorimeters. In principle this would be unavoidable only for 10%, while the charged particles could benefit from the much superior resolution of the tracking system. Theoretically, resolutions better than $20\%/\sqrt{E}$ would be achievable with typical detectors if each particle would be reconstructed individually and the optimal device be used.

In practice, this requires the energy depositions of the individual particles in the calorimeters to be disentangled from each other, such that those from neutral hadrons can be isolated. The particle-flow method tackles this task with a combination of very finely segmented calorimeters and sophisticated pattern recognition algorithms to analyse the event topology. The performance is then limited by imperfections in this separation, the so-called *confusion*. If neutral particles (photons or hadrons) are not resolved from charged ones, their energy is wrongly considered as measured with the tracker and thus not accounted for. On the other hand, the energy of shower fragments erroneously classified as extra neutral particles is double-counted.

The concept of particle-flow calorimetry has implications for the overall detector design (see Chapter 13). It requires a large-volume, high-efficiency, low-density tracking system, a strong magnetic field and calorimeters inside the solenoid. For the calorimeters, particle-separation and pattern-recognition capabilities are emphasised even more than energy resolution. This is sometimes misunderstood. In particular the hadron energy resolution is still very important not only for ensuring good precision for neutrals but also for assisting the pattern recognition.

However, the calorimeters must be very dense, compact and extremely finely segmented in both the transverse and longitudinal dimensions. The preferred ECAL absorber material is tungsten with its small radiation length and Molière radius of 3 mm, in order to maximise the separation between electromagnetic showers. For the HCAL steel is preferred for cost reasons, but tungsten is also considered for highest energies. In order to resolve the shower topology, not only the ECAL segmentation but also that of the HCAL must be of the order of X_0, and not λ_I, like in classical hadron devices. Transverse cell sizes are 0.5–1 cm for the ECAL and 1 or 3 cm for the HCAL, depending on the readout technology. The enormous channel density (of the order of 100 million for the ECAL and 10–100 million for the HCAL), each read out individually, represents a tremendous technological challenge.

The jet energy performance of a particle-flow detector depends not on the calorimeter alone, but also on the tracking system (mainly its efficiency) and very much on the quality of the reconstruction software for tracking, calorimeter clustering and combination of information. The most advanced algorithm to-date is PandoraPFA [18]. The algorithm performs clustering, using track-based seeds and then uses cluster topological information such as directionality and also energy-momentum matching to merge clusters or remove fragments.

On the basis of a rather detailed and realistic detector simulation and reconstruction, the performance has been analysed and understood in terms of different contributions to the resolution. For the ILD model (see also Chapter 13), it can be parametrised as

$$\frac{\text{rms}_{90}}{E} = \left[\frac{21}{\sqrt{E}} \oplus 0.7 \oplus 0.004\,E \oplus \left(\frac{E}{100}\right)^{0.3} \right] \% \quad (E \text{ in GeV}),$$

where rms_{90} is the r.m.s. of the smallest range containing 90% of the events. The individual terms represent contributions from the intrinsic calorimetric resolution, tracking imperfections, leakage and confusion. This energy dependence is illustrated in Figure 15.5 and compared with the resolution obtained with a traditional approach from calorimetric information alone. Even at jet energies as high as 500 GeV, where confusion (shown separately in the figure) becomes dominant, particle flow brings a significant improvement. Note that the degradation at high energies is also due to leakage, which affects the purely calorimetric measurement much more severely.

The simulations with PandoraPFA have been used to optimise the ILD global detector parameters as well as calorimeter design parameters, in particular the granularity, and confirmed the choices above. Despite the notorious imperfections in hadronic-shower simulations the results are remarkably stable under variations of the particular model used. Nevertheless, over the past years a considerable effort was invested into the experimental verification of these ideas. The CALICE collaboration has developed and built highly granular prototypes of calorimeters for a linear collider and exposed them to test-beams. First studies have been made with a very compact tungsten ECAL [19] instrumented with 30 layers of 1×1 cm^2 silicon pad diodes and a scintillator steel HCAL [20] with 3×3 cm^2 tiles. A hadron show-

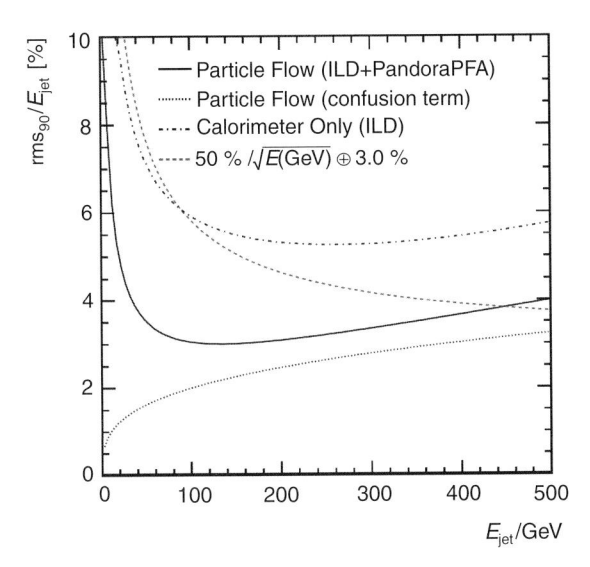

Figure 15.5 Jet energy resolution vs. energy using the particle-flow method (for the ILD detector concept). The contribution due to confusion is shown separately. For comparison also the resolution obtained with the calorimeter alone and a naive parametrisation are shown (adapted from [18]).

er event is depicted in Figure 15.6. The combined set-up has a hadronic energy resolution of $49\%/\sqrt{E}$, where the fine granularity has been exploited for a simple software compensation method [21].

It is practically impossible to verify the jet energy resolution of a particle-flow detector with artificial jets produced in test-beams, since this would require magnetic momentum spectroscopy, large acceptance as well as simulations to account

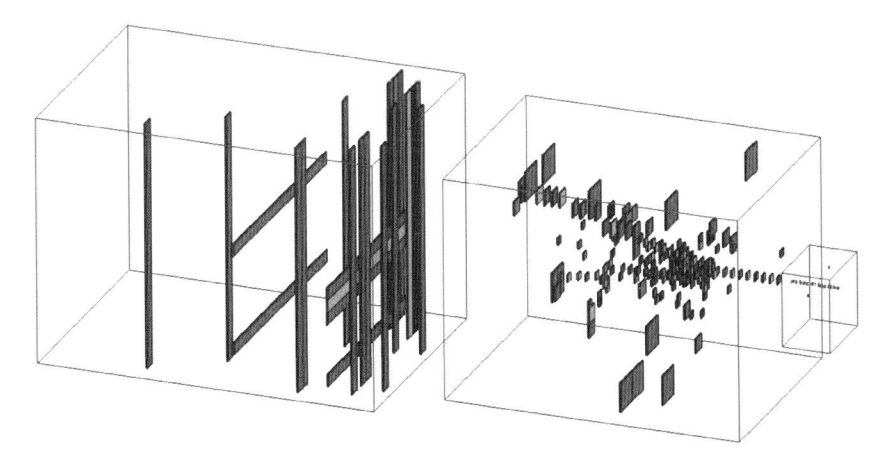

Figure 15.6 A reconstructed hadron in the CALICE prototype module. The hadron enters the detector from the right (Source: DESY).

for target and acceptance losses. However, thanks to the high granularity, the cell occupancies are small and event overlay techniques can be used to study the particle separation power of detectors and particle-flow algorithms with real data. First applications of PandoraPFA to CALICE data have confirmed the simulations [22]. Also shower shapes and internal topologies like charged track multiplicity inside the shower are reasonably well modelled [21].

Future R&D concentrates on coping with the technical realisation of the large channel density and the related integration issues for a compact detector, like for example power supply and dissipation issues. Also, the merits and drawbacks of different readout technologies for particle-flow calorimetry still have to be explored. Alternative ECAL schemes include scintillators or monolithic active pixel sensors; for the HCAL different gaseous techniques are being developed (see also Chapter 14). These offer cost-effective ways to even finer transverse segmentation, 1×1 cm^2, if combined with simplified readout electronics. In this so-called *digital* or *semi-digital approach* to calorimetry the energy measurement is replaced by hit counting which also provides an estimate of the number of particles in the shower and thus the primary particle's energy. Prototypes are being built and will be put into test-beams in the forthcoming years.

15.6
Summary

Calorimeters in high energy physics are used in a wide variety of applications, from high-precision homogeneous calorimeters for electrons to luminosity detectors and high particle flux beam monitors for heavy ions. Very different technologies, depending on the requirements on resolution, compactness, signal speed and particle flux have been utilised. In order to meet the challenges of collider physics at the Terascale, new ideas have been put forward and partially been established. On the one hand they hold the potential to study the new physics domain with unprecedented precision at the International Linear Collider, and on the other hand they allow the increased luminosity of the planned upgrade of the LHC to be fully exploited.

References

1 Particle Data Group, Amsler, C. et al. (2008) Review of Particle Physics, *Phys. Lett. B*, **667**, 1, doi:10.1016/j.physletb.2005.04.069

2 Wigmans, R. (2000) Calorimetry – Energy Measurements in Particle Physics, Oxford University Press.

3 Wigmans, R. (1987) On the energy resolution of uranium and other hadron calorimeters, *Nucl. Instrum. Meth. A*, **259**, 389–429, doi:10.1016/0168-9002(87)90823-0.

4 Aubert, B. et al. (2005) Development and construction of large size signal electrodes for the ATLAS electromagnetic calorimeter, *Nucl. Instrum. Meth. A*, **539**, 558–594, doi:10.1016/j.nima.2004.11.005.

5 Pinfold, J. *et al.* (2008) Performance of the ATLAS liquid argon endcap calorimeter in the pseudorapidity region $2.5 < |\eta| < 4.0$ in beam tests, *Nucl. Instrum. Meth. A*, **593**, 324–342, doi:10.1016/j.nima.2008.05.033.

6 NA48 Collab., Fanti, V. *et al.* (2007) The beam and detector for the NA48 neutral kaon CP violations experiment at CERN, *Nucl. Instrum. Meth. A*, **574**, 433–471, doi:10.1016/j.nima.2007.01.178.

7 Adzic, P. *et al.* (2007) Energy resolution of the barrel of the CMS electromagnetic calorimeter, *JINST*, **2**, P04004, doi:10.1088/1748-0221/2/04/P04004.

8 Gabriel, T.A., Groom, D.E., Job, P.K., Mokhov, N.V., and Stevenson, G.R. (1994) Energy dependence of hadronic activity, *Nucl. Instrum. Meth. A*, **338**, 336–347, doi:10.1016/0168-9002(94)91317-X.

9 ZEUS Calorimeter Group, Andresen, A. *et al.* (1990) Response of a uranium scintillator calorimeter to electrons, pions and protons in the momentum range $0.5\,\mathrm{GeV}/c$ to $10\,\mathrm{GeV}/c$, *Nucl. Instrum. Meth. A*, **290**, 95, doi:10.1016/0168-9002(90)90347-9.

10 ZEUS Calorimeter Group, Behrens, U. *et al.* (1990) Test of the ZEUS forward calorimeter prototype, *Nucl. Instrum. Meth. A*, **289**, 115–138, doi:10.1016/0168-9002(90)90253-3.

11 Abramowicz, H. *et al.* (1981) The response and resolution of an iron scintillator calorimeter for hadronic and electromagnetic showers between 10 GeV and 140 GeV, *Nucl. Instrum. Meth.*, **180**, 429, doi:10.1016/0029-554X(81)90083-5.

12 H1 Collab., Abt, I. *et al.* (1997) The tracking, calorimeter and muon detectors of the H1 experiment at HERA, *Nucl. Instrum. Meth. A*, **386**, 348–396, doi:10.1016/S0168-9002(96)00894-7.

13 ATLAS Collab., Aad, G. *et al.* (2008) The ATLAS Experiment at the CERN Large Hadron Collider, *JINST*, **3**, S08003, doi:10.1088/1748-0221/3/08/S08003.

14 Wigmans, R. (2009) Recent results from the DREAM project, *J. Phys. Conf. Ser.*, **160**, 012018, doi:10.1088/1742-6596/160/1/012018.

15 Brient, J.-C. and Videau, H. (2002) The calorimetry at the future e^+e^- linear collider, http://arxiv.org/abs/hep-ex/0202004.

16 Morgunov, V.L. (2003) Calorimetry design with energy-flow concept (imaging detector for high-energy physics), Prepared for 10th International Conference on Calorimetry in High Energy Physics (CALOR 2002), Pasadena, California, 25–30 Mar 2002.

17 Magill, S.R. (2007) Innovations in ILC detector design using a particle flow algorithm approach, *New J. Phys.*, **9**, 409, doi:10.1088/1367-2630/9/11/409.

18 Thomson, M.A. (2009) Particle Flow Calorimetry and the PandoraPFA Algorithm, *Nucl. Instrum. Meth. A*, **611**, 25–40, doi:10.1016/j.nima.2009.09.009.

19 Adloff, C. *et al.* (2009) Response of the CALICE Si-W electromagnetic calorimeter physics prototype to electrons, *Nucl. Instrum. Meth. A*, **608**, 372–383, doi:10.1016/j.nima.2009.07.026.

20 Adloff, C. *et al.* (2010) Construction and commissioning of the CALICE analog hadron calorimeter prototype, *JINST*, **5**, P05004, doi:10.1088/1748-0221/5/05/P05004.

21 Simon, F. for the CALICE Collab. (2010) Particle Showers in a Highly Granular Hadron Calorimeter, http://arxiv.org/abs/1008.2318.

22 Markin, O. for the CALICE Collab. (2001), presented at CALOR2010, Beijing, 2001; to appear in the proceedings.

16
Muon Detectors: Catching Penetrating Particles

Kerstin Hoepfner and Oliver Kortner

16.1
Sources of Muons

As already discussed in Part One, many physics processes which were out of reach until now will become accessible at the LHC, such as the production of the Higgs and new gauge bosons or of potential superpartners of the known fundamental particles. Such processes are, however, highly obscured by Quantum Chromodynamics (QCD) reactions, creating multiple jets and leptons of low transverse momenta. The interesting non-QCD processes often manifest themselves with leptons of high transverse momenta in the final state. Muons are of particular interest, as they lose very little energy during their passage through matter and therefore are the only charged particles traversing the entire detector. Muonic final states provide the cleanest signatures for the detection of new physics processes.

Figure 16.1a shows the transverse momentum distributions of muons from the decays of Z bosons, a potential Standard Model Higgs boson H with an assumed mass of 130 GeV, supersymmetric Higgs bosons A with a mass of 300 GeV, and Z' bosons with a mass of 2 TeV. As the masses of these resonances cover a wide range, muons from their decay have transverse momenta between a few GeV and 1 TeV, which have to be detected and identified by the LHC detectors.

The main source of muons, however, are meson decays, as can be concluded from the inclusive muon cross sections in Figure 16.1b. Most of the muons with transverse momenta $p_T^\mu < 10$ GeV originate from the decays of charged pions and kaons produced in proton–proton collisions. Between $p_T^\mu \approx 10$ GeV and 50 GeV, decays of b and c mesons are the dominant source of muons; for higher p_T^μ their inclusive cross sections are comparable to the ones from W-boson and Z-boson decays. Muons from t-quark decays are much rarer than those from b and c decays, as the production cross section of $t\bar{t}$ pairs is small due to their high mass. Additional charged pions are produced when final-state hadrons are absorbed in the calorimeters. Since these pions have even lower transverse momenta than the primarily produced ones, muons from their decays are too soft to escape the calorimeters and mostly remain undetected. Although muons from all of the above sources can

---

Physics at the Terascale, First Edition. Edited by Ian C. Brock and Thomas Schörner-Sadenius
© 2011 WILEY-VCH Verlag GmbH & Co. KGaA, Weinheim.
Published 2011 by WILEY-VCH Verlag GmbH & Co. KGaA.

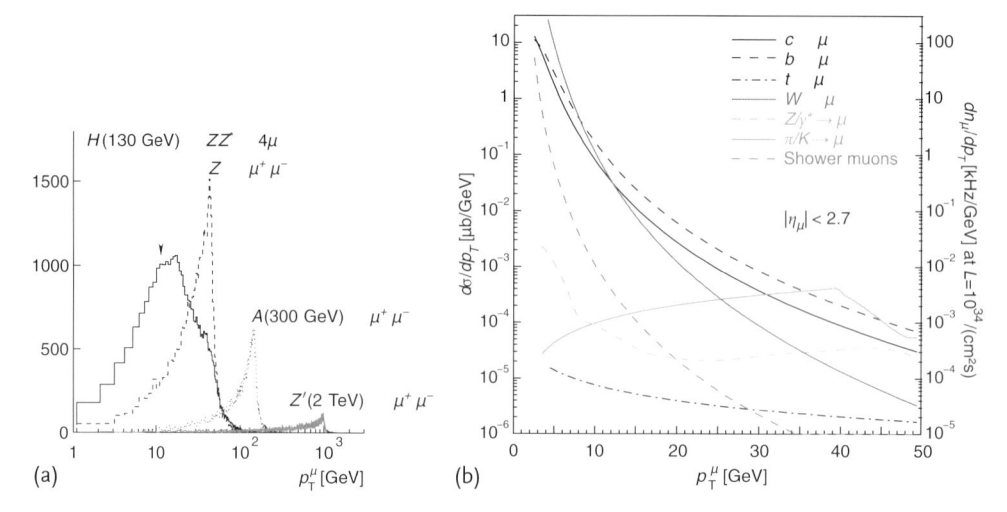

Figure 16.1 (a) transverse momentum distribution for muons from H, Z, A, and Z' decays as predicted by [1]; (b) transverse momentum dependence of inclusive muon cross sections for ATLAS [2].

be detected, at the LHC one concentrates on those muons which do not originate from pion and kaon decays or hadronic showers.

Both the ATLAS and CMS detectors have excellent muon detection and identification capabilities, as will be explained in the following sections.

16.2
Energy Loss of Muons and Muon Identification

The fact that muons lose little energy when passing through matter is essential for muon identification. There are four basic processes of energy loss for muons: ionisation and excitation of atoms, direct electron–positron pair production, bremsstrahlung and nuclear interactions. Figure 16.2 shows the contributions of these processes to the specific energy loss of muons as a function of the muon energy.

Up to muon energies of 1 TeV, *nuclear interactions* contribute so little to the specific energy loss that they can be neglected. *Bremsstrahlung processes* deflect a muon by the electric field of a nucleus under emission of a bremsstrahlung photon. In *direct electron–positron pair production*, this photon is virtual and disintegrates into an electron–positron pair. Because of the large mass of the muon, both the direct pair production and the bremsstrahlung process only play a role above muon energies of 100 GeV, where their contributions to the energy loss are of similar size and rise linearly with the muon energy. Below 100 GeV, muons lose their energy predominantly by *excitation and ionisation* of atoms.

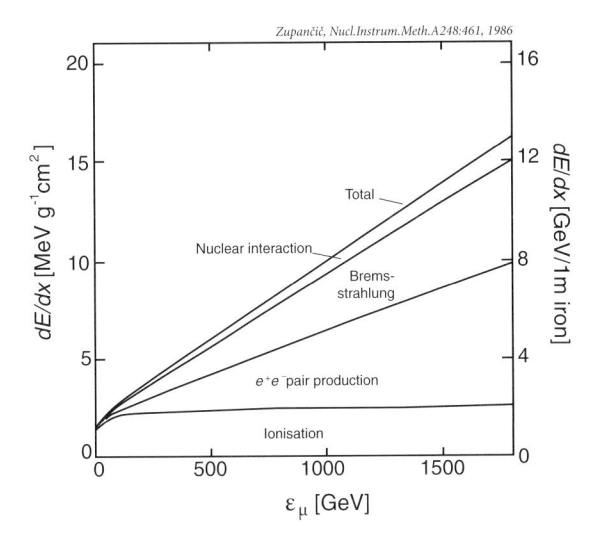

Figure 16.2 Contributions of the individual primary processes to the specific energy loss of muons (adapted from [3]).

According to Figure 16.2, muons with energies below 100 GeV lose less than 2 GeV of energy when passing through 1 m of iron, while a 1 TeV muon loses about 8 GeV. The total energy loss of muons in the calorimeters of ATLAS and CMS is approximately the same as in 3 m of iron, such that muons need at least 4–5 GeV of energy in order to escape the calorimeters.

Highly energetic muons lose only a small fraction of their energy in the detector, while other charged particles are absorbed in the calorimeters. Since electrons are 200 times lighter than muons, they lose most of their energy by emission of bremsstrahlung. Charged hadrons undergo strong interactions, resulting in their absorption in the hadron calorimeters. Consequently, all charged particles which escape the calorimeters are usually classified as muons. For a high muon-detection efficiency and a low misidentification rate, muons should be isolated from jets, and their track segment in the muon system should be matched with one reconstructed in the inner tracking detectors. Traditionally, a layer of muon detectors is mounted outside the calorimeters, and the muon momentum is measured from the curvature of its trajectory reconstructed in the inner detector.

The LHC experiments ATLAS and CMS go beyond the traditional muon identification, as they have the capability to measure the muon momentum stand-alone in their muon systems. The implemented technological choices differ because the designs of the detectors follow different strategies. In order to explain these strategies, the principles of the momentum measurement of charged particles are briefly described in the next section.

16.3
Measurement of Muon Momenta

The charge and the momenta of charged particles are deduced from the curvature of their trajectories in the presence of a magnetic field. When a charged particle of charge q and momentum p travels an infinitesimal distance dl in a uniform magnetic field B orthogonal to its path, it is deflected by the angle

$$d\alpha = \frac{q}{p} B dl . \tag{16.1}$$

If the magnetic field along the full path S of the particle is orthogonal to S, but varying in size, the total deflection angle α becomes

$$\alpha = \frac{q}{p} \int_S B dl . \tag{16.2}$$

The sign of the deflection angle distinguishes positively from negatively charged particles; its magnitude is inversely proportional to the particle momentum. Assuming that the field integral $\int B dl$ is known with higher precision than the deflection angle and applying Gaussian error propagation, (16.2) leads to the following expression for the fractional momentum resolution, δp:

$$\frac{\delta p}{p} = \frac{1}{\int_S B dl} \cdot \frac{p}{q} \cdot \delta \alpha , \tag{16.3}$$

with $\delta \alpha$ being the measured precision of α which has two contributions: the constant angular resolution $\delta \alpha_0$ of the tracking system; and a multiple-scattering term which is inversely proportional to the momentum and proportional to a constant, c_{MS}, which takes into account the amount of material traversed by the particle. Overall, the fractional momentum resolution is given by

$$\frac{\delta p}{p} = \frac{1}{\int_S B dl} \cdot \sqrt{\left(\frac{p}{q}\right)^2 \delta \alpha_0^2 + \left(\frac{c_{MS}}{q}\right)^2} . \tag{16.4}$$

Equation 16.4 shows that the fractional momentum resolution is limited by the ratio of the multiple scattering to the field integral. At large momenta, it is dominated by the angular resolution of the tracking detector. ATLAS and CMS adopt different approaches to optimise the momentum resolution. In order to maximise the field integral in the inner detector, CMS uses a magnetised iron muon spectrometer while ATLAS uses an air-core magnet system instrumented with high-resolution muon chambers.

16.4 Muon Identification in ATLAS and CMS

The muon system of ATLAS [2] is designed to measure muon momenta with high accuracy independently of the inner detector. It uses a system of eight superconducting air-core toroid coils, producing a toroidal magnetic field of $B = 0.4\,\text{T}$ on average in the muon system. The air-core structure minimises the amount of material for traversing muons, and hence their multiple scattering. Deflections of the muons in the spectrometer's barrel section are measured by three layers of tracking chambers with a spacing of 2.5 m. The inner and outer layers are attached to the toroid coils, while the middle layer is mounted in the middle of the coils. The endcap section consists of three disks of tracking chambers with 6 m spacing with the two inner disks enclosing the endcap toroid system. The sagittae of bent muon trajectories in the spectrometer are of the order of $0.5\,\text{m}/p\,[\text{GeV}]$, that is, 500 µm for muon momenta, $p^{\mu} = 1\,\text{TeV}$. To measure these with an accuracy of 10%, ATLAS uses *monitored drift-tube chambers* (MDT) which have a spatial resolution of better than 40 µm. An optical *alignment system* is used to monitor the geometry of the muon spectrometer with 30 µm accuracy. Since the monitored drift-tube chambers have a maximum response time of 700 ns and are therefore slow compared to the 25 ns bunch-crossing frequency of the LHC, *resistive-plate chambers* (RPC) in the barrel part and *thin-gap chambers* (TGC) in the endcap are used for triggering. With a response time of $< 10\,\text{ns}$, these triggering chambers are capable of assigning muons to the correct bunch crossing. In addition, the resistive-plate chambers provide the third coordinate, as the precision chambers measure only one two-dimensional projection. The muon spectrometer is instrumented up to pseudorapidities of 2.7.

In the design of the CMS detector, great emphasis is given to a very good momentum resolution of the inner tracking detector. As momentum resolution is inversely proportional to the bending power, a superconducting solenoid coil provides a high magnetic field strength of $B = 3.8\,\text{T}$ to the inner tracking detector[40]. To close the field lines, the CMS solenoid needs to be complemented with an iron yoke, which is instrumented with four layers of muon chambers. As the momentum resolution of the CMS muon system is dominated by multiple scattering in the iron yoke, tracking chambers with a spatial resolution of $\approx 100\,\text{µm}$ are sufficient in the muon system. Drift-tube chambers (DT) with a maximum drift time of up to 380 ns instrument the barrel part of the CMS muon system and *cathode-strip chambers* (CSC) with 50 ns response time the endcaps. All three chamber types are able to identify the correct bunch crossing with on-chamber electronics and contribute to the first-level muon trigger. Fast RPCs are used in the barrel and the endcap sections of the muon spectrometer, providing a measurement complementing the three-dimensional track segment and triggering information obtained from the precision chambers. The CMS muon system has a pseudorapidity coverage up to 2.4.

---

40) For comparison: as ATLAS does not have an iron yoke, the magnetic field in the inner detector of ATLAS is only 2 T.

The muon momentum resolutions achievable in stand-alone mode of the ATLAS and CMS barrel muon systems are shown in Figure 16.3. The ATLAS muon spectrometer has a very good transverse momentum resolution of 2–4% for transverse muon momenta $p_T^\mu \leq 300\,\text{GeV}$; the resolution reaches 12% for $p_T^\mu = 1\,\text{TeV}$. Energy-loss fluctuations only influence the momentum resolution at very low p_T^μ. The resolution has a lower limit of 2% caused by multiple scattering in the muon spectrometer. For $p_T^\mu > 300\,\text{GeV}$, the curvatures of the muon trajectories in the muon spectrometer are so small that the intrinsic spatial resolution of the chambers and the limited accuracy of the chamber alignment dominate the momentum resolution of the spectrometer.

In the CMS muon system, the transverse momentum resolution is limited by the chamber resolution only for high values of p_T^μ, while multiple scattering in the iron yoke is the dominant contribution up to $p_T^\mu \approx 300\,\text{GeV}$. For muon momenta below 1 TeV, the larger amount of multiple scattering in CMS is partly compensated by the five times larger bending power of the CMS magnetic field such that the stand-alone transverse momentum resolution for soft muons is only three times worse than in ATLAS. For TeV muons both detectors have a similar momentum resolution of about 10%.

The situation improves significantly when combining the momentum measured by the muon system with the one of the inner detector (see Figure 16.4). CMS achieves an overall p_T^μ resolution between 0.5 and 2% for $p_T^\mu \leq 300\,\text{GeV}$, due to the high bending power and the high resolution of the inner detector system. The bending power in the ATLAS inner detector is about half as large as in CMS, and the combined muon momentum resolution is about two times poorer than in CMS. The bending power of the CMS magnet is much smaller in the endcaps than in the barrel part of the muon system which makes the stand-alone muon momentum resolution worse. For muons measured by the endcap detectors, the precision is totally determined by the inner detector, while ATLAS profits from its high stand-alone performance.

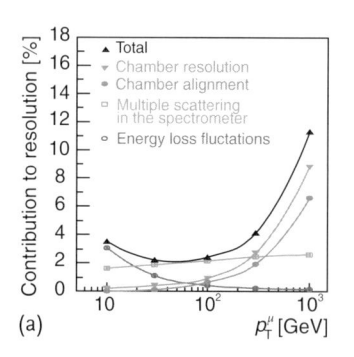

Figure 16.3 Momentum resolution of the barrel muon systems in stand-alone mode. (a) ATLAS [2]; (b) CMS [4]. (The curves are the result of a fit of the function $\sqrt{a_{MS}^2 + (a_{chamber} \cdot p_T)^2}$ to the published resolution points.)

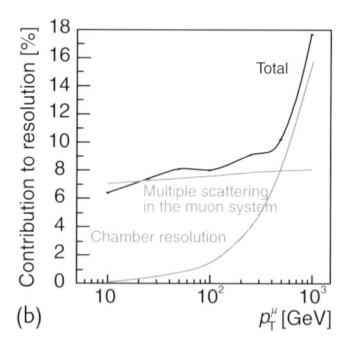

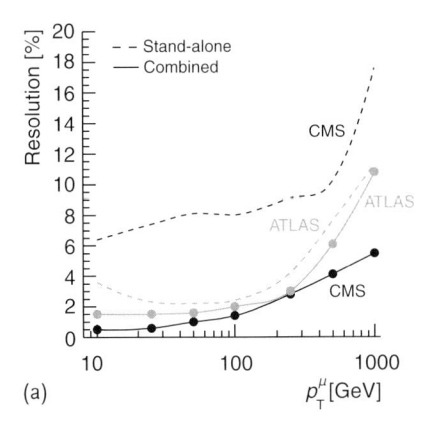

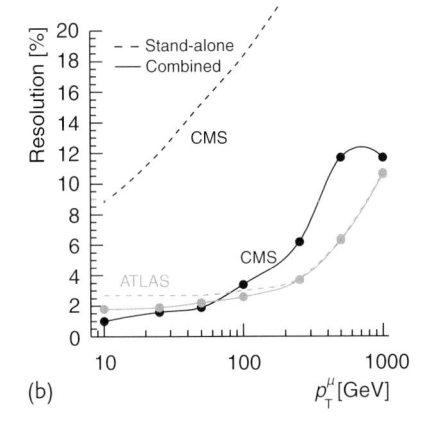

Figure 16.4 Momentum resolution of the ATLAS and CMS detectors after combining the momentum measurements of the inner detectors and the muon systems [2, 4]. (a) momentum resolution for muons in the barrel region; (b) momentum resolution for muons in the endcap region.

16.5
ATLAS and CMS Muon Chambers

The muon systems of ATLAS and CMS are both based on gaseous detectors for precision tracking and triggering. Given the large surfaces to be covered, this is the only detection technology besides scintillator plates providing a reasonable cost-per-channel. Scintillators have frequently been used for muon detection in the past, a good choice if muon identification is the primary task. However, they provide a lower spatial resolution than gaseous detectors. Small cell sizes of $\mathcal{O}$ (mm–cm) are achievable with gaseous detectors and further enhanced with methods such as drift-time readout or operating at overpressure. Being the oldest particle detection technology, by now a large variety of gaseous detectors have been developed, notably for applications in muon systems.

In CMS the barrel muon technologies are precision drift-tube chambers and resistive-plate chambers, while the endcap region is instrumented with cathode-strip chambers for precision tracking complemented by resistive-plate chambers [4, 5]. There are two reasons for the choice of two different technologies for precision tracking in the barrel and endcaps:

- the presence of a magnetic field would affect the drift of primary and secondary electrons, thus requiring a field-free environment or sufficiently small drift distances. Chamber slots without magnetic field are only available in the barrel where the uniform magnetic field is largely contained in the iron. This is not the case in the endcaps where the field is inhomogeneous, hence requiring a detection technology insensitive to magnetic fields such as CSCs;

- the barrel rates of $\mathcal{O}$ (1 Hz) are moderate enough to essentially avoid multiple hits in the long drift tubes even at LHC design luminosity, while the forward region can experience about an order of magnitude higher rates due to the solenoidal field. This argument is even stronger for chambers at high pseudorapidity where rates of $\mathcal{O}$ (1 kHz) may occur. CSCs have a better rate capability due to their shorter drift distances. An alternative would have been drift tubes of smaller diameter which increases the number of channels and, hence, the cost.

The ATLAS muon system [2] is characterised by the toroidal magnetic field and no material between the chambers, hence a very high resolution is beneficial. The precision chambers are all based on drift tubes, in the barrel as well as most of the forward region. Only very high pseudorapidities are instrumented with CSCs, providing a better handling of the higher rates. Both areas are in addition instrumented with RPCs, mainly providing the third coordinate. The endcap region is also instrumented with thin-gap chambers for triggering purposes.

16.5.1
Drift-Tube Detectors

Drift tubes with either round, square or rectangular cross section and a central anode wire operating in avalanche or drift mode are very common in many muon systems. In muon spectrometers tubes of $\mathcal{O}$ (cm) diameter are typically arranged in chambers of $\mathcal{O}$ ((1–5) m^2) area.

The rectangular cross section of the CMS barrel drift-tube chambers allows a full coverage with sensitive material. Dead spaces are only caused by the 1 mm thin walls between cells of 4×1 cm^2 cross section. ATLAS monitored drift tubes are based on round cells of 3 cm diameter stacked in layers.

Because of the low occupancy in muon systems, the chambers in both detectors can operate with a relatively slow standard drift gas, for example Ar/CO$_2$. For example, 85% Ar with 15% CO$_2$ has a drift velocity of $\sim$55 μm/ns at nominal pressure, yielding a maximum drift time for the CMS drift chambers of 380 ns. This corresponds to 16 bunch crossings at the LHC, which is acceptable as the occupancy in the muon system is very low. In addition, the bunch can be identified with electronics based on the "mean-timer" method where the drift times of four staggered cells are combined to a unique track segment.

Measuring the drift time, a method pioneered by central drift chambers, yields a resolution which is about 50 times better than a pure "digital" readout, where only the signal wire is identified. Such a resolution can be improved by almost an order of magnitude when operating at overpressure as done with the ATLAS monitored drift tubes. Their point resolution at 3 bar is about 80 μm [2] to be compared to 250 μm at nominal pressure.

The ATLAS tubes are assembled and tested individually before being glued together to form either rectangular or trapezoidal chambers (by varying the length of the individual drift tubes). Two groups of three individual layers form a tracking station, as shown in Figure 16.5. Each drift-tube station measures the muon tracks

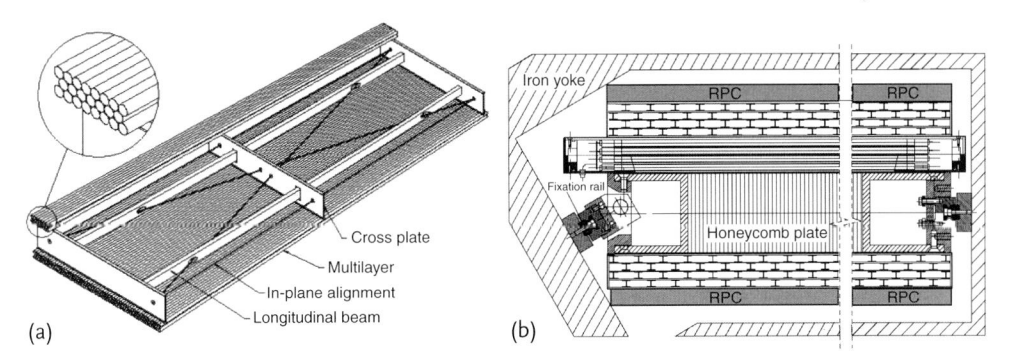

Figure 16.5 Implementations of drift tubes in muon systems. (a) ATLAS monitored drift tubes [2] combine two groups of three individual layers measuring the transverse projection. (b) CMS barrel drift-tube chambers combine three superlayers of four individual layers and measure both projections. The drift-tube chambers are enclosed between RPC chambers [5].

in the bending plane of the toroidal magnetic field while the trigger chambers measure the muon tracks in the orthogonal plane.

CMS uses $4 \times 1 \, \text{cm}^2$ drift cells with a 50-μm steel wire, operated at ambient pressure. These cells are not pre-fabricated as individual tubes but are formed by gluing together large aluminium plates with a set of spacers. This design was first used for UA1 with cell sizes of $10 \times 3 \, \text{cm}^2$, almost one order of magnitude larger than the CMS cells, reflecting the increasing resolution requirement for muon systems. Most of the CMS muon barrel stations are built from three superlayers, each consisting of four individual layers of drift tubes, as seen in Figure 16.5. While two such superlayers (separated by 30 cm) measure the $r\phi$ projection, the third one, rotated by $90°$, determines the rz projection, such that a muon station provides a three-dimensional track segment.

16.5.2
Resistive-Plate Chambers

Resistive-plate chambers were developed from spark chambers and provide a large signal amplitude while being relatively simple in their construction. They are one of the few implementations of gaseous detectors without anode wires. A thin (2 mm) gas gap is enclosed between highly resistive plates (either Bakelite with $\rho \approx 10^9$–10^{11} Ω cm or glass with $\rho \approx 10^{13}$ Ω cm) covered with a conductive graphite coating on the outside. The movement of the charge in the gas-filled gap induces a signal outside the plates, which is subsequently picked up by external readout strips (or pads), which usually cross each other and thus provide a 2-dimensional readout. Combined with the small thickness an RPC station thus provides a 3-dimensional space point. Figure 16.6 shows a double-gap RPC with a common readout strip enclosed between both chambers.

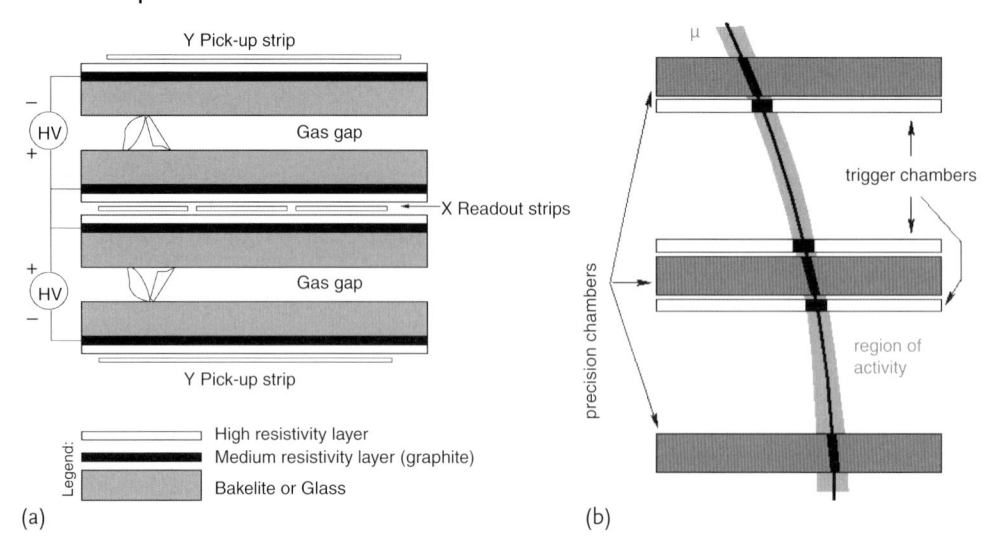

Figure 16.6 (a) principle of a double-gap RPC; (b) track reconstruction in the barrel of the ATLAS muon spectrometer.

ATLAS and CMS instrumented 3650 m² and 8000 m² respectively, both operating in avalanche mode.

16.5.3
Cathode-Strip Chambers

The conceptual design of a cathode-strip chamber is shown in Figure 16.7. The anode wires are arranged in planar layers spaced by 1–2 mm, thus determining the spatial resolution. The layer of wires is enclosed between two cathodes, and the gap is filled with an appropriate gas mixture for generating primary electrons and amplifying the signal near the anode wires, just like with drift tubes.

The anode signal is fast, determined by the drift velocity, but the drift velocity of the positively charged ions to the cathode is a 1000 times slower. While the previously described detectors do not use the cathode signal, CSCs with a segmented cathode do. The charge spreads over several strips and charge interpolation provides a resolution better than (strip width)/$\sqrt{12}$. Combining the information from the cathode strips and anode wires provides a three-dimensional space point. Such cathode-strip chambers are the precision detector component of the CMS muon endcaps and instrument the regions of high pseudorapidity of the ATLAS muon endcaps. Their main advantages in these forward regions are their capability to handle larger occupancies than drift tubes and that they are rather insensitive to the magnetic field due to their short drift distances. Most of the ATLAS endcap region is instrumented with thin-gap chambers – multi-wire proportional chambers with a signal from the anode wires.

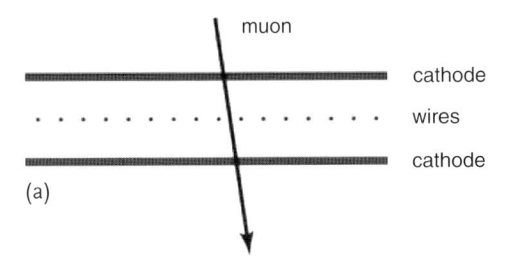

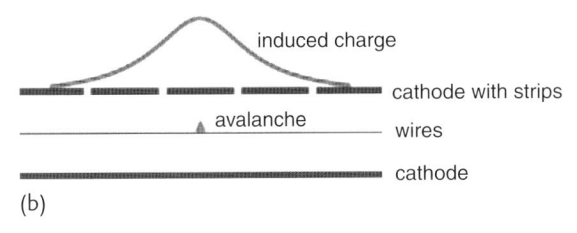

Figure 16.7 Working principle of a cathode-strip chamber. (a) Passage of a muon through a layer of a cathode strip chamber. (b) Distribution of the charged induced on the cathode strips after avalanche formation at the wire.

16.6
Muon Track Reconstruction and Identification

All high-momentum tracks that are recorded by the muon system of ATLAS and CMS are muon tracks. Both experiments follow the same track-reconstruction strategy.

Figure 16.6b illustrates the track-reconstruction steps in case of a muon with $p_T^\mu > 10\,\text{GeV}$ in the barrel part of the ATLAS muon spectrometer. As the trigger chambers have a much lower granularity and simpler geometry than the precision chambers, their hits can easily be used to define a so-called "region of activity" through which the muon must have passed. In the second step of the track reconstruction, straight segments are reconstructed in the precision chambers within the region of activity. The local straight segments of the region of activity are then combined to a curved candidate trajectory which is finally refined by a global refit of the hits lying on the candidate trajectory.

The combination of the trajectory found in the muon system with a matching trajectory in the inner detector serves two goals: (1) the improvement of the momentum resolution, especially in the case of CMS where the stand-alone muon momentum resolution is poor compared to the momentum resolution of the inner detector; (2) the rejection of muons from pion and kaon decays. The mother pions and kaons of the muons have decay lengths of a few metres. Thus, most of them leave a long track in the inner detector before they decay. Since a considerable fraction of the energy of the mother pion or kaon is carried away by neutrinos, the energy of the decay muon is significantly lower than the energy of its mother.

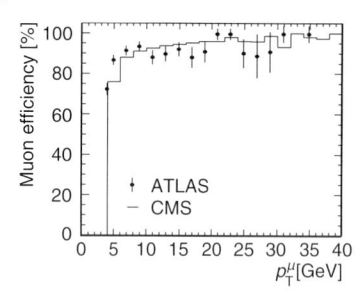

Figure 16.8 Muon-identification efficiency of ATLAS [7] and CMS [6].

Consequently the momentum of the inner-detector track does not match that of the muon-system track. Therefore, muons from pion or kaon decays can be rejected by requiring a good match between the inner-detector and muon-spectrometer trajectories.

The procedure described above cannot be applied to muons of $p_T^\mu < 10\,\text{GeV}$ because their trajectories do not extend to the outermost layers of the muon systems. For CMS such muons are absorbed by the iron yoke; in the case of ATLAS, the magnetic field in the spectrometer is strong enough to bend them away from the outermost layers. Therefore, the reconstruction starts with a low-momentum inner-detector track in the same solid angle as the segments found in the muon system, extrapolates it to the muon system, and the segments lying on the extrapolation are selected. The number of these segments, their length and position must be compatible with the momentum of the inner-detector track. This requirement rejects muons from pion and kaon decays. Pions and kaons decaying in the calorimeters deposit much more energy there than muons which are minimum ionising particles. One therefore also requires that the energy deposited along the extrapolated inner-detector trajectory in the calorimeters is compatible with the muon hypothesis. Monte Carlo studies [6, 7] show that an efficiency $> 80\%$ and a fake rate $< 0.5\%$ can be achieved for muons of $p_T^\mu > 5\,\text{GeV}$ in both experiments.

Figure 16.8 summarises the muon-identification efficiency of ATLAS and CMS. The two experiments achieve the same track-reconstruction efficiency, namely about 90% for $5 < p_T^\mu < 20\,\text{GeV}$ and $> 96\%$ for $p_T^\mu > 20\,\text{GeV}$ with a fake rate of $< 0.5\%$. This result is particularly difficult to achieve in the case of ATLAS. Simulations [8] show that, different from CMS, a sizable amount of low-energy neutrons leaks out of the calorimeters into the muon spectrometer. These neutrons excite nuclei in the entire experimental hall of ATLAS, such that the muon spectrometer is operated in a huge neutron and γ background. This background leads to occupancies of up to 20% in the precision chambers, but less than 0.3% in the trigger chambers due to their much shorter response time. The low occupancy of the trigger chambers is responsible for a reliable determination of the regions of activity in the track reconstruction in the muon spectrometer of ATLAS.

References

1 Sjostrand, T. *et al.* (2006) PYTHIA 6.4 physics and manual, *JHEP*, **05**, 026, doi:10.1088/1126-6708/2006/05/026.
2 ATLAS Collab. (1997) The ATLAS Muon Spectrometer – Technical Design Report, Tech. Rep., CERN/LHCC 97-22. http://cdsweb.cern.ch/record/331068.
3 Zupancic, C. (1985) Physical and Statistical Foundations of TeV Muon Spectroscopy. http://cdsweb.cern.ch/search?sysno=000073705CER, Contribution to Workshop on Muon Detection at SSC/LHC, Madison, WI, 4–6 Apr 1985.
4 CMS Collab. (1997) The Muon Project – Technical Design Report, Tech. Rep., CERN/LHCC 97-32. http://cmsdoc.cern.ch/cms/TDR/MUON/muon.html.
5 CMS Collab. (2008) The CMS experiment at the CERN LHC, *JINST*, **3**, S08004, doi:10.1088/1748-0221/3/08/S08004.
6 James, E. *et al.* (2006) Muon identification in CMS, http://cdsweb.cern.ch/record/927392, CMS Note 2006/010.
7 ATLAS Collab., Aad, G. *et al.* (2008) The ATLAS experiment at the CERN Large Hadron Collider, *JINST*, **3**, S08003, doi:10.1088/1748-0221/3/08/S08003.
8 Baranov, S. *et al.* (2005) Estimation of Radiation Background, Impact on Detectors, Activation and Shielding Optimization in ATLAS, http://cdsweb.cern.ch/record/814823, ATLAS Note ATL-GEN-2005-001.

17
Luminosity Determination: Normalising the Rates

Ian C. Brock and Hasko Stenzel

17.1
Outline

Luminosity is a common ingredient to any cross-section measurement through the basic relation

$$\sigma = \frac{N}{L}, \tag{17.1}$$

where N is the experimentally determined number of events and L the absolute integrated luminosity for the data-taking period leading to the observation of N. Experimentally the luminosity determination is often split into the *relative* monitoring of the luminosity during a run or fill of the accelerator and the *absolute* calibration of the integrated luminosity, which is used offline in the physics analyses.

Several methods for luminosity measurements are proposed for Terascale colliders, which can be classified into two groups: accelerator-type determinations based on beam properties measured with appropriate machine instrumentation devices and rate measurements of selected physics processes. The latter are typically obtained with dedicated small-angle subdetectors of the experiments.

In this chapter the relative and absolute luminosity measurements using physics processes are discussed, while the accelerator methods are developed in Chapter 12. The relative luminosity monitoring requires rate measurements of the selected processes with sufficiently large acceptance, good linearity and time stability. The absolute calibration involves an accurate determination of the acceptance and efficiency of the monitor and a cross section of the process in question which has to be experimentally measured or theoretically calculated with good precision.

Note that two values for the luminosity are usually needed. The machine wants to know how much luminosity is delivered to the experiments. The experiment wants to know what is the integrated luminosity which corresponds to the data they have collected. It is therefore necessary to know the dead-time of the experiment (see Chapter 18.2.4) and correct the numbers accordingly. How this correction is done depends on whether the events are simply counted or whether the information in the luminosity detectors is read out as part of the normal data chain.

17.2
Luminosity Determination in e^+e^- Machines

In e^+e^- colliders, *Bhabha* scattering at small angles, $e^+e^- \to e^+e^-(\gamma)$, is usually used to measure the luminosity. Its cross section can be calculated with high precision from Quantum Electrodynamics (QED) and depends only weakly on the properties of the Z boson, even when running at centre-of-mass energies close to the Z pole. The cross section for the process is large, which leads to a small statistical uncertainty on the measurement, and there is usually no statistical limitation when studying the experimental systematic uncertainties. Colliders running at energies below the Z pole usually used small-angle Bhabha scattering to measure the relative luminosity and large-angle scattering for the absolute luminosity.

The discussion here concentrates on the measurements made at the Z pole, as these impose the toughest demands on the measurement precision. The ratio $\Gamma_{\text{inv}}/\Gamma_\ell$, that is, that of the decay width of the Z into invisible particles (neutrinos) over the decay width to charged leptons can be calculated very accurately in the Standard Model as discussed in Chapter 3. It is also a crucial value for the determination of the number of light neutrino families, which was discussed in the same chapter. For both of these measurements the error on the integrated luminosity is one of the larger systematic uncertainties.

To lowest order, the small-angle Bhabha cross section (integrated over the azimuthal angle, ϕ) in a detector with a polar-angle coverage from θ_{min} to θ_{max}, is given by

$$\sigma = \frac{16\pi\alpha^2}{s}\left(\frac{1}{\theta_{\text{min}}^2} - \frac{1}{\theta_{\text{max}}^2}\right), \tag{17.2}$$

where α is the fine-structure constant and s is the square of the centre-of-mass energy.

The most important features of a detector for the accurate determination of the luminosity are:

- well-known geometry;
- very high and well-known trigger efficiency;
- small and well-understood backgrounds;
- full coverage of the azimuthal angle.

The last item ensures that small transverse offsets of the detector with respect to the beam have a negligible effect on the luminosity determination.

At the time of the original design of the LEP detectors, a goal of a luminosity measurement with $\approx 1\%$ precision was set and was also achieved during the first few years of running. However, when running at or close to the Z peak, the efficiency of the detectors for the process $Z \to$ hadrons could be measured with a systematic uncertainty of about 0.1%. The Bhabha cross section had also been recalculated with higher precision, so that a measurement of the luminosity with an accuracy of about 0.1% was desirable. More detailed information on how this precision was achieved can be found in [1] and references therein.

The LEP experiments took somewhat different approaches to achieve this goal. The L3 experiment installed a silicon tracker in front of the BGO calorimeter, while the OPAL and ALEPH collaborations built new luminosity monitors using a combination of tungsten absorbers and silicon wafers. DELPHI used a calorimeter of lead and scintillating tiles together with two layers of silicon detectors close to the shower maximum. With the new detectors, it was possible to achieve an uncertainty on the luminosity measurement that was very similar to the precision with which the efficiency for detecting hadronic Z decays could be determined.

The L3 luminosity monitor consisted of a calorimeter made of bismuth germanate (BGO) crystals, which provided excellent energy resolution and a very efficient trigger, complemented by a tracker made of single-sided silicon wafers (SLUM). The wafers had a very high intrinsic geometrical precision (1–2 μm) and could be accurately positioned and measured (6 μm). An overview of the position of the detector in the L3 experiment is shown in Figure 17.1. Bhabha events were selected using the calorimetric measurement in the BGO to provide a background-free sample of events, while the silicon tracker was used to select only those Bhabhas that were contained in a precisely defined fiducial volume.

For both experimental and theoretical reasons it was preferable to select events inside a narrow fiducial volume (N) on one side of the detector and a wider re-

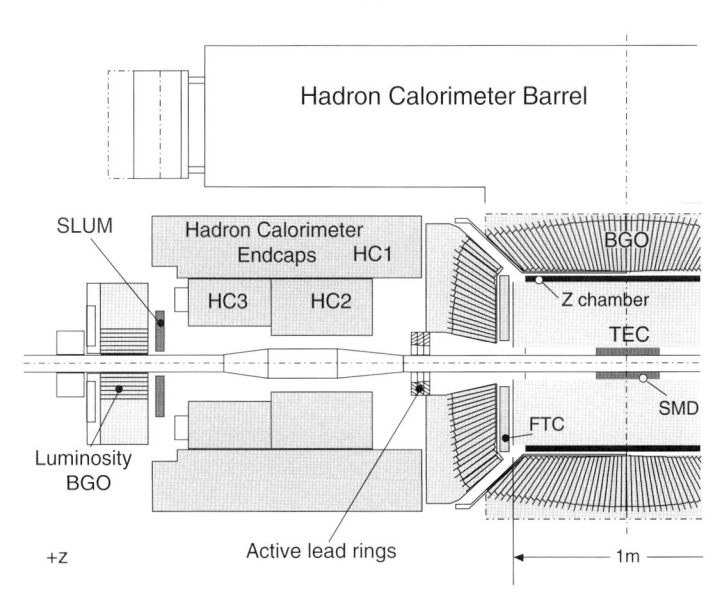

Figure 17.1 Side-view of the central region of the L3 experiment on the $+z$ side, showing the beam-pipe and the position of the luminosity BGO calorimeter and the silicon tracker (SLUM).

Table 17.1 Systematic uncertainties on the luminosity measurement at L3 in the 1994 data. The second column shows the uncertainty using only the BGO calorimeter and the third column shows that for the combined BGO and silicon analysis.

Source	Contribution to $\Delta L/L$ (%)	
	BGO analysis	BGO + silicon analysis
Trigger	Negligible	Negligible
Event selection	0.3	0.05
Background	Negligible	Negligible
Geometry	0.4	0.03
Monte Carlo statistics	0.06	0.06
Theory	0.11	0.11
Total	0.5	0.14

gion on the other side (W). Such a selection also accepts radiative events where the photon remains undetected in the beam-pipe (initial-state radiation). In addition, it is better to select the scattered electron and positron using a calorimeter measurement, because events with photon radiation close to the scattered electron or positron (final-state radiation) are also accepted due to restrictions in spatial resolution. At the end, the average value obtained from using a narrow selection on one side and then on the other side was taken.

The systematic uncertainties on the measurement for the L3 detector are summarised in Table 17.1.

ALEPH and OPAL both built silicon-tungsten calorimeters. The use of tungsten limited the size of the electromagnetic showers and allowed very compact calorimeters. As the measurement of the radial coordinate was most important, the pads were curved and subdivided much more finely in radius than in azimuth. The acceptance was defined using the silicon layers close to the shower maximum (about $7X_0$). For the layers that defined the fiducial volume, a very accurate survey of their geometry was necessary. Even things like thermal expansion of the detector (about $2\ \mu m/°C$) had to be taken into account in order to achieve the required level of accuracy.

DELPHI followed a somewhat different approach. Their calorimeter was made of $27X_0$ of lead equipped with 47 layers of scintillating tiles. In addition silicon pads were placed at depths of $4X_0$ and $7X_0$. On one side of the experiment they placed a carefully machined tungsten mask, which was used to define the inner radius cut. This mask was surveyed and aligned to an accuracy of $20\ \mu m$.

All four experiments ultimately achieved experimental uncertainties on the luminosity measurement below 0.1% – quite a remarkable achievement for a cross section that changes so fast with the scattering angle!

In parallel to the improvement in the detectors there was also rapid progress in the theoretical calculations. This was to a large extent due to the fact that two independent groups provided Monte Carlo generators that were used by the experiments. This meant that the generators could be cross-checked – there was also a healthy competition to provide the best generator [2]. The OLDBIS generator (a revision of the BABAMC generator [3, 4]) was ultimately used as a cross-check, while BHLUMI [5] was used as the main generator. Monte Carlo events were generated for the process $e^+e^- \to e^+e^-(\gamma)$ at $\sqrt{s} = 91.25$ GeV using BHLUMI V2.01 and then passed through the detector simulation programs[41]. With the computing power available at the time (early 1990s), it was quite a challenge to generate enough Monte Carlo events to keep the statistical error on the MC below that of the measurements.

At a future e^+e^- linear collider such as the ILC, the required precision on the measurement of the luminosity is very similar to that at LEP. Similar calorimeters to those used by ALEPH and OPAL are planned in order to achieve this goal. Studies show that such detectors would also work well at the ILC and that in particular the background levels can be kept under control.

17.3
Luminosity Determination at HERA

The original aim at HERA was to measure the luminosity with a precision of about 5%. After the start of data-taking, it fairly quickly became clear that it would be possible to achieve a higher precision, and ultimately an uncertainty of 1–2% was achieved. Like at LEP, at HERA it is possible to use a simple QED process with high rate to measure the luminosity – essentially bremsstrahlung radiation off the incoming electron in the field of the proton, $ep \to e'p\gamma$. The process is called the *Bethe–Heitler* process, and its rate was first calculated in 1934 [7]. Neglecting the spin and finite size of the proton the cross section is given by

$$\frac{d\sigma_{BH}}{dE_\gamma} = 4\pi r_e^2 \frac{E_{e'}}{E_\gamma E_e} \left(\frac{E_e}{E_{e'}} + \frac{E_{e'}}{E_e} - \frac{2}{3} \right) \left(\ln \frac{4 E_p E_e E_{e'}}{M_p M_e E_\gamma} - \frac{1}{2} \right). \quad (17.3)$$

As is usual for a bremsstrahlung process, the photon and electron emerge at a very small angle, θ_γ, with respect to the incident electron direction:

$$\frac{d\sigma}{d\theta_\gamma} \sim \frac{\theta_\gamma}{((m_e/E_e)^2 + \theta_\gamma^2)^2}. \quad (17.4)$$

The initial idea at HERA was to measure both the scattered electron and the radiated photon. The scattered electron is bent in the field of the beam magnets, and, since it has a lower energy than the beam electrons, it will at some point leave the beam-pipe. It can then be detected, and due to the relation between particle momentum and bending radius in a magnetic field, the position of the detector will then define the electron energy. In this process the recoil of the proton can be

---

41) The L3 detector simulation is based on GEANT Version 3.15, see [6].

neglected to a very good approximation, so that the following relation holds:

$$E_\gamma + E_{e'} = E_e \, . \tag{17.5}$$

This, in turn, means that if one can simultaneously detect the photon and the electron, it is possible to precisely calibrate the photon energy. The background conditions at HERA were such that it was actually possible to only measure the radiated photon, while the electron detectors were used for calibration purposes.

The rate of Bethe–Heitler events is so high that events (photons) were just counted, rather than being recorded. This also meant that it was straightforward to provide both a luminosity for the machine (without dead-time) and one that took the dead-time into account, by simply gating the scalars which count the photons with the experiment live-time.

Challenges to a precise measurement of the luminosity include: good control of the geometrical acceptance; pile-up effects; a good absolute energy calibration; non-Gaussian proton-beam tails (*satellite bunches*) and suppressing synchrotron radiation without adding too much dead material in front of the detector.

For the HERA 1 running period an experimental uncertainty of $\approx 1\%$ was achieved. Multiple interactions from a single bunch were not too much of a problem and the size of the satellite bunches was quite small. In addition there is a theoretical uncertainty of between 0.5% and 1% depending on the selection criteria.

With the HERA 2 upgrade in 2002 (see Chapter 12), the rate of interactions increased to more than 1 per bunch crossing – an effect which required significant corrections. In addition, due to changes in the machine optics and other effects, the synchrotron-radiation background increased substantially. To compensate for this, the ZEUS collaboration increased the thickness of the carbon absorber in front of the photon detector to about $4X_0$. A second independent luminosity measurement was made by using a window of well-defined thickness (85% aluminium, 11% silicon plus small amounts of copper, iron and magnesium) which converted about 10% of the Bethe–Heitler photons into e^+e^- pairs. These were then bent in a dipole field and detected in a separate detector, the so-called spectrometer. The reduction of the rate of bremsstrahlung photons by a factor of 10 for this measurement also meant that multiple interactions were less of a problem. Ultimately the spectrometer had a lower systematic uncertainty than the photon calorimeter and was used as the main device to measure the luminosity.

The H1 collaboration also used a photon detector placed around 100 m from the experiment to measure the photons from the Bethe–Heitler process during both HERA running periods. In addition they used photons from Compton scattering at large angles ($\theta > 100$ mrad) to determine the integrated luminosity for longer running periods. The latter process gave a statistical precision of around 1% using several months of data.

17.4
Luminosity Determination at Hadron Colliders

In this section, three methods for luminosity determination at hadron colliders are presented, together with their implementation at the Tevatron and the plans for the LHC.

The traditional method for beam luminosity determination at hadron colliders relies on the counting of inelastic events with small-angle detectors with a large acceptance for inelastic processes placed at large pseudorapidity. Alternatively, if the experiment is equipped with *roman pots* (detectors mounted inside the beam-pipe and operated in vacuum that are moved close to the beam once stable conditions have been declared) at even larger pseudorapidities, then the well-known Coulomb interaction contributing to elastic scattering can be exploited to obtain an absolute luminosity calibration. Recently, also high-p_T processes have been proposed to calculate the luminosity for large datasets; in particular the inclusive production of W and Z bosons is a good candidate, because it profits from small theoretical systematic uncertainties of the next-to-next-to-leading order (NNLO) Quantum Chromodynamics (QCD) calculations and a steadily improving PDF (parton distribution function) precision.

17.4.1
Luminosity from Counting Inelastic Events

The rate of inelastic events as measured by a given luminosity monitor (LM) is related to the absolute luminosity by

$$\mathcal{L} = \frac{dN}{dt} \frac{1}{\sigma_{LM}}, \tag{17.6}$$

where $\sigma_{LM} = \sigma_{in}/\epsilon_{LM} a$ is the observable cross section seen by the luminosity monitor, ϵ the detection efficiency and a the acceptance. Because of limited geometrical acceptance and efficiency losses, σ_{LM} is smaller than the total inelastic cross section, σ_{in}. The rate $\frac{dN}{dt}$ in (17.6) can be expressed as the mean number of interactions, μ, times the beam-revolution frequency f:

$$\mathcal{L} = \frac{\mu f}{\sigma_{LM}}. \tag{17.7}$$

The challenge for the luminosity monitor is to provide a good measurement of μ and a precise estimate of the acceptance for inelastic events, while the value for the inelastic cross section entering σ_{LM} typically must be taken from other detectors or experiments or from theoretical calculations.

The inelastic cross section can be decomposed into the several contributions,

$$\sigma_{in} = \sigma_{nd} + \sigma_{dd} + \sigma_{sd}, \tag{17.8}$$

where the single-diffractive (sd) and double-diffractive (dd) processes are described by the colour-neutral exchange of a pomeron; colour-exchange processes are classified as inelastic non-diffractive (nd) (see also Chapter 11). The observable cross section for the luminosity monitor is given by

$$\sigma_{LM} = \epsilon_{LM} \left(a_{nd} \sigma_{nd} + a_{dd} \sigma_{dd} + a_{sd} \sigma_{sd} \right), \tag{17.9}$$

where ϵ_{LM} denotes the overall hardware efficiency of the monitor, a_{nd} is the acceptance of the monitor for non-diffractive events, and a_{dd} and a_{sd} are similar terms for the diffractive processes.

The most precise determination of the inelastic cross section does not consist (as (17.8) might suggest) of a measurement of the individual contributions. Instead, the total cross section, σ_{tot}, is measured via the optical theorem and the elastic contribution, σ_{el}, is subtracted in order to obtain σ_{in} via $\sigma_{in} = \sigma_{tot} - \sigma_{el}$. The optical theorem relates the total cross section, σ_{tot}, to the forward elastic scattering amplitude,

$$\sigma_{tot} = \frac{16\pi(\hbar c)^2}{1+\rho^2} \frac{dN_{el}/dt|_{t\to 0}}{N_{in} + N_{el}}, \quad \rho = \left. \frac{\Re(F_{el}(t))}{\Im(F_{el}(t))} \right|_{t\to 0}, \tag{17.10}$$

where N_{in} and N_{el} are the numbers of inelastic and elastic events, respectively, and F_{el} denotes the amplitude of the elastic scattering cross section. This method of using the optical theorem has the advantage of being independent of the actual machine luminosity and was used by the Tevatron experiments CDF [8] and E811 [9], which simultaneously measured the total inelastic rate and the differential elastic rate including the forward scattering region at small t. Then the elastic t spectrum has to be extrapolated to $t \to 0$ to give the optical point needed in (17.10). In practice a fit is applied to the t spectrum, in the simplest case an exponential, to extract the optical point and total number of elastic events, N_{el}. The inelastic cross section according to (17.10) can be re-written in terms of the ratio $R = N_{in}/N_{el}$ of inelastic to elastic events,

$$\sigma_{in} = \frac{16\pi(\hbar c)^2}{1+\rho^2} \frac{dN_{el}/dt|_{t\to 0}}{N_{el}} \frac{R}{(1+R)^2}, \tag{17.11}$$

which facilitates the combination of the two measurements from CDF and E811. The combined value of the inelastic cross section at the Tevatron is $\sigma_{in} = 59.3 \pm 2.3$ mb at $\sqrt{s} = 1.8$ TeV [10]. Assuming a $\ln^2 s$ energy dependence of the cross sections, one obtains for Run 2 ($\sqrt{s} = 1.96$ TeV) $\sigma_{in} = 60.7 \pm 2.4$ mb. This value of σ_{in} is used in the luminosity calibration for all Tevatron experiments.

The relative contributions of diffractive and non-diffractive processes needed for the acceptance in (17.9) are determined by tuning the Monte Carlo simulation to specific distributions measured by the monitor [11] and supplemented by direct measurements of the single diffraction cross section [12]. It is important to distinguish between these contributions since their acceptance for typical luminosity monitors is very different: close to 100% for non-diffractive, about 70% for double-diffractive and as low as 15% for single-diffractive events [11].

An illustrative example of a traditional luminosity monitor is given by the DØ luminosity monitor [13]. It consists of annular scintillator counters surrounding the

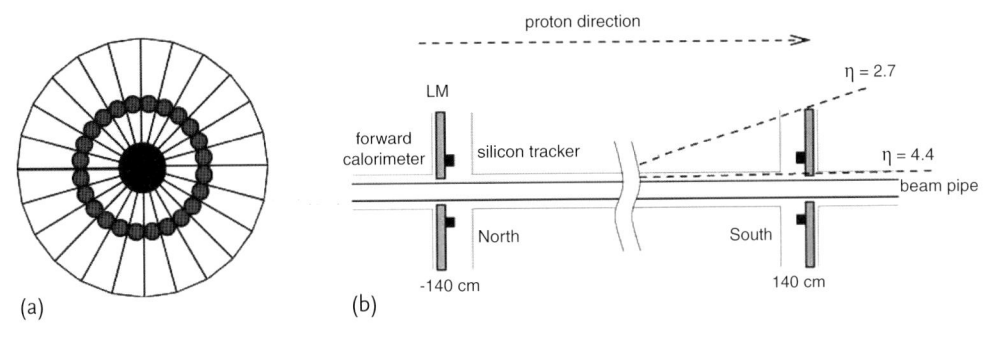

17.4 Luminosity Determination at Hadron Colliders

Figure 17.2 Experimental setup of the luminosity monitor for the DØ experiment. (a) r-φ-view of annular scintillators, (b) r-z view showing the location of the Luminosity Monitor (LM) (adapted from [13]).

beam-pipe and covering a pseudorapidity range of $2.7 < |\eta| < 4.4$ and measures a large fraction of inelastic events with good acceptance. The experimental setup is shown in Figure 17.2.

A convenient method to measure the mean number of interactions per bunch crossing, μ, (called the *zero-counting* method) consists of a measurement of the probability of having no interaction per bunch crossing. In general, the probability, P_n, for n interactions in a given crossing follows Poisson statistics,

$$P_n = \frac{\mu^n}{n!} e^{-\mu}. \tag{17.12}$$

P_0 is easily obtained by taking the ratio of the number of bunch crossings with no interaction over the total number of crossings. In practice the measurement is performed over a period of about one minute in order to achieve, at the same time, a statistical uncertainty for P_0 below 1% with approximately constant instantaneous luminosity. Then μ is simply given by

$$\mu = -\ln P_0, \tag{17.13}$$

which can be substituted into (17.7) and yields the expression for the luminosity,

$$\mathcal{L} = -\ln P_0 \frac{f}{\sigma_{LM}}. \tag{17.14}$$

The final result for the integrated luminosity of Run 2a [13] used the value $\sigma_{LM} = 48.0$ mb and has an uncertainty of 6.1%, which receives equal contributions of 4% from the systematic uncertainties for the inelastic cross section and for the μ determination.

Luminosity determination in CDF relies upon the *Cerenkov Luminosity Counter* (CLC) [14], which consists of conical tubes of gaseous Cerenkov detectors arranged in pointing geometry (i.e. they point back to the interaction point) and located in the forward region of CDF. The tubes have a length of up to 180 cm, are arranged in three concentric layers and cover a pseudorapidity range from 3.75 to 4.75. The counters use isobutane at about atmospheric pressure as the Cerenkov radiator, which gives a Cerenkov angle of 3.2° and a light emission momentum threshold of 9.3 MeV for electrons and 2.6 GeV for pions. Particles from the interaction point traversing the entire tube produce about 100 photoelectrons, detected with photomultipliers connected via light-collecting cones to the tubes. The advantage of this design over annular scintillator counters is the increased radiation tolerance of the Cerenkov counters and their ability to discriminate primary particles from the interaction point from secondary particles, the latter being in part below threshold or traversing the detector at large angles and thus producing smaller signals. Furthermore, in the absence of Landau tails, the number of particles can be reconstructed from the signal amplitude [15]. The single-particle peak can be determined at low luminosity (where the occupancy is low) and can be used to calibrate the amplitude spectrum, after applying isolation cuts to reduce secondary showers produced in the surrounding material.

The detector's capability to measure the number of particles is an asset at a higher Tevatron luminosity of $\mathcal{L} \approx 10^{32}\,\text{cm}^{-2}\,\text{s}^{-1}$ with about 6 $p\bar{p}$ interactions per bunch crossing, where the number of empty bunch crossings becomes small and the statistical uncertainty of the luminosity measurement with the zero-counting method larger. Therefore, as an alternative method, the mean number of interactions can be measured directly by counting the number of particles per bunch crossing, N_B, divided by the mean number of particles at low luminosity, N_0, where only one interaction per bunch crossing is observed,

$$\mu = N_B/N_0 . \tag{17.15}$$

In principle, this method can also be applied to hits rather than to particles, but the hit rates tend to saturate at higher luminosity and deteriorate the linearity, requiring model-dependent Monte Carlo corrections [15].

The CLC design was ported to the LHC and taken up by ATLAS for the *LUminosity Cerenkov Integrating Detector* (LUCID) [16]. LUCID is located 17 m from the interaction point and embedded in an absorber shield protecting machine elements from secondary particles. The layout of LUCID is shown in Figure 17.3 together with other ATLAS forward detectors.

LUCID's present implementation consists of only 20 Cerenkov tubes of length 150 mm with a diameter of 15 mm, coupled in part directly to photomultipliers and in part via quartz fibres to distant multi-anode PMTs. The challenge for LUCID compared to CLC is the much higher radiation level and the more severe background conditions. For the high-luminosity operation, an upgrade of LUCID is planned with an increased number of tubes for better acceptance, and quartz-fibre readout throughout the system to cope with the anticipated radiation level. At low luminosity, the zero-counting method can be employed and the mean number of hits and particles for single interactions per bunch crossing can be calibrated. At high luminosity, with about 25 interactions per bunch crossing, the particle- and hit-counting methods must be used to measure μ. Monte Carlo simulations indicate that saturation effects introducing non-linearity become important at such high luminosities, and the challenge will be to find reliable correction methods.

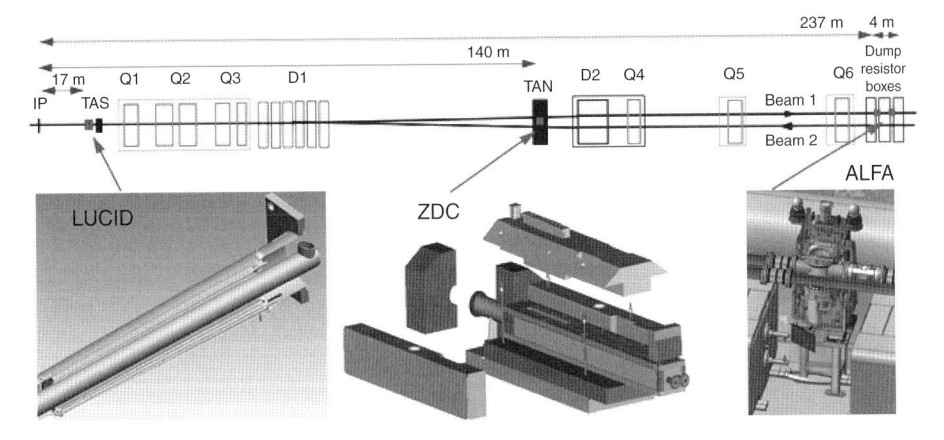

17.4 Luminosity Determination at Hadron Colliders

Figure 17.3 The ATLAS luminosity and forward detectors LUCID, ZDC and ALFA (adapted from [16]).

CMS has chosen the Hadron Forward (HF) calorimeter [17] as its luminosity monitor. The main task of the HF, which covers the pseudorapidity range from 3 to 5, is the measurement of jets. The HF has steel plates as absorber and uses quartz fibres embedded in grooves in the absorber structure as active material. This technology has been selected because of its radiation tolerance – in the endcap a dose of 10 Mrad is anticipated in one decade of LHC operation. Given the primary calorimetry purpose of the HF, counting of particles is not possible. However, the rather fine granularity allows new luminosity-sensitive observables to be derived. In particular the total sum of transverse energy, E_T, calculated in rings of constant η exhibits a clear correlation with the luminosity and a linear behaviour over a large luminosity range.

At the LHC the total cross section will be measured by the dedicated TOTEM experiment [18] using the luminosity-independent method (17.10) with an expected precision of a few percent. This measurement can be used to calibrate the various luminosity monitors of the LHC experiments as well as the beam instrumentation monitors. In addition, TOTEM can provide a direct measure of the absolute luminosity by an inversion of the procedure to measure the total cross section,

$$\mathcal{L} = \frac{1+\rho^2}{16\pi(\hbar c)^2} \frac{(N_{\text{in}} + N_{\text{el}})^2}{\mathrm{d}N_{\text{el}}/\mathrm{d}t|_{t\to 0}}, \qquad (17.16)$$

independent of the total cross section itself. This might be very useful for a calibration of the CMS luminosity monitors, which share the same interaction point with TOTEM, provided they can operate at the low luminosity of about $\mathcal{L} \approx 10^{28}\,\text{cm}^{-2}\,\text{s}^{-1}$ – these are the anticipated conditions with a high β^* optics enabling the total cross section measurement for TOTEM.

At the start-up of the LHC the absolute luminosity was obtained from a Monte Carlo calibration of the luminosity monitors, which suffered from a systematic uncertainty of $\pm 20\%$ due to the model dependence of diffractive processes. During the course of 2010 this calibration was significantly improved by a so-called van der Meer scan [19]. This improved the luminosity accuracy to ±11%.

17.4.2
Luminosity from Elastic Scattering

Elastic scattering alone provides another source of luminosity calibration if the Coulomb–nuclear interference (CNI) region at very small scattering angles can be accessed. At that point the well-known Coulomb amplitude brings an additional handle to extract from the t spectrum both the luminosity and the total cross section. The method was pioneered by UA4 at the Sp$\bar{\text{p}}$S and consists of a fit of the following expression to the t spectrum:

$$\frac{dN}{dt} = \mathcal{L}\frac{1}{16\pi}|F_C(t) + F_N(t)|^2 , \tag{17.17}$$

$$F_C(t) = -8\pi\alpha\hbar c \frac{G^2(t)}{t} \exp(i\alpha\Phi(t)) , \tag{17.18}$$

$$F_N(t) = (\rho + i)\frac{\sigma_{\text{tot}}}{\hbar c} \exp\left(\frac{-Bt}{2}\right) , \tag{17.19}$$

where $G(t)$ is the proton form factor and $\Phi(t)$ is the Coulomb phase. To measure elastic scattering an experimental setup similar to the one used for the total cross section has to be realised, that is, detectors have to be placed at a large distance from the interaction point and operated close to the beam in order to track elastically scattered protons. Special beam conditions are required such as a high β^* optics with parallel-to-point focusing in order to relate directly the measured impact position at the detector to the scattering angle at the interaction point. However, while for the total cross section only the nuclear part of the process has to be covered at $t > 10^{-2}$ GeV^2, for the luminosity even smaller scattering angles must be reached to access at least the Coulomb–nuclear interference point where the Coulomb amplitude equals the nuclear amplitude,

$$F_C(t) = F_N(t) , \quad t \approx 6.5 \times 10^{-4} \text{ GeV}^{-2} . \tag{17.20}$$

This corresponds to a scattering angle of 3 μrad at the LHC. The ALFA [20] subdetector of ATLAS is set up to exploit this method in order to provide an absolute luminosity calibration point for LUCID and the other luminosity monitors. ALFA is composed of four roman pot stations, two at each side of the ATLAS interaction point at a distance of 240 m, as shown in Figure 17.3. Each station contains two vertically movable roman pots, which are instrumented with a scintillation fibre tracker. Under ideal beam conditions the detectors can approach the orbit as close as 1.5 mm, and the CNI region can be accessed with a good acceptance of about 65%. In practice the luminosity will be extracted from a fit of (17.17) to the t spectrum, and systematic studies indicate that a precision of 3% can be obtained with this method.

17.4.3
Luminosity from W/Z Production

Lately, the accuracy of theoretical calculations for selected high p_T processes has become competitive for luminosity calibration. This is due to two effects: (i) the accuracy of structure function data from HERA steadily improved, resulting in an increasing precision of the global PDF fits; (ii) considerable progress was made on the partonic predictions for hard scattering processes at NNLO. The most promising *candle process* is the inclusive production of W and Z bosons with leptonic decay modes [21], which profit from clean experimental signatures and (at least at the LHC) from large production rates. From a given number of events, N_{obs}, observed in a time interval corresponding to the data-taking, the integrated luminosity is determined according to

$$\mathcal{L} = \frac{N_{obs} - N_{bg}}{\sigma^{th}_{acc} \epsilon_{exp}}, \qquad (17.21)$$

where N_{bg} is the number of background events. σ^{th}_{acc} is the theoretical cross section within the experimental acceptance cuts, typically applied on η and p_T of the leptons which are required to be isolated and $\not{E}_T$ for the W. ϵ_{exp} is the detector efficiency to observe the events within the cuts. The analysis to obtain the number of events as well as the efficiency and background is essentially the same as for the cross-section measurement, except that for the cross section an acceptance correction is needed to extrapolate to the full acceptance while for the luminosity the theoretical cross section within the acceptance is required. In a recent analysis CDF [22] obtained an experimental accuracy of better than 2% for both efficiencies and acceptances and about 1% for the backgrounds. The prospects for the LHC look similar according to simulations [23]. The key issue is thus the accuracy of the calculation for the theoretical cross section within the cuts, which is more challenging than the acceptance, where a number of uncertainties cancel to some extent in the ratio of accepted to total cross section. The fact that cuts are imposed affects the convergence of NNLO calculations, compared to fully inclusive observables. Another limitation is the lack of a combined EW × QCD calculation: QED final-state radiation affects the observables and some simplified Monte Carlo simulations explicitly assuming factorisation must be applied for corrections. At present, an accuracy of about 3% for the calculation of σ^{th}_{acc} is obtained for W and Z production [24, 25].

In conclusion, the luminosity determination at the Tevatron has reached a final precision of 6% and after one year of running 11% has been achieved at the LHC. For Standard Model cross-section measurements at LHC the luminosity uncertainty will quickly become dominant. A better calibration of the beam current transformers will allow the precision to be improved by a factor of two within a year. Further improvements to about 3% accuracy can be expected at a longer time-scale from elastic scattering and from W/Z production.

References

1 Dallavalle, G.M. (1997) Review of precision determinations of the accelerator luminosity in LEP experiments, *Acta Phys. Polon. B*, **28**, 901.

2 Jadach, S. et al. (1996) Event Generators for Bhabha Scattering, http://arXiv.org/abs/hep-ph/9602393.

3 Bohm, M., Denner, A., and Hollik, W. (1988) Radiative Corrections to Bhabha scattering at high-energies (I): Virtual and soft photon corrections, *Nucl. Phys. B*, **304**, 687, doi:10.1016/0550-3213(88)90650-5.

4 Berends, F.A., Kleiss, R., and Hollik, W. (1988) Radiative corrections to Bhabha scattering at high-energies. (II). Hard photon corrections and Monte Carlo treatment, *Nucl. Phys. B*, **304**, 712, doi:10.1016/0550-3213(88)90651-7.

5 Arbuzov, A. et al. (1996) The present theoretical error on the Bhabha scattering cross section in the luminometry region at LEP, *Phys. Lett. B*, **383**, 238, doi:10.1016/0370-2693(96)00733-2.

6 Brun, R. et al., "GEANT 3", CERN DD/EE/84-1 (Revised), September 1987.

7 Bethe, H. and Heitler, W. (1934) On the stopping of fast particles and on the creation of positive electrons, *Proc. Roy. Soc. Lond. A*, **146**, 83. http://www.jstor.org/stable/2935479.

8 Abe, F. et al. (1994) Measurment of the antiproton–proton total cross section at $\sqrt{s} = 546$ and 1800 GeV, *Phys. Rev. D*, **50**, 5550, doi:10.1103/PhysRevD.50.5550.

9 E-811 Collab., Avila, C. et al. (1999) A measurement of the proton-antiproton total cross section at $\sqrt{s} = 1.8$ TeV, *Phys. Lett. B*, **445**, 419, doi:10.1016/S0370-2693(98)01421-X.

10 Klimenko, S., Konigsberg, J., and Liss, T.M. (2003) Averaging of the inelastic cross section measured by the CDF and E811 experiments, http://www.slac.stanford.edu/spires/find/hep/www?r=FERMILAB-FN-0741, FERMILAB-FN-0741.

11 Bantly, J. et al. (1997) DØ luminosity monitor constant for the 1994–1996 Tevatron run, http://www.slac.stanford.edu/spires/find/hep/www?r=FERMILAB-TM-1995, FERMILAB-TM-1995.

12 CDF Collab., Abe, F. et al. (1994) Measurement of $\bar{p}p$ single diffraction dissociation at $\sqrt{s} = 546$ GeV and 1800 GeV, *Phys. Rev. D*, **50**, 5518, doi:10.1103/PhysRevD.50.5535.

13 Andeen, T. et al. (2007) The DØ experiment's integrated luminosity for Tevatron Run IIa, http://www.slac.stanford.edu/spires/find/hep/www?r=FERMILAB-TM-2365.

14 CDF Collab., Acosta, D. et al. (2001) The CDF Cherenkov luminosity monitor, *Nucl. Instrum. Meth. A*, **461**, 540, doi:10.1016/S0168-9002(00)01294-8.

15 Acosta, D. et al. (2002) The performance of the CDF luminosity monitor, *Nucl. Instrum. Meth. A*, **494**, 57, doi:10.1016/S0168-9002(02)01445-6.

16 ATLAS Collab., Aad, G. et al. (2008) The ATLAS Experiment at the CERN Large Hadron Collider, *JINST*, **3**, S08003, doi:10.1088/1748-0221/3/08/S08003.

17 CMS Collab. (2008) The CMS Experiment at the CERN LHC, *JINST*, **3**, S08004, doi:10.1088/1748-0221/3/08/S08004.

18 TOTEM Collab., Anelli, G. et al. (2008) The TOTEM experiment at the CERN Large Hadron Collider, *JINST*, **3**, S08007, doi:10.1088/1748-0221/3/08/S08007.

19 Van der Meer, S. (1968) CERN-ISR-PO-68-31.

20 ATLAS Collab. (2008) ATLAS Forward Detectors for Measurement of Elastic Scattering and Luminosity Determination, http://cdsweb.cern.ch/record/1095847.

21 Dittmar, M., Pauss, F., and Zurcher, D. (1997) Towards a precise parton luminosity determination at the CERN LHC, *Phys. Rev. D*, **56**, 7284, doi:10.1103/PhysRevD.56.7284.

22 CDF Collab., Abulencia, A. et al. (2007) Measurements of inclusive W and Z cross sections in $p\bar{p}$ collisions at $\sqrt{s} = 1.96$ TeV, *J. Phys. G: Nucl.*

Part. Phys., **34**, 2457, doi:10.1088/0954-3899/34/12/001.
23 ATLAS Collab. (2008) Expected Performance of the ATLAS Experiment – Detector, Trigger and Physics, Tech. Rep. CERN-OPEN-2008-020, CERN, http://arxiv.org/abs/0901.0512, arXiv:0901.0512.
24 Adam, N.E., Halyo, V., Yost, S.A., and Zhu, W. (2008) Evaluation of the theoretical uncertainties in the $W \to l\nu$ cross sections at the LHC, *JHEP*, **0809**, 133, doi:10.1088/1126-6708/2008/09/133.
25 Adam, N.E., Halyo, V., and Yost, S.A. (2008) Evaluation of the theoretical uncertainties in the $Z \to \ell^+\ell^-$ cross sections at the LHC, *JHEP*, **0805**, 062, doi:10.1088/1126-6708/2008/05/062.

18
Trigger Systems in High Energy Physics Experiments

Eckhard Elsen and Johannes Haller

In spite of the tremendous progress in data processing capability, HEP experiments generally cannot afford to cope with the large data volume generated in the detectors in real time. The quest to measure rare events necessitates an interaction rate and data volume that challenges the technical capabilities for data storage and analysis. A real-time *trigger* has to be introduced in order to select the interesting interactions for permanent storage and to discard the background.

The key concepts of a trigger system are introduced in the following: *timing, clocked readout, dead-time, trigger logic* and *buffering*. By way of example these concepts are illustrated for the trigger systems of the H1 experiment at HERA, ATLAS at LHC and for a detector at a future International Linear Collider (ILC). The relevant methods for tackling the trigger effects and biases introduced by a trigger system in an analysis of the recorded data are briefly explained.

Evidently the level of detail must be restricted here and only concepts are outlined. For a broader discussion the reader is referred for example to [1], where interesting aspects of trigger systems are discussed at a general level, and to the detailed description of the trigger systems of the experiments operating at the high energy frontier [2–7]. More details on the accelerators can be found in Chapter 12 and the general layout of the experiments is discussed in Chapter 13.

18.1
Introduction

The definition of a trigger and the distinction between a trigger and a *readout system* is not well-defined and has changed over time. A typical scintillator coincidence experiment of the early days of particle physics initially constituted the entire readout of the experiment (Figure 18.1a). The coincidence rate gave a good indication of the particle flux, the essential number for a cross-section measurement.

The same scintillator setup can be used to trigger a drift chamber readout (Figure 18.1b). The fast coincidence of the scintillator signals defines the *time of passage/interaction* (T_0) and initiates the *readout phase* to record the charge signals from

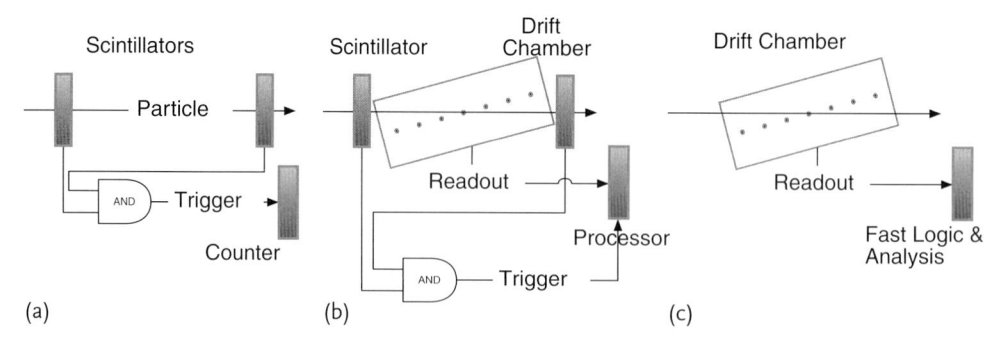

Figure 18.1 Evolution of level-1 trigger schemes: (a) temporal and spatial constraints derived from the scintillator signal increment a coincidence counter; (b) the scintillator coincidence triggers the readout of a drift chamber; (c) the fast logic synchronously matches the recorded drift chamber hits to valid track patterns in real time and the track is reconstructed with high precision in a subsequent analysis.

the drift chamber. In this case the scintillator coincidence (the trigger) selects horizontal tracks only and defines the time of passage. This is a general feature of a trigger system: provision of experimental (here: spatial) constraints and definition of the T_0, with the goal of selecting interactions of interest and reducing background. In this experimental setup the data path is split in two parts: the *trigger path* that provides a fast and often coarse signal, and the *readout path* that provides the detector response with full granularity and resolution. For the combination of trigger signals from different detectors (here: two scintillators) a central component, often called *central trigger*, is needed.

Progress in the development of high-speed logic can be used for a further simplification of the scheme of Figure 18.1b: the continuous sampling of the signals of the drift chamber and the online identification of the coincidence pattern that corresponds to the charges generated by the passage of a real particle (Figure 18.1c) can be dealt with simultaneously. The T_0 and the spatial origin of the track can be deduced in *real time* from the redundancy of the drift chamber alone. The transfer to permanent storage for offline analysis can be initiated by such a coincidence pattern. The chamber has become *self-triggering*. There is no difference between trigger and readout path so that the full resolution of the detector is available for the trigger decision. The large-scale experiments of today typically implement a mix of the variants shown in Figure 18.1b,c, depending on the properties of the detector component.

The example of Figure 18.1b exhibits another important feature: *latency*. The temporal pattern of drift chamber hits caused by a track is only available after the relatively long drift time has elapsed – the charge has to drift to the wire. Consequently the trigger signal can only be formed after a certain period, the *trigger latency*. This latency may vary from detector to detector, and appropriate *delays* have to be inserted in the coincidence evaluation. Similarly a *readout latency* may be encountered which may differ from the trigger latency due to for example different electronics. In order to compensate for these variations, appropriate *readout pipelines* have to be implemented to align the relevant detector information in time. Note that the delays and pipelines have been omitted for simplicity in Figure 18.1.

The term *level-1 trigger* will be used in the following for a prompt signal that is based on the identification of *regions of interest* (see also Section 18.3.3) in a synchronous[42] fashion, which typically initiates the readout, that is, the transfer to the next trigger level or to permanent storage.

Hadron colliders brought a new challenge: the hadronic interaction rate alone is so high that it would easily saturate processing power and storage capabilities. An additional degree of selection becomes mandatory. The trigger has to separate interesting physics processes (signal) from well-understood processes (background). Such a distinction is only possible if the trigger is aware of physics features in the event. The trigger has to evaluate the lepton content, momentum balance, individual particle momenta and so on in real time, which implies that data from calorimeters, muon chambers and tracking detectors have to be used in the trigger.

Typically a respectable number of preferably *orthogonal* trigger conditions[43] are required to satisfy all physics goals of the experiment. The complete set of conditions defines the "trigger mix" of an experiment.

18.2
Elements of a Trigger System

Modern HEP experiments are omni-purpose detectors. They pursue several physics goals at the same time. Consequently, the trigger system has to accommodate varying physics demands and event rates. Precision measurements may employ triggers running at a rate of 1 kHz or more, whereas small-scale physics rates may be as rare as one event per month. Evidently such demands require the set up of various trigger conditions, the trigger mix, that are simultaneously evaluated. The conditions may hence require a *jet* candidate OR an *electron* candidate, to give but one example. Such *multi-purpose* trigger systems process the data from several detector components. Table 18.1 gives an indication of how various detectors provide the information for a trigger condition. The result of the trigger evaluation in a subsystem is made available to the central trigger, which executes the *global trigger decision* and retains the event if at least one condition is fulfilled. The trigger decision for each condition is often recorded in the event itself for later analysis (see Section 18.4) in binary format using *trigger bits* which encode the information.

---

42) Here synchronous means synchronous with the clock of the experiment, cf. Section 18.2.1.
43) The coincidence rate of orthogonal triggers is given by their accidental coincidence rate unless there is a physics signature that synchronously sets the individual components.

Table 18.1 The relevant subsystems and trigger conditions employed in typical omni-purpose detectors.

Trigger	Trigger condition
Calorimeter	energy, total energy, jets from calorimeter
Electron	localised shower in the first layers of the calorimeter
Muon	penetrating tracks in the muon detector
Track	tracks and momentum (threshold)
Vertex	vertex position from tracking detector
Central	coincidences and thresholds from various subsystems

18.2.1
Clocked Readout and Triggering

Modern colliders are clocked by the radio frequency (RF) of the accelerating fields, typically some 100 MHz. The bunch-crossing time (BC) is an integer multiple of the period, typically 25 to 100 ns. Linear colliders envisage using RF of 1.3 GHz and up to 12 GHz to achieve much shorter bunches with bunch crossings every 300 ns or ~ 1 ns, respectively. The same BC time reference drives the transfer of signals from the detectors, their digitisation and the placement in the front-end pipelines. Collider experiments use a *discrete* time; they are clocked. The trigger timing function (i.e. the T_0 mentioned before) is hence reduced to the *bunch identification*, that is, the identification of the bunch in which the interaction occurred.

Typically not all bunch locations are filled in a collider. Gaps in the fill pattern are used to safely extract the beam, to give but one example. The fill pattern is a priori known or can be derived from beam pick-up signals. The simplest trigger, called *minimum bias trigger*, is hence the one that reads out the detector for each filled bunch.

It should be noted that the temporal alignment of the signals from different detector components on the discrete time, as needed to evaluate the coincidence of signals from the same bunch crossings, can be a formidable task. With sometimes thousands of channels and their individual signal propagation distances[44], the delays of all electronic channels must be adjusted for *optimal timing* within a bunch crossing to ensure full detector efficiency. In high-luminosity colliders the propagation delays often exceed the intra-bunch distance so that the next bunch collision occurs before the particles of the preceding interaction have reached the far detectors and signals have been collected. The trigger delays hence have to be adjusted so that a slow detector is channelled through a short delay and the signals of fast detectors travel through a correspondingly longer delay before the coincidence is formed (Figure 18.2). This adjustment of delays is an important step of the trigger commissioning, that is, the "timing-in". Initially all signals of one detector compo-

---

44) The propagation speed of an electronic signal in a cable is about 5 ns/m. Hence even small variations in the detector location must be taken into account.

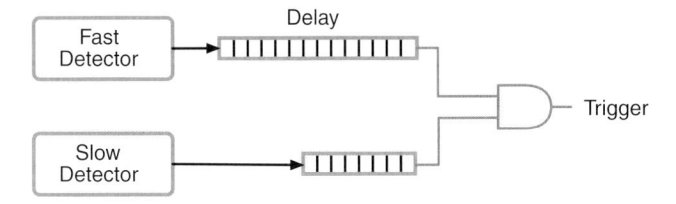

Figure 18.2 Fast and slow detectors require adjustable delays to form a proper coincidence of the signals in the trigger.

nent are brought into coincidence. Following that internal timing-in, overall delays are adjusted so that all detectors are synchronised. The latter task is typically realised in the central trigger.

In the case of the setup of Figure 18.1b a separate timing-in of the pipelines of the various channels of the readout path is needed in order to allow a proper alignment and combination of all event fragments in the readout in a given event.

18.2.2
Central Trigger Logic

It is rare that the signal from one detector alone is used to initiate the readout of the entire detector. Typically much lower detector thresholds can be realised when employing coincidences of two different signals, and the logical AND of two signals is probably the most frequently realised *trigger logic*. However, there are many more logic conditions that are routinely applied in the central trigger:

- AND and OR of two or more signals;
- majority logic: (n out of m detectors have responded, with $n \leq m$);
- threshold: n lines encode a digital value, for example energy E, and the condition $E > E_{thr}$ is used to trigger;
- topological patterns with a well-known signal profile.

The trigger circuitry must provide a synchronous decision within the duration of the bunch-crossing interval. Since general-purpose computers would not guarantee the response in real time and for every bunch, the logic has to be embedded in custom-made electronics. There are several means of realising a trigger decision using modern electronics.

Arbitrary logic for a finite number of logic inputs can best be realised with a *lookup table* (LUT): the trigger inputs drive the address lines of a memory while the memory cell contains the encoded logic condition, that is, *Yes/No*. The output of such a lookup table consequently defines the logic response. Evidently any logic can so be encoded. However, depending on the memory chip used and the physics requirement, the address space may be large, so that encoding/loading of the memory cells may take a disproportionately long time or even becomes prohibitive.

Whenever the response to a given trigger input pattern is to return a quantitative value and not only a *Yes/No* decision, *Contents Addressable Memories* (CAMs) may be adequate. These multi-bit memory cells are packed with ancillary comparator logic so that when a given pattern is presented to the logic the address or index of the memory cell(s) appears at the output. An example for CAMs is the momentum reconstruction of a charged track in a magnetic field. The measured coordinates (the input pattern) are presented to the CAM, which returns the momentum of the particle from its pattern library.

When the number of address lines becomes prohibitive, *Field Programmable Gate Arrays* (FPGAs) provide a solution. These high-density chip architectures combine logic cells with storage cells. Their logic function is programmed at startup so that they maintain much of the flexibility of a lookup table. Nowadays FPGAs can be clocked at GHz rate, so that serial computations can be envisaged even in between bunch crossings. The same feature is supported by programmable *Digital Signal Processors* (DSPs), which are commonly used. Manufacturers offer highly integrated components with embedded DSPs and custom programmable gate arrays. In these cases the user profits from the high-speed interconnects inside the same logic chip.

18.2.3
Buffering

The cross section, σ, of a physics process defines a probability, ν, for an interaction per time interval: $\nu = \sigma \mathcal{L}$, where $\mathcal{L}$ is the instantaneous luminosity. The rate of events follows a *Poisson distribution* with mean ν. The trigger and readout path hence have to cope with the large ensuing instantaneous rate variations. In order to de-randomise the arrival time, events are stored in a *buffer* once they have been triggered by the level-1 trigger system and readout. The next event is deposited in a subsequent buffer. The advantage is that the further processing capability (i.e. the next trigger level or transfer to storage) no longer has to be adjusted to fit in the smallest possible time period between triggers (i.e. the next bunch crossing). Rather, the processing time, T_p, has to satisfy the constraints on average, $\langle \nu \rangle \leq 1/\langle T_p \rangle$. Otherwise the number of buffers would eventually be exhausted. Such a scheme is commonplace, although in all practical cases the number of buffers is limited to a finite number. Once all buffers are filled, triggering must be inhibited in order to avoid overwriting of events recorded earlier[45].

18.2.4
Dead-Time

Electronic components have a finite reaction time, for example due to the discharging of capacities or the reestablishment of the initial conditions after processing

---

45) It should be added that in practical applications T_p also follows a distribution, since the demand on serial processing typically depends on the complexity of the events. The layout of the processing capability of a trigger system is hence not trivial.

one event. While the level-1 trigger electronics itself is especially designed for processing with the maximum possible rate (the clock frequency), the readout electronics of the detector channels may have been laid out for a considerably lower rate, since only the events accepted at level-1 must be read out. High trigger rates may hence saturate the readout capacity of the detector, for example by exhausting the buffer pool (cf. Section 18.2.3). If this happens, further level-1 triggers must be inhibited by a *feedback system*. Consequently *dead-time* arises when the system is not able to accept further events until buffers have been freed.

The dead-time fraction, f_D, is defined as the fraction of the time of active bunch crossings for which the triggering is inhibited over the total live-time. It is a key ingredient in the measurement of the cross section: $\sigma = \nu/(1 - f_D)/\mathcal{L}$. In a non-buffered system dead-time arises whenever the next event $n+1$ follows the event n at a time $t_{n+1} < t_n + T_p$, where a constant processing time, T_p, has been assumed for simplicity. In a multi-buffer environment the variations are drastically levelled out and the dead-time is much reduced. With n buffers available in the system, the dead-time fraction can be computed from the sum of the individual Poisson distributions,

$$f_D = 1 - \sum_{k=0}^{n} \frac{\mu^k}{k!} e^{-\mu}, \tag{18.1}$$

with $\mu = \nu T_p$. The expected behaviour of f_D as a function of μ is shown in Figure 18.3 for $n = 1, 3$ and 5 buffers. With more buffers available higher input rates can be realised: a dead-time of 50% is reached at 1.7 kHz for one buffer and 3.7 kHz and 5.7 kHz for 3 and 5 buffers, respectively for an assumed processing time, T_p, of 1 ms. The examples demonstrate the importance of tackling second ($n = 2$) and higher order ($n > 2$) dead-time which arises when two or even more events have to be dealt with simultaneously.

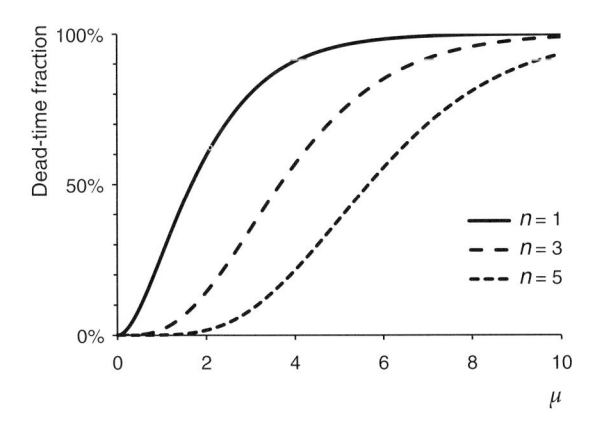

Figure 18.3 The development of the dead-time fraction, f_D, as a function of the input rate, $\mu = \nu T_p$, for 1, 3 and 5 buffers in the input chain. T_p is the processing time and ν the input rate for a given cross section.

Since dead-time is so pivotal to all measurements of cross sections, it is worthwhile to mention a subtlety arising in collider experiments: the dead-time may become a function of the bunch number in the fill pattern of the collider. For example, it is conceivable that the first bunch of a train of particles causes a smaller dead-time than a later bunch in the train, since the remainders of the signals caused by earlier bunches may incur additional dead-time. While in principle the above concept of dead-time can be extended easily, it is advisable from the point of view of physics analyses to retain the simplicity of an overall dead-time. Nevertheless, the capability of measuring a *bunch-specific dead-time* has to be designed into the trigger hardware components.

18.2.5
Multi-level Trigger Systems and Event Building

The level-1 trigger retains those events that satisfy the physics selection criteria implemented in the *synchronous* level-1 logic. Given the timing constraints, the information may be coarse and lead to inefficiencies compared to the precise information reconstructed offline. A *multi-level* trigger may arrive at more sophisticated decisions. Following the buffering concept introduced before, events accepted by the level-1 trigger are placed in intermediate buffers and *asynchronously* analysed by algorithms of the next trigger level, called *level-2*. Obviously, following this principle, systems can be designed using even more levels. Typically the algorithms of the higher levels are run in commercial computer farms since processing time constraints are relaxed. Evidently dead-time will arise when all intermediate buffers have been filled. This may happen when the processing speed of the next filtering stage is inadequate as in the example of Figure 18.4, where the multiplexing[46] introduces dead-time. The resulting dead-time may be tolerable if the level-1 trigger rate is known to remain small.

Nowadays higher-level trigger systems are generally designed to have the full information of a detector component available for the trigger decision starting at level-2. However, the event information may be restricted to specific detector components since the information of the various detectors has not been merged yet. For example, at level-1 the muon trigger may indicate a pattern match for penetrating tracks in the muon system. At level-2 the detailed information provides tracks with a precise momentum measurement. This information suffices to implement a muon momentum threshold in the trigger, but cannot be combined with the signature of a recoiling jet, since the calorimeter information has not yet been combined with the muon information at this stage. The combination of data fragments from the various detector components is the task of the *event builder*. It is a fundamental design choice at what stage the full event information is made available. Evidently, the earlier the combination is done the more sophisticated trigger decisions can be implemented, with the drawback of a potentially much higher load due to data

---

46) The multiplexer sequentially channels the data from several readout pipelines to one high-resolution ADC, a design chosen for example for cost reasons.

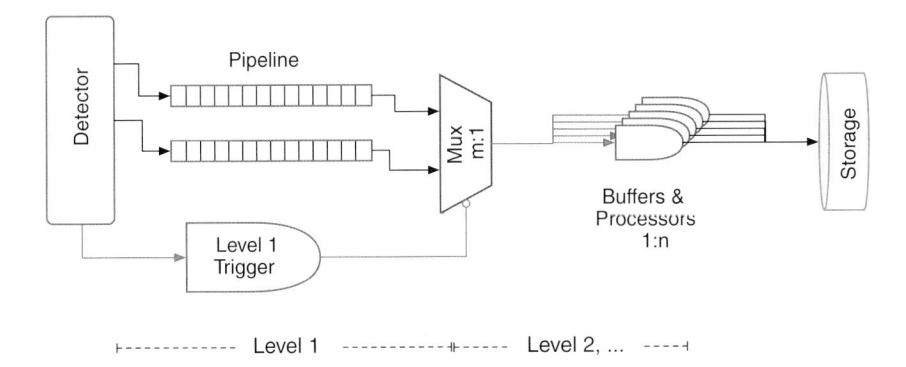

Figure 18.4 The level-1 trigger enables the readout and passes the events to 1 of n buffers. The level-1 trigger may introduce latency when the readout is slower than the collider clock cycle. This may happen for example when the pipelines are multiplexed $m:1$, as in this example. Dead-time may also arise when all processor buffers are filled.

traffic in the network connecting the various parts of the system. It is interesting to note that the ATLAS [6] and CMS [7] experiments have implemented different integration levels of this scheme in their trigger systems: while in ATLAS the events are only partially built after level-1 and completed after level-2, the CMS system combines the data fragments already after level-1, that is, all data are transferred with the full level-1 rate.

18.2.6
Trigger Rates and Downscale

The various trigger conditions designed for different physics goals must be optimised such that the total recording rate of the experiment does not exceed its technical rate capability (*bandwidth*). For a multi-purpose detector the overall rate must be appropriately shared amongst the trigger conditions, that is the different physics goals of the experiment. Since the latter and the conditions of data-taking (e. g. instantaneous luminosity, background rates) change over time, multi-purpose trigger systems require flexible adaptations of the trigger setup to control the rate of accepted events. Values of threshold parameters may be changed (e. g. the E_T threshold for electromagnetic objects could be increased in order to reduce the rate of the corresponding trigger condition) or even the logic combinations in the central trigger may be modified. Evidently the trigger logic must make provision for such flexibility, which can be realised with the programmable logic mentioned before.

The bandwidth restrictions at the different trigger levels may prevent the recording of all events fulfilling high-rate trigger conditions. Such conditions are nevertheless useful since they may for example serve as *monitor triggers* for the evaluation of the *trigger efficiency* or allow physics studies in special regions of the phase space (e. g. measurements at low p_T). An artificial but well-defined *prescaling* of such high-rate triggers helps to contain the rate. The corresponding *downscale factor*, d, is included in the trigger condition, such that on average only every dth selected event is recorded. The choice of the downscale factor is hence an important step in setting up the trigger mix for a multi-purpose experiment. The experimental simplicity of a trigger condition with lowest possible physics bias has to be contrasted with the statistical losses introduced by a high downscale factor. The optimal mixture is often not a priori known.

18.3
Trigger Systems in Modern HEP Experiments

All high energy collider experiments [2–7] of the past two decades have installed a multi-level trigger system to select the interesting physics from for example beam-gas induced backgrounds or to discard the less interesting interactions that otherwise would have swamped the storage capacity. Typically the rate has to be reduced by several orders of magnitude. Such a severe selection can only be achieved in a multi-level system where a single trigger condition fulfilled (out of potentially many conditions) suffices to transfer the event to the next trigger level. The coarser decisions at the lower level are implemented in custom-made hardware that processes the signals from prompt detectors. The higher-level triggers are realised in computer farms that have the full event information available for the most sophisticated and least biased decision. The technological trends push the most general and flexible computer farm selection to the lowest trigger levels [6, 7].

18.3.1
Trigger Strategies

A trigger condition, particularly when based on coarse or incomplete detector information, may introduce an unwanted bias into the physics selection, for example via inefficiencies. The goal of each experiment is hence to run with the least stringent trigger conditions. This desire has to be contrasted with the realities of the operating parameters of the collider (beam current, interaction rate and background level) and the recording capability of the data-acquisition and storage system. Table 18.2 summarises the parameters that are most relevant for colliders operating at the energy frontier. The e^+e^- collider LEP is seen to be a notable exception: with typically 4 (and later 8) bunches placed on the circumference of the machine, the time between bunch crossings is tens of microseconds. Much higher interaction rates are realised in two-ring colliders where the time interval between bunch crossings drops to 25 ns as in the LHC. In addition, the primary interaction at LEP involved an *electroweak* vertex which resulted in a small interaction rate. Consequently the LEP experiments were able to record basically all physics interactions. Triggering is substantially more challenging at hadron colliders where the strong interaction causes a huge inelastic event rate while other physics events, such as electroweak processes, events with high momentum transfer or new physics processes at high

Table 18.2 Accelerator parameters relevant for trigger and data-acquisition systems. For HERA and LHC the *Number of bunches* indicates the used bunches out of the possible bunch locations. For Tevatron the two values refer to initial and final number of filled bunches for Run 2.

Accelerator	Number of bunches	Crossing interval [ns]	Total inel. event rate [Hz]	Trigger rate [Hz]	Event size [kB]
LEP (26.7 km)	4 (later 8)	22 000	< 6	< 6	~100
HERA (6.3 km)	180 of 220	96	~10^5	~20	~200
Tevatron 2 (6.3 km)	36, 103	132–396	~10^7	~40	~350
LHC (26.7 km)	2835 of 3564	25	~10^9	~200	~1500

mass are rare. For these experiments highly selective trigger systems have been designed since only a small fraction of the total event rate can be recorded.

In general, studies of the above-mentioned rare processes are limited by the available statistics and therefore the relevant trigger conditions must avoid the application of downscale factors in order to record the maximum number of events. Whenever the properties of the events of interest are well known (e.g. known processes of the Standard Model, such as Z, $W^{\pm}$ or $t\bar{t}$ production in $pp/p\bar{p}$ collisions or neutral-current and charged-current reactions in deep inelastic scattering at HERA) specialised *exclusive* trigger conditions are used for the selection. However, if the signature of new physics is not known, the events have to be selected more *inclusively*. In these cases the presence of a number of candidates of a particular particle type exceeding a (transverse) energy threshold (e.g. more than one electron with $p_T > 50$ GeV), may suffice. With this inclusive selection strategy even events with unexpected topologies at high energy are likely to be retained. Such an inclusive trigger strategy is typically pursued at the start of the data-taking at a new collider, when the properties of the events are partially unknown and the physics analyses have to be prepared for the unexpected. The LHC experiments will follow this inclusive trigger strategy. Since the energy spectrum of particles in background events is usually steeply falling, the threshold value applied is generally a compromise between available bandwidth and coverage of the phase space.

For the study of the strong interaction, for the analysis of background processes outside of the signal region of new physics or for the monitoring of the detector performance, the experiments are also interested in events with lower trigger thresholds for which the rates are high. Such events can be recorded within the bandwidth constraints by the usage of trigger conditions requiring *multiple* particle candidates or by the application of downscale factors (see Section 18.2.6) which artificially reduce the recorded data volume.

18.3.2
Example HERA: the Trigger System of the H1 Experiment

Analyses at newly commissioned hadron colliders are often concerned with large transverse momenta where new physics is expected, and the H1 experiment [2] at the *ep* collider HERA was no exception. HERA collided 27.6 GeV electrons with 920 GeV protons with a bunch spacing of 96 ns. Its primary physics goals were the measurements of the proton structure over a wide kinematic range in the Lorentz-invariant variables x and Q^2, electroweak physics and the search for new phenomena at large momentum transfer. At large p_T the known cross sections fall quickly and hence large integrated luminosities are required.

The H1 experiment employed an energy trigger using the liquid argon calorimeter and a track trigger derived from the drift chamber supplemented by a muon trigger. The fast signals from proportional chambers near the beamline were used to recognise the bunch crossing. In order to trigger on for example deep inelastic events the total energy contained in the 45 000 individual channels of the calorimeter had to be summed quickly and reliably. With the analogue electronics placed outside the cold calorimeter, signal integration times of 1 μs were necessary to contain the electronic noise, which otherwise would have dominated the energy measurement. The technique chosen for the energy measurement was a *sample-and-hold* circuit for which sampling and readout cannot proceed simultaneously. Consequently the H1 calorimeter was either sampling or in readout mode, the latter consuming a period of 1 ms. This feature immediately implied that every accepted trigger induced 1 ms of dead-time.

H1 used a three-level trigger system: an accept condition at level-1 (L1Accept) had to be issued after 2.5 μs; once issued dead-time started accumulating. The level-2 decision had to be issued after the fixed period of 21 μs. A reject decision at this stage incurred an overall dead-time of 35 μs in order to allow the pipelines to be cleared. At level-3, following the level-2 accept condition (L2Accept), more sophisticated trigger algorithms could be run to arrive at a physics-related decision, for example a D^* selection. With the readout of the detector proceeding, a *fast clear* signal could be issued at any moment to resume data-taking. The level-3 accept signal was given at the latest after 800 μs.

At the time of designing the electronics (late 1980s) microprocessor and data-acquisition systems were orders of magnitude slower than nowadays. Memory easily became a cost issue. As a result the calorimeter DSPs reserved only two buffers to place the event data. Once both buffers were filled second-order dead-time[47] was encountered. The effect of a finite buffer depth is illustrated in Figure 18.3.

The multi-level trigger system of the H1 experiment [8] constantly evolved over the lifetime of the experiment, with the biggest steps taken during the HERA upgrade shutdown in 2001. Prior to this major upgrade the level-2 trigger was augmented by a sophisticated topological trigger and a neural network trigger [9]. After

---

47) Obviously in such an architecture a second event can quickly be buffered. A third event cannot be registered unless the buffers are cleared.

the upgrade shutdown a multi-level *fast track trigger* [10] was introduced with the primary goal of triggering on heavy-quark physics at small transverse momenta.

Given the sample-and-hold architecture of the calorimeter readout system, it was not conceivable to upgrade the trigger system such that the higher-level trigger could utilise the full event information. Consequently at each level an independent and parallel path had to be implemented to the standard readout. While this was technically possible it led to a tremendous effort in the implementation of a higher-level trigger as is exemplified by the fast track trigger [10]. H1 implemented the variant shown in Figure 18.1b at all three stages: the trigger information was derived independently from the high-resolution readout of the experiment.

18.3.3
Example LHC: the Trigger System of the ATLAS Experiment

The LHC has been designed to collide protons at a centre-of-mass energy of 14 TeV with luminosities of up to 10^{34} cm^{-2} s^{-1}. Proton bunches intersect with a rate of 40 MHz, corresponding to a time interval between bunch crossings of 25 ns. At nominal luminosity ~25 interactions per bunch crossing are expected. In this challenging environment the trigger system of the ATLAS experiment [6] has to reduce the rate to below 200 Hz, the maximum rate that can be processed by the offline computing facilities, while making every attempt to record previously undetected and rare physics processes. For example a Standard Model Higgs boson with a mass of 120 GeV, decaying into two photons, is expected to occur in one out of 10^{13} interactions.

The ATLAS trigger is composed of three levels: level-1 (L1), level-2 (L2) and the Event Filter (EF). A schematic overview of the system is given in Figure 18.5. It implements several of the basic elements discussed before and was designed to select events containing e, γ, μ, $\not{E}_T$, τ and hadronic jets.

The first level is a hardware-based system, implemented in electronics and firmware. Its decision is based on rather coarse data from the calorimeters and dedicated fast muon chambers. L1 reduces the event rate to below 75 kHz (upgradeable to 100 kHz) within a fixed latency of less than 2.5 μs, during which the data of each readout channel are retained in on-detector pipelines. Hence the general setup of Figure 18.1b is implemented in ATLAS. The higher trigger levels consist almost entirely of commercially available computers and networking hardware. After buffering they access more detector information to reduce the event rate to approximately 3.5 kHz after L2, and about 200 Hz during the selection of the EF which runs after the event building.

The L1 system is composed of three parts (cf. Figure 18.5 insert): the Calorimeter Trigger, the Muon Trigger and the L1 event-decision part implemented in the *Central Trigger Processor* (CTP).

Calorimeter Trigger In ATLAS, calorimetry is provided by lead and copper liquid-argon sampling calorimeters (LAr) and an iron scintillator-tile sampling calorimeter (TileCal) for hadron calorimetry in the barrel part. In order to reduce the chan-

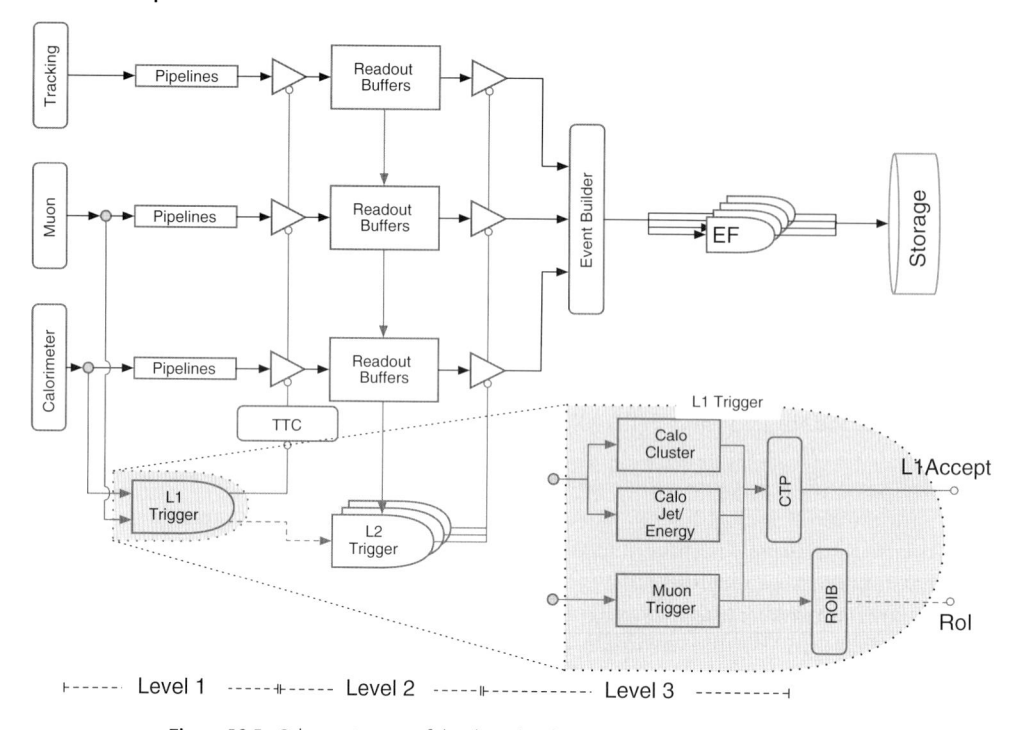

Figure 18.5 Schematic view of the three-level ATLAS trigger system. The insert displays the processing of calorimeter and muon trigger information at level-1.

nel count in the trigger decision, signals of neighbouring calorimeter cells are summed to form ∼7200 rather coarse signals, called trigger towers. These signals are transmitted via long cables to the processing electronics of the calorimeter trigger system, which is installed in the ATLAS electronics cavern and thus shielded by concrete walls from the main tunnel in order to protect the electronics from the high radiation environment. The trigger tower signals are analysed in various programmable hardware components (FPGAs): first the data are digitised and the exact bunch crossing is identified (preprocessor); subsequently candidates for highly energetic electron/γ and τ/hadron objects are identified in the electronic modules of the cluster processor (Calo Cluster) and their transverse energy is discriminated against programmable thresholds. In parallel the jet/energy processor (Calo Jet/Energy) identifies jet candidates, discriminates their transverse energy value, E_T, against programmable thresholds and evaluates several global energy sums using the digitised trigger tower signals. Synchronously with the 40 MHz machine clock (i.e. for each bunch crossing) the computed multiplicities of e/γ, τ/hadron and jet candidates as well as the global energy information are transmitted via short parallel cables to the CTP where the global event decision is taken.

Muon Trigger The other source of information available in the L1 trigger decision are the fast muon chambers which are installed in three layers around the interaction region. The algorithms of the muon trigger are based on hit coincidences in different detector layers within geometrical windows, whose width is matched to the transverse-momentum threshold applied and hence profits from the deflection of muons in the magnetic field. The number of muon candidates for each threshold is transferred to the CTP.

Central Trigger Processor The CTP is composed of several electronic modules housed in a single electronic crate. It evaluates the L1 conditions for each bunch crossing, that is, with a rate of 40 MHz based on the information received on parallel cables from the muon and calorimeter triggers and ancillary sources (scintillators, random triggers, etc.). While the timing-in of the various channels inside the muon and calorimeter triggers is done in the subsystems, several delay pipelines are used to align the combined signals from these trigger parts at the input of the CTP. The CTP logically combines the signals of up to 256 trigger conditions according to a programmable trigger setup. For instance one of these trigger conditions could implement the selection of events containing more than two jet candidates with transverse energies above 10 GeV and more than two muon candidates with transverse momenta above 12 GeV. The rate of each trigger condition can be artificially reduced in the CTP using a downscale factor. The level-1 accept signal (L1Accept), which is calculated as the logical OR of all configured trigger conditions, is then transmitted for each event (or rather bunch crossing) to each of the thousands of readout channels of the detector via a tree-like system of optical fibres in the *Trigger Timing and Control* system (TTC) [11] so that the L1Accept is available at the detector front-end to initiate the readout.

Buffering The *L1 latency* in ATLAS is defined as the fixed time interval between the actual crossing of the bunches at the interaction point and the arrival time of the L1Accept signal from the CTP at the detector front-end. During this time all data are stored in pipeline memories. With no L1Accept signal the data are discarded. In case of a positive L1Accept the front-end electronics of the detector components transmit their data to computers where the data fragments are buffered (in so-called *Readout Buffers*) for later use (e. g. by L2 or for event building) following the concept introducing in 18.2.3. A tree of signals from each detector component to the CTP enables the sub-detectors to inhibit the generation of further L1Accepts if so required – at the cost of global dead-time of the experiment. In addition automatic algorithms are applied inside the CTP which inhibit further L1Accept signals in case of too high readout rates.

L2 Trigger For accepted events, the L1 systems send additional data via dedicated optical fibres to the second-level trigger (via the so-called *Region-of-Interest Builder* (RoIB), cf. Figure 18.5). For the muon and calorimeter triggers these data fragments contain the identified geometrical position (η and ϕ) of the object candidates. For each event these *Regions of Interest* (RoI) are transmitted via the L2 supervisor to one CPU of the L2 computer farm where the L2 selection is carried out. Via a large networking system the L2 CPU has access to the data stored in the Readout Buffers at full granularity and precision around the RoI (approximately 2% of the total event data). RoIs identified at L1 are used to seed the L2 selection, which consists of a series of software algorithms. At L2 several trigger conditions are run in parallel. An event is selected if one of the trigger conditions is fulfilled. Otherwise, events are rejected and their data are discarded in the Readout Buffers.

Event Filter For events accepted in the L2 selection all event fragments stored in the Readout Buffers are retrieved via the network and combined in a single data fragment (*event building*), which is subsequently transmitted to one of the CPUs of the event filter processor farm. Hence the EF selection procedure can make use of the full event data and can apply offline-like event selection algorithms. If an event fulfils one of the configured trigger conditions, the data are transmitted to permanent storage. Otherwise they are discarded.

18.3.4
Example LC: Trigger-Less Data Acquisition at the ILC

The International Linear Collider [12] (ILC) is being designed to collide electrons and positrons of 250 GeV each accelerated in linear accelerators. The ILC features a duty factor[48] of only 0.5% due to its special bunch structure that arises from a train of 3000 bunches separated by 300 ns followed by a break of 199 ms. It is hence conceivable to record and buffer all bunch crossings of a bunch train in the readout electronics. The subsequent 199 ms gap can then be used to empty the buffers of the readout electronics, filter the events according to their physics properties and transfer the events to permanent storage. Hence the pulsed mode allows the usage of a trigger system similar to the setup described in Figure 18.1c. A dedicated trigger system with fast processing and intermediate pipeline storage of the data read (cf. Figure 18.1b) is not needed as long as the transfer and reconstruction capability is adequate. In contrast, the LHC will eventually reach almost a 100% duty cycle. Together with the enormous physics background in pp collisions, the level-1 trigger has to be highly selective not to saturate the subsequent readout bandwidth.

In both cases the demands on the trigger/filter system arise from satisfying the equation: $\langle \nu \rangle \langle T_P \rangle < 1$, where ν is the input rate and T_p the processing time at this stage. Both at the LHC and the ILC the average processing power, which scales with the number of installed processors, has to be chosen to satisfy the input rate, cf. Figure 18.3.

---

48) The duty factor of a collider is the fraction of filled bunch positions amongst the possible positions.

18.4
Trigger Systems and HEP Data Analysis

The action of the trigger has implications for the physics analysis. The shortcuts and approximations introduced in quickly evaluating physics quantities at the early trigger levels affect the boundaries of the phase space due to resolution effects. Offline, these inefficiencies must be corrected for to obtain a fair representation of the original event distributions. In addition the correction for the well-quantified event losses in the trigger downscale procedure has to be evaluated and applied. The correction procedure will be outlined below using a single-level trigger, which explains the basic principle [13]. The same reference offers an extension of the concept to an analysis using several trigger conditions at various trigger levels.

The correction for the trigger implications in an analysis comprises three steps: (i) the choice of trigger conditions from the complete trigger mix that promises high efficiency for the physics process under study[49]; (ii) the evaluation of the trigger efficiency preferentially from the data itself to safeguard from possible inadequate descriptions in the Monte Carlo simulation; and (iii) the application of the correction for inefficiencies together with the effects of the downscale procedure.

The efficiency of a trigger condition is naturally defined as the ratio of the events accepted by the trigger condition to the number of events that would be included in the offline analysis without applying the condition to an unbiased event sample. Inevitably this definition depends on the phase space defined in the analysis offline. The additional inefficiency of the offline selection with respect to for example the Monte Carlo simulation will hence have to be treated in the correction of the offline selection itself.

The trigger efficiency correction using data itself necessitates the availability of an independent data sample collected with a reference trigger condition orthogonal to the studied trigger. Since the efficiency typically varies as a function of the kinematic variable, q, the efficiency has to be determined differentially:

$$\epsilon(q) = \frac{\text{number of events selected by both trigger conditions}}{\text{number of events selected by reference trigger condition}}, \quad (18.2)$$

for each bin. Note that only events fulfilling the offline selection are used. An example for the procedure would use a trigger condition solely based on signals in the tracking detectors as a reference for a calorimeter trigger condition. Often the data-taking rates for the reference triggers are artificially reduced (by prescaling) so that the statistics is limited. In these cases the data-driven procedure provides a consistency check of the validation of the trigger efficiency as derived from Monte Carlo simulations.

The effect of a trigger threshold on a quantity measured with finite resolution can be seen in the "turn-on curve" of the corresponding trigger efficiency of Figure 18.6. The event variable, q, may correspond to for example the E_T of the electron. The quality of the trigger selection, that is, the resolution, σ_q, is reflected in

---

[49] A preparatory step in the analysis consists, of course, of designing an adequate trigger condition before data-taking that will retain the interesting events with high efficiency.

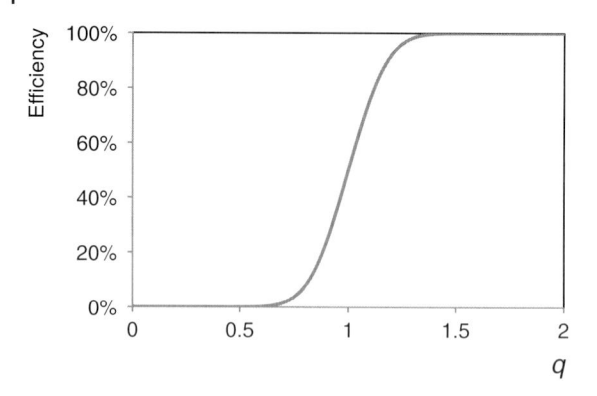

Figure 18.6 Typical efficiency turn-on curve of a trigger condition applying a threshold criterion on a single event quantity as a function of the corresponding value, q, determined offline (assumed threshold $q = 1$, trigger resolution $\sigma_q = 0.2$).

the steepness of the turn-on curve. The curve also displays the events lost above $q = 1$ and the additional events that enter the sample due to resolution effects for $q < 1$.

Quite often the event sample used in the analysis is composed of events with a variety of trigger conditions. The efficiency correction can hence not be applied as a simple factor in the cross-section calculation. Instead, each event has to be considered individually with its corresponding weight. The weight for event j depends on the trigger condition i in run[50] k and can be computed as

$$w_j = \frac{d_{ik}}{\epsilon_{ik}(q_j)}, \tag{18.3}$$

where d_{ik} is the downscale factor for trigger i in run k and $\epsilon_{ik}(q_j)$ is the efficiency of this trigger in this run as a function of a set of event parameters, q_j. If the downscale factor, d, and the efficiency, ϵ, do not vary over the phase space of the physics analysis, the weights of the events $j = 1,\ldots, N$ passing the offline selection criteria, including the trigger requirement, are computed as $w_j = d/\epsilon$, and the original number of events[51] can be calculated as $N_{\text{orig}} = \sum w_j = N \times (d/\epsilon)$.

Large variations in the downscale factors from run to run may lead to results that are statistically not very significant, since large factors dominate the result (cf. (18.3)). Under these circumstances it is advantageous to resort a luminosity weighted average weight calculated over all runs. This average weight is calculated

---

50) Data at collider experiments are usually collected in event samples of separate "runs", in which stable detector performance and steady running conditions are maintained. The trigger setup, in particular the downscale factors, are kept constant within one run, but may vary from run to run as a response to changing conditions, for example instantaneous luminosity and background rates.

51) Further corrections for detector efficiency, acceptance and so on must be taken into account. These corrections are outside the scope of the present trigger-related discussion.

from the number of selected events, N, and the original events, N_{orig}:

$$w_j = \frac{N_{\text{orig}}}{N} = \frac{N_{\text{orig}}/\sigma}{N/\sigma} = \frac{\sum_{k=1}^{N_{\text{runs}}} L_k}{\sum_{k=1}^{N_{\text{runs}}} L_k \frac{\epsilon_{ik}(q_j)}{d_{ik}}} \ . \tag{18.4}$$

Here N_{runs} and L_k represent the total number of runs and the dead-time corrected integrated luminosity of run k, respectively. As expected for a given number of original events, that is, for a given integrated luminosity, the so averaged weight for a trigger depends only on the total number of events collected via this trigger, N. An optimisation of the downscale factors during data-taking increases the statistics of the sample, reduces the weights and increases the statistical significance – as desired.

18.5
Summary and Outlook

Trigger systems for the large multi-purpose HEP experiments have seen a remarkable development over the past two decades. Not only was it necessary to separate noise signals from genuine physics interactions, but hadron machines immediately demanded the separation of interesting physics events from less interesting signatures. The trigger had to be "physics-aware". This demand could only be fulfilled exploiting the tremendous progress in electronics integration. The LHC experiments demonstrate already today that following the first-level trigger the selection can be made in a trigger farm that is based on commercial processors and fast network switches. The examples discussed suggest that a purely software-based trigger could soon become a reality.

References

1 Frühwirth, R. et al. (2000) Data Analysis Techniques for High-Energy Physics, Cambridge University Press.
2 H1 Collab., Abt, I. et al. (1997) The H1 detector at HERA, Nucl. Instrum. Meth. A, 386, 310, doi:10.1016/S0168-9002(96)00893-5.
3 ZEUS Collab. (1993) Status Report, Chap. Trigger, http://www-zeus.desy.de/bluebook/bluebook.html.
4 CDF-II Collab., Blair, R. et al. (1996) The CDF-II detector: Technical design report, http://www-cdf.fnal.gov/upgrades/tdr/tdr.html, FERMILAB-PUB-96-390-E.
5 DØ Collab., Abazov, V.M. et al. (2002) Run IIb Upgrade Technical Design Report, http://www.slac.stanford.edu/spires/find/hep/www?r=FERMILAB-PUB-02-327-E, FERMILAB-PUB-02-327-E.
6 ATLAS Collab., Aad, G. et al. (2008) The ATLAS Experiment at the CERN Large Hadron Collider, JINST, 3, S08003, doi:10.1088/1748-0221/3/08/S08003.
7 CMS Collab., Adolphi, R. et al. (2008) The CMS experiment at the CERN LHC, JINST, 0803, S08004, http://dx.doi.org/10.1088/1748-0221/3/08/S08003.
8 Sefkow, F., Elsen, E., Krehbiel, H., Straumann, U., and Coughlan, J. (1995) Experience with the first level trigger of

H1, *IEEE Trans. Nucl. Sci.*, **42**, 900, doi:10.1109/23.467771.

9 Nicholls, T. *et al.* (1998) Concept, design and performance of the second level trigger of the H1 detector, *IEEE Trans. Nucl. Sci.*, **45**, 810, doi:10.1109/23.682642.

10 Newman, P.R. *et al.* (2007) Results of the H1 Fast Track Trigger, http://www.slac.stanford.edu/spires/find/hep/www?r=RT2007-TDAQ-TRIG03, presented at 15th IEEE Real Time Conference 2007 (RT 07), Batavia, Illinois, 29 Apr–4 May 2007.

11 Ashton, M. *et al.* (2000) Timing, trigger and control systems for LHC detectors, http://ttc.web.cern.ch/TTC/RD12status99.pdf, CERN-LHCC-2000-002.

12 Brau, J. (ed.) *et al.* (2007) International Linear Collider Reference Design Report. 1: Executive summary. 2: Physics at the ILC. 3: Accelerator. 4: Detectors, http://www.linearcollider.org/about/Publications/Reference-Design-Report, ILC-REPORT-2007-001.

13 Lendermann, V. *et al.* (2009) Combining triggers in HEP data analysis, *Nucl. Instrum. Meth. A*, **604**, 707, doi:10.1016/j.nima.2009.03.173.

19
Grid Computing in High Energy Physics
Wolfgang Ehrenfeld and Thomas Kreß

19.1
Introduction

At the start of the LEP particle collisions in 1989, the computer hardware used for high energy physics (HEP) data analysis consisted mainly of big mainframe machines. A short time later workstations became available at a reasonable price and were used more and more. Around the beginning of the new millennium the era of *personal computers* (PCs) in HEP started. Nowadays PCs and for some aspects also laptops play an important role in the experiments' computing and end-user data analysis.

Although distributed computing was already partly used to process data in the high energy physics experiments of the LEP and HERA eras, the LHC experiments fully rely on it for almost all aspects. During the planning for the LHC it became evident that concentrating most of the computer hardware and manpower needed for data-taking and analysis at CERN alone was not possible. For LHC computing, large building capacities hosting the IT devices are necessary, and there is also a high demand for electricity and cooling power. In addition, the funding agencies clearly prefer a scheme where a substantial fraction of the money spent stays in the home country to operate hardware bought from local companies and to finance local operators and expert manpower. In a lot of countries large computing facilities and expert knowledge existed already, which can be used for the benefit of LHC computing. Therefore, there was a need for a distributed computing concept where critical services like data recording, a first fast data processing and the storage of the recorded data are provided at CERN, while a large fraction of the resources are maintained all over the world connected by fast and reliable networks, which are operated and controlled by dedicated software, the *Grid middleware*.

The basic principles of a computing Grid were introduced in the late 1990s by I. Foster and C. Kesselman in their famous *Grid Blueprint* [1]. The *Worldwide LHC Computing Grid* (WLCG) project [2] was approved by the CERN Council (see Sections 22 and 21) on 20 September 2001 to develop, build and maintain a distributed computing infrastructure for the storage and analysis of data from the LHC exper-

Figure 19.1 The metallic globe in the CERN Computing Centre illustrating the Worldwide LHC Computing Grid (WLCG) (Source: CERN).

iments. The WLCG collaboration combines the expertise and computing power of more than 140 IT centres situated in over 30 countries. Figure 19.1 shows exemplarily the CERN IT centre as one example of the WLCG member centres distributed over the world.

In the past few years the four large LHC collaborations ALICE, ATLAS, CMS and LHCb have published detailed technical design reports for their computing models [3–6] with more than 100 pages each. This article concentrates on the implementations by the ATLAS and CMS collaborations and focuses on the aspects relevant for user analysis of experimental data.

The LHC computing model manifests itself in a so-called *Tier architecture* illustrated in Figure 19.2. At CERN, the Tier-0 (T0) centre, raw data collected from the detectors are promptly reconstructed and stored. A copy of the data is then distributed to the Tier-1 (T1) sites. The T1 centres provide a wide range of computing services for the collaborations, with high reliability and throughput, at large sites – ten for ATLAS, seven for CMS – distributed all over the world. Tier-2 (T2) centres – 50–100 per experiment usually situated at laboratories or bigger institutes – provide flexible resources with large processing power, but with looser storage, availability and connectivity requirements than a T1 centre. Tier-3 (T3) is a term not rigorously defined technically. It may offer some or even all Grid and experiment-specific analysis functionality, but is characterised by the fact that the control of the resources is completely in the hands of the local site, while for the first three Tier layers *Memoranda of Understanding* (MoUs) exist which guarantee firm processing, storage, network and manpower resources to the collaborations. At the end specialised interactive clusters (like the *National Analysis Facility* (NAF) at DESY for German users, see Section 19.9), desktop clusters or standalone PCs or even laptops are used to perform the final steps of the analysis and to produce the physics results.

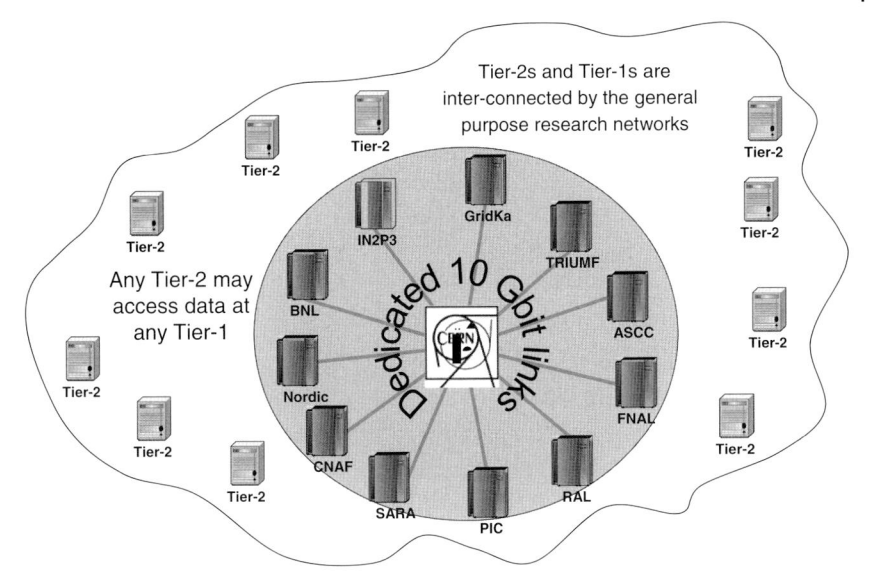

Figure 19.2 LHC Grid Tier layer architecture (adapted from [2]).

19.2
Access to the Grid

The LHC collaborations distribute their recorded data for analysis on their Grid Tier sites all over the world. To access the data, users have to obtain (and to renew regularly) a valid *Grid certificate* from an accredited *Certificate Authority* (CA), for example at GridKa at the Karlsruhe Institute for Technology for users affiliated to a German university or science laboratory. Using these certificates short-term (typically of the order of a few days) proxies are created by the users and used throughout for authentication (the proof of the user's identity) and authorisation (the permission to use the requested services). The experiments use the *Virtual Organisation Management Service* (VOMS) to administer the certificates of the group of users which is allowed to utilise the various Grid resources provided for the experiment. To define for example a national community to provide their members with higher CPU resources at their national T2 centre, a VOMS group can be set up to which people can be assigned. Furthermore, if a user has to perform certain privileged tasks (like that of an experiment's MC production manager), he/she can use the same certificate but authenticates himself/herself with a specific VOMS role extension to gain special privileges for performing the central task. Thus, VOMS groups and roles can be used to provide a fine-grained privilege control system to allow an unequal utilisation of the resources.

19.3
Tier-0 Grid Layer

The main purpose of the Tier-0 centre at CERN is to quickly, completely and safely store the raw and the first reprocessed data delivered by the experiments' data acquisition systems and to provide a first fast event reconstruction, together with alignment and calibration constants. From the T0 a second copy of the data is transferred to the Tier-1 centres. The T0 workflows are operated by a handful of data management experts from the experiments, in close collaboration with the CERN IT experts for hardware, software and the Grid middleware.

19.4
Tier-1 Grid Layer

The Tier-1 centres are mainly used for centrally organised collaboration workflows. The collaborations' T1 centres share a second copy of the initial LHC datasets for custodial storage on their tape systems. They also collect and save the Monte Carlo simulation events produced mainly at the Tier-2 sites. Another main purpose of the T1s is a repeated re-reconstruction of the data with improved detector alignment and calibration constants and new features implemented in updated releases of the experiment software, and the extraction of data samples like event selections and skims to be further analysed by the users at the T2 sites. Using the Tier-1 centres for user analysis in parallel to the organised workflows is possible to some extent. However, some collaborations do not allow massive user analyses at their T1s, with the exception of some privileged users controlled by the VOMS group and role mechanism.

The majority of the Tier-1 centres provide resources for more than one LHC collaboration. All T1s are equipped with very reliable CPU, disk and tape storage hardware and with excellent and redundant wide-area network connections. The core services are monitored by the computer centre's staff in a full-time 24/7 maintenance mode.

19.5
Tier-2 Grid Layer

ATLAS and CMS operate around 50–100 Tier-2 sites each. Although used differently in detail, a common feature is that these sites provide the major resources for user analyses and large-scale Monte Carlo simulation.

Typically, the relevant data for a user analysis are stored at one or more Tier-2 sites. In a Grid job, the user specifies his analysis computing code as well as the desired datasets to analyse and submits the job to the Grid. The Grid middleware finds one or more appropriate locations where the desired job input datasets are located, runs the jobs and returns the output to the user.

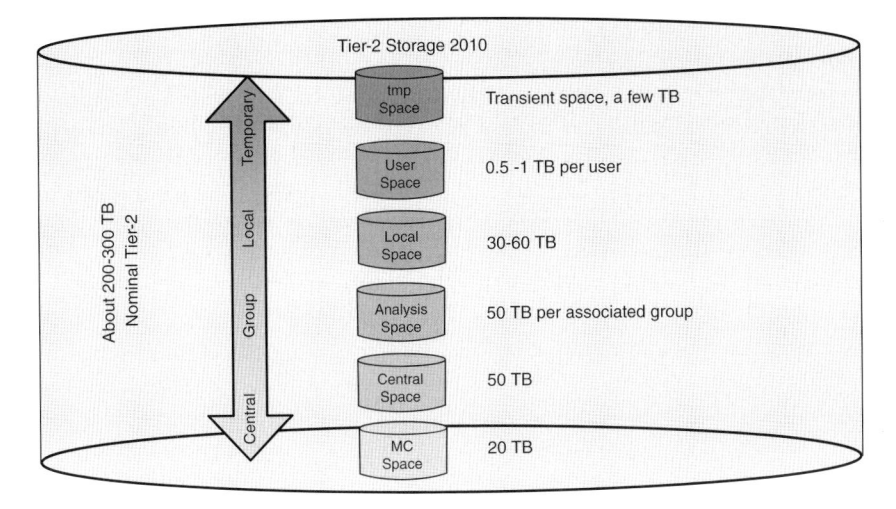

Figure 19.3 Storage layers at a nominal CMS Tier-2 site.

At a CMS Tier-2 centre, the CPU resources pledged in the WLCG MoU to the collaboration are divided equally among central operation and user analysis. The CMS T2 storage resource allocation is shown in Figure 19.3. For the year 2010 a nominal CMS Tier-2 provides about 200–300 TeraBytes (TB) of disk storage. A few TB are needed as transient space. Besides a smaller area at CERN, the T2s are the only locations where users are provided with a guaranteed amount of disk resources. A CMS user gets in the order of 0.5 to 1 TB for private data to serve as a Grid home directory, in most cases at a national Tier-2. Some 30–60 TB are used to store datasets of particular interest or for a high-throughput access for the associated local or national users. CMS supports about 20 detector performance and physics groups with their own storage resources at the T2s to host datasets or selections of particular interest for the group. A typical group owns storage resources at about 3–5 Tier-2 sites. The group organises and controls the transfers and production and deletion of dedicated datasets stored at this location. 50 TB for datasets heavily accessed by the whole collaboration are managed by central operation teams. A further 20 TB are used to temporarily host the produced Monte Carlo simulation events before the files are merged and finally moved to a Tier-1 tape system for custodial storage.

At some Tier-2 sites, beside these WLCG-pledged storage resources, the local and national users are provided with additional disk space from additional funds.

19.6
Tier Centres' Hardware Components

For the HEP Grid Tier layers 0 to 2, commodity hardware is not reliable enough to operate high-quality services. For a collection of identical computing nodes running the analysis jobs (*worker nodes*, WN) and for Grid middleware control machines, mostly robust hardware in blade technology or dedicated server rack systems are used these days. In most cases the computer mainboard of such a node is equipped with two or even more physical CPUs. In 2010, each CPU has 2–8 processing cores. Usually an analysis job runs on one core and can address of the order of 1–2 GB of virtual memory. A blade or server rack system is usually connected with 1–2 Gbit/s Ethernet to the local area network. The trend is now to move from copper-based Ethernet to shared 10 GBit/s optical network links. Apart from the access to the storage, the analysis jobs on the WNs do not have to pass information to each other; consequently, more advanced inter-network connections like "InfiniBand" or "Myrinet" with low latency are not necessary. A data sample to be analysed is split into several parts with a few thousand events each, such that an average analysis job takes at most a few hours to be finished on one WN core. The crash of a worker node running an analysis job is not a real problem, since the job is either automatically re-sent by the Grid middleware or can be resubmitted.

Tier-0 and Tier-1 centres have to safely store the data. For this purpose, huge tape systems are used which are slower than disk systems but more reliable. If data on tape have to be processed, they are read back to disk buffers and analysed from there. Almost all of the Tier-2 sites do not provide tape storage, since in the LHC experiments' computing models, data at a Tier-2 site can be re-created or re-transferred from a T1 in a reasonable time, and the necessary drives and robot systems to access tape media are quite expensive. In storage systems redundant power supplies are usually used for the disk boxes and storage control servers. The disks are protected by Raid systems to prevent data loss in case of a failure of single disks. A typical disk server with several dozen TB of capacity is equipped with CPU power and memory similar to a worker node's, but usually with a higher bandwidth connection to the network backbone. The Tier-1 sites and CERN are interconnected with redundant 10 Gbit/s links even across the Atlantic; the wide-area network connection of the Tier-2s vary in the range of 1–10 Gbit/s.

19.7
Tier-3 Grid layer

As mentioned in Section 19.1, a MoU regulates the processing, storage and manpower resources for the first Tier layers 0–2. The use of these resources is governed by the collaboration's computing model, as discussed in the previous sections. These models also allow for additional computing and storage resources not covered by the MoU regulations and call their aggregation Tier-3. In general T3 resources are local resources provided either by an individual institute, or by a regional or national community. For simplicity these communities are called "local community" throughout the rest of this section. As these resources are not MoU-regulated, their owners have the exclusive right of use, but also need to install, operate and support their resources themselves, usually only with support from the collaboration experiment on a *best effort* basis.

How the access to Tier-3 computing and storage resources is provided is defined by the local community. It can range from local to Grid-based access, even including a mixture of both. Different aspects need to be considered. Clearly, local resources are much easier to maintain than Grid-based ones; on the other hand Grid-based resources allow the usage of certain services provided by the experiments, which can reduce the amount of administration. Furthermore, it must be guaranteed that the T3 resources and their access to the experiment's data and software are retrievable by the local community. The prime example for this kind of discussion is storage. Grid-enabled storage clearly needs a higher amount of maintenance because an additional Grid software and protocol layer is needed on top of the local file system. On the other hand one gets for free all the experiment's tools to administrate the data. These tools can include automatic data import and export and data bookkeeping. Global bookkeeping is of particular advantage if more than one user is interested in the same dataset and when data need to be deleted. Similar arguments hold for the question if computing access should be local or Grid-enabled, although Tier-3 sites tend to operate local batch systems. There are two reasons for Grid-enabled T3 computing resources. Either the Grid service nodes can be operated at no additional cost as the site already provides these, or the computing resources need to be shared between a larger community due to funding constraints. In the latter case, user administration based on Grid VOMS groups and roles is usually easier than local user administration.

At the beginning of LHC data-taking all experiments provided guidelines for a typical Tier-3 setup to guide the interested groups. The idea is to provide a common installation, which can be set up and operated with minimal manpower. As an example, the current ATLAS T3 centres can be grouped into the following categories:

- *Local Tier-3*
  Computing and storage resources are only available to local users. The main focus of the Tier-3 is the support of the users from the local institute.
- *Grid Tier-3*
  CPU and some storage resources are Grid-enabled and therefore available for a larger community. Sites can also contribute to official Monte Carlo event simulation. One reason for this setup is that part of the hardware and operational costs are supported either by regional or national funding agencies.
- *Tier-3 associated to Tier-1 or Tier-2*
  The institute already operates a T1 or T2 centre and local resources are added to the existing Grid worker nodes or storage system. Operation and maintenance costs for additional T3 resources are usually quite low. The VOMS system can be used to grant access to the T3 CPU and storage resources to only users from the local or national community. Alternatively, for the computing resources a dedicated fair share and higher priority on the batch resources can be provided.

The boundaries between the different categories are not very strict. A special case of the last category is the German National Analysis Facility (NAF) which is discussed in Section 19.9. The idea of the NAF is to provide additional computing and storage resources to all members of the national community for data analysis. This includes additional Grid-based resources for early data analysis steps on the Grid and interactive resources for efficient and fast end-user data analysis. This means that the size of the NAF resources need to be larger than the Tier-3 resources from a single local institute.

19.8
User Analysis on the Grid

Most Grid sites are optimised for CPU-intensive tasks and the overall bandwidth to the data based on the requirements of standard experiment's jobs like data reconstruction and Monte Carlo event generation, detector simulation, digitisation and reconstruction. Here, the requirements for these use cases are well known and the experiment's workflow and production system is optimised to maximise throughput while not overloading the different components of a site. For example, an ATLAS simulation job has moderate requirements on memory (1 GB), but processing time can be as long as several minutes per event. A reconstruction job has higher demands on memory (2–3 GB), but has a much higher processing rate (order of 10 s). This implies that the necessary I/O rate must be higher. In order to increase the failure resistance of the different involved components, input files are downloaded from the local Grid storage elements (SE) to the local disks of the worker node before processing; output is written to the local disk and uploaded to the local Grid SE after the event processing has finished. This implies that the I/O rate during event processing is provided by the local disk, which is usually quite high. However, in many cases the available disk capacity per job is limited. For production jobs moderate download and upload rates are acceptable. This approach makes particular sense if the full event information is processed.

In a broad view there are only two necessary workflows for user data processing: data analysis and private Monte Carlo production.

Private Monte Carlo production is needed for those analyses where the standard Monte Carlo samples provided by the experiment are not sufficient and dedicated Monte Carlo samples are needed for physics interpretation or study of systematic uncertainties. Not all of these samples can be produced within an experiment's production system using the experiment's resource share. Hence, it is the task of the user to do this. The processing requirements are moderate and described above.

In a very simplistic view data analysis is done in only three steps: selection of an interesting dataset, measurement of the interesting quantities and calculation of statistical and systematic errors. Access to the full dataset is usually possible but takes time due to the dataset size. In order to finish this process in a given time and to be able to adjust the event selection from time to time, data analysis is done in steps or cycles, mostly in two: a pre-selection step and a final selection step, but more steps can be added. In each step, the overall data volume one works with in the next step is reduced. A reduced dataset implies that the processing time will be shortened, increasing the overall turnaround. This strategy will only improve processing time if modifications and thus reprocessing of an earlier step are less frequent than at a later step. Data volume reduction can either be done by *event skimming*, the selection of certain events, or by *event slimming*, that is by storing only parts of the event data, or by a combination of both. In most cases only event skimming is applied in order to be more flexible in the event analysis later on. For example, the pre-selection is usually based on trigger items, detector quality and some basic kinematic quantities like the number of physics objects. More advanced selections will be done in the next analysis step.

The processing requirements for data analysis are quite broad and depend on the analysis step. The pre-selection should mainly reduce the data volume, and typically not much event processing is done. If event slimming is applied, the event data can be reduced down to a small percentage. This step requires a very high I/O rate and only little computation. If one moves further up in the chain towards the final measurement, the dataset is much smaller but the information density is much higher and much more event analysis is done. The processing requirements shift from high I/O rate to more processing power. The estimate of systematic errors is usually done via the variation of one single parameter within the analysis chain. This is repeated for a few event quantities. This does not add any new performance requirements for the single job, but on the overall processing power: the number of systematic variations could be of the order of a few tens, and thus enough processing power is needed.

Finally, the experiment's workflow adds another level of complexity to this discussion. In general, there is more than one way to arrive at a reduced dataset suitable for user analysis. There are different choices in accessing the data and how to process them. Some of the steps include already event skimming and event slimming. This usually reduces computing time downstream in the analysis chain but increases the necessary I/O bandwidth at the same time.

Despite the diversity of the given examples, some common user analysis characteristics can be extracted: the amount of input data is rather large while the event computation can be quite moderate, therefore demanding a high I/O throughput and large storage with high performance for input and output files.

Furthermore, the processing paradigm changes from average to overall. Standard production like Monte Carlo event generation and reconstruction aims at the optimisation of the average processing time of a job. If one of the jobs in a batch takes much longer it is not so important. This is different in user analysis where the next step can only be processed if the previous one finished fully. Also, sometimes for a measurement the relevant data and Monte Carlo samples and events have to be analysed completely. Therefore, the overall turnaround for user analysis should be fast.

19.9
National Analysis Facility

As discussed in the previous section individual user analysis requires a different infrastructure and setup of a computing facility from the sequential central workflows. The setup needs to achieve high data throughput while allowing for the required flexibility in the analysis workflow.

The concept of a *National Analysis Facility* (NAF) was introduced to provide a generic, large-scale computing facility optimised for individual user data analysis complementary to the standard Grid Tiers. A similar facility has been set up at CERN: the CERN Analysis Facility (CAF), which has been designed for time-critical analysis work related to data calibration and commissioning. The NAF provides computing and storage resources with the following characteristics:

- enough disk space to host all interesting input data for user analysis;
- dedicated disk space for analysis output;
- analysis disk space optimised for I/O-intensive jobs;
- exclusive Grid computing resources for standard Grid applications;
- interactive cluster for fast analysis of large datasets and computing-intensive jobs.

In total a fast turnaround time is expected while some degree of flexibility is needed to quickly adapt to changing analysis workflows.

From the operations point of view, a NAF of substantial size should be placed either at a Tier-1 or large Tier-2 centre in order to gain from the existing expertise and from already available resources. It is straightforward to add additional computing and storage resources to the existing Grid resources at low maintenance cost. The already existing Grid storage reduces the amount of additional storage needed to host the input data. Furthermore, it is expected that the administrators have enough expert knowledge in running the large batch and file systems needed for efficient user analysis.

In the following the German NAF at DESY [7] will be used as an example setup of such a concept. One reason to establish the NAF at DESY was the already existing Tier-2 for the ATLAS and CMS experiments there. The different building blocks of local and Grid computing and storage resources are shown in Figure 19.4.

The data import, export, and storage areas are handled by the Tier-2 Grid storage systems. A major part of the NAF storage resources is added to the existing Grid storage systems to provide sufficient space to host all data relevant for analysis and to provide enough space for user data (1 TB/user). The DESY Grid storage systems are visible to all NAF computing components. As some part of the user analysis is expected to run at Tier-2 centres, the NAF provides additional computing resources which enlarge the DESY Tier-2 Grid cluster. NAF users have a dedicated CPU share with higher priority to these resources, depicted as the NAF Grid cluster in Figure 19.4. This allows NAF users to get a faster access to the Grid resources than other users from their experiment. For I/O-intensive user analysis jobs, local

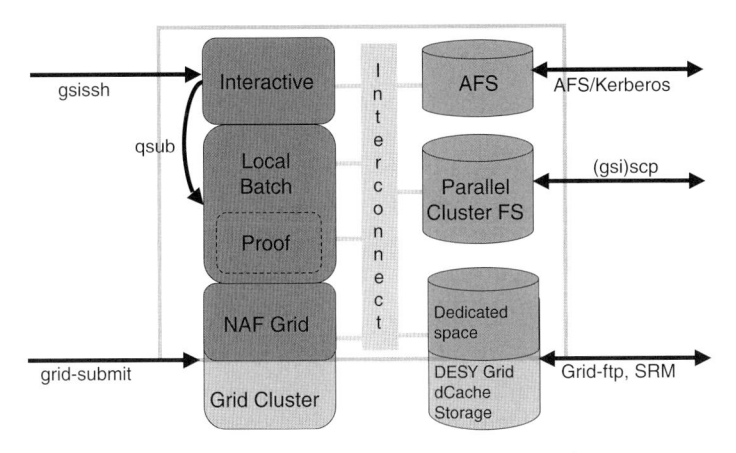

Figure 19.4 The building blocks of the German National Analysis Facility (NAF) at DESY. The NAF-specific part (upper box) is built on top of the DESY Grid Tier-2 infrastructure.

computing and storage resources are provided. The access to them is provided via dedicated workgroup servers. Authentication is based on standard security techniques using the user's Grid proxy. On the workgroup servers and the local batch machines the user can find a dedicated AFS home directory, which can be accessed directly from outside from an AFS client installation, for example at an institute's or laboratory's cluster PC.

For non-interactive tasks, a batch queueing system is available, which favours short jobs, but also allows for jobs running as long as one week. On the workgroup servers and local batch machines, the user has access to a global, high performance clustered file system using "Lustre". The connection between this storage system and the worker nodes is based on InfiniBand, providing a very high I/O bandwidth suitable for dedicated end-user analysis.

As an example of the size of such a facility, the ATLAS NAF resources for 2010 are given. The ATLAS share of the DESY Tier-2 consists of 650 cores and 740 TB of Grid storage. In comparison, the ATLAS NAF contribution on the Grid level is 80 cores and 500 TB disk storage and for the interactive/local part 160 cores and 60 TB of high performance disk space are available for the about 200 registered ATLAS users.

19.10
Last Steps of a Typical HEP Analysis

The typical waiting time to obtain the Grid jobs' output is in the order of 1–2 days per input dataset. In most cases the jobs run over the recorded collision data as well as the signal and all relevant background Monte Carlo event samples, in order to obtain efficiencies and purities and for systematic studies. Usually a user runs his/her analysis jobs on the Grid a few times per month only. Specifying soft selection criteria he/she obtains interesting signal and background events in a convenient data format and sufficient number to be analysed quickly in more depth from a facility which supports interactive work like the NAF, a desktop PC or even a laptop. If the Tier-2/3 does not provide a direct (non-Grid) access to the user files one can (with some restrictions on the local Linux operating system) install a selection of Grid middleware (*user interface* (UI)) which provides tools to copy datasets from Grid sites to local storage.

Using a popular HEP analysis framework like ROOT [8], publicly available for all important operating systems and easy to install, or appropriate software packages of the experiment the selection cuts can be fine-tuned, output histograms of the relevant physical values can be produced, and fit routines can be run to obtain for example cross section numbers or limits, or quantities like the mass of a particle. To run through the events on the local disk with a new set of cuts is usually feasible in the order of minutes.

Only if relevant event object quantities were not fully considered for the local sample extracted by the Grid jobs, or if the experiment's software or detector alignment and calibration constants got updated and the events were re-reconstructed, or if at the end in a selection all the available statistics have to be processed, does one have to resubmit the Grid jobs.

19.11
Cloud Computing – the Future?

During the last few years large companies realised that the concept of *buying and selling* computing power and resources *on demand* is very attractive. The commercial provider builds and provides big computing centres, preferentially in an area with low development costs for land, electricity and for cooling capacity. Sometimes also a significant reduction of taxes is granted as a compensation for substantial new employment possibilities. Big IT providers have to plan their resources such that they can cope with peak demand; therefore selling surplus resources in calmer times is a very sensible approach. A customer can now buy for a specified time a specific amount of CPU and disk resources and, if necessary, other services like databases and support. After their utilisation he finally gets a bill or his credit card is debited. For a user this is similar to renting web space – a concept which became popular in the last years. If the provider is able to provide flexibly also large resources, the concept can be very interesting for other sectors of industry which then do not have to plan, construct and maintain their own computing facilities with all the risks of under- or overestimating their needs. Usually cloud services from different vendors are not easily interchangeable, since the vendor wants to bind the customer closely to the company. Whether cloud computing will become important also for high energy physics will probably become clearer in the next few years. A likely scenario would be to buy Monte Carlo event simulation CPU power in addition to the major resources from the Grid to handle peak demands.

19.12 Data Preservation

In the past data preservation was not much of a concern. New high energy physics experiments came along in a regular drumbeat, regularly superseding one another in terms of what could be done with the data they produced. Nowadays, experiments are getting bigger, more complex and much more expensive. The time interval between experiments is increasing. Data from current experiments are collected with significant financial and human effort and are mostly unique. Currently, full data preservation and long-term analysis is rarely established, although some effort [9] has been made in different past and current experiments. Nevertheless, the importance of data preservation is commonly appreciated.

The value of old data was recently demonstrated by the high energy physics group at the Max Planck Institute for Physics in Munich. The group resurrected old data from the JADE experiment at DESY taken in the 1980s and combined them with more recent data taken by the OPAL experiment at CERN to improve the measurement of the strong coupling constant, α_s [10]. Applying new theoretical input and new experimental insights and methods, the old JADE data provided new results on α_s and its energy dependence, in an energy range which today is not otherwise accessible. This re-analysis of JADE data was not straightforward, as some part of the data was only available in the form of computer printouts.

In the following different aspects of data preservation will be discussed.

Documentation Within the collaborations each analysis is rather well documented, but most of the information is restricted to members of the collaboration. The general public only has access to the published article, which describes the analysis method and discusses the results. The available space for published articles is usually limited, especially for letters, which are only a few pages long. Hence not all analysis details can be discussed or all results be presented in full detail. This limits the usefulness of published data for later interpretation in the context of new theoretical models or combination with measurements from other experiments.

A first model of data preservation would be to provide additional documentation. This can include more information associated with publications (extra data tables, high-level analysis code, ...), internal collaboration notes, general experimental studies (for example on systematic correlations) and so on.

Data An economic way of preserving real and simulated data without the need for any experiment-specific software would be to just preserve the basic event-level four-vectors describing the detected particles. Clearly, this simple storage model is not sufficient for full physics analysis, except for a few particular cases. Nevertheless, it would be a useful model to roughly check new models and for outreach and education purposes.

Full Analysis Framework The preservation of data and the full analysis framework is needed for real data analysis. For example a re-interpretation of existing measurements in terms of new models usually requires Monte Carlo simulations of these models.

Preservation of data and long-term access to it is usually a financial matter. For archiving purposes data are stored on tapes and need to be replicated to new media from time to time. This is necessary to detect damaged or lost files or if tape technology changes.

Supporting long-term analysis is much more difficult and challenging since it not only involves the experiments' computing framework but also external software like ROOT or CERNLIB. For example, some of the software frameworks of the LEP experiments were written in FORTRAN making extensive use of the CERN FORTRAN library (CERNLIB). While FORTRAN compilers will stay around for some time, the support of CERNLIB was dropped a few years after the end of the LEP experimental programme. There are two ways to proceed: one can either run the analysis code on an outdated operating system, which might not be supported by current hardware, or one migrates to more modern libraries and operating systems. This usually involves significant manpower to migrate code and, more importantly, to validate the correct functionality.

Still, the mere preservation of data and the analysis framework might not be enough to perform data analysis. Coming back to the first point, documentation of expert knowledge is extremely important. For example expert knowledge, like information related to the quality of the Monte Carlo simulation and certain Monte Carlo corrections, is not always documented. Further examples are the knowledge about quantities which are less sensitive to calibration or alignment effects.

As discussed data preservation is important for the scientific reach of an experiment and its importance is commonly appreciated. On the other hand it involves substantial additional work. Only the future can tell if the LHC experiments will undertake the necessary efforts.

References

1 Foster, I. and Kesselman, C. (1998) *The Grid: Blueprint for a New Computing Infrastructure*, Morgan Kaufmann Publishers, Inc.

2 The LCG TDR Editorial Board (2005) LHC Computing Grid, Technical Design Report, http://cdsweb.cern.ch/record/840543.

3 ATLAS Computing Group (2005) Atlas computing, Technical Design Report, http://cdsweb.cern.ch/record/837738.

4 ALICE Collab. (2005) Alice computing, Technical Design Report, http://cdsweb.cern.ch/record/832753.

5 CMS Computing Project (2005) The computing project, Technical Design Report, http://cdsweb.cern.ch/record/838359.

6 LHCb Collab. (2005) LHCb computing, Technical Design Report, http://cdsweb.cern.ch/record/835156.

7 Haupt, A. and Kemp, Y. (2010) The NAF: National analysis facility at DESY, *J. Phys.: Conf. Ser.*, **219**, 052007, doi:10.1088/1742-6596/219/5/052007.

8 Brun, R. and Rademakers, F. (1997) ROOT: An object oriented data analysis framework, *Nucl. Instrum. Meth. A*, **389**, 81, doi:10.1016/S0168-9002(97)00048-X, http://root.cern.ch/.

9 Study Group for Data Preservation and Long Term Analysis in High En-

ergy Physics (2009) Data Preservation in High Energy Physics, http://arxiv.org/abs/0912.0255.
10 JADE Collab., Pfeifenschneider, P. et al. (2000) QCD analyses and determinations of α_s in e^+e^- annihilation at energies between 35 and 189 GeV, *Eur. Phys. J. C*, **17**, 19–51, doi:10.1007/s100520000432.

Part Three The Organisation

20
The Sociology and Management of Terascale Experiments: Organisation and Community

R. Michael Barnett and Markus Nordberg

20.1
Introduction

This chapter focuses on the more sociological aspects related to experiments for Terascale physics. It addresses the impact of the type of physics, technologies used, and available funding (and other resources). These affect the organisation and life of the physics community engaged in constructing and operating large detector apparatus for Terascale physics endeavours. Experience gained from work on the ATLAS experiment at CERN is used in addressing these issues, but the comments and observations are intended as a more general description of the subject matter.

Terascale physics, as described well elsewhere in this book, implies three key aspects relevant for an experiment: performance required by the energy and luminosity; technology (choice of detectors); the time frame (and budget) available. Each of these three aspects will now be examined in more detail. Figure 20.1 summarises the line of thought.

20.2
Performance and Instruments of Funding

First, consider the performance and its implications. According to the laws of physics, the higher the collision energies, the larger the detector has to be in size (or density). More specifically, a larger radius of the detector allows a much improved resolution of momentum transverse to the beam, p_T, as needed in the experiment. Higher luminosity (higher rate of collisions) increases the rate of radiation damage to the inner parts of the detector. This implies the use of state-of-the-art technologies that have the best resistance to radiation damage.

In simple terms, this means that Terascale physics experiments are huge in size, even for the standards of the particle physics community. They stand tall as a multi-storey building, weigh many thousand tons (all the way up to the equivalent of the weight of an Eiffel tower or two) and contain many millions of functional elements.

Physics at the Terascale, First Edition. Edited by Ian C. Brock and Thomas Schörner-Sadenius.
© 2011 WILEY-VCH Verlag GmbH & Co. KGaA, Weinheim.
Published 2011 by WILEY-VCH Verlag GmbH & Co. KGaA.

20 The Sociology and Management of Terascale Experiments: Organisation and Community

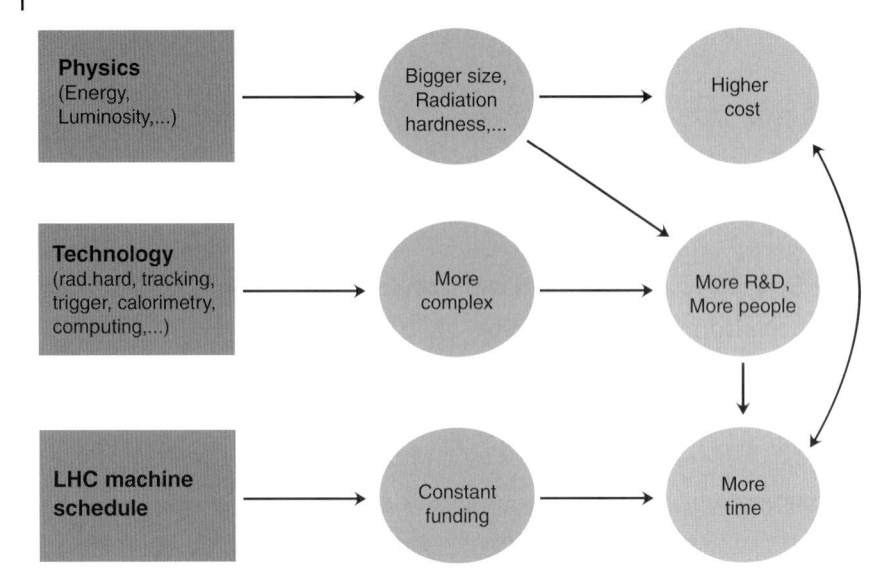

Figure 20.1 Line of thought in this chapter: three key aspects for Terascale experiments – performance, technology and time scales/budget.

The capital investment runs in the multi-hundred-million to multi-billion Euro/US dollar range, depending on the selection of detector technologies and the adopted accounting methodology. It is therefore clear that the construction of such devices requires significant resources both in terms of money and people.

No single country has sufficient intellectual and financial resources to realise such large and challenging Terascale projects and to exploit effectively the expected scientific output. International collaboration is the only way to accomplish such goals.

Experiments of this scale require substantial organisation, both in terms of organising the project itself and in the way resources are collected, shared and reported back to the funding agencies.

The underlying instrument for achieving such ambitious goals is a loose contractual framework, which in experimental particle physics is generally accomplished as a "Memorandum of Understanding" (MoU). Typically, there is a MoU for the construction phase and a separate one for the operations phase. It is, in most cases, a legally non-binding document that describes the focus and aims of the experiment and how the resources are shared, all on a best-effort basis. The MoU explains what is being built and who does what; it describes how costs are determined, shared and reported; it describes how the project is run and governed (e.g. the election procedure for project managers and different board members); it also defines the role of and relationship with the host laboratory. The MoUs are signed either by the participating institutes (universities and laboratories) or by their supporting funding agencies (often a mix of the two).

The signing counterparty is often the host lab (CERN, DESY, FNAL, and so on), in particular when the project itself is not a legal entity. Host labs hosting Terascale physics experiments usually apply the so-called ICFA (International Committee for Future Accelerator) guidelines, where the host lab covers all direct costs up to the point of producing particle collisions in a cavern area where the experiment is located (see Chapter 22 for a discussion of the large laboratories and their role in particle physics). According to these guidelines, the experiment is then to contribute in some way, specifically laid out in the MoU, to the costs of facilitating the experiment. As a rule of thumb, the annual running costs are gauged at 5–10% of the capital cost of the experiment itself, depending whether – and to what extent – manpower costs are included.

This latter point makes a big difference as to how the experiment is actually integrated into the host lab structures. In Europe, in places such as CERN, manpower costs are handled separately whereas in the US, they are an integral part of the project resources allocation. The host lab responsibilities are thus differently defined across the continents, and that is an important parameter while designing the governance structures of the collaboration.

In the US, the experimental projects are traditionally more integrated into the departmental structures of the host labs than is the case for laboratories in Europe. The experimental projects are considered part of the responsibility of the lab (even if a large share of the total resources comes from outside), and thus the projects are integrated within its departments, in order to ensure that they get the resources needed and that the host lab policies are followed.

The funding agencies signing the MoU with the aim of channelling resources both centrally for the experiment and to the participating institutes in their countries need to agree to a common cost accounting and reporting mechanism. National accounting standards differ significantly from one country to another. For example, in Italy a financial commitment made by the funding agency for the future is technically a Euro already spent.

In Japan and the US, a foreign lab or a project within that foreign country is to be treated as a (industrial) supplier and therefore is subject to the same rules and procedures that also apply to industrial, profit-making organisations. The foreign lab has to submit an offer, structured along the ministry (funding agency) guidelines.

In the US, the budget contingency is given to the project management, whereas in Europe the funding agency does not give the project any contingency and asks the project to come back to ask for contingency funding, if needed.

Some countries allow, or even insist on, annual cost indexation, in others the indexation has to be fixed *ex-ante* (in advance). "Cost indexation" means the country applies annual inflation indices on their contributions that were originally agreed in fixed prices at the beginning of the project (typically agreed for the first year of construction).

All such aspects need to be considered and dealt with in the MoU. Typically, a dedicated "Finances Board" or "Resources Board" is set up for following and allocating common resources to such experiments, consisting of members from

Table 20.1 The size of typical Terascale experiments, including estimates for completed upgrade work (*). The members in the table include those contributing to scientific publications. The construction budgets assume 1 US$ = 1.2 MCHF.

Experiment	Members	Institutes	Countries	Construction (M$)
CDF	600	60	15	390*
DØ	540	90	18	180*
ALICE	1000	110	30	130
CMS	2900	180	38	440
ATLAS	3000	170	38	450
LHCb	730	50	16	60

each funding agency, the management of the experiment and (typically) of the host lab.

The apportioning of costs also presents issues that must be handled effectively. There is no absolute mechanism or formula for that. One attempts to negotiate a fair sharing of the costs between the host lab and the participating funding agencies.

In some cases, such as at CERN, the host lab has a dual role, acting both as a lab and as a participating institute and thus (in effect) as a funding agency. CERN scientists are a group in the experiments, and the CERN PH department signs the MoU as a participating institute.

Large teams from countries with wealthy economies are usually expected to carry a larger share of the costs than small teams from countries with still developing economies. Consideration of the possible contributions of teams/countries (while agreeing to the sharing of MoU responsibilities) is a central part of the process of forming the collaboration.

A Terascale experiment has several thousand members from several hundred institutes (universities and laboratories), from all around the world. ATLAS and CMS have about 3000 members each (including graduate students), whereas ALEPH (at LEP) had about 400 and CDF (at the Tevatron) has grown above 600. ATLAS currently has 174 institutions from 38 countries. Table 20.1 summarises the size of typical Terascale experiments.

Like the related funding agencies, also the institutes need to get organised. Usually, they form a "Collaboration Board" where each institute has representation. In large collaborations, project managers (or coordinators) are often elected to their posts by majority vote with fixed mandates in time. This is done in order to ensure rotation and offer chances for younger members of the collaboration to take wider responsibilities. The Collaboration Board, or equivalent, needs to establish policies on matters such as the mechanism for institutes wishing to join – or leave – the experiment, crafting an authorship policy, sharing of construction and operation tasks, definition and election of people for managerial tasks. The MoU needs to incorporate all these considerations.

20.3
Technology, Project Structures and Organisation

The choice of technology has an impact on the project structure and the organisation. Technology plays a key role in the sociology of experiments. This is because the scientific and technological expertise of the collaboration will cluster around the planned and finally selected technologies. As Terascale physics experiments push the frontiers of both physics and technology by an order of magnitude or so, also the level of complexity increases correspondingly. An experiment such as ATLAS or CMS at CERN has at least 10 million components someone has touched by hand several times during the construction and integration phase. That is roughly three times more than the components of a Boeing 747 or almost twice as much as on an Apollo Moon Lander.

A dedicated Terascale experiment is built only once, not in series. By definition, it's one of a kind, a prototype that has to be built to work on the first go. Terascale experiments belong to the category of "Big Hairy Objects" that will radically change our perception of the world.

Handling this level of complexity requires setting up dedicated teams around the key technological challenges. Before full-scale production can start, the participating research teams need to demonstrate to their peer-review committees – typically formed by the host lab with international, uninvolved experts – that the detector will work. This is established by setting technical milestones, which are then reviewed. The approved technical solutions are usually recorded and reported in documents called "technical proposals" or similar.

Despite the detailed planning and careful evaluation of technical or technological solutions, adequate flexibility is required to allow the teams to dynamically change their composition as unforeseen technological problems or hurdles need to be overcome. Outside help is always welcome.

Unlike in industry, where product development is more incremental rather than radical and where the work can usually be contracted out to well-defined entities either inside or outside the firm, Terascale projects cannot be planned top-down, *ex-ante* (in advance). That is, all possible "what-if" situations cannot be defined before launching the project. Because of the high level of complexity, it would otherwise never converge. Problems are thus faced when they occur; some detailed technology decisions are left to the last possible moment given the overall construction schedule that is determined by the related accelerator system.

Not only do such complex undertakings require dedicated pools of scientific and technical expertise from all around the world, they also need some level of hierarchy that provides adequate steering without compromising the individual freedom or institutional constraints imposed by the hundreds of collaborating institutes behind the common effort.

Let us start with leadership. In most cases, large experiments elect their own leaders bottom-up. Candidates for a particular leadership position (such as project leader for a particular subsystem) are proposed by the groups working on that system. Candidates for the "spokesperson" and "Collaboration Board chair" are typically proposed by the membership of the whole collaboration. The leadership is then elected, and their proposed management teams are endorsed. The "electing bodies" are the institutions relevant to the particular position, for example the entire collaboration for the Collaboration Board chair or spokesperson.

Managing a large and far-flung collaboration requires a well-structured organisation. Each detector subsystem often has its own management team. At the same time, the project management, possibly supported by an "Executive Board" or equivalent, maintains general oversight of the project.

In most cases, the spokesperson has a "Technical Coordination Team" to help him/her and to make sure that all the separate subsystems can fit together. In parallel, there are national representatives whose function is to oversee the distribution of resources from each participating country to all the collaborating groups from that country, and to make sure that these resources are well-used. With work being done in so many different locations world-wide and many subsystems being designed and assembled independently, one needs to assure that all detector pieces fit and work together. In many cases, the Technical Coordination Team works with all the subsystem groups to ensure that the separate pieces will fit together without interference, and that the full detector can be assembled in its interaction region.

The spokesperson may also have a "Resources Coordinator" or "Manager" for help in preparing annual budget reports to the funding agencies, including the host lab. Depending on the agreement with the host lab concerning the use of financial services, the Resources Coordinator may also be responsible for the related collaboration accounts, expenses and cash flows. The spokesperson often also has a deputy or deputies to share the work and, depending on the collaboration rules, their responsibilities may also be rotated.

Apportioning tasks fairly and effectively across a large collaboration is essential. One tries to match the interests and resources of the participating teams to the tasks. This can succeed only if everyone is also willing to share the less interesting but necessary tasks, the so-called "no-fun jobs", such as pulling cables or labelling electronics modules. This in turn works because the physicists are motivated by the prospect of the exciting results to be obtained, and know that these depend on having a complete working detector system. Of course it is not always easy to arrive at an optimal sharing of tasks such that everyone is satisfied, and that all tasks are assigned.

Figure 20.2 shows an example of the organisation structure of a large Terascale collaboration (ATLAS, 2010).

Although leadership structures and the sharing of tasks are important for the success of a Terascale experiment, the role of the individual expert is even more important. After all, in practice such state-of-the-art projects are built by a collaboration of thousands of people with pieces constructed at several hundred different institutions around the world. It involves more than just following a sheet of instructions possibly used to draft the MoU. In plain reality, the construction of a large experiment is a collaboration of many dedicated individuals who have to work together as a team while often being thousands of miles apart.

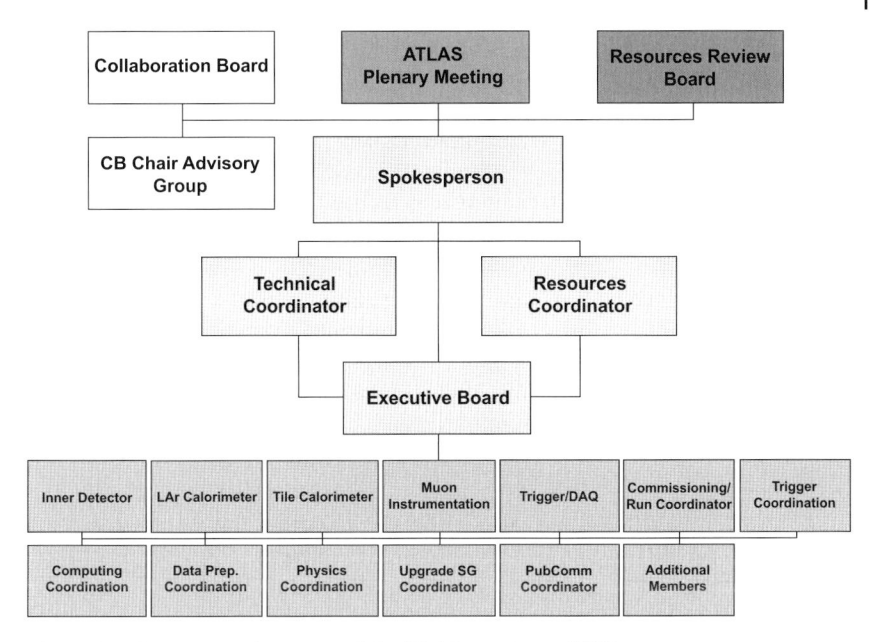

Figure 20.2 Organisational structure of the ATLAS experiment (2010).

Thus a Terascale experiment brings together people of different cultures, religions, races and economic backgrounds, and asks them to work together on a massive project when the typical person cannot know the large majority of people in the experiment.

How can this then possibly work when there is not necessarily an imposed top-down hierarchy and when many of the people and institutions have perhaps not worked together before? Part of the answer lies in the balance between individual creativity and the process of being part of a large collaboration. The successful design and construction of a large and complex detector requires the creative participation of a very large number of people. It is not the collaboration that is creative, but the sum of its individual members. There are numerous sub-subsystems in such large experiments so that people mostly work in small groups and contribute creatively.

The fact that all the systems must fit and work together, and be affordable, necessarily imposes some limits on the creative directions that people can take. It also means that people learn to work together independent of backgrounds.

Many important decisions need to be made, but for the process to be successful the voices of individuals not in the leadership need to be heard. This is achieved by splitting the overall experiment into smaller subsystems where decisions can be made more rapidly and independently, naturally in consultation with the project management (typically the Technical Coordination). The pros and cons are initially discussed in the subsystem plenary meetings. The term "plenary" here implies that all collaborators working within the specified group (for example a subsystem) can participate in the meetings and make their voices heard.

Recommendations can then be discussed in bodies such as the Executive Board or equivalent and presented in the plenary meetings that play a primary role in forming a large consensus about issues for which decisions are required. The leadership can only "lead" the collaboration to decisions that are understandable to all, or at least to a large majority. Directed by common sense rather than dogma, practical constraints may influence decisions, like costs, schedule, and the availability of teams to take responsibilities for the execution. Procedurally there is often a clear sequence of steps to formally make decisions, using a hierarchical structure within the subsystems. Ultimately, matters can be directed up for a vote in the Collaboration Board, which can have the final say in the matter.

The global spread of Terascale collaborations with many physicists working in countries as far apart as the US, Spain, Russia, China and Japan has a concrete impact on how people work together. It implies that factors such as the transport of components need to be taken into account during the construction, and that communication and logistics play a major role. Modern forms of communication such as video conferencing and the advanced usage of the World Wide Web (WWW) have greatly facilitated long-distance communication and enabled distant participants to be actively involved.

Nonetheless, this global spread also implies that scientists from other continents have to travel long distances to participate in the discussions and meetings, in the detector assembly and in testing or operational activities. They may have to spend extended periods away from their homes and home institutions. However, all scientists are after the same goal, namely doing frontline physics, and they are willing to endure the above inconveniences to achieve that goal.

More generally, communication amongst thousands of people is critical for the success of a large experiment. Bridging the large distances and time differences presents special challenges. Electronic communication plays a major role (e-mail, WWW, telephone, video conferencing). However, regular direct human contacts are crucial elements in the communication. Traditional face-to-face meetings still play an important role in the modern life of a Terascale physicist.

20.4
From Data Analysis to Physics Publications – the Case of ATLAS

Although all the above-mentioned factors shape the sociology of a Terascale collaboration, it goes without saying that the long-awaited physics analysis and final results are the fundamental driving force behind any scientific collaboration. This aspect also requires consideration.

How to organise this? Dividing the analysis of data among thousands of people may seem an impossible task. Well, pragmatism rules here, as well. The scientists will pursue these research areas mostly in small groups working at their home institutions, but in close contact with each other. Typically, all collaborators are invited to analyse the data by being part of analysis teams, which consist of physicists from many institutions. The final results are discussed amongst these individuals and groups which then develop a consensus.

In some respects, data analysis by individual physicists can be compared to data analysis by astronomers using the Hubble Space Telescope (although particle physics looks at the microcosmos). Both groups of scientists can freely choose the research areas and data that interest them most.

Agreeing on results and publishing papers in a giant collaboration does present challenges. Generally the data analysis work will be done in small groups pursuing different research directions. The analysis is then presented in a draft which will be subject to comments by all collaborators, to careful review and eventually to discussion within a plenary meeting. Ordinarily, two or more groups have done an analysis and differences will need to be resolved. Eventually, this process leads to a consensus, and the agreed-upon paper is submitted for publication.

As an example, ATLAS physics analyses are carried out within physics groups and within related combined performance groups. Each of these groups has two convenors, who each normally serve for a two-year term, with staggered appointments. Within each group, there are usually subgroups – for the combined performance groups these are typically arranged on an *ad hoc* basis to give short-term flexibility.

For the physics groups a more concrete subgroup structure has been defined. This substructure of dedicated smaller groups has been introduced into physics working groups in order to be able to deal efficiently with the large number of topics. The work in the subgroups is organised by sub-convenors, who work in close collaboration with the respective physics group convenors.

Table 20.2 gives an overview of the current working groups within ATLAS.

In ATLAS, physicists undertaking an analysis present it to the relevant analysis group from the beginning of their studies. When an analysis begins to reach maturity, the convenors of the analysis group inform the "Physics Coordinator" who selects a team of 2–3 people acting as "Analysis Reviewers" (AR). They include both

Table 20.2 Overview of the current working groups within ATLAS.

Physics groups	Combined performance groups	Other groups
B physics	e/gamma	Trigger
Top quark	Flavour tagging	ATLAS luminosity
Standard Model	Jet and $\not{E}_T$	
Higgs boson	Tau lepton	
Supersymmetry	Muon	
Exotic particles	Tracking	
Heavy ions		
Monte Carlo		

senior and younger people. Some or all of them will become members of the "Editorial Board".

When a physics analysis is nearing completion, the working group convenors are notified, as well as the Physics Coordinator and the "Publications Committee" (PubComm). The PubComm chair, in consultation with others, sets up an Editorial Board which follows the process to the end. Closely related analyses are encouraged to combine efforts before coming forward with a first draft.

The first step in the approval for publication of any analysis will be its approval by the relevant physics analysis group. After approval, it must be presented at dedicated plenary physics meetings, held once every two weeks, and announced to the collaboration. A draft of the paper has to be made available to the collaboration at least one week before. The collaboration will be invited to submit comments within two weeks. However, under exceptional circumstances, this time could be shortened to as little as three days. Upon receiving the comments by the collaboration, the editors modify the paper correspondingly and comment in writing on the suggestions and questions received.

When the Editorial Board believes that the paper is close to completion and the comments are accounted for, there will be a presentation to an open meeting. Here, any significant comments are dealt with. When the Editorial Board is satisfied that all comments have been dealt with appropriately, the final draft will be posted. Collaboration members will have 3–7 days for a last check of the paper.

After a final open meeting step was passed, the PubComm Chair in consultation with the Editorial Board Chair will recommend to the spokesperson that the paper should be approved for publication. Drafts are sent to the CERN directorate for approval as CERN preprints.

Although the above process is specific to ATLAS, all Terascale projects have set up similar processes in order to ensure quality and to enable participation of individuals inside the collaborations.

For large collaborations, of course a major question is how any one collaborator gets credit for his/her contribution. Internal publications within the collaboration, usually with one or a few authors, will document the individual contributions. Also, leading contributions are often recognised by asking the person in question to present results at conferences. However, the large collaborations still have to learn better how to handle this question in a fair way. Often major results are obtained in a collective way, because people are willing to share the tasks. A typical Terascale physics publication starts with some 20 pages of names. In contrast, in the fields of economics and sociology, three authors already make a crowd.

20.5
Budget and Time Considerations

Further elements that have a fundamental influence on the design and sociology of Terascale experiments are money and time. Not surprisingly, in this world, which economists define as a "land of scarce resources" – in particular in recent times –

both the technological and financial implications of reaching Terascale physics are far-reaching.

First, developing state-of-the-art scientific equipment, and beyond, requires constant R&D efforts by the particle physics institutes around the world, as well as by the related high-tech industries. Supporting such global, coordinated initiatives before signing formal MoUs is not easy and requires strong leadership from big countries and research institutions. In practice progress is made, but at a pace that is measured in years rather than in months. This means that large collaborations get going slowly, over a time span of many years. Given that governments come and go, obtaining a long-term commitment with aggressive funding levels is unlikely. Instead, progress will have to be made assuming more or less fixed (or even declining) annual budgets, thus stretching out the life span of a large collaboration to several decades.

Once the initial R&D and prototyping period reaches a stage of a better understanding of the implied technologies, the actual construction planning can get under way. Typically, the planned budget expenditure profile is bell-shaped over time, peaking a couple of years before the scheduled construction completion. Together with the funding agencies, a common, agreed annual budget allocation mechanism, following from the agreed MoU as described above, is set up. In the case of ATLAS, such budget allocation and funding reporting mechanisms were put in place in the "Resources Review Board" (RRB). The process is summarised in Figure 20.3.

The construction MoU establishes a planned project payment profile. Using this as the baseline, regular information on the payment profiles are collected from each subsystem and then rolled up. The actual annual payments are then compared

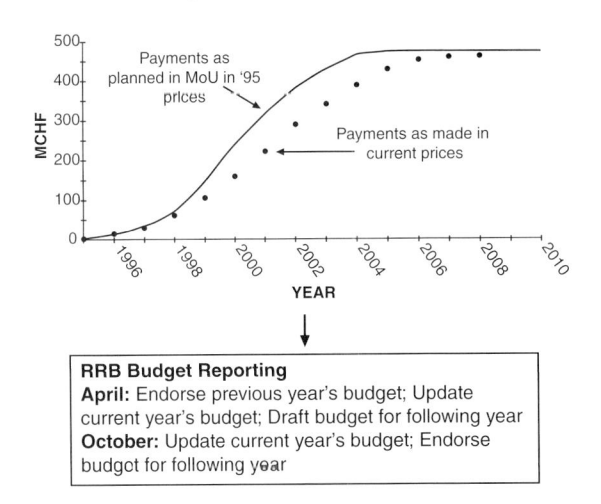

Figure 20.3 Comparison of budget allocation and funding for ATLAS.

with the original baseline and reported back to the funding agencies. In the case of ATLAS, the RRB meets twice a year, each April and October. A status report is thus provided twice a year and all the contributions, normalised to the pledged amounts in the MoU, get reported and updated. It should be noted that such simple reporting does not incorporate schedule milestones against resources consumed as is the case, for example, in "Earned Value Management" (EVM) schemes. The latter is well established and widely used for example in the US.

Second, and related to the above, constructing advanced devices of the size of apartment blocks requires skilled, dedicated people. Institutes have them, but only a limited supply, and it takes time to get teams of world-leading experts up and running. New generations of people need to be trained and fostered. Meanwhile, old timers will retire. It may well be that in the future, a Terascale physicist will only belong to one single collaboration during his/her professional career to reap the fruit of a long desire. Because of these long lifetimes, many physicists will join more than one experiment at the same time to make full use of opportunities to contribute to exciting science.

As an example of how collaborations form and of the timescale needed for a Terascale project, consider the following timeline of the ATLAS experiment:

1984 – First workshop discussions of a hadron collider in the not-yet-built LEP tunnel.
1987 – First workshop discussions of multi-purpose detectors at the LHC.
1989 – First steps to setting up a collaboration – ECFA Study week.
1992 – Expression of Interest presentations by ASCOT and EAGLE (amongst others).
1992 – ATLAS (combined from ASCOT and EAGLE) submit Letter of Intent (birth of ATLAS) with 88 institutions.
1994 – Submission of the ATLAS Technical Proposal.
1997 – Start of construction.
2003 – Start of installation in the ATLAS cavern.
2008 – First LHC beam in ATLAS.
2009 – First 0.9 TeV collisions.
2010 – First 7 TeV collisions.

20.6
Conclusions

The management and sociology of Terascale physics follows from three key design parameters: performance required, technology and the budget (and time frame) available. Terascale experiments are very focused enterprises and, despite of their well-defined scope, require large financial and human resources to make them happen. The community behind such endeavours is driven by a vision of new, exciting physics. They engage in long-term commitments (extending in many cases over decades) and benefit from and enjoy working together with people from all around the world.

Such collaborations are based on individual freedom, creativity and a shared sense of fairness. This is supported by project management and administrative structures where informal, legally non-binding MoUs are typically put in place. The participating institutes, including thousands of members, commit to construct and operate such devices over decades, while sharing all results as equal partners in the subsequent physics papers. In science, Terascale physicists build a world of relative harmony.

Further Reading

Boisot, M. *et al.* (2010) *Collisions and Collaboration: The Organization of Learning in the ATLAS Experiment at the LHC*, Oxford University Press, Oxford.

Merali, Z. (2010) Physics: the large human collider, *Nature*, **464** 482–484.

Lillestol, E. (2008) CERN: a unique experience, http://public.web.cern.ch/public/en/People/UniqueExperience-en.html.

Liyanage, S., Wink, R., and Nordberg, M. (2007) *Managing path-breaking innovations*, Greenwood Publishing, Westport.

Marchand, D. (2009) *The ATLAS and LHC Collaborations at CERN*, Case Study IMD-3-2015, IMD, Lausanne.

Wegener, D. (2008) Sociology of the ARGUS Collaboration, http://argus-fest.desy.de/e301/e306/ARGUS-wegener-socio1.pdf.

Shrum, W., Genuth, J., and Chompalov, I. (2007) *Structures of Scientific Collaboration*, The MIT Press.

Morrison, D.R.O. (1978) in: *Physics from Friends*, (R. Armenteros, A. Burger, Y. Goldschmidt-Clermont, J. Prentki eds.), Papers dedicated to C. Peyrou on his 60th birthday, 351–365, Geneva.

Knorr Cetina, K. (1999) *Epistemic Cultures*, Harvard University Press, Cambridge/MA; K. Knorr Cetina (2007) *Interdiscipl. Sci. Rev.*, **32** 361–375.

Traweek, S. (1998) *Beamtimes and Lifetimes*, Harvard University Press, Cambridge/MA.

Tuertscher, P. (2008) *The Emergence of Architecture in Modular Systems: Coordination across Boundaries at ATLAS, CERN*, Dissertation, Universität St. Gallen.

21
Funding of High Energy Physics

Klaus Ehret

21.1
Outline

This chapter describes general aspects of the funding of high energy physics (HEP), with special emphasis on the German funding system. Considering that details of the funding are different for different countries and that several different sources of funding are used, a comprehensive summary is out of the scope of this review. The first Section 21.2 provides a survey of the field. It briefly reviews the need of high energy physics, describes how the work and responsibilities are organised and explains the resulting demands on funding. The need of large research facilities causes a close interplay between politics and governmental authorities for the funding of high energy physics, which also affects the constitution and management of accelerator laboratories and large-scale experiments. Several of these aspects are also discussed in more detail in Chapters 20 and 22. Hence the review of accelerator laboratories in Section 21.3.1 and of collaborations in Section 21.3.3 and their funding is kept short within this chapter.

Section 21.4 describes the federal structure of the funding of basic research in Germany and provides a survey of the most prominent national funding resources for HEP. They are complemented by funds provided by the European Commission (EC), which are sketched in Section 21.5. Several general aspects of strategic decision-making are reviewed in Section 21.6. As a prominent example several aspects of the LHC funding are discussed in the last Section 21.7.

21.2
High Energy Physics – an International Effort between Accelerator Laboratories and Universities

Before explaining the complex funding structure of HEP, it is useful to briefly recall how high energy physicists organise their work. High energy physics requires large research infrastructures (RI). HEP is the prime example for the common interna-

Physics at the Terascale, First Edition. Edited by Ian C. Brock and Thomas Schörner-Sadenius.
© 2011 WILEY-VCH Verlag GmbH & Co. KGaA, Weinheim.
Published 2011 by WILEY-VCH Verlag GmbH & Co. KGaA.

tional construction and usage of large research infrastructures and therefore also a prime example for the funding of large RI. In a simplified view there are, on the one hand, accelerator laboratories that construct and operate particle accelerators (see Chapter 22). On the other hand, there are international HEP collaborations which construct the detectors, operate the experiments and exploit the physics data (see Chapter 20). The success of high energy physics relies on a close collaboration of accelerator laboratories and physicists from the particle physics community. Particle physicists are not only users of particle accelerator laboratories. Their science vision is also the driving force for the development and construction of accelerators. Accelerator laboratories are either national or international institutions. Details of their *institutional* funding depend on the constitution of the laboratories and on individual national regulations. HEP collaborations usually organise themselves around a detector which exploits the beams of a particle accelerator. The larger fraction of the physicists usually comes from universities, besides that there are also collaborators from the host laboratory and other national laboratories. The accelerator laboratory is the host laboratory for the collaboration. It is interested in the successful construction and operation of the experiment, provides technical and administrative support for the collaboration, and is quite often an active member of the collaboration. University groups usually acquire additional *project funds*, often referred to as *third-party funds*, for their participation in HEP collaborations.

The complexity and challenges of the machines and detectors usually require a long phase of research and development (R&D) before the construction can start. Typically several years are needed for the construction followed by the operation period of the experiment which quite often lasts for more than one decade[52].

Finally it takes several years to finalise the data analysis and to disassemble the setup. Therefore the total lifetime of HEP accelerators and their experiments easily expands to a few decades. This complete period has to be taken into account to estimate the "total cost of ownership" (TCO) of an experiment. Throughout this long period, funding of the project must be secured – and the integrated costs over the complete lifetime are significant. Therefore, the approval of new projects relies – besides excellent science scope and the prospects of success – also on secure funding over a long period, including the construction cost and the expenses necessary for operation and physics analysis.

21.3
Funding and Interplay with Politics

The need for large and costly infrastructure together with the long operation periods and the international collaborations makes the funding of HEP a political and governmental affair, with a close attendance of national governments and funding

---

52) The progress of high energy physics depends strongly on the integrated collected luminosity. Large statistics is essential both for the search for new "rare" processes and the improvement of the accuracy of existing measurements. Experience shows that it typically takes several years to reach the maximum luminosity of a machine.

agencies already in the early planning phase of new projects. One distinguishes usually between laboratories and university groups. The basic funding of laboratories is institutional, although they may also acquire project funds designated for well-defined projects. In contrast, the institutional funding of the university provides just the basis for the work of the university groups. For their participation in particle physics experiments, additional project funds are essential.

The challenge of HEP funding and its close interaction with political decision-taking becomes even more evident if one recalls the usual schemes and handling of public finances. In parliamentary democracies, parliaments are sovereign in budgetary matters. National budgets are quite often planned by the government and enacted by the parliament on an annual basis. In Germany and other countries this budget planning is complemented by a mid-term financial planning for several years which, however, is still much shorter than the typical lifetime of high energy physics experiments. One has to keep in mind that a well-accepted and commonly used principle in politics and for budget decisions is that the complete funding of a project must be secured before approval. Both physicists and governmental authorities are interested in the successful construction and operation of effective research projects with reliable funding schemes. This common interest is essential for their joint planning and cooperation.

21.3.1
Accelerator Laboratories for High Energy Physics

Large laboratories provide a long-term support for large research infrastructures. The foundation is usually arranged within the constitution or convention of the laboratory, which is signed by governmental authorities, thus rendering the laboratory a separate legal entity[53]. These constitutions usually control the strategy and policy of the laboratory including its administration and committees and the representation of the governmental authorities which supervise the laboratory. The particle physics laboratories are usually financed by state governments or state funding agencies. Important decisions are taken in accordance with authorities from the governmental funding agencies, especially decisions on new projects and new priorities. Budgetary issues depend, among other things, on national regulations, resulting for example in annual budget negotiations or budget periods of several years.

The formation of big accelerator laboratories contributed significantly to the success of high energy physics in the last half century. However, there is of course competition between the best proposals for different projects from various research fields for funding and the construction of large RI. National authorities decide on the priorities and the scope of governmental support, taking into account new developments and demands. The programmes of the national laboratories are reg-

---

53) A prime example for these legal constitutions are the CERN conventions, http://council.web.cern.ch/council/en/Governance/Convention.html.

ularly evaluated and adapted according to scientific needs and governmental priorities. Projects may be stopped, or even complete institutions may be closed, as happened to the SSC project which was cancelled in 1993[54]. In this respect CERN has a privileged position. Since CERN is an international organisation for particle physics, the research focus and the funding is secured by the convention which was signed by the individual member states. In the following, the institutions DESY and CERN are described as two representative examples. For more details see also Chapter 22. Other laboratories have similar organisational structures.

DESY was founded in 1959 in Hamburg by means of a treaty signed by the federal minister for atomic energy and the Hamburg mayor. DESY has two locations, one in Hamburg and a second in Zeuthen near Berlin. The laboratory has about 2000 staff members and more than 3000 visiting scientists each year. DESY develops, runs and uses accelerators and detectors for photon science and particle physics. The annual budget of about 190 million Euro is funded jointly by the German federal government (90%), the city of Hamburg and the German state Brandenburg[55]. The research centre belongs to the Helmholtz Association (HGF) and is financed according to the programme-oriented funding concept of the HGF, see Section 21.4.2. DESY's governmental structure is based on: (i) a Board of Directors which manage the laboratory; (ii) the Scientific Council and scientific advisory committees, which advise DESY on important scientific matters, and (iii) the Administrative Council with representatives from the national government and federal states, which decides on budget and administrative issues.

CERN[56], the European Organisation for Nuclear Research located along the Franco-Swiss border near Geneva, is the world's largest and most important laboratory for high energy particle physics. CERN was founded in 1954 as an international scientific organisation for the purpose of collaborative research. It was one of Europe's first joint ventures and now has 20 European member states[57]. The CERN laboratory operates expressly for research in high energy particle physics of a purely scientific and fundamental character[58]. Member states contribute to the capital and operating costs of CERN's programmes of around one billion Swiss Franc (CHF) per year. They are represented in the Council, which is responsible for all important decisions about the organisation and its activities.

CERN employs around 2500 people, and approximately 8000 visiting scientists (half of the world's particle physicists from some 580 institutes and universities around the world) use CERN's facilities. Physicists and their funding agencies from

---

54) The Superconducting Super Collider (SSC) project was cancelled by the US government after having already spent two billion US dollar, see http://www.hep.net/ssc/.

55) http://www.research-in-germany.de/research-landscape/rpo/research-infrastructures/41800/11-1-desy.html

56) Conseil Européen pour la Recherche Nucléaire

57) Some states or international organisations for which membership is either not possible or not yet feasible are observers: The European Commission, India, Israel, Japan, Russian Federation, Turkey, UNESCO and USA.

58) http://public.web.cern.ch/public/en/About/About-en.html, http://council.web.cern.ch/council/en/EuropeanStrategy/ESConvention.html

both member and non-member states are responsible for the financing, construction and operation of the experiments on which they collaborate. CERN spends much of its budget on building new machines (such as the Large Hadron Collider) and only partially contributes to the cost of the experiments.

The CERN Council[59] is the highest authority and has responsibility for all important decisions. It controls CERN's activities on scientific, technical and administrative matters. The Council approves programmes of activity, adopts the budgets and reviews expenditure. Each member state has two official delegates to the CERN Council. One represents his or her government's administration; the other represents national scientific interests. The Director-General, appointed by the Council usually for a period of five years, manages the laboratory. He is assisted by a Directorate and runs the lab through a structure of departments. The Council and Directorate are assisted by the Scientific Policy Committee, evaluating CERN's scientific programme, and the Finance Committee, dealing with all issues related to financial contributions by the Member States and to the organisation's budget and expenditure.

The CERN convention bestows upon the organisation two missions, namely the operation of the CERN laboratory in Geneva and the organisation and sponsoring of international cooperation in the field. These missions have been actively addressed since the adoption of the *European Strategy for Particle Physics* in July 2006[60], where general and organisational issues addressing the central role of CERN and the CERN Council for the strategic planning of particle physics are described in Articles 1, 2, 11 and 12. The CERN Council sets up working groups and held a special *European Strategy Session* with the purpose of fulfilling this function, namely to update the medium- and long-term European strategy for particle physics and to follow up its implementation.

21.3.2
Constitution of International Collaborations

Particle accelerators for high energy physics are utilised by international collaborations. They play the key role from the first ideas of new experiments, during the R&D and construction period up to the operation and physics phase of these challenging devices, see also Chapter 20. The cooperation inside HEP collaborations is usually based on a legally non-binding "Memorandum of Understanding" (MoU) under the auspices of the host laboratory. This has been proven to be a very successful arrangement for research collaborations with non-commercial goals. The MoU concept merges, under moderate administrative regulations, the freedom to the individual researcher with a common shared responsibility for the project, including financial matters. Engagement in and the success of HEP projects depends essentially on the motivation and ambitions of the researchers, for which the MoU provides a well-coordinated framework without restrictive regulations. Financial

---

59) http://council.web.cern.ch/council/en/Welcome.html
60) http://council.web.cern.ch/council/en/EuropeanStrategy/MandateES.html

commitments inside a MoU are usually arranged with the appropriate funding agencies. For larger enterprises the MoUs are also signed by governmental representatives[61]. As a legally non-binding commitment it is also a helpful concept on a "best-effort basis" to deal with the contrast between the necessity for a secured planning for long-term projects and annual national budget decisions.

21.3.3
Project Funds for University Groups

Besides the education of young academics, universities provide the essential basis of the high energy physics "community" and play a key role in bringing in new ideas and doing physics research. For their participation in high energy experiments, university groups use their institutional funds and infrastructure, but additional project funds are indeed vital. Dedicated public project funding programmes devoted to universities, with funding periods of a few years, usually follow a common procedure: (i) a funding call is issued by the funding agency that sets the objectives and priorities of the intended funding like the science case, political goals and administrative boundary conditions; (ii) project proposals are submitted by research groups, describing the project with its goals and the requested resources; (iii) (peer-) review of the proposal, leading to a recommendation which provides the basis for the funding decision; (iv) approval of high-ranked proposals which are then financed (maybe at a reduced level) within the available budget frame. Groups may apply again, in a sequential funding call, for public money for their participation on the same experiment, but with a new project, implying different work packages and new goals. Several countries have dedicated funding lines for particle physics, which helps to secure and finance the essential long-term contributions and commitments of university groups to HEP experiments.

Usually there are different complementary sources for third-party funding, for example different funding agencies, national and regional grants, research foundations, funds from topical initiatives, grants from industry and so on. Additionally in Europe funding for basic research through the EC has become very important within the last decade, see Section 21.5.

Details about the structure of funding agencies, about how funding is provided and how decisions are taken varies within each country. The complete funding of HEP is based on different complementary pillars. Several aspects are explained in the next section which describes the German funding system.

---

61) This applies also to the construction and the operation of MoUs for the LHC experiments.

21.4
Federal Structure of Science Policy and Funding in Germany

Germany has a broad variety of funding and support programmes, with public and private sources of funding for basic and applied research[62]. The federal government system in Germany with its federal states (referred to in the following as "Länder") defines also the support and funding of basic research in Germany, which follow the organising principle of subsidiarity. This means that matters ought to be handled by the smallest, lowest or least centralised competent authority. Supporting basic research is therefore primarily a task of the Länder. Institutions and projects of scientific research with supra-regional importance are co-funded by the federal government. The Basic Constitutional Law ("Grundgesetz") of the Federal Republic of Germany defines, in Article 91b, the legal framework for a cooperation of the federal government and the Länder in the funding of research.

High energy physics in Germany is funded by several different bodies with defined complementary responsibilities. Because of the need for large accelerator facilities and international cooperations, the Federal Ministry of Education and Research (BMBF) plays an active role in the funding of particle physics. BMBF provides a significant fraction of the funding for particle physics in Germany and accompanies the strategic discussions in the field.

21.4.1
Federal Ministry of Education and Research – BMBF

BMBF supports research topics of supra-regional, international or of fundamental significance or which are specific to large-scale research and require an institutional structure through institutions such as the Helmholtz Association or the Max Planck Society. The support for scientific basic research is focused on large-scale research infrastructures, international coordination and international organisations related to large-scale facilities. This denotes the federal interest and therefore BMBF provides for example the annual German share to CERN of around 200 million CHF.

Large-scale facilities are a fundamental component of Germany's research infrastructure. Construction and operation is carried out by the Helmholtz Research Centres as well as the institutes of the Leibniz Association, the Max Planck Society or by international research organisations like CERN, ESRF[63], ESO[64] and ILL[65]. Research at large-scale facilities is promoted, among others, by excellent research groups at German universities. This concurrence and cooperation of external research groups and the large-scale facilities in national and international centres is supported financially by the BMBF in the framework of the so-called "Verbund-

---

[62] http://www.research in germany.de/dachportal/en/research-funding/2868/research-funding.html
[63] European Synchrotron Radiation Facility, www.esrf.eu.
[64] European Southern Observatory, www.eso.org.
[65] Institut Laue-Langevin, www.ill.eu.

forschung" (Collaborative Research). With specific funds, primarily addressed to universities, scientifically excellent projects at large-scale facilities are funded.

During the last decade BMBF spent around 250 million Euro per year on particle physics, the main fraction being used for the institutional funding of CERN and DESY. The sum also includes the BMBF share for German Research Foundation (DFG) (see Section 21.4.3.1) and Max Planck Society (MPG) (see Section 21.4.2.3) activities in particle physics and BMBF project funds for university groups. One should keep in mind that the number contains some uncertainties, demonstrating the general difficulties of global statistical numbers on funding. One has to define whether activities, or which fraction of them, belong to particle physics or to other research fields such as nuclear physics, astrophysics or applied science. This is quite often ambiguous. Also for the accelerator operation and accelerator research one has to define which fractions of the expenditures are accounted for by particle physics activities. One has to do the same, for example, for the expenditures of a laboratory on general infrastructure and administration. Another difficulty is the accounting of the work of individual persons, for example a professor who works on a research project and gives lectures at the university.

Within the next two sections a brief survey of basic research institutions and funding suppliers in Germany is given, with special emphasis on the relevance for high energy physics.

21.4.2
Institutions for Basic Scientific Research in Germany

21.4.2.1 Universities
Responsibility for education and research in Germany lies with the 16 Länder. They run most of the German universities, which constitute the largest and most important body for basic scientific research in Germany and provide the basis of scientific education and research. In Germany there are 395 universities (including 190 universities of applied sciences) with more than 500 000 staff members, 180 000 scientists and over two million students. The annual budget is around 9 billion Euro, provided mainly by the Länder, but additional third-party funding, for example from the DFG or the BMBF, is crucial, especially for forefront research activities.

21.4.2.2 HGF – Helmholtz Association
The Helmholtz Association is the largest scientific institution in Germany[66]. It comprises 16 scientific-technical and biological-medical research centres with approximately 28 000 staff, of which 2000 are employed by DESY and 9000 are scientists. The annual budget is approx. 2.8 billion Euro, with 90% coming from the federal government. Particle physics is mainly performed at DESY, one of the world's leading centres. But also the Grid Computing Centre Karlsruhe (GridKa) at KIT[67] and the John von Neumann Institute of the Forschungszentrum Jülich contribute

66) http://www.helmholtz.de/en/, www.terascale.de, http://www.helmholtz.de/en/about_us/programme_oriented_funding/
67) Karlsruhe Institute of Technology http://www.kit.edu/kit/english/index.php

to this field[68]. Moreover, the Helmholtz Association provides project funding by means of Helmholtz Alliances. The Alliance "Physics at the Terascale" is a prime example for these structured research networks.

Research centres belonging to the Helmholtz Association are financed according to a concept called "programme-oriented funding". Financing is oriented towards long-term research programmes in which several research centres participate, with the aim to retain the autonomy of the individual centres. The research programmes for five-year periods are evaluated together with external reviewers in a competitive process.

21.4.2.3 MPG – Max Planck Society

The Max Planck Society is the research organisation for basic research outside the universities in Germany. It performs research in natural sciences, life sciences, social sciences as well as in the humanities. Institutes study self-defined core themes and take up new and innovative research areas which are beyond the scope or capability of universities. More than 13 000 people, including 4800 scientists, are employed at 80 MPG research institutes. The annual budget of more than one billion Euro is provided in equal shares by the federal government and the Länder. The Max Planck Institute of Physics in Munich and the Max Planck Institute for Nuclear Physics in Heidelberg are active in the field of particle physics.

21.4.2.4 Other Research Institutions

Other research institutions like the Leibniz Association or Fraunhofer Gesellschaft (FhG) do not perform research in the field of particle physics directly. However, they may contribute as potential partners for dedicated projects, for example for technology development. An important example is the FhG institute IZM Berlin, where the bump-bonding for the ATLAS pixel detector was carried out.

21.4.3
Project Funding for Basic Research

21.4.3.1 German Research Foundation – DFG

The DFG is the central, self-governing research funding institution in Germany[69]. It promotes basic research mainly at universities via projects like Research Training Groups ("Graduiertenkollegs"), Collaborative Research Centres ("Sonderforschungsbereiche") or the Excellence Initiative. The DFG has an annual budget of around 1.3 billion Euro and funds more than 20 000 research projects each year. The budget is provided in approximately equal parts by the federal government and the Länder, and the funds are distributed according to demand.

Proposals are peer-reviewed by selected members from the scientific community, ensuring the quality of the decision. The final funding decision, including the level of funding to be awarded, is made by the DFG's Joint Committee or a Grants

---

68) The GSI in Darmstadt is a centre for hadron and nuclear physics.
69) http://www.dfg.de/en/index.jsp

Committee, consisting of scientists and representatives from the federal and Länder governments.

Amongst others, the funding of basic particle physics theory research at universities is provided by the DFG.

21.4.3.2 BMBF Project Funding of Basic Research – Collaborative Research

Research at large-scale facilities is promoted by excellent research groups at German universities. This fosters an active and fruitful exchange between the operators of large-scale facilities and the external user community, which leads to the conception of new experiments and to the development of innovative instruments and research methods. A particular priority of the BMBF is the so-called Collaborative Research ("Verbundforschung"), which enables excellent research groups, primarily at universities, to collaborate with the outstanding experimental large-scale facilities at national and international research centres. This funding is complementary to DFG funding and provides essential resources for the participation of German university groups in large international collaborations. High energy physics is a prime example for federal project funding. It has a long tradition (since the 1970s) of supporting especially university groups with the focus on experimental groups participating at experiments located at DESY and CERN. Proposals are peer-reviewed by a BMBF evaluation board, and funding is usually provided for a period of three years. Besides excellent science and high prospects for success, an obvious federal interest is mandatory for the approval of funds. This instrument was strengthened in 2008 by a dedicated initiative of the federal ministry to improve the utilisation of CERN by German scientists.

Figure 21.1 summarises the funding since 1997. The participation of German university groups in HEP experiments at HERA, LEP, Tevatron, SPS and LHC

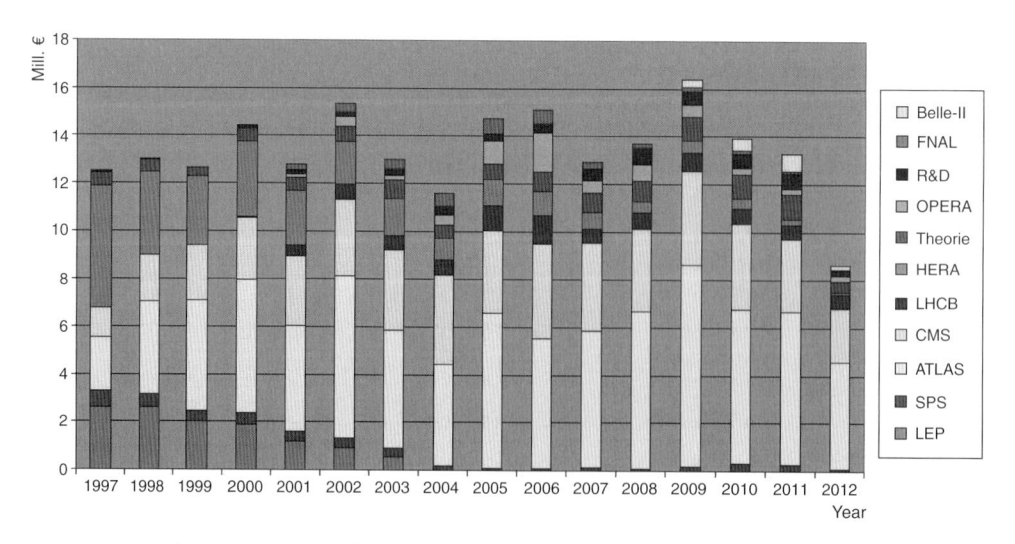

Figure 21.1 BMBF Collaborative Research funding for particle physics in Germany from 1997 to June 2012. The different shades indicate the amount spent on each of the projects supported by the BMBF.

Table 21.1 Funding and size of the four LHC collaborations together with the German share (status as of spring 2010). M&O means funds for "Maintenance and Operations".

Experiment	ATLAS	CMS	LHCb	ALICE
Initial MoU funding (MCHF)	468.5	450.1	70.3	118.8
Requested funding 2007 (MCHF)	540.9	566.3	75.3	138.8
BMBF MoU funding (MCHF)	32.5	17	3.8	12.6
BMBF construction funding incl. cost to completion (MCHF)	37.1	22.5	4.4	17.9
BMBF funding 1997–6/2009 (M€)	64.0	37.6	6.2	23.4
Authors total/D	2762/348	1859/142	506/38	890/103
Countries and institutions	38/172	32/182	15/54	32/114
Number of German institutes	13 + 2	5 + 1	2 + 1	8 + 1
M&O 2009 MCHF	20.6	17.6	3.9	6.1
M&O 2009 MCHF – BMBF	1.55	0.87	0.13	0.35

is supported with an annual budget of around 12.5 million Euro; this also includes R&D activities on detectors and accelerators for future experiments and phenomenological theory research to support the experimental activities. Approximately half of the money is spent on research associates and young academics. The contribution of German university groups to the construction and operation of the LHC experiments is also provided through these funds, see Table 21.1.

With the constitution of BMBF research centres ("Forschungsschwerpunkte", BMBF-FSP) in 2006, a new instrument was introduced to support basic research and provide sustainable long-term support for nationwide excellent research networks working within a large-scale infrastructure environment. The aim is to improve coordination and cooperation and to combine the usage of resources. LHC groups are a prime example. The constitution of the ATLAS and CMS FSPs was an important measure to strengthen the German groups at the transition from the construction phase to the running and physics period of the LHC experiments.

21.4.3.3 The Helmholtz Alliance "Physics at the Terascale"

The Helmholtz Alliance "Physics at the Terascale" is a structured research network comprising all high energy physics institutes at German universities, the two Helmholtz centres DESY in Hamburg and KIT in Karlsruhe and the Max Planck Institute of Physics in Munich. The Alliance acts as a tool for a more effective collaboration, in particular between experimentalists and theorists. With its focus on infrastructure and comprehensive tasks the Terascale Alliance is complementary to the BMBF collaborative research funding.

21.4.3.4 Additional Support for Scientific Research

Germany has a large number of funding programmes for scientific research and science careers. The diversity of sponsoring options is unique. Besides large organisations such as the German Research Foundation and the German Academic

Exchange Service (DAAD), scientists are also supported by numerous prominent foundations such as the Alexander von Humboldt Foundation or the Volkswagen Foundation. Furthermore, there are individual programmes provided by universities, research institutions like the MPG or HGF, by the Länder and by industry[70].

21.5
European Research Area and EC Funding

Funding provided by the EC has become very important over the last decade. Building a European Research Area was of vital interest for the European governments and the EC. In 2000 the European Union (EU) decided to create the *European Research Area* (ERA). This unified Europe-wide area should enable researchers to move and interact seamlessly, and benefit from world-class infrastructures and working with excellent networks of research institutions. It should optimise European, national and regional research programmes in order to support the best research throughout Europe and to coordinate these programmes to address major challenges together. Especially in the field of research infrastructures the EC is very active. An exhaustive summary is beyond the scope of this article, which concentrates on a schematic overview and selected items with relevance to particle physics[71].

21.5.1
European Research Council – ERC

The *European Research Council* (ERC) was launched in 2007 with a budget of 7.5 billion Euro up to 2013 as a part of the Seventh Research Framework Programme (FP7). It is supposed to be a core of the European Research Area, focusing solely on fundamental investigator-driven frontier research[72]. The ERC supports science and scholarships carried out by individuals or individual teams led by established Principle Investigators competing at the European level, within and across all fields of research. It operates according to the principles of scientific excellence, efficiency and accountability. The ERC complements other funding activities in Europe such as those of the national research funding agencies. Being "investigator-driven", or "bottom-up", in nature, the ERC approach enables researchers to identify new opportunities and directions for research, rather than being led by priorities set by politicians. This approach aims to channel funds

---

[70] The interested reader may find more information about research funding in Germany in http://www.research-in-germany.de/dachportal/en/research-funding/2868/research-funding.html

[71] The interested reader may find additional information for example in http://ec.europa.eu/research/era/index_en.htm, http://cordis.europa.eu/home_en.html and http://en.wikipedia.org/wiki/Seventh_Framework_Programme.

[72] http://erc.europa.eu, http://en.wikipedia.org/wiki/European_Research_Council

into new and promising areas of research with a greater degree of flexibility. ERC grants are awarded through open competition to projects headed by young and established researchers, irrespective of their origins, who are working in Europe – the sole criterion for selection is excellence.

21.5.2
Seventh Research Framework Programme – FP7

The research funding programme of the European Union FP7 (Seventh Framework Programme for Research and Technological Development of the European Union)[73] will last for seven years from 2007 until 2013. The programme has a total budget of over 50 billion Euro. This is a substantial increase compared with the previous Framework Programme, FP6, and reflects the high priority of research in Europe. It is the largest funding programme for research projects in the world. Its main strategic objectives are the construction and strengthening of the European Research Area. Specific goals are to gain leadership in key scientific and technology areas and to stimulate the creativity and excellence of European research. In order to complement national research programmes, activities funded from FP7 must have a European added value such as research projects carried out by consortia from a minimum of three member states. They should address challenges that are so complex that they can only be addressed at European level.

Four specific programmes were created to address the corresponding objectives: (i) Ideas, managed by an autonomous entity, the ERC; (ii) Capacities, targeted on research infrastructures (RI); (iii) Cooperation, funding of transnational scientific research activities within ten defined thematic priorities; (iv) People, focus on supporting the training, the mobility and the career development of European researchers, mainly through the expansion of Marie Curie Actions[74]. Grants are determined on the basis of calls for proposals and a peer-review process. The RI part of the FP7 Capacities programme provides 1.8 billion Euro to optimise the use and development of the best research infrastructures existing in Europe. It supports a wider and more efficient access to, and use of, existing RI by means of "Integrated Infrastructure Initiatives" (I3). Furthermore, it supports new RI of pan-European interest in all fields of science and technology by funding of design studies and the preparatory phase for the construction of new infrastructures. With this comes a strong interconnection to the European Strategy Forum on Research Infrastructures (ESFRI) roadmap process[75].

There are many HEP-related FP7 projects. For example, among the projects with DESY participation[76] one may find the EGEE III (Enabling Grids for E-sciencE) project, the ILC-HIGrade (International Linear Collider and High Gradient Superconducting RF-Cavities) project or the SLHC-PP (Preparatory phase of the Large Hadron Collider upgrade) project.

---

73) http://cordis.europa.eu/fp7/home_en.html
74) http://ec.europa.eu/research/mariecurieactions/index.htm
75) http://ec.europa.eu/research/infrastructures/index_en.cfm?pg=esfri
76) http://eup.desy.de/fp7

21.6
Strategic Decision-Making

No unique scheme for the approval of research projects exists, but similar procedures and criteria are applied commonly. The complexity of approval processes grows with the size and budget needs of the project. Several HEP-related aspects are discussed in Chapter 22. This section highlights some general aspects and lists bodies relevant for decision-making in the field of particle physics.

Ideas and proposals for new projects come from the scientific community. Essential for any approval is a convincing scientific case. For smaller national projects the scientific excellence is often the decisive point for the funding agencies. Laboratories usually receive advice from their scientific committees and take their decisions in agreement with the governmental authorities. A sketched review of a common procedure for the approval of new experiments in particle physics helps to depict essential steps and the interaction with the different entities involved in the approval process. The physics case and ideas for the realisation are submitted in the form of an "Expression of Interest" (EoI)[77] to the directorate of a laboratory. These ideas are reviewed and if the EoI is supported by the laboratory, the authors are encouraged and supported to prepare a proposal. This implies an R&D phase with detailed studies of the physics case and of technical solutions as well as a compilation of the costs. Along with this, the collaboration is formed and funding issues are addressed. Common criteria for the review and the decision about the proposal are scientific objectives, its excellence, the capability of the collaboration, the assessment of project costs, the prospects of success, the international competition and recognition, a comparison with alternative solutions as well as funding issues like available support, secured resources and prospects for funding. A positive decision usually implies the conditional approval of the project with the request to prepare a detailed technical design report (TDR) and to find solutions for the open and weak points of the proposal. If all this is settled the proposed experiment might finally be approved and the construction can start.

For large-scale research infrastructures or international projects the decision process might be more complex, but it usually follows comparable criteria and a similar procedure[78]. In the last decade the importance of large RIs has been realised and has received special attention. Governments and funding agencies also cooperate in the early planning phase before approval and funding decisions are taken. Therefore, road-map and strategy-building processes have been established at various national and international levels to prioritise different projects as a sound basis for decision-taking. Examples are the ESFRI process in Europe or the CERN Council Strategy Group. More details are given in Chapter 22.

---

77) This is often referred to as "Letter of Intent" (LoI).
78) Proposals for new projects are inspired by the science case, followed by an R&D phase with a design study leading to a preparatory phase before the final approval.
79) http://public.web.cern.ch/public/en/lhc/Milestones-en.html

21.7
Funding of the LHC and Its Experiments

The Large Hadron Collider (LHC) is the world's largest and highest energy particle accelerator. It is presumably the largest and most expensive research enterprise for basic research on earth. The estimated expenditure for the construction of around 3 billion Euro was mainly provided by CERN, but also non-member states like the US and Japan contributed[79]. The LHC was originally conceived in the 1980s and approved for construction by the CERN Council in late 1994. A symposium in 1984 in Lausanne, Switzerland, is considered as the official starting point for work on the LHC. R&D working groups were set up to study the various aspects of the physics and to investigate the broad variety of technical challenges. The formation of the LHC experiments started in spring 1992, when the "Expressions of Interest" were made. After more than ten years of construction the machine was finally completed in 2007. An incident a few days after the first beam operation in September 2008 required substantial repairs. At the end of March 2010 the LHC started delivering collision data using two 3.5 TeV proton beams colliding in the experiments. After a long shutdown for further improvements and consolidation of the machine in 2012 it is planned to increase the energy of the LHC to its design energy of 7 TeV per beam. There exist detailed upgrade and consolidation plans of the CERN accelerator complex and plans for the modification of the interaction regions to improve the performance (i.e. higher luminosity) of the LHC until the year 2020.

There are six experiments at the LHC run by international collaborations. ATLAS and CMS are the two largest experiments, based on general-purpose detectors. The two medium-size experiments, ALICE and LHCb, have specialised detectors: ALICE for heavy ion collisions, LHCb for b physics. TOTEM and LHCf located close to CMS and ATLAS, respectively, are much smaller and focus on "forward physics".

Table 21.1 provides an overview of cost, funding and size of the collaboration for the four large LHC experiments and the related German share. It first lists the initial MoU construction cost and the final construction cost followed by the German share provided by the BMBF through Collaborative Research funds. BMBF funding in the fifth row refers to the Collaborative Research funds from the year 1997 up to June 2009, including personnel and travel expenses. Row six lists the number of authors together with the fraction coming from Germany, followed by the number of countries and institutions of the collaborations. The number of German institutions gives the number of university institutions plus the number of groups from research centres. DESY participates in ATLAS and CMS, GSI is active in ALICE and MPG institutes contribute to ATLAS and LHCb. In the last two rows, the numbers for the operation costs (maintenance and operation, M&O) for the year 2009 and the BMBF share are listed. This annual operation cost of the LHC experiments is nearly 50 million CHF, of which the BMBF provides around 3 million CHF.

Due to the outstanding importance of the LHC and its experiment, the German federal ministry signed the construction MoUs as well as the MoUs for M&O of

---

80) http://committees.web.cern.ch/committees/all/welcomeLHCRRB.html

the four large LHC experiments. The new challenge for the construction, besides the large amount of required money, was the long-term commitment over more than a decade. BMBF requested in advance of the construction MoU signature commitment letters from the universities which planned to contribute. The universities provided and maintained the infrastructure for R&D and for the detector construction, while BMBF supported the groups and provided the funds for the construction expenditures. The LHC Resources Review Boards (LHC-RRB), with representatives of each experiment's funding agencies, meet twice per year to discuss the status and funding issues of the LHC experiments[80].

It soon became obvious that computing for the LHC experiments is enormously challenging. Therefore the "Worldwide LHC Computing Grid", WLCG, was launched in order to accomplish this crucial task. The WLCG combines the computing resources of more than 100 000 processors from over 130 sites in 34 countries. Funds are acquired from different sources – besides the contribution provided by national funding agencies and laboratories there is an essential support from the EC.

21.8
Summary and Outlook

Particle physics studies the structure of the universe at its most fundamental level. In order to obtain new insights about the elementary particles and fundamental forces of our world and to understand the development of the early universe, large research infrastructures with reliable long-term support are essential. Especially the high energy frontier requires huge and expensive accelerators, making HEP an international or even global enterprise and also a political issue: not only funding and decision-taking but already the strategic planning takes place in close coordination with governmental authorities. As described in this chapter, the funding of particle physics is a quite complex affair, with numerous complementary funding sources.

Besides the basic scientific findings, there are economic and societal benefits of particle physics research, like education and technology transfer[81].

There is a competition for public funding. Like other research communities, particle physicists have to explain why it is useful for the society to fund their activities. Outreach activities (see Chapter 23) to foster public understanding of science are of major importance to secure public support and funding of particle physics. Together with coherent planning on a global scale, particle physics is in a good position to continue the successful story of the last half century with its fundamental findings and is now entering new and exciting eras of discovery to resolve the mysteries of our universe.

---

81) This is not explained in this chapter, but the interested reader may find information about this topic for example in http://public.web.cern.ch/public/en/About/Fundamental-en.html.

22
The Role of the Big Labs
Albrecht Wagner

Particle or high energy physics seeks answers to basic questions such as: what is our world made of and how does it work? The most important tools in this quest are accelerators and their associated detectors. Since about 1950 many such devices have been built, mainly by individual countries or regions (America, Asia and Europe). In the following, a number of questions will be discussed concerning the role of the large laboratories in particle physics. The examples given in this section mainly come from one laboratory, DESY, a large national laboratory located in Hamburg, Germany, providing facilities for scientists from around the world. DESY was founded 50 years ago as a research centre for particle physics open to German universities and thus exhibits all the characteristic features of a large laboratory.

22.1
Why Does Particle Physics Need Large Laboratories?

Historically, science was done by one or a few scientists pursuing their ideas, in a small laboratory if needed. Also the first experiments exploring the inner structure of matter, one of the key questions of particle physics, were performed by Ernest Rutherford in 1909 in this way. He and his collaborators used a radioactive source emitting α particles and measured the angular distribution of these particles after having been scattered off the atoms in a thin gold foil. The only equipment needed to perform this experiment was an α source, a collimator, a gold foil and foils of scintillating material to detect the scattered particles.

Since the times of Rutherford, the need and desire to probe nature with ever-increasing accuracy has forced scientists to develop more and more complex measurement devices, accelerators and their detectors, in the same way as astronomers have built telescopes of increasing size and improved performance to explore the universe.

Over the last 60 years particle physics has been pushing the energy frontier by a factor of approximately 10 every ten years, in the desire to probe nature with ever-

Physics at the Terascale, First Edition. Edited by Ian C. Brock and Thomas Schörner-Sadenius
© 2011 WILEY-VCH Verlag GmbH & Co. KGaA, Weinheim.
Published 2011 by WILEY-VCH Verlag GmbH & Co. KGaA.

increasing accuracy. Simultaneously, the available luminosity increased in a similar way.

The increasing size of the facilities, their growing technical complexity, the need to develop new technologies to solve the scientific questions, the increasing length in time during which the facilities are used by scientists and the growing demands on new computing infrastructure have created the need to pool skills and resources in national and international laboratories. This process started in Europe in the field of basic science in the 1950s, most notably through the foundation of the European Centre for Nuclear Research, CERN.

22.2
Examples of Large Laboratories

A few examples of major particle physics laboratories in the three world regions mentioned above should help to illustrate the spectrum of their work and funding.

In Europe CERN, the European Organisation for Nuclear Research, was founded in 1954. The laboratory is located near Geneva in Switzerland. It was one of Europe's first joint scientific projects and now has 20 member states. It is one of the world's largest centres for scientific research, its sole mission being to use its complex scientific instruments to study the basic constituents of matter, the fundamental particles. CERN's annual budget amounts to 750 M€, provided by the member states proportional to their GNP (the Gross National Product, a measure of the value of goods and services a country produces in a given year). It has 2250 staff, serving 10 000 users.

DESY, the Deutsches Elektronen-Synchrotron, is a leading centre for the investigation of the structure of matter and has locations in Hamburg and Zeuthen (near Berlin). DESY develops, runs and uses accelerators and detectors for photon science and particle physics. DESY is a national research centre supported by public funds (90% from the federal republic and 10% from the federal states in which DESY is located) and a member of the Helmholtz Association. DESY was founded in 1959 as a research laboratory for German universities active in the field of particle physics. Over the years it has placed an increasing emphasis on using its accelerators for photon science. In 2007, with the end of operation of the electron–proton collider HERA, DESY has stopped running in-house experiments in particle physics, but continues its involvement in the field through a tight collaboration with CERN at the Large Hadron Collider, the provision of test-beams and through an active role in the development of the International Linear Collider. The annual budget of DESY is 190 M€, and it has 1800 staff serving more than 3000 users.

The Fermi National Accelerator Laboratory (Fermilab) is the only remaining laboratory in the United States dedicated solely to particle physics. Located near Chicago, it is a US Department of Energy national laboratory. Fermilab operates the Tevatron, which until the start-up of the LHC was the world's most powerful accelerator. Fermilab also hosts some smaller experiments and neutrino experiments. The annual budget of Fermilab is 300 M€, and it has 1930 staff serving 2300 users.

The Stanford National Accelerator Laboratory (SLAC) at Stanford University operates a two-mile linear accelerator and has been a pioneer in the development of electron–positron colliders. Originally a particle physics research centre, SLAC is now a multipurpose laboratory for astrophysics, photon science, accelerator and particle physics research. The annual budget of SLAC is 220 M€, and it has 1700 staff serving 2500 users.

In Asia the leading accelerator laboratory in particle physics is KEK, the High Energy Research Organisation in Japan, near Tsukuba. It operates an electron–positron colliding-beam accelerator (KEK-B) on its Tsukuba Campus and is strongly engaged in the construction and operation of the J-PARC (Japanese Proton Accelerator Research Complex) project at the Tokai Campus. KEK also operates accelerators for synchrotron radiation. The annual budget of KEK is 327 M€, and it has 700 staff serving approximately 3000 users. In relation to the other laboratories, the staff of KEK is smaller and the budget larger. This is due to the fact that in Japan more work is outsourced to industry.

A direct comparison of the budgets of the different laboratories is not possible as the annual budgets sometimes do not include the money for major new projects. However, the given numbers reflect the size of the operations.

A comparison of these examples of accelerator laboratories shows that only two of them (CERN, Fermilab) focus solely on particle physics while the other three operate accelerators also for other scientific applications, such as material science and biology. All of them, however, place major emphasis on all aspects of accelerator and detector science.

22.3
Complementarities of Universities and Large Laboratories

Setting up large laboratories did not mean that universities lost their role in particle physics. On the contrary, universities and laboratories developed a very fruitful symbiosis in which the partners complement each other in their skills and strengths.

Universities typically develop and operate special infrastructure, such as detector laboratories, and foster specialised expertise, scientific diversity, and above all, the young people who get their education at universities.

Research centres typically operate large unique facilities (such as accelerators), developed with and for the scientific users, the vast majority of which come from universities. They have the know-how, the skills and the engineering capacities necessary to build large and complex facilities, they play a key role in the strategic development of the field (due to the long lead times to develop, approve, build and operate new facilities) and they provide the long-term technical and management support not only for accelerators, but also the detectors. They also have the long-term funding necessary for large projects. The typical lifecycle of today's accelerators and detectors is well beyond 10–20 years and will be discussed later.

In the following these aspects are discussed in more detail.

22.4
Key Functions and Assets of Large Laboratories

Large laboratories fulfil two major roles: they create new research facilities to advance their field of science and they help the users from universities to make best use of the facilities such as accelerators, detectors and computer systems. In addition to their goal of doing high-level research, laboratories are therefore also service providers for the users. In that role the management of the laboratory carries responsibility not only for its own staff and infrastructure, but also for providing the best environment for external users, frequently from abroad. In addition it has to guarantee long-term support of the technical devices, support staff, funding and political support.

One of the most important roles of large laboratories is their ability to support projects of a very long lifecycle, such as the large projects in particle physics. A typical lifecycle can be illustrated using the example of the electron–proton collider HERA at DESY in Hamburg: at the end of 1979, after several years of discussion among particle physicists, workshops and preparatory development work, the detailed plans for this collider were presented for the first time.

Six years later, in 1984, the agreement to build HERA was signed by the federal and state ministers funding the project. The civil construction and the work on all components started shortly after project approval. In October 1991 the first electron–proton collisions were observed. Data-taking started one year later and continued for 15 years, including a major upgrade phase, until June 2007.

This example, which is typical for projects of similar size in particle physics, illustrates that such an endeavour needs the long-term stability and support of a large laboratory. In contrast, the typical duration of project funding in universities is around three years.

22.4.1
Research and Development of Accelerators and Experiments

Accelerators and detectors are mostly unique, specially designed devices combining many new features of high technology and meeting challenging demands concerning performance, reliability and radiation tolerance. Reaching the performance goals typically requires many years of R&D at the technology frontier.

Technology examples in the field of accelerator R&D are the development of superconducting magnets for high magnetic fields or of superconducting radio-frequency (SCRF) structures to generate very high electric fields. Also new accelerator concepts have been developed, such as colliders: during the first decades of particle physics the beam of the accelerator was directed at a target in which the desired interaction took place ("fixed-target experiments"). The collider concept, in which two bundles of particles are first accelerated in opposite directions and then brought to collision, was only developed later.

Accelerator development requires a broad expertise in accelerator physics, magnet technology, material science, vacuum technology, radio-frequency technology, mechanical engineering, electronics and control software, to name just a few. This is the reason why nearly 50% of the staff in the large laboratories works in the accelerator division.

A recent example is the development of superconducting radio-frequency structures reaching unprecedented high accelerating fields (gradients). This work started around 1990 in the framework of a collaboration of world experts in the field of SCRF. They formed the TESLA collaboration, which later became the TESLA Technology Collaboration. The mission of this collaboration is to advance SCRF technology R&D and related accelerator studies across the broad diversity of scientific applications and beyond the borders of individual projects. It supports and encourages free and open exchange of scientific and technical knowledge, expertise, engineering designs, and equipment. About 50 institutions from 12 countries around the world are members of this collaboration, half of them being large laboratories providing the necessary infrastructure.

In order to test their new concepts and ideas the TESLA collaboration built jointly the TESLA Test Facility (TTF) at DESY. This superconducting electron linear accelerator proved the soundness of the common design and the feasibility to reach high gradients. It now also serves as the basis for the Vacuum Ultraviolet Free Electron Laser, FLASH. Likewise, other laboratories around the world (such as Fermilab, Jefferson Laboratory, or KEK) built test facilities for new developments and experiments in SCRF technology, beam and light physics, as well as associated developments such as instrumentation and diagnostics.

Joining forces and combining the world know-how has led to a major increase of the achievable gradients, from 5–7 MV/m in 1990 to over 40 MV/m in 2010. At the same time the cost per unit length could be decreased by a factor of 5, thus making the technology usable on a large scale.

The development of detectors follows a similar pattern, although in this field the university-based groups play a much stronger role in the R&D of new detector concepts. The reason is that detector development is smaller in size and requires smaller infrastructures than accelerator development.

22.4.2
Construction

The construction phase of both accelerators and detectors requires major technical infrastructure, skilled engineers, technicians, project managers, civil engineers and purchasing experts. These can typically only be found in large laboratories.

The accelerator complex includes a large number of systems, prototypes of which are developed, designed and tested in the laboratory and then built by industry. Components frequently are the result of new ideas and developments and therefore a first of their kind. Therefore the production process involves close interaction between the developing engineers and technicians and the manufacturer. During the construction phase in industry, the progress of the work at the companies has to be closely monitored by the project leaders. Upon delivery a detailed quality control follows, assuring that the necessary specifications are being met.

The basic subsystems of an accelerator are the source (the device where the particles are produced), magnets which guide and focus the charged particles along their flight path, structures by which the particles are accelerated in electrical fields, vacuum pipes through which the particles move and sensors which monitor the flight path. Each of these subsystems is driven and controlled by electronics and computers. The overall operation is assured by combining, monitoring and analysing all the relevant data in a central control room. This broad variety of systems requires a similar breadth of skills among the staff, in order to develop, build and maintain them properly. This includes also the need to survey the position of all accelerator elements, in the case of HERA thousands of components at a length of 15 km. A key challenge during accelerator construction is the logistics of installation, interconnection, local and systems testing, which all require adequate expertise.

Laboratories normally outsource the civil engineering work to specialised companies, but they need among their staff highly qualified experts for the planning stage and to critically supervise the civil construction, especially the accelerator tunnel, the experimental areas and other buildings.

This brief description of the elements of accelerator construction illustrates, with a few examples, the broad spectrum of scientific and technical skills needed to build and assemble an accelerator, requiring the staff of a large laboratory.

The detector construction follows a similar, but more decentralised pattern, as many components are built by university institutes, making use of their local infrastructure. For the large detector components the collaborations frequently make use of the construction capabilities of the host laboratory. As the detectors are tightly linked to the accelerator, both in space and in function, the work on the so-called machine–detector interface requires the close collaboration of the experimental team with the responsible engineers of the laboratory. Together they have to develop ways to assemble the various detector elements, to service them in beam position, to supply them with electricity and gas, to synchronise accelerator and detector electronics, and to solve the safety issues.

22.4.3
Operation

Also the operation of complex research infrastructure such as accelerators and detectors requires a staff having various skills. At DESY, the accelerator department has 21 groups with responsibilities ranging from the individual accelerators to subsystems such as power, water and cooling, controls and diagnostics, software development, vacuum, cryogenic systems, radio frequency, operation, and personal safety and interlock systems, just to mention a few. The long lifetime of the projects requires maintaining the know-how of all systems through persons and documentation. Most of the major laboratories have a similar mix of skills in their staff. Some, however, have dedicated operation teams, focusing entirely on the operation of accelerators.

The operation of detectors is mostly done by the staff from the participating universities with support from the host laboratory mainly in the area of support infrastructure, such as power, cooling, gas supply, and safety. The reason lies in the fact that the detectors are built by many partner institutions, each carrying the responsibility for one or only a few subsystems. This responsibility is maintained for the entire life cycle of the experiment. Participating in the detector operation and monitoring the quality of the data is also an important educational element for the students working on the experiment, as it provides a vital element in the understanding of the meaning and quality of the recorded data.

Increasing performance of networking has stimulated a development towards more decentralised operation of detectors, where laboratories from around the world and also universities have "virtual" control rooms, allowing them to monitor from home the status of the accelerator, the detector and its subsystems. Remote operation has also been tested successfully for accelerators, but is not yet used in a routine fashion (see also Section 22.10).

22.4.4
Computing

Particle physics has pioneered the use of computers for extensive data collection, online and offline analysis, simulation studies, and online instrument control. Both accelerators and detectors contain a large number of subsystems. Transferring efficient and reliable information to and from these subsystems requires sophisticated distributed computing systems. These systems involve not only new hardware but also advanced, specially developed software.

Computer simulation has become an essential part of experimentation in particle physics (and in other fields of science). Its applications range from the simulation of particle trajectories in the accelerators or storage rings and the performance of complex detectors to the interaction of particles in matter and the design of radiation shielding.

To deal with the great amount of data arising in the course of the Large Hadron Collider (LHC) experiments, the Grid computing concept has been developed at CERN (see Chapter 19). It combines the computer resources from multiple domains for a common task, the analysis of the LHC data and the corresponding simulations which all require unprecedented computing power as well as the capability to store and process large amounts of data.

An efficient collaboration requires an optimised exchange of information, data and software. This need led to the development of the WWW software technology by CERN.

The design, assembly, maintenance and development of the large computing facilities and associated data storage and networks, the guaranteed availability of the computing services around the clock and a rapid intervention in case of failures requires skilled teams of scientists and technicians. These tasks are therefore typically performed by the large laboratories.

Computing services could in principle be outsourced, at least as long as one deals with routine operations. But past experience has shown that particle physics with its particular demands had frequently to invent new schemes to match its needs. This requires close and flexible interaction between the scientists and the computer experts and is a typical task for large laboratories.

22.4.5
User Support

In order to enable users to work efficiently at a large laboratory, as a member of one of the experiments and coming frequently from abroad, the host laboratory has to provide the users with substantial support. This ranges from a housing service to providing office space, computer access, and solving visa issues. All large laboratories therefore have established a user office to help with these matters.

A large number of outside users is regarded as a measure of the attractiveness of the scientific and technical programme of a laboratory. At the same time it creates a considerable financial burden, as users typically do not pay for the service they get.

22.4.6
Management

The management of large scientific laboratories such as the ones mentioned above involves many aspects, which are beyond the scope of this review. They range from defining and implementing the research strategy, setting up the adequate organisational structures, managing the financial and human resources, monitoring the efficiency of the facilities, to integrating the views and responding to the needs of the users[82].

Management of large scientific projects needs to strike a balance between the necessary rules of project management and the consensus-based decision-taking of large collaborations. Therefore, the large collaborations have developed their own management structures which usually differ from the more hierarchical structures of the laboratories (see Chapter 20).

22.4.7
Scientific Staff at Large Laboratories

While the need for large laboratories is hardly questioned when it comes to the construction and operation of facilities, the scientific staff of these laboratories is also vital for their success, for several reasons.

A strong in-house group of scientists, both experimental and theoretical, has proven to enhance the scientific output and increase the efficiency of operation,

---

[82] A more detailed discussion can be found in C. Gelès et al., *Managing Science*, John Wiley & Sons, 2000.

because of the close interaction and common goals of the outside users and the in-house staff. The scientists play also a leadership role in the strategic planning of their field.

In order to fully exploit the data taken with the help of the accelerators and detectors, a strong theory group that is able to develop the necessary theoretical concepts and tools is a key asset. At the same time it fosters a quick and effective interaction with the scientists stationed at the laboratory.

In short, the quality of a large laboratory and of its scientific staff are strongly correlated.

22.5
Collaborations and Their Individual Members

The field of experimental particle physics is characterised by the large and growing size of the teams joining forces to build, operate and analyse a joint experiment. These teams are called "collaborations" (see Chapter 20). In order to cope with the increased size and complexity of the experiments in an adequate way, collaborations have grown with the energy of the accelerators, by about a factor of six every ten years. They were able to cope with this growth by applying the lessons learnt during one step to the next one. Nevertheless, doing science in very large collaborations remains a challenge.

Collaborations are formed and start working together during the phase when new accelerators are being developed and constructed. The key ingredients for a successful collaboration are: sharing a common scientific goal and vision and having complementary competences, both scientific and technical (for the building of detectors). But similarly important are the human chemistry and trust as they are the basis for long-term success.

Even collaborations with well over 2000 people, like those at the Large Hadron Collider, have a very light, non-hierarchical management structure. There is no clear command structure and the decisions, physics output and planning are based on teamwork. Most of the budget is held by the individual university groups, but all the funding agencies involved in an experiment meet once or twice per year to control the budget flow.

In order to understand the role of individuals in an experiment, one has to realise that the word "experiment" is misleading. An "experiment" in particle physics is an ensemble of highly complex subdetectors, each fulfilling a certain purpose in the observation of particles generated in the accelerator. This ensemble of components records everything that happens and resembles a photographic plate in astronomy with which one records all signals from the sky. These data are then used to address many different scientific questions. Therefore, an "experiment" is a device to pursue hundreds of different physics questions.

Three aspects of the work in large collaborations are of great importance for young scientists: scientific independence, competition and visibility.

Independence: many scientists contribute to the construction of the subdetectors which are frequently assembled at the universities; each scientist participates in the data-taking by taking shifts, as accelerators and detectors operate around the clock. Each scientist in a collaboration is then free to use all the recorded data to pursue the scientific question he or she is interested in. This is the prerequisite to cover all the scientific aspects of an experiment and to make the best use of the data.

Competition: there is competition between scientists within one collaboration and competition between collaborations. Both can be compared with soccer. The first resembles the competition between players within one team and the second the one between teams.

Visibility: the visibility of scientists within a collaboration is straightforward: good young scientists are quickly recognised for their individual contributions, are given more and more responsibilities, are proposed as speakers at international conferences, and so on. The bigger and partially unsolved problem is the visibility from the outside, which plays a key role for later appointments. At the moment there is an intense discussion trying to assure this external visibility even for collaborations of 2000 persons.

In summary, the work in very big collaborations requires on the one hand the willingness to work as a team, but at the same time gives a lot of freedom for scientific creativity and independence within the framework of the possibilities provided by the large research infrastructure.

22.6
Organisational Models for Particle Physics Facilities

In 1993, the International Committee for Future Accelerators (ICFA) produced a classification of different organisational models for the construction and operation of particle physics accelerators and experiments:

- National or regional facilities: they are built and operated by the host country or host region. The planning, project definition and the choice of parameters are internationally coordinated. Examples for such facilities are DESY, SLAC and KEK.
- "Larger" facilities which cannot be funded by one country or region: here the host country or region seeks contributions for the construction from other countries or institutes. These contributions will preferably be provided in components or subsystems. The facility is again planned and defined internationally. The operation is the responsibility of the host country. This model has become known as the so-called "HERA model".
- Very large projects needing a collaboration of several countries with comparable share of the total construction and operation cost: in this model the participating countries should make their contributions again through components or subsystems in a similar way as large collaborations building jointly a major detector facility. A facility under this model would be a common property of the

participating countries or laboratories. They would also share the responsibility and the cost for operation. The staff for the operation would also come from the participating countries. To set up this type of collaboration, governmental agreements are expected to be needed. The first accelerator being built so far according to this model is the linear accelerator driving the European XFEL in Hamburg. The International Linear Collider might be built following this model.

- Very large projects in the framework of an existing international organisation: This type of collaboration is characterised by common funding for the construction and for operation of large projects. CERN is the prominent example, including the LHC experiments.

22.7
Access to Large Laboratories and Their Facilities

Science has benefited enormously from the fact that the results of basic science are published and available to all. The selection of scientific projects at a specific facility is therefore based only on the scientific merit of individual proposals. This principle of open access is a central element of the guidelines defined by ICFA. Host institutes benefit as much from their users' intellectual and technical capabilities as the users benefit from the facilities at the laboratories. Some aspects might need reconsideration now that the number and geographical distribution of sites is quite different from what it was in the past. This will be discussed in Section 22.10.

22.8
Strategic Planning for Different Laboratories and Regions

The strategic planning for future projects of the large laboratories is increasingly being done in a globally coordinated process. This process are coordinated by regional and international committees, such as ACFA (Asian Committee for Future Accelerators), ECFA (European Committee for Future Accelerators) and HEPAP (High Energy Physics Advisory Panel in the US).

ECFA was set up in 1963 in order to help coordinate the European high energy physics programme[83]. Participating physicists are from the 20 member countries of CERN. ECFA is an advisory body to the CERN management, the CERN Council and its committees, to the DESY management and its Scientific Council, and to various other organisations, both national and international. All major facilities constructed in Europe are discussed in ECFA. Special workshops are organised to investigate the details of the projects, so as to obtain a regional consensus. For instance, ECFA supported the development of LEP, HERA and LHC in this way. ECFA is actively participating in the definition of the European Strategy for Particle

83) Guidelines for ECFA Terms of Reference, http://ecfa.web.cern.ch/ecfa/en/termsofref.html.

Physics in the CERN member states and monitors its implementation. ACFA is set up in a similar way to ECFA.

HEPAP is an advisory panel to the US funding agencies, the Office of Science of the Department of Energy, and the Mathematical&Physical Sciences Directorate of the National Science Foundation. It discusses the national high energy physics programme, which encompasses the conduct of experimental and theoretical high energy physics research and accelerator R&D.

ICFA, the International Committee for Future Accelerators, was created to facilitate international collaboration in the construction and use of accelerators for high energy physics. It was created in 1976 by the International Union of Pure and Applied Physics. Its purposes are: to promote international collaboration in all phases of the construction and exploitation of very high energy accelerators; to regularly organise world-inclusive meetings for the exchange of information on future plans for regional facilities and for the formulation of advice on joint studies and uses; to organise workshops for the study of problems related to high energy accelerator complexes at the energy frontier and their international exploitation and to foster research and development of the necessary technology.

The directors of the large laboratories are members of the regional and international committees.

22.9
Decision Process and the Role of Politics

The decision process for each major project in particle physics has its own characteristic features. But each time three elements have to be fulfilled in order to move ahead: the formation of a scientific consensus; a technical solution; a political decision. The example of the International Linear Collider (ILC) is shortly reviewed as the most recent and not yet completed process.

The linear collider activities started with a convergence towards a common project: in 2001 road-map discussions about the future of particle physics were held in the three regions America, Asia, Europe, leading to an overwhelming world-wide consensus about scientific priorities and for a project in which positrons collide with electrons at energies up to 500 GeV, with luminosity above 10^{34} cm^{-2} s^{-1}. Substantial overlap in running with LHC was recommended.

The reason for reaching such a consensus was based on the shared understanding that the LHC will lead the way and has a large energy reach for quark–quark, quark–gluon and gluon–gluon collisions at 0.5–5 TeV, thus covering a broad spectrum of physics. On the other hand, the ILC would provide a second view with very high precision in electron–positron collisions with fixed energies, adjustable between 0.1 and 1.0 TeV and a well-defined initial state. Together, LHC and ILC would be complementary tools for the exploration of the Terascale.

The document summarising the scientific case, entitled "Understanding Matter, Energy, Space and Time", was signed by more than 2700 scientists from all around the world.

The scientific strategy is presently still being reviewed, following its own pattern in each region. For example, in 2006 a panel, set up by the US National Research Council of the National Academies, was charged to identify and prioritise the scientific questions and opportunities that define elementary particle physics and to recommend a 15-year implementation plan. One of its recommendations was that the US should announce its strong interest to become the host country for the ILC and should undertake the necessary work to provide a viable site and mount a compelling bid.

In Europe, in 2005 the CERN Council, as the body in which all European funding agencies engaged in particle physics are represented, launched a process to define a European strategy for particle physics. One of the resulting recommendations approved by the Council in 2006 was: "It is fundamental to complement the results of the LHC with measurements at a linear collider. In the energy range of 0.5 to 1 TeV, the ILC, based on superconducting technology, will provide a unique scientific opportunity at the precision frontier; there should be a strong well-coordinated European activity, including CERN, through the Global Design Effort, for its design and technical preparation towards the construction decision, to be ready for a new assessment by Council around 2010."

Also in Asia similar priorities were set.

Since the early 1990s several solutions for solving the technical challenges of a linear collider were pursued world-wide, converging in a common review process to two relatively mature solutions, a room-temperature linear accelerator based on copper structures, mainly developed by SLAC and KEK together with international partners, and one which was based on superconducting radio-frequency structures developed by the international TESLA collaboration and led by DESY.

As it became increasingly clear that only one technical solution could be pursued with the necessary strength, ICFA in 2003 established the "International Technology Recommendation Panel" (ITRP), charging it to recommend one technology. The recommendation was to be based on all relevant scientific, technical, schedule and cost considerations. The recommendation of ITRP was to pursue the superconducting solution. It was presented to ICFA in August 2004 which unanimously endorsed this recommendation.

After this decision had been taken a lot of enthusiasm, willingness to self-organise and a strong sense of initiative was visible in all participating laboratories. This was remarkable as many people had previously invested considerable effort and time to develop other technologies. But the ITRP decision was accepted and already a few months later a first workshop in Japan helped to advance the global collaboration and established well-defined work packages.

In 2005 ICFA appointed a director for the Global Design Effort (GDE), and the regions (Asia, Europe and the North America) nominated their regional directors. A second ILC workshop took place, leading to a baseline configuration for the ILC. From then on the project proceeded steadily, guided by the GDE mission: (i) produce a design for the ILC that includes a detailed design concept, performance assessments, reliable international costing, an industrialisation plan, analysis of possible sites where the ILC could be built, as well as detector concepts and scope;

(ii) coordinate world-wide prioritised proposal-driven R&D efforts (to demonstrate and improve the performance, reduce the costs, attain the required reliability, etc.).

The third element, the political decision process, is still ongoing. This process is quite complicated and slow, due to its global nature. From the beginning, many reasons spoke for a truly global project: the necessary sizable funds, the scientific and technical challenges, the political climate concerning basic research and the growing time gap between new projects.

The community took many steps to help this process: it reached a scientific consensus, made a difficult technology choice and created a world-wide organisation for coordinated accelerator and detector work.

On the political side, the German Science Council evaluated the TESLA proposal in 2002 and gave a very strong recommendation to the funding agency, the Federal Ministry. However, the ministry was not ready to propose Germany as host country, but encouraged DESY to continue to work on the necessary R&D in international collaboration. On the other hand, the ministry approved the construction of an "X-ray Free Electron Laser" (XFEL) using the superconducting technology, which could be considered as a pilot project for the ILC.

In 2003 the first meeting of representatives of all funding agencies interested in participating in a linear collider (FALC) took place and the OECD set up a consultative group which in 2004 lead to an OECD Ministerial Statement supporting the ILC[84]. FALC continues to meet on a regular basis.

In summary, impressive and steady progress has been made in the past 10 years to generate a strong basis for the decision to go ahead with the construction of an International Linear Collider. But for a final decision the scientists and politicians are eagerly awaiting the first indicative results from the LHC.

22.10
Possible Future Developments

During the past 50 years, high energy accelerators have not only become major research tools for nuclear and particle physics, but also influenced many other fields of science and industry by providing powerful sources of synchrotron radiation and other beams. New accelerator concepts have been the key to both an increased understanding of nature via fundamental research and the growing application of accelerators and accelerator techniques in other fields. It is therefore important to continue to develop new accelerators and to maintain accelerator expertise worldwide.

However, the size and cost of future large accelerators will most likely outstrip the resources of a single region, and building them will require a new approach. One way is via the framework of an international collaboration. A collaboration for a major accelerator facility must meet the following two challenges: (i) maintain

---

84) OECD Global Science Forum, Report by the Consultative Group on High-Energy Physics, http://www.oecd.org/dataoecd/2/32/1944269.pdf. This report was completed in 2002 and endorsed by OECD Science Ministers in 2004.

and nurture the scientific culture of all participating laboratories; (ii) maintain the visibility and vitality of each partner.

Furthermore, all participating countries must be willing to invest and to commit themselves through long-term agreements. A possible solution would be a "Global Accelerator Network" (GAN) in which scientists and engineers from laboratories and research centres around the world could form a network to integrate their scientific and technical knowledge, ideas and resources, and focus them on a common project – a merger of world-wide competence.

The GAN would allow participating institutes to continue important activities at home while being actively engaged in a common project elsewhere. All of the participants could demonstrate a visible level of activity, thus maintaining a vital community of scientists and engineers, and attracting students to the field of accelerator research and development. Last but not least, the network approach could facilitate the thorny problem of site selection for new large accelerator facilities.

The approach is based on the substantial experience gained in the construction and operation of large particle physics experiments at the LHC and LEP (CERN), HERA (DESY) and Fermilab's Tevatron. In these projects, multinational teams, motivated and united by a common research goal, share the responsibilities of a large experiment. In this way, many groups, mostly from universities, become technically and financially responsible for the design, construction, operation and understanding of parts of the detector, which may be small but are nevertheless vital to the success of the experiment. Much of the work would be done at the home institutes.

On the other hand, most accelerators so far have been built and are operated by only one laboratory. An important exception is the HERA electron–proton collider at DESY, where major accelerator components were designed and built in laboratories in other countries. However, once installed, responsibility for their operation and maintenance was handed over to the host laboratory, DESY. The LHC at CERN has also been built in a very similar way with major contributions from outside CERN.

In the GAN framework, new accelerator facilities, as well as experiments and beamlines for synchrotron radiation, would be designed, built and operated by an international collaboration of "partner" laboratories and institutes.

The machine would be built at an existing laboratory – the "host" – to capitalise on available experience, manpower and infrastructure. The host state would have to underwrite a major part of the finance and to make a clear commitment to support the project throughout its duration. (In the case of CERN, the host states are not the principal sponsors of international facilities built on their territory – the organisation as a whole is responsible.)

Each partner would take responsibility for certain components of the project, designed, built and tested at home before being delivered to the host site. This responsibility would be maintained even after delivery. Component maintenance, operation and development would be carried out as much as possible from the home institutes, using modern communications technology. For this the partners would need to maintain duplicates of accelerator components for testing, checking and development. In some institutions, "copies" of the accelerator control room could even provide for highly efficient round-the-clock operation. At the host site, a core team, under guidance from all partners, would provide the necessary on-site technical support.

With GAN, major capital investment and operation funding would be taken up inside the partner states. Operational costs (mainly electricity), excluding manpower, would be shared by all partners according to a predefined arrangement. Most manpower would remain in the partner institutions, except during periods of installation and overhaul, and during collaboration meetings.

Details of the collaboration and management structures, together with the exact sharing of responsibilities between partners and the host, have yet to be worked out, but examples can surely be found within existing arrangements.

Remote control and diagnostics, allowing off-site partners to participate on site actively, are the key GAN features. While this would be an innovation for accelerators, there already exists substantial experience world-wide in the remote operation of large technical installations.

In major particle physics experiments, subdetectors are frequently monitored and run remotely. Large telescopes for astronomy are operated remotely – experiments on satellites and on distant planets are routinely operated from control centres on Earth. In industry, remote diagnostics and operation have become standard, even in nuclear power plants.

Several human aspects, such as national visibility and political and public identification with the project are also involved. How can the desired "corporate identity" be attained? How much manpower is needed at the host site and at the partner institutes? What scientific sociology will emerge? Many of these issues resemble those that have already been encountered in large experiments, which will serve as useful role models.

Whatever the challenges, a GAN could provide the framework for the construction and operation of future large accelerators, which would otherwise be impossible to realise.

22.11
Summary and Outlook

Particle physics can look back on a unique scientific success story of sixty years. The success is owed to an extensive sharing of the use of existing facilities, of ideas and of openness to international cooperation. This has enabled the field to gain an unprecedented insight into the innermost structure of matter and the universe, the Standard Model and its experimental verifications being a striking example.

The success would have not been possible without the very close collaboration between large laboratories and their users from universities around the world. Co-operation in particle physics has been initiated and continuously nourished by scientists themselves, the leading players invariably being physicists or engineers, and the basic motivation has always been the best use of available facilities and resources for the progress of science. Only the close interplay between all partners with their complementary skills has enabled this progress.

Large laboratories have not only been the "home base" for the development of science, they have also played an important political role. This started with the foundation of CERN in an attempt to bridge the gaps between countries after World War 2. It continued successfully by working across the "Iron Curtain" and by integrating scientists from developing countries.

A successful past does, however, not guarantee an equally successful future. The development of particle physics has arrived at a crucial crossroads. The immediate path into the future will most likely be charted by the results from the LHC and the physics results at the Terascale. Should the results confirm important new physics at this energy scale, then a Linear Collider will be the obvious next step. Its realisation will require a truly global collaboration, models for which are being developed. Should the LHC results show unexpected features, the future strategy will possibly have to be revised.

But one thing is clear: whatever the outcome, the field will continue to rely on the strength, inventiveness and power of its large laboratories, as well as of its individual scientists.

23
Communication, Outreach and the Terascale

James Gillies and Barbara Warmbein

Although this chapter is the last in this book, it should be something that you bear in mind constantly, from the start to finish of any project at the Terascale, because good communication underlies any modern endeavour: it is part of physics at the Terascale whether we are communicating to our peers, our neighbours or the public at large. In the past, finding a communication chapter in a physics textbook was unheard of, so what's new?

After the LHC's first beams in 2008, one newspaper declared that CERN is the new NASA, ready to inspire a new generation of scientists just as the moon shots of the 1960s had done. Collider physicists are becoming the new rocket scientists. People not only have greater interest in what was once an extremely remote subject, but they are also hungry for quick, direct and accurate information. This is a unique opportunity for scientists, and for the communication experts who work with them, to put science back where it belongs: at the heart of culture in the broadest sense. It is a great chance to streamline core messages and tell the story that particle and machine physicists have known all along: it is exciting to seek answers to fundamental questions about the universe, because curiosity is part of what makes us human. Not everyone can have the excitement of sitting in a control room late at night and getting to know something new for the very first time, but good communication is the next best thing. It can help put science back into the heart of popular culture, giving everyone the chance to experience the sense of awe and humility that research at the cutting edge brings.

Our curiosity about nature is why we do science at the Terascale, and that has to be the first key message of the field's communication and outreach. But there are many more positive messages to deliver. While working on the question what makes the universe tick, Terascale physicists are solving hugely complex technological tasks on a daily basis – "accelerating science and innovation", as CERN likes to put it. By working towards shared scientific goals, they are bridging cultural and linguistic divides and are uniting thousands of people without hierarchies or formal agreements: it is no accident that the rest of the world is increasingly looking at management structures in particle physics, hoping to learn some valuable lessons.

Physics at the Terascale, First Edition. Edited by Ian C. Brock and Thomas Schörner-Sadenius
© 2011 WILEY-VCH Verlag GmbH & Co. KGaA, Weinheim.
Published 2011 by WILEY-VCH Verlag GmbH & Co. KGaA.

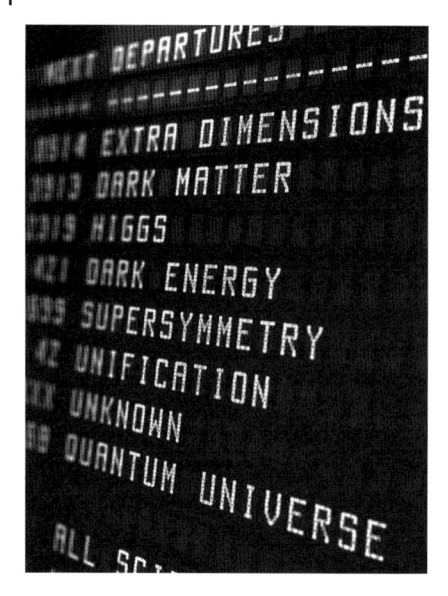

Figure 23.1 Communicating all the exciting discoveries that wait at the Terascale needs strategy and vision (Source: ILC).

Science and communication have to work together to harness the new image of particle physics – and the particle physicist; this chapter tells you how, what and, first of all, why.

23.1
Why Communicate?

Science is an integral part of modern culture. It pervades almost every aspect of our daily lives from the food that we eat to the technology in our living rooms. Yet the scientific process is becoming ever more detached from people's everyday lives, leaving them ill-equipped for the modern world, and without the necessary tools to make informed decisions on a wide range of subjects. How do we solve the energy crisis? Is climate change real? Should we vaccinate our children? The list is extensive, and the answers that individuals and society as a whole come up with will have a profound impact on all our futures. This alone is reason enough for scientists to engage the public with what they do – but there are other, less altruistic, reasons as well.

In the field of basic research, it is largely public money that is being spent, and that comes with an obligation to communicate. The public, after all, have a right to know where their money is going, and indeed to take decisions on where that money should be spent. "Is fundamental science a worthwhile use of public money," is a question we can add to the list above, and if we want the answer to be "yes", then we have to give opinion formers, and the public at large, appropriate information to reach an informed conclusion.

There are many reasons why basic physics research has a very powerful case to communicate. Curiosity about the universe is part of the human condition. It is one of the things that distinguish human beings from other animals, and it is what has allowed us to progress culturally and technologically. Curiosity is why astronomy is repeatedly ranked as one of the most popular scientific subjects, ranking as highly as health issues. In other words, human beings would appear to rank understanding of nature as highly as personal well-being.

For most people, this curiosity can be sated by the media, books and films, but for those who choose to pursue a career in science, it is a driving force that makes basic science not only a force for knowledge, but also for technological innovation. If scientists in pursuit of the Higgs boson find a technological hurdle in their way, more often than not they will find a solution. This has brought about immediate benefit in domains ranging from information technology to medical imaging and taking in solar energy collection along the way. Basic science has a proven track record of providing the necessary raw material – knowledge – for applied scientists to turn into applications that improve the human condition. The World Wide Web is but the tip of the spin-off iceberg.

Finally, although no scientist working in particle physics today would claim that the discovery of the Higgs boson will produce immediate tangible benefits for society, few would rule out that possibility. Scientists have learnt the lesson of history. Human ingenuity being what it is, most of basic science's discoveries tend to find applications sooner or later. One famous episode from history has Michael Faraday telling his finance minister, in response to a question about the utility of research into electricity, that one day the government might be able to tax it. Faraday is remarkable in his vision, and he is an exception to a more pessimistic rule. The history of science is populated more by quotes along the lines of Rutherford's famous claims that ideas of energy from the atom were moonshine. Today, we know better than to be so categorical.

Coupled with the clear benefits to society that research brings is the fact that the gap between reality and public perception when it comes to basic science can be vast. Particle physics laboratories are frequently perceived to be closed, secretive and potentially dangerous. The reality could not be more different: CERN, for example, is open to the public six days per week, makes the results of its research as widely available as possible and actively courts public attention.

Fear of science is also a phenomenon that gains ground when the gap between science and culture widens. It is increasingly common to hear science named as a culprit for society's ills from climate change to Franken foods. Again, the reality could not be more different – science has extended life expectancy and improved the quality of life in much of the world. Rather than driving climate change, it is science that gives us the means to understand the phenomenon and address it. And rather than creating monster foods, science allows us to improve efficiency in food production. There is no doubt that the fruits of science can be misapplied, but in these cases it is often the way society chooses to use science rather than the

Figure 23.2 Visitors in DESY's HERA tunnel during the Open Day 2009. Public events like Open Days attract huge crowds to every lab (Source: DESY).

science itself that leads to misapplications. This is all the more reason for scientists to engage with society, to play their part in ensuring that science is put to work for the good of society.

Science is an area in which humans have shown that they are able to cooperate – whatever cultural background they have. It has always been this way. Science attracts those people whose sense of curiosity about their surroundings is the most developed. Its basic empirical approach and its language of mathematics are universal, transcending any man-made barriers of culture or language. For this reason, it is entirely natural that when nations wish to work together, science is often the first step. That was the case for CERN, established in the aftermath of a particularly bruising period in European history after World War 2 to build bridges between nations through science. And it's the case today in the Middle East where the SESAME lab is being established in Jordan along much the same lines as CERN before it, although with very different scientific goals.

Finally, physics is perceived to be a difficult subject, and indeed it is for those who wish to understand fully all its arcana and to contribute to its development. Conceptually, however, it is within the grasp of anyone with a modicum of curiosity. Tell people that we understand no more than about 4% of the universe, and in general they'll want to know more and are prepared to make the necessary effort to further their knowledge.

For all these reasons, communication of basic science is important. Clarity, openness and transparency are the keys to making that communication effective.

23.2
The Place of Communication Within an Organisation

"We need a brochure" is perhaps the most common thing a communications professional in a scientific organisation will hear in the course of their career. Perhaps we do need a brochure, but there are many steps to be taken before arriving at such a conclusion. What strategic goals is our communication supporting? What message are we aiming to deliver, and to whom? Only after questions like these have been thought through can we say whether a brochure is the best way to make sure our message reaches its target.

The task of communication is frequently misunderstood, or taken for granted, when things are going well. In times of crisis, communications professionals all too often find themselves fire-fighting on behalf of their organisations. It doesn't have to be that way. Communication is a strategic and many-faceted activity, encompassing internal communication, corporate communication, stakeholder relations, education and outreach, among others. Communicators are often, by necessity, good at fire-fighting. But with proper strategic thinking, fire-fighting should rarely be necessary.

The objective of communication is to help an organisation or project to achieve its strategic goals: it is a management function. Do we need a brochure? If that brochure is going to achieve a desired outcome with a key audience, then yes, we do need a brochure. If it is going to contain baffling information not appropriately packaged for the intended audience, and consequently finds itself dumped in a rubbish bin at the earliest opportunity, then perhaps the brochure is not the ideal vector for communicating with that audience.

Scientists have a tendency to communicate for altruistic reasons, sharing their enthusiasm for what they do. As communicators, we have to ask ourselves whether our initiatives are satisfying the strategic goals of our organisations. Thankfully, many of our activities satisfy strategic goals. At many laboratories, public visits pass a message of openness, teacher programmes address the strategic objective of increasing the number of students studying science at university, and initiatives with local schools help to embed the lab as a positive actor in its host region. These are all strategic goals of the organisation, supported by communication activities.

Communication should be part of the day-to-day management process of any organisation and be built into any project from the start. Communication takes as input the strategic goals of the organisation or project. The questions a communicator then needs to ask are: who do we need to be talking to in order to achieve our goals; what are our key messages; and what are the tools we have at our disposal for delivering those messages to our audiences. This forms the basis of a communication plan (see for example the cover page of CERN's 2009–2013 communication plan in the annex). Terascale physics projects of the future should include communication in the planning process, and in this respect the International Linear Collider is setting a strong example. The ILC project has included communication tools and activities at the highest level of management from a very early project stage.

In the world as a whole, the communication function is becoming increasingly professionalised with many universities offering postgraduate studies in the field. A recent study carried out by the European Association of Communication Directors showed that around 85% of people working in communication have some form of university or professional qualification in communication management, with over 70% being qualified to at least Masters degree level[85].

The same study shows that in 85% of cases, communications professionals believe their job is to support the organisation's goals, with 60% of respondents saying that they play a role in defining business strategies. At the CEO level, increasing numbers of directors define their roles in terms of communication, recognising that good communication underlies everything from public image to worker relations. All this shows that the rightful place for communication within a project is as an integral part of the project from the start, and the place for the communication function within an organisation is at the top level of management.

23.3
Audiences and Tools – the Basics

It would be interesting to jump 15 years into the future and take a peek at what audiences will look like then: in communication, things are changing very fast. We live in exciting times where "twitter" is the top word of the year 2009[86] and communication tools are in a state of flux. Boundaries between who communicates and who receives the communication are blurring, sometimes even disappearing, and we are moving towards an unknown, albeit lively, entertaining and challenging communication future.

Press offices at labs and communication teams for projects are only human: they enjoy trying out new toys and testing new tools to find out how and whether they work. The new social media offer a plethora of new tools – the basics, however, will always remain: we need to target our messages at specific audiences using tailor-made tools. The concept of audiences and tools is universal to all areas of communication and is as essential for communicators as mathematics is for physicists.

Traditionally and theoretically, audiences for communication activities are well defined. If you follow a communication, which you are best advised to do, you begin with your organisation's strategic goals, define the key messages that support those goals and then choose the tools best adapted to convey those messages accurately to your target audience.

---

85) European Communication Monitor 2009, ISBN 978-3-9811316-208.
86) See the Global Language Monitor's annual global survey of the English language, http://www.languagemonitor.com/news/top-words-of-2009.

Some of the people who can influence the present and future of your project or lab will value the classic brochure. Your neighbours and fans will flock to open days and subscribe to your twitter feed. The people who work on the project need internal newsletters and magazines to stay informed and involved. And journalists need professional press releases, enthusiastic and coherent scientists for interviews and one-to-one personal relationships so that they can write well-informed, up-to-date articles or prepare a piece for the news. Then there's industry – the people who build big or small parts for the lab, and who can influence the first target audience: the decision-makers. They need targeted communication too. And we should never forget the upcoming generation of citizens and researchers – schools and universities, the students who will form the workforce of the future and the teachers who inspire and train them.

The communication landscape is a complex one, the Rockies rather than the prairies, and it is full of grey areas. Do journalists constitute a target audience of their own, or are they a vector? Is it the people who read their stories who are the true target audience, or rather the people they vote for? The answer is "all of the above": journalists, decision-makers, members of the public, corporate managers hoping to land a contract with a major lab are all audiences for our communication, and all expect and need communication tailored to them. Then there's the new media: should bloggers receive the same treatment as journalists? Who can and cannot subscribe to press releases? Does someone living next door to one of our facilities have the same needs in terms of communication as someone living far away? How important are the social media? Should we place as much faith in Wikipedia as in Britannica? At least two experiments would suggest not. According to the Irish Times[87], a fake quote attributed to Maurice Jarre by a Dublin student in Wikipedia rapidly found its way into mainstream media. Similarly, when someone gave a new German minister an extra name in Wikipedia, it was duly picked up by the press[88]. We need to exercise extra caution when evaluating social media sources.

All these questions need consideration in any communication plan, so how do you give each target audience what it needs while at the same time pursuing your organisation's goals? A variety of tools – some tried and tested, some new and exciting – are at the disposal of the world's particle physics communication experts.

23.4
How to Engage with Your Audience

In the modern communication landscape, if you have messages you want to pass, it's increasingly important to be as visible as possible in as many places as possible for as much of the time as possible. Coupled to that is the fact that communication is no longer a one-way street: as communicators, we are no longer in complete

---

87) http://www.irishtimes.com/newspaper/ireland/2009/0506/1224245992919.html
88) http://www.bildblog.de/5704/wie-ich-freiherr-von-guttenberg-zu-wilhelm-machte/

Figure 23.3 Fermilab has an active Citizens' Task Force, putting people from the towns and villages surrounding the lab in regular contact with the lab for an exchange of ideas and advice (Source: Fermilab).

control of information about our organisation. People are talking about us, and increasingly they expect to engage in a dialogue with us. In this world, a smart communication team will make full use of the social media phenomenon engaging as far as possible, and in effect turning its audiences into vectors for their messages: if people like what you're saying, they'll rebroadcast it for you, ensuring that your messages are everywhere, all the time. The corollary, of course, is that those who don't like your messages will do what they can to ensure that their messages are louder than yours. Figuring out how to avoid that has become part of the communication planning process.

But let's get down to basics. One of the first rules of science communication – and all other communication for that matter – is: no tool without a strategy. All communication should pass this three-question test: What is the strategic goal we want to achieve? Who is our target audience? What's the best tool to reach that audience?

"We need a brochure" may be the answer to the third of those questions, but "we need a twitter feed" is equally likely. Brochures, however, do have an important place in the communicator's armoury, along with many other tools. Here's a selection of tools that you may find useful, along with some of their typical uses.

- *Brochures* Skilfully written and well-designed brochures are very powerful tools: holding all your key messages neatly on paper, they are easy to ship and distribute, put into press kits, give to important visitors and use as a reference. A disadvantage is that they can be expensive to produce and it is difficult to keep them, and all their translated versions, up to date. In the days of Web 2.0 and carbon footprints they can seem old-fashioned, but they continue to be a staple of any project's communication diet because virtual paper isn't yet up to the task.
- *Websites* Websites are a one-stop information opportunity for many different target audiences, which make them both extremely useful but at the same time a difficult tool to master. Everybody uses the web – from primary schools to fund-

ing agencies. Publicly funded top-level research institutes have the obligation to communicate, and a good website will not only explain the whats, the whys and wherefores of research, but will also offer regular updates, images and videos for download, clear contact details and possible interaction points. For scientists working in the lab or on the project, the intranet is perhaps the most important tool for internal communication, containing technical information, an events calendar, document servers, software or simulation tools and the latest news for staff.

- *Press releases* The most respected, and perhaps most abused, tool of communication between organisations and the media are press releases. Press releases must come from an established source, contain information that is relevant for the world outside the organisation, and be written according to established guidelines. Journalists use them to produce a story or clip (or blog entry), often accompanied by a more in-depth interview with an expert. Journalists receive hundreds of press releases every day and generally take little more than a few seconds to decide whether to read or discard them. The more established a lab is as a source, and the more responsible its use of press releases, the more effective they are likely to be.
- *Personal relationships* Communication is built on trust, and the best way to build trust is to get to know someone. Know a journalist? Give them a ring from time to time, even if there's nothing in particular to say about your project or organisation. Relationships of trust between scientists, communicators and journalists are not only the bedrock of effective media relations, but are also vital for institutional communication with all stakeholders.
- *Newsletters* Newsletters mainly serve an essential target group: the community. Information is a currency, and knowing what's going on not only gives the newsletter recipient a certain sense of wealth, it can also help them to go about their everyday science business. Newsletters keep collaborators up to date on new projects, upcoming or achieved milestones, the thoughts and plans of their directors – anything from how to get a visa for the next important collaboration meeting to the lab's 50th anniversary celebrations. However, they are also read by fans and by the media and can be a useful tool to spread news that doesn't merit a press release. A good newsletter should be fast, easy and fun to read and as entertaining as a good magazine.
- *Events and exhibitions* Events are a great opportunity to touch base with your neighbours, friends and sceptics. Some are small and reach a very limited audience, but an open day at a major lab can draw crowds of several tens of thousands – what better way to share some of the excitement of science than letting people take a glimpse behind the curtains and witness the enthusiasm of scientists at work. Events are also a way to give members of the public the chance to interact with real science and real scientists.

Exhibitions are another tried-and-trusted way of telling the taxpayers what happens to their money. Sometimes localised versions of exhibitions really take off. An exhibition in a subway station in Berlin about CERN, the LHC and its experiments and Germany's role in the project was turned into a travelling exhibition and has attracted more than 100 000 visitors in a little over a year, travelling around the country. Other countries have similar projects, and there's of course CERN's own travelling exhibition. People in Europe do not have to go to Switzerland to learn more about mankind's greatest experiment, and future projects are already dipping their toes into the world of science fairs and European conferences to reach the audiences they need to have on their side: fellow scientists and decision-makers. An all-important ingredient for all these: the real scientist who answers questions and talks about his research and visions. Events and exhibitions also serve the strategic goal of demonstrating openness.

- *Educational programmes* No field can survive if it isn't kept alive by fresh brains. With student numbers in decline, scientists at the Terascale must be competitive against other academic disciplines and fields of physics, and a number of programmes and projects for school children and students exist to inspire them to become those scientists who will run the machines people plan today.
- *Social Media* Blogs, twitter, facebook. The language of communication has suddenly become a lot richer, as has the number of information sources available to all our audiences. The key difference between social media and traditional media is that anyone can publish, and anyone can engage. This democratises communication, but it also makes the reliability of information harder to evaluate. In the past, people would instinctively know how to evaluate information from FAZ, Liberation, The Guardian or El Pais against other offerings. In the social media world, it will take a while for people to get to grips with what to expect from a Quantum Diarist as opposed to someone writing on Cosmic Variance or scienceblogs.de. In the meantime, social media offer a real opportunity to engage with our audiences.

Whatever tools we choose, evaluating their impact is an important part of the communication process. It is not always easy to evaluate the success of communication activity. Some metrics are obvious, usually the quantitative ones like queues at an open day, number of visitors for a show or hits on a website. These give hard numbers, but do not give qualitative information on the impact. Most press offices collect and file articles on their organisation and field of research, but qualitative media monitoring is harder to do, and is very labour intensive. Metrics such as advertising value equivalent produce some impressive figures about how much you would have had to spend to get the same amount of visibility, but give no information about how your messages are being received. Some of the more sophisticated, and expensive, media monitoring services offer a qualitative analysis of coverage. Social media are increasingly being covered by such services, and automated tone analysis (ranking articles as positive, negative or neutral) is starting to appear. More difficult to evaluate is the success of an educational event, of a phone call between a journalist and a communicator, or of a brochure. The influence can be enormous, yet the connection cannot always be made. Evaluation is, nevertheless, an essential part of the communication function. We need to know that our communication activity is supporting the long-term strategic interests of our organisations. Questionnaires, focus groups and interviews all offer access to information, and should be built into communication plans. Cross correlations can also provide useful metrics: does increased coverage of a lab correlate with increased clicks on the job vacancies site, for example, and does that feed through to high-quality applicants.

23.5
Communication at the Terascale

It is a phenomenon universally acknowledged that people are first and foremost interested in – people. It explains the success of facebook, the popularity of gossip magazines and the fact that stories that make it into the science section of general-interest newspapers are almost always about medicine or human biology. Human evolution, cures for cancer, diseases and psychology make for good reading because they are scientific and people can relate to them. However, in a sales study made by New Scientist magazine a few years ago the bestselling issue of the magazine, based on shop sales and thus the cover, was not the one about the human brain but instead the one about Einstein. People may love medicine and biology because it's close to home, but they also like big science – astronomy and particle physics. Maybe it's because big science is so far from home. Maybe it's because we've always been curious about our universe. Ever since we first appeared in the Rift Valley all those thousands of millennia ago, our curiosity has marked us apart and driven our species' success.

Mankind's fascination with space and the enormously big is a connection that particle physicists have harnessed to engage their audiences with the microcosm: it works in our favour that the science of the extremely big and the science of the extremely small address the same fundamental questions, the same fundamental curiosity. The superlative projects and record-breaking machines of today and tomorrow generate fascination simply because of the sheer scale of the human endeavour involved. With that comes a hunger for information that can be overwhelming, and an opportunity to engage broad audiences with the science we do and the innovation we foster.

That fascination is ultimately good for the field, but it brings exposure: if things go wrong, the world will know about it. In a social media world, however, exposure is a fact of life, not an option, and it is far better to be part of the conversation than on the sidelines. When CERN first switched on the LHC in 2008, it was perhaps the most visible scientific event in history. Given that the LHC broke down shortly afterwards, would CERN have been better off playing it low-key? A better question might be "did CERN have the option", and to that the answer is no. Big science projects are in the spotlight, and the coverage CERN had following the LHC break down and through the recovery was positive in tone, thanks in part to the good will established by bringing the machine to life in full public view.

The LHC is not the only example. In December 2003, a British-led Mars probe, Beagle 2, was scheduled to reach the red planet. The probe was successfully ejected from Mars Express on 19 December, and nothing has been heard from it since. Throughout the month the project leader was constantly on the news, explaining

Figure 23.4 A live radio interview during CERN's First Physics Day in 2010. Many communication offices offer media training for scientists to help them prepare for situations like this one (Source: CERN).

Figure 23.5 A picture of people that went around the world – first beam splashes at ATLAS in November 2009 (Source: CERN).

his project and baring his soul. The mission was a failure, yet public support for space science rose, which brings us back to where we began this section: it is a phenomenon universally acknowledged that people are first and foremost interested in – people. If you can bring together a great human story with great science, you are likely to be on to a winner.

The invention and reinvention of the web has led to a number of new phenomena that makes things much easier for communicators – but also much more complicated. It was hard to miss the start of the LHC on 10 September 2008, and the incident that brought the LHC to a standstill nine days later, because it was the major news item all over the world in newspapers, radio stations, online magazines, blogs, websites and TV. Part of the reason for this is that what began with a handful of people escalated through Web 2.0 social media channels into perhaps the most overhyped story of 2008: the LHC, they said, could produce earth-eating black holes. Of course there is no possibility of that happening, and it is somewhat ironic that the product of a particle physics lab should become the tool of favour for those who would detract from particle physics.

While CERN was preparing to switch on LEP in 1989, Tim Berners-Lee was penning the proposal for what would become the World Wide Web. Progress always attracts merchants of doom, and the start-up of LEP was no exception. Without a web, however, the doom merchants found no traction. A couple of years later, Fermilab suffered protesters at the gate for the Tevatron, but there was little echo. By 1999, however, when Brookhaven switched on RHIC, the web was well established, and the doom merchants managed to generate some coverage. By 2008, the web had become Web 2.0, and the communication landscape had changed forever. There's no doubt that the black-hole story contributed to the LHC's notoriety, and there's equally no doubt that the overhyping of the black-hole story is no more than a Web 2.0 phenomenon.

The challenge for communicators is to harness new tools to their advantage. The 2008 start-up of the LHC was possibly the first accelerator start-up to be tweeted live. When the LHC restarted in November 2009, twitter was a well-established part of CERN's communication strategy. The number of followers reached around 16 000 in 2008, and has climbed to over 130 000 through the restart process. That's a social media sign that people are interested in particle physics. Traditional market research also shows that while people may not know about quarks and leptons, bosons and fermions, they do think that curiosity-driven research is worthwhile. Put another way, the world is interested, ready to listen, and we are the happy storytellers of the Terascale.[89]

---

89) **Recommended as essential reading and links:**

CERN Courier: http://cerncourier.com,

symmetry magazine: www.symmetrymagazine.org,

symmetry breaking: http://www.symmetrymagazine.org/breaking/,

CERN bulletin: http://bulletin.cern.ch,

ILC NewsLine: http://www.linearcollider.org/newsline,

Fermilab Today, SLAC Today, DESY inForm: http://www.fnal.gov/pub/today/, http://today.slac.stanford.edu, http://www.desy.de/inform,

interactions.org, particle physics news and resources: www.interactions.org,

European Particle Physics Communication Network (EPPCN): https://espace.cern.ch/forum-EPPCN,

European Particle Physics Outreach Group (EPPOG): http://eppog.web.cern.ch.

Appendix
CERN Strategic Communication Plan 2009–2013, Summary

Scope

This plan covers the period 2009–2013, during which the first results from the LHC will be published. Its main objective is to build on the platform created by the public impact of the LHC start-up in 2008 to position CERN's message of basic science as a key knowledge and innovation driver. Particular attention will be given to internal communication, recognising both the need for effective communication to a diverse community, and that every member of that community is a potential ambassador for CERN and particle physics. It covers work to be carried out mainly in the Communication group (DG-CO), but also in the Education group (PH-EDU), working closely with the InterActions network of laboratory communicators, the European Particle Physics Communication Network (EPPCN), the European Particle Physics Outreach Group (EPPOG) and the LHC Outreach Group (LOG). Resource plans for individual actions will be developed where required following management approval.

Main Elements Are

- Communications through the LHC repairs and restart
- Protocols for announcements
- Internal communication
- Local communication
- Conditions of use for CERN image and information
- Brand management
- Web 2
- Provision of a quality visit experience at CERN
- Exploitation of CERN's scientific heritage

Objectives

- To foster support for CERN and particle physics
- To position CERN as a global laboratory and a world leader in fundamental physics research
- To position fundamental science as a driving force for innovation

Key Messages

- Fundamental science satisfies the basic human instinct to explore
- Fundamental science is a driving force for technical innovation, collaboration and scientific education: without fundamental science, there is no science to apply
- CERN is a world leader in fundamental research
- The LHC will launch a new era of discovery and understanding of fundamental questions about the universe
- LHC technologies create benefits for society
- CERN and the LHC are excellent examples of science transcending barriers of age, religion, gender and nationality

Target Audiences

- Science and technology opinion formers
- The CERN community (including scientific, administrative and professional staff, contractors working on site and users)
- The local community
- The broader scientific community
- Media
- Educational systems, primarily at high school level
- The general public

Index

a
accelerator   243–262
  – beam–beam effect   246, 248
  – beam emittance   245
  – beam protection system   258
  – beamstrahlung   246
  – bunch trains   249
  – bunches   245
  – charge exchange injection   252
  – CLIC   259, 261
  – collider   243
  – damping time   248
  – emittance   245
  – energy   244
  – fixed target   243, 434
  – HERA   253–255
  – ILC   259, 261
  – interface to detector   266
  – LEP   244–250
  – LEP 2   249
  – LHC   245, 255–258
     linear collider   245, 258–262
  – Livingston plot   244
  – luminosity   245, 248, 251, 255
  – muon collider   259
  – muon cooling   259
  – particle source   250, 252–253
  – performance   248, 251, 255
  – pinch effect   246
  – Pretzel scheme   248
  – PS   258
  – separation scheme   248
  – SPS   250, 253, 258
  – storage ring   243
  – superconducting magnet   244, 250
     synchrotron radiation   244, 246
  – Tevatron   250–253
  – tune shift parameter   248
  – two-beam acceleration   259

accelerator laboratories, *see* big labs   417
active pixel detector   300, 303
ADD model   215
ALEPH   272
ALICE   19, 280, 296, 384
alignment   291, 309–310
α_s 8, 38, 41, 45, 63, 75
alternative Higgs   209–211
analysis framework   394
astroparticle physics   16–19
  – air showers   18
  – cosmic rays   17
  – dark matter   18
  – multi-messenger approach   18
  – neutrinos   16
  – photons   17
  – point sources   17
  – weakly interacting massive particle (WIMP)   18
asymptotic freedom   74, 93
ATLAS
    269, 281, 296, 299, 320, 325, 335, 337, 363, 371, 375, 384
automated calculation   97
avalanche photodiode   322
axino   149

b
B factories   173
B-meson mixing   174–178
  – measurements   176–178
  – mixing phenomenology   174–176
  – time-dependent decay rates   176
bandwidth   371
baryon number   44
baryon parity   146
baryon triality   147
beam–beam effect   246, 248
beam emittance   245

Physics at the Terascale, First Edition. Edited by Ian C. Brock and Thomas Schörner-Sadenius
© 2011 WILEY-VCH Verlag GmbH & Co. KGaA, Weinheim.
Published 2011 by WILEY-VCH Verlag GmbH & Co. KGaA.

beam-pipe 266
beamstrahlung 246
beta function 75
Bethe–Bloch formula 295, 314
Bethe–Heitler process 351
BFKL evolution 230
Bhabha scattering 348
big labs 417–419, 431–447
– access 441
– assets 434–439
– budget 433
– collaborations 439–440
– decision process 442–444
– examples 432–433
– facilities 432
– functions 434–439
– future 444–446
– history 431
– organisation 440–441
– politics 442–444
– staff 438–439
– strategic planning 441–442
– universities 433
Born level 97–98
branes 213–216
Breit frame 84
bremsstrahlung 314, 334, 351
brochure 456
BSM 209–221
– ADD model 215
– alternative Higgs 209–211
– branes 213–216
– compactification 213
– composite Higgs 211–213
– compositeness 211–213
– 5D warped models 213
– duality 214
– extra dimensions 213–216
– extra gauge bosons 218
– flavour changing neutral currents 169–172
– grand unified theory 216–218
– heavy quarks 169–172
– hidden sector 220
– hidden-valley models 220
– Kaluza–Klein 213
– Kaluza–Klein excitations 215
– leptoquarks 218–219
– little Higgs model 211
– model-independent search 220–221
– partial compositeness 211–213
– quirk model 220
– Randall–Sundrum model 216

– sequential standard model 218
– strings 213–216
– technicolour 211–213
– top quark 189, 204–205
– unparticle model 220
budget
– big labs 433
– collaborations 410–412
buffer 363, 368
bump-bond 302
bunch identification 366
bunch trains 249

c
CALICE 328
calorimetry 267, 313–330
– active medium 313
– bremsstrahlung 314
– compensation 324
– confusion 327
– conversion 315
– digital approach 330
– dual readout 326–327
– e/h ratio 323
– e/mip ratio 320
– electromagnetic energy loss 314–315
– electromagnetic shower 316–318
– hadronic energy loss 315–316
– hadronic shower 318–319
– homogeneous calorimeter 267, 319
– ion/mip ratio 323
– ionisation 314
– minimum ionising particle 315
– mip 315
– Molière radius 317
– nuclear interaction length 315
– PandoraPFA algorithm 328
– particle detection 313–316
– particle flow 284, 327–330
– particle shower 316–319
– radiation length 314
– readout 322
– resolution 319–325
– response 319–325
– sampling calorimeter 267, 319
– sampling fraction 320
– shower shape 317
– spallation process 315
– total absorption calorimeter 327
cascade decay 151–152
CASTOR 237
cathode-strip chamber 296, 337, 342
CDF 188, 278

central trigger   364, 367, 371
CERN   432
CERN Council   383, 419, 428, 441, 443
charge exchange injection   252
charge trapping   305
charged current   36
charged-current interaction   47
chargino   145
charm quark   5
chirality   47
CKF, see combinatorial Kalman filter   306
CKKW algorithm   109
CKM matrix   35, 43, 164–166, 188, 199, 201
  – CP violation   166
  – limits on new physics   168
  – parameters   166
  – unitarity   165
  – unitarity triangle   166–168
  – Wolfenstein parametrisation   166
CLIC   259
clocked readout   363
cloud computing   394
cluster hadronisation   114
CMS
  266, 282, 296, 299, 322, 335, 337, 371, 384
collaborations
  270, 401–413, 419–420, 439–440
  – budget   410–412
  – Collaboration Board   404
  – communication   408
  – contractual framework   402
  – cost indexation   403
  – data analysis   408–410
  – decisions   408
  – funding   401–404
    host lab   403
  – international   402, 419–420
  – leadership   405
  – manpower   403
  – Memorandum of Understanding
    402, 420
  – organisation   406
  – publication process   408–410
  – resources   401
  – role of individuals   407
  – size   404
  – spokesperson   405
  – technical coordination   406
  – technology   405
  – time scale   410–412
  – working groups   409
collider   243
collinear safety   78

colour   28, 74
colour charge   75
colour glass condensate   19
combinatorial Kalman filter   306
communication   408, 449–461
  – audience   454–455
  – brochure   456
  – culture   450
  – curiosity   451
  – event   458
  – exhibition   458
  – fear of science   451
  – newsletter   455, 457
  – press release   457
  – social media   458
  – strategic goals   453
  – tools   454–455
  – website   457
Compact Linear Collider, see CLIC   259
compactification   213
compensation   324
  – hardware compensation   325
  – software reweighting   325
computing
  – analysis framework   394
  – cloud computing   394
  – data preservation   395–396
  – distributed computing   383
  – event skimming   391
  – event slimming   391
  – Grid   383–396
  – ROOT   394
  – Tier-0   386
  – Tier-1   386
  – Tier-2   386
  – Tier-3   388
  – Tier architecture   384
  – user analysis on the Grid   390–391
  – Worldwide LHC Computing Grid   384
confinement   41, 74, 93, 112, 220
cosmic microwave background   14
cosmic rays   17
CP violation   6, 43, 166, 178–179, 185
CSC, see cathode-strip chamber   296
culture   450

d

DAF, see deterministic annealing filter   308
dark matter   18
data analysis   408–410
data-driven procedure   379
dead-time   294, 363, 369
  – bunch-specific dead-time   370

DELPHI 273
DEPFET 303
depletion 300
DESY 432
detector 265–287
 – beam-pipe 266
 – cables 269
 – calorimeter 267
 – collaborations 270
 – concept 265–268
 – control systems 269
 – cooling 269
 – cryogenics 269
 – electromagnetic calorimeter 267
 – gas 269
 – hadron calorimeter 267
 – infrastructure 269–270
 – interface to accelerator 266
 – missing energy 268
 – muon detector 268
 – number of channels 269
 – omni-purpose detector 265
 – operation 270
 – organisation 270–271
 – pixel detector 266
 – push–pull operation 286–287
 – safety 270
 – solenoid magnet 267
 – strip detector 266
 – subdetector 265
 – vertex detector 266
 – wire chamber 266
deterministic annealing filter 308
DGLAP evolution 229
diamond 305
diffraction 225–238
 – BFKL evolution 230
 – CASTOR 237
 – DGLAP evolution 229
 – diffractive final states 233
 – diffractive parton distribution function 234
 – diffractive scattering 227
 – diffractive structure function 234
 – elastic scattering 226
 – energy dependence 226, 228
 – ep scattering 225–228
 – gluon density 230, 232
 – hard diffraction 234
 – Higgs boson 235
 – impact parameter 226
 – instrumentation 237–238
 – LUCID 238
 – multiple scattering 235–237
 – new physics 235
 – optical theorem 226
 – parton distribution function 228–231
 – parton dynamics 228–231
 – pomeron 226, 233
 – pp scattering 225–228
 – QCD 226
 – rapidity gap 225, 234
 – rapidity gap survival probability 235
 – saturation 231–233
 – small x 228–231
 – soft diffraction 234
 – t dependence 226, 228
 – total cross section 226, 228
 – TOTEM 238
 – underlying event 235–237
distributed computing 383
Drell–Yan process 79, 107, 116, 130, 195, 233
drift-tube detector 340–341
DSP, see logic implementation
 – Digital Signal Processor 368
dual readout 326–327
duality 214
DØ 188, 267, 279

e
EC funding 426–427
effective theories 44–45
elastic scattering 226, 358
electroweak interaction 23, 28–37, 39–40, 47–69, 123
 – charged-current interaction 47
 – chirality 47
 – measurements 51–60
 – neutral-current interaction 47
 – theoretical predictions 61
electroweak precision measurements 188
electroweak symmetry breaking 123, 164, 189
electroweak unification 4, 8
elliptic flow 19
emittance 245
ENC, see equivalent noise charge 293
energy loss 308
energy resolution 319–325
 – constant term 320
 – electromagnetic sampling calorimeter 320–321
 – hadron calorimeter 322–325
 – homogeneous calorimeter 322
 – noise term 320
 – sampling fraction 320

– standard parametrisation   320
– stochastic term   320
equivalent noise charge   293
European Research Area   426–427
event builder   370, 378
event shape   84
event skimming   391
event slimming   391
extra dimensions   154, 213–216
　– compactification   213

f

facilities   432
factorisation   76, 97
fast clear   374
fear of science   451
Fermi constant   39, 51
Fermilab   432
fermion masses   34–36
FF, see Monte Carlo
　– fragmentation function   77
fine-structure constant   51, 61
fine tuning   12
fixed-order calculation   97–98
fixed target   243, 434
flavour changing neutral currents   5, 169–172, 205
flavour mixing   6
flavour oscillation   171
flavour tagging   292, 310–311
forward-backward asymmetry   48, 54, 58
forward–backward charge asymmetry   195
forward physics   225–238
fourth generation   67
FPGA, see logic implementation
　– Field Programmable Gate Arrays   368
fragmentation function   77, 83, 112
fragmentation, see Monte Carlo
　– hadronisation   112
funding   401–404, 415–430
　– BMBF   421–422
　– decision-making   428
　– EC funding   426–427
　– European Research Area   426–427
　– European Research Council   426
　– federal structure in Germany   421–426
　– laboratory budgets   417–419
　– LHC   429–430
　– organisation of HEP   415–416
　– politics   416–420
　– project funds   416, 423–425
　– public finances   417
　– research infrastructure   415

– research institutions in Germany   422–423
– third-party funds   416, 420
– time scale   416
– universities   420

g

Gargamelle   4
gas electron multiplier   298
gas gain   295
gas mixture   295–296
gaseous detector   266, 291–299
gauge symmetry   147
gauge theory   24–27
Gaussian sum filter   308
GEM, see gas electron multiplier   298
generalised parton distribution   21
ghost hit   302
GIM mechanism   5, 170, 205
global trigger decision   365
gluon   6–7, 74
gluon density   230
$g_\mu - 2$ 60
Goldstone boson   32, 42
GPD, see generalised parton distribution   21
grand unified theory   216–218
Grid computing   383–396
GSF, see Gaussian sum filter   308
GUT, see BSM
　– grand unified theory   216

h

H1   8, 21, 276, 325, 363, 374
hadronisation   98, 111–115
　– cluster hadronisation   114
　– independent fragmentation   112
　– Lund string fragmentation   113
hard diffraction   234
heavy ion physics   19–20
　– abundances   20
　– colour glass condensate   19
　– elliptic flow   19
　– final-state properties   20
　– gluon density   19
　– jet quenching   20
　– QCD   19
　– quark–gluon plasma (QGP)   19
　– RHIC   19
　– saturation   19, 233
　– small x 19
heavy quarks   163–185
　– asymmetric B factories   173

- *B* factories   173
- *B*-meson measurements   172–185
- *B*-meson mixing   174–178
- *B*-meson mixing measurements   176–178
- *B*-meson production   173
- CKM matrix   164–166
- CP violating phase   178–185
- CP violation   166, 179
- flavour oscillation   171
- flavour violation   166
- generations   163
- limits on new physics   168
- mass   164, 166
- mass eigenstates   164
- mixing   164, 166
- mixing phenomenology   174–176
- neutral meson mixing   170
- new physics   163, 169–172
- null tests   172
- parameters   166
- rare decays   184–185
- Standard Model   163–168
- time-dependent decay rates   176
- unitarity triangle   166–168
- Yukawa interaction   164

Helmholtz Alliance   XIX, 425
HERA   253–255, 275–278
- parton distribution function   9
HERA-B   276
HERA model   440
HERMES   276
hidden sector   220
hierarchy problem   12
Higgs   11, 123–140
- alternative Higgs   209–211
- composite Higgs   211–213
- coupling   140
- decay   126
- decay to $\gamma\gamma$ 134, 136
- diffraction   235
- electroweak fit   139
- electroweak precision observables   125
- electroweak symmetry breaking   123
- exclusion   131–132, 134–135, 138
- gluon fusion   129, 134
- Higgs-strahlung   127, 130, 132
- LHC prospects   136–139
- LHC sensitivity   138
- LHC vs. Tevatron   136
- little Higgs model   154, 211
- mass   123
- mass determination   139
- mass limit   123, 134–135, 139
- production at hadron colliders   129–131
- production at LEP   127–129
- production at LHC   136
- radiation off top quarks   130
- reconstruction   132
- search implications   139–140
- searches at LEP   132–134
- searches at Tevatron   134–135
- selection strategies   131
- signal-to-background ratio   131
- vacuum stability   125
- VBF   134, 136
- vector boson fusion   134, 136
- *WW* fusion   127, 130
- *ZZ* fusion   130

Higgs boson   123–140
Higgs mechanism   31–34
Higgs-strahlung   127, 130, 132
homogeneous calorimeter   319
host lab   403, 416
hybrid pixel detector   299, 302–303

i
ideogram method   202
ILC   259, 283–287, 303
ILD   284–286, 311
impact parameter   226
inelastic cross section   353
infrared safety   78
International Linear Collider, *see* ILC   259

j
jet   86
jet algorithm   86
jet quenching   20
jet shape   91
jet structure   91, 99

k
Kalman filter   308
Kaluza–Klein   154, 213
Kaluza–Klein excitations   215
KEK   433

l
L3   274
laboratories, *see* big labs   431
lattice gauge theory   75
left-right asymmetry   54
LEP   8, 244–250, 271–275
LEP 2   249
lepton number   44
lepton parity   147

leptoquarks   218–219
LHC   245, 255–258, 280–283
LHCb   280, 296, 384
LHCf   280
lightest supersymmetric particle   148
linear collider   245
little Higgs model   154
Livingston plot   244
logic implementation
  – CAM   368
  – Contents Addressable Memories   368
  – Digital Signal Processor   368
  – DSP   368
  – Field Programmable Gate Arrays   368
  – FPGA   368
  – lookup table   367
  – LUT   367
LSP   148
LUCID   238
luminosity   245, 248, 251, 255, 347–359
  – Bethe–Heitler process   351
  – Bhabha scattering   348
  – bremsstrahlung   351
  – counting of inelastic events   353–358
  – e^+e^- machines   348–351
  – elastic scattering   358
  – hadron colliders   353–359
  – HERA   351–352
  – ILC   351
  – inelastic cross section   353
  – LEP experiments   349–350
  – optical theorem   354
  – parton luminosity   80
  – total cross section at the LHC   357
  – TOTEM   357
  – W/Z production   359
  – zero-counting method   355
Lund string fragmentation   113

m

Majorana neutrino   148
manpower   403
MAPS, *see* monolithic active pixel sensor   303
mass scheme   202
matrix element method   203
matter parity   146
MC, *see* Monte Carlo   97
MDT, *see* monitored drift-tube chamber   337
Memorandum of Understanding   402, 420
micro-pattern gas detector   293, 298–299
Micromegas   298
microstrip gas chamber   298
Millepede program   309

minimal supergravity   147
minimum bias   116, 366
minimum ionising particle, *see* mip   300
mip   300, 315, 320, 323
missing energy   268
MLM algorithm   109
Molière radius   317
momentum resolution   336, 338
monitored drift-tube chamber   337
monolithic active pixel sensor   303
Monte Carlo   97–118
  – automated calculation   97
  – Born level   97–98
  – CKKW algorithm   109
  – cluster hadronisation   114
  – factorisation   97
  – fixed-order calculation   97–98
  – fragmentation function   112
  – hadronisation   98, 111–115
  – independent fragmentation   112
  – infrared singularities   97
  – jet structure   99
  – large logarithms   97
  – Lund string fragmentation   113
  – matching   108–111
  – MLM algorithm   109
  – next-to-leading order calculation   98–99
  – next-to-next-to-leading order calculation   99
  – parton shower   99–103
  – phase-space integral   97
  – regularisation   98
  – renormalisation   97
  – singularities   97
  – subtraction scheme   98
  – tree level   97
  – ultraviolet singularities   97
  – underlying event   115–118
MoU, *see* Memorandum of Understanding   402
MPGD, *see* micro-pattern gas detector   298
MSSM   147
MSSM18   159
mSUGRA   147
multiple scattering   235–237, 308, 336
multivariate analysis   198
muon collider   259
muon cooling   259
muon detector   268, 333–344
  – ATLAS   337
  – cathode-strip chamber   296, 337, 342
  – CMS   337

– combination with tracking information   338, 343
– drift-tube detector   340–341
– gaseous detector   339
– magnetic field integral   336
– momentum resolution   336, 338
– monitored drift-tube chamber   337
– multiple scattering   336
– muon chamber   339–342
– muon energy loss   334–335
– muon identification   337–338, 343–344
– muon momentum   336
– region of activity   343
– resistive-plate chamber   296, 337, 341–342
– scintillator   339
– solenoid magnet   337–338
– thin-gap chamber   296, 337
– toroid magnet   337–338
– track reconstruction   343–344
muon energy loss   334–335
muon identification   334–335, 337–338, 343–344
muons   268, 333–334
  – bremsstrahlung   334
  – direct electron–positron pair production   334
  – excitation of atoms   334
  – identification   334–335
  – ionisation   334
  – nuclear interactions   334
  – sources   333–334

n

NA48   322
NAF   392
National Analysis Facility   392
neural network   202
neutral current   36
neutral-current interaction   47
neutralino   145
neutrino   13–16
  – accelerator-based experiments   14
  – atmospheric   14
  – β decay   13–14
  – cosmic microwave background   14
  – Dirac vs. Majorana   14
  – double-β decay   14
  – Majorana   148
  – mass   13
  – mass hierarchy   14
  – oscillations   14–16
  – solar   15
  – SUSY   148

neutron   316
new physics, see BSM   50, 60, 69, 163, 169, 172, 209
newsletter   455, 457
next-to-leading order calculation   98–99
  – real corrections   98
  – virtual corrections   98
NLO   98–99
NNLO   99
noise   293
November revolution   4
nuclear interaction length   315
nuclear tomography   21

o

OPAL   274
optical theorem   61, 226, 354
outreach, see collaborations
  – communication   449

p

particle detection   313–316
particle flow   284, 327–330
particle shower   316–319
  – electromagnetic shower   316–318
  – hadronic shower   318–319
  – Molière radius   317
  – shower shape   317
parton   73
parton density function, see HERA
  – parton distribution function   77
parton distribution function   9, 77, 80, 117, 228
  – BFKL evolution   230
  – DGLAP evolution   229
  – diffraction   228–231
  – diffractive   234
  – evolution   81
  – generalised   21
  – HERAPDF   9
  – splitting function   81
parton luminosity   80, 190
parton shower   99–103
  – angular-ordered shower   104–105
  – antenna dipole shower   107–108
  – approximations   99
  – implementation   104–108
  – leading-order parton shower   99–103
  – partitioned dipole shower   106–107
  – shower evolution   100–101
  – shower scheme   104–108
  – shower time   99
  – splitting operator   102–103
  – unitarity condition   99

pattern recognition 306–307
PDF, see HERA
  – parton distribution function 77
phase-space integral 97
photomultiplier 322
pinch effect 246
pixel detector 266, 299
pn-junction 300
Poisson distribution 368
politics 416–420, 442–444
pomeron 226
prescaling 371, 379
  – downscale factor 372–373, 377, 380
press release 457
Pretzel scheme 248
project funds 416, 423–425
proton decay 146–148
proton hexality 147
proton structure 8
public relations, see collaborations
  – communication 449
publication process 408–410

q
QCD 27–28, 41–42, 73–94, 226
  – α_s 8, 38, 41, 45, 63, 75
  – asymptotic freedom 74, 93
  – beta function 75
  – confinement 41, 74, 93, 112, 220
  – diffraction 226
  – heavy ion physics 19
  – jet structure 99
  – parton luminosity 190
  – running coupling 38, 41, 45, 75
  – strong coupling 8, 38, 75
QCD structure constant 75, 93
QED 24
Quantum Chromodynamics, see QCD 73
Quantum Electrodynamics, see QED 24
quark 73
quark flavour physics 163–185
quark–gluon plasma (QGP) 19
quarks 5, 28
quirk model 220

r
R-parity 146
radiation damage 304–305
radiation length 314
radiative return 58
Randall–Sundrum model 216
rapidity gap 225, 234
  – survival probability 235

rare decays 169, 184–185
readout
  – buffers 377
  – latency 364
  – path 364
  – phase 363
  – pipelines 365
  – system 363
real time 364–365, 367
region of interest 365, 377
regularisation 98
renormalisability 4
renormalisation 37, 97, 202
renormalisation equation 75
resistive-plate chamber
  294, 296, 337, 341–342
response 319–325
ROOT 394
RPC, see muon detector
  – resistive-plate chamber 294
running coupling 38, 41, 45, 75

s
sample-and-hold 374–375
sampling calorimeter 319
saturation 19, 231–233
  – gluon density 232
  – heavy ion collisions 233
scale uncertainty 78
scaling violation 74
sea-saw mechanism 148
selectron 144
self-triggering 364
semiconductor detector 291, 299–305
shower evolution 100–101
SiD 284
silicon 300–301, 305
single top quark
  – cross section 196–197
  – non-SM production 204
  – observation 197–199
  – production 195–199
  – production at the LHC 199
singularities 97
SLAC 433
soft diffraction 234
solenoid magnet 267, 337–338
sphericity 84
spin crisis 20
spin-off 451
spin physics 20–21
  – generalised parton distribution 21
  – structure functions 21
  – transverse structure 21

spin puzzle 20
splitting function 81
spontaneous symmetry breaking 32
SPS 250, 253, 258
Standard Model 4, 7–8, 23–45, 47–69
– asymmetries 48, 52
– charged-current interaction 47
– CKM matrix 164–166
– constraints 60–69
– electroweak interaction 28–37, 39–40
– electroweak observables 39–40
– electroweak parameters 39–40
– ε parameters 66
– extensions 13, 65–67
– fermion masses 34–36
– fine-structure constant 61
– flavour changing neutral currents 169–172
– forward-backward asymmetry 48, 54, 58
– fourth generation 67
– $g_\mu - 2$ 60
– heavy quarks 164–168
– Higgs mechanism 31–34
– left-right asymmetry 54
– lineshape 53
– low energy data 60
– measurements 51–60
– neutral-current interaction 47
– new physics 50, 60–69
– parameter fits 52, 65, 68
– precision measurements 49–51
– problems 12–13, 24
– STU parameters 66
– theoretical predictions 61
– top quark 187
– two-fermion process 58
– W-boson couplings 57
– W-boson mass 55–57
– weak mixing angle 62
– Yukawa interaction 34–36, 164
storage ring 243
straw-tube detector 296
strings 213–216
– duality 214
strip detector 266, 299, 301–302
strong coupling 8, 38, 75
structure function, see HERA
  – parton distribution function 8
subjet multiplicity 91
Sudakov exponent 101
Sudakov exponential 80
Sudakov factor 108
Sudakov logarithm 80

SUGRA 147
superconducting magnet 244
superfield 145–146
supergravity 147
superpotential 145–146
supersymmetry, see SUSY 11, 65, 143
SUSY 143–159
– at the LHC 151–152
– axino 149
– baryon parity 146
– baryon triality 147
– broken symmetry 147
– cascade decay 151–152
– chargino 145
– coupling determination 155
– discrete gauge anomaly-free 147
– discrete gauge symmetry 147
– E_6 151
– gravity 147
– lepton parity 147
– LHC mass reach 152
– lightest supersymmetric particle 148
– LSP 148
– mass eigenstates 145
– mass reconstruction 152, 154, 156–157
– mass spectrum 148, 158
– matter parity 146
– measurements 151–159
– minimal supergravity 147
– MSSM 147
– MSSM18 159
– neutralino 145
– neutrino mass 148
– parameter determination 157–159
– particle masses 144
– phenomenology 149–151
– Poincaré algebra 143
– properties at the ILC 155–157
– properties at the LHC 152–155
– proton decay 146–148
– proton hexality 147
– R-parity 146
– sea-saw mechanism 148
– search strategy 151
– selectron 144
– signature 151
– spin measurement 154, 157
– SPS1a benchmark point 158
– stransverse mass 153
– superfield 145–146
– supergravity 147
– superpartner 144

- superpotential   145–146
- transformations   143
symmetries   42–44
synchrotron radiation   244, 246

t
tau lepton   5–6
tau neutrino   11
technical coordination   406
technicolour   211–213
technology   405
template method   202–203
Tevatron   250–253, 278–280
TGC, *see* muon detector
   – thin-gap chamber   296
thin-gap chamber   296, 337
third generation   5
third-party funds   416
thrust   84
thrust axis   84
Tier architecture   384
time of passage/interaction   363
time projection chamber   296
time scale
   – collaboration   410–412
   – funding   416
timing   363
   – delay   364
   – L1 latency   377
   – latency   364
   – optimal   366
top quark   187–205
   – CKM matrix   188, 199, 201
   – decay   189, 192, 199–201
   – decay rate   199
   – discovery   10, 188
   – electroweak precision measurements   188
   – electroweak symmetry breaking   189
   – extensions of the Standard Model   189
   – flavour changing neutral currents   205
   – forward–backward charge asymmetry   195
   – ideogram method   202
   – LEP indications   187
   – lepton+jets channel   203
   – lifetime   188
   – mass   189, 201
   – mass measurement   188, 202–204
   – mass scheme   202
   – matrix element method   203
   – measurement of R_b 201
   – modified minimal subtraction scheme   202
   – neural network   202
   – new physics   204–205
   – non-SM single top-quark production   204
   – non-SM top-quark decays   205
   – on-shell scheme   202
   – pair-production cross section   190–192
   – pair production in hadronic collisions   190–195
   – pair-production measurement   192–193
   – polarisation   189, 193
   – pole-mass scheme   202
   – properties   188
   – QCD background   192
   – quasi-free quark   189
   – renormalisation scheme   202
   – single top-quark cross section   196–197
   – single top-quark multivariate analysis   198
   – single top-quark observation   197–199
   – single top-quark production   195–199
   – single top-quark production at the LHC   199
   – size   187
   – spin correlations   193–194
   – Standard Model   187
   – $t\bar{t}$ final states   192
   – template method   202–203
   – Tevatron   188, 192
   – top–antitop resonances   204
   – V–A structure   189
   – W-boson helicity   199–201
top-quark mass   188–189, 201–204
toroid magnet   337–338
total absorption calorimeter   327
total cross section   226, 228, 357
TOTEM   238, 280, 357
Townsend coefficient   293
TPC, *see* time projection chamber   296
track fit   306–309
track reconstruction   306–309
tracking
   – alignment   291, 309–310
   – combinatorial Kalman filter   306
   – flavour tagging   310–311
   – hit residual   309
   – Kalman filter   308
   – local reconstruction   306
   – parameter resolution   306
   – pattern recognition   306–307
   – track fit   306–309
   – track reconstruction   306–309
   – vertex reconstruction   306

tracking detector 291–305
- active pixel detector 300, 303
- amplification 293, 295
- annealing 305
- bump-bond 302
- bunch-crossing tagging 296
- cathode-strip chamber 296, 337
- charge trapping 305
- dead-time 294
- DEPFET 303
- depletion 305
- diamond as detector material 305
- drift velocity 295–296
- electron multiplication 293
- equivalent noise charge 293
- flavour tagging 292
- gas electron multiplier 298
- gas gain 293, 295
- gas mixture 295–296
- gaseous detector 266, 291–299
- Geiger–Müller mode 293
- ghost hit 302
- hybrid pixel detector 299, 302–303
- MAPS 303
- micro-pattern gas detector 293, 298–299
- Micromegas 298
- microstrip gas chamber 298
- monolithic active pixel sensor 303
- noble gas 295
- noise 293
- particle identification 296
- pixel detector 266, 299
- position resolution 295, 301
- proportional mode 295
- quench gas 295
- radiation damage 304–305
- radiation tolerance 304–305
- resistive-plate chamber 294, 296, 337
- semiconductor detector 291, 299–305
- signal shape 293
- silicon as detector material 300–301, 305
- space–charge effect 293
- straw-tube detector 296
- strip detector 266, 299, 301–302
- thin-gap chamber 296, 337
- time projection chamber 296
- Townsend coefficient 293
- track reconstruction 291
- triggering 293–294, 296
- vertex detector 266
- working principle 292, 300
tree level 97

triangle anomaly 38, 43
trigger 363–381
- bits 365
- efficiency 371, 379
- exclusive trigger 373
- inclusive trigger 373
- latency 364
- level-1 logic 370
- level-1 trigger 365
- level-2 trigger 370
- logic 363, 367
- minimum bias 366
- mix 372
- monitor 371
- multi-level 370
- multi-purpose 365
- orthogonal conditions 365
- path 364
- T_0 363
two-fermion process 58

u
underlying event 115–118, 235–237
unitarity triangle 166–168
universities 415, 420, 422, 433
unparticle model 220

v
vertex detector 266, 292, 300, 310

w
W-boson couplings 57
W-boson helicity 199–201
W-boson mass 55–57
W/Z production
- luminosity 359
weak gauge bosons 7
weak mixing angle 4, 33, 62
weak neutral currents 4
website 457
wire chamber 266
WLCG, see Worldwide LHC Computing Grid 384
Wolfenstein parametrisation 166
working groups 409
Worldwide LHC Computing Grid 384

y
Yukawa interaction 34–36

z
Z lineshape 53
ZEUS 8, 21, 277, 324